WITHDRAWN

D1712939

SYNTHETIC FUELS HANDBOOK

Properties, Process, and Performance

James G. Speight, Ph.D., D.Sc.
University of Trinidad and Tobago
O'Meara Campus
Arima, Trinidad and Tobago

New York Chicago San Francisco Lisbon London Madrid
Mexico City Milan New Delhi San Juan Seoul
Singapore Sydney Toronto

The **McGraw·Hill** Companies

Library of Congress Cataloging-in-Publication Data

Speight, J. G.
   Synthetic fuels handbook : properties, process, and performance / James G. Speight.
     p.   cm.
   ISBN 978-0-07-149023-8 (alk. paper)
   1. Synthetic fuels. I. Title.
   TP360.S675   2008
   662'.66—dc22
                                                                                                      2008009846

Copyright © 2008 by The McGraw-Hill Companies, Inc. All rights reserved. Printed in the United States of America. Except as permitted under the United States Copyright Act of 1976, no part of this publication may be reproduced or distributed in any form or by any means, or stored in a data base or retrieval system, without the prior written permission of the publisher.

1 2 3 4 5 6 7 8 9 0   DOC/DOC   0 1 4 3 2 1 0 9 8

ISBN 978-0-07-149023-8
MHID 0-07-149023-X

Printed and bound by RR Donnelley.

McGraw-Hill books are available at special quantity discounts to use as premiums and sales promotions, or for use in corporate training programs. To contact a special sales representative, please visit the Contact Us page at www.mhprofessional.com.

This book is printed on acid-free paper.

Sponsoring Editor
   Larry S. Hager
Production Supervisor
   Richard C. Ruzycka
Editing Supervisor
   Stephen M. Smith
Project Manager
   Vastavikta Sharma,
   International Typesetting and Composition
Copy Editor
   Priyanka Sinha,
   International Typesetting and Composition

Proofreader
   Upendra Prasad,
   International Typesetting and Composition
Indexer
   Broccoli Information Management
Art Director, Cover
   Jeff Weeks
Composition
   International Typesetting and Composition

Information contained in this work has been obtained by The McGraw-Hill Companies, Inc. ("McGraw-Hill") from sources believed to be reliable. However, neither McGraw-Hill nor its authors guarantee the accuracy or completeness of any information published herein, and neither McGraw-Hill nor its authors shall be responsible for any errors, omissions, or damages arising out of use of this information. This work is published with the understanding that McGraw-Hill and its authors are supplying information but are not attempting to render engineering or other professional services. If such services are required, the assistance of an appropriate professional should be sought.

CONTENTS

Preface   ix

Chapter 1. Fuel Sources   1

1.1 Conventional Fuel Sources   /   4
1.2 Nonconventional Fuel Sources   /   5
    1.2.1 Tar Sand Bitumen   /   6
    1.2.2 Coal   /   8
    1.2.3 Oil Shale   /   10
    1.2.4 Natural Gas   /   14
    1.2.5 Gas Hydrates   /   16
    1.2.6 Biomass   /   16
    1.2.7 Bioalcohol   /   18
1.3 Synthetic Fuels   /   19
1.4 Production of Synthetic Fuels   /   22
    1.4.1 Thermal Decomposition   /   22
    1.4.2 Gasification   /   23
    1.4.3 The Fischer-Tropsch Process   /   24
    1.4.4 Bioprocesses   /   25

Chapter 2. Natural Gas   27

2.1 History   /   27
2.2 Formation and Occurrence   /   28
2.3 Conventional Gas   /   31
    2.3.1 Associated Gas   /   31
    2.3.2 Nonassociated Gas   /   32
    2.3.3 Liquefied Natural Gas   /   32
2.4 Unconventional Gas   /   33
    2.4.1 Coalbed Methane   /   33
    2.4.2 Gas Hydrates   /   34
    2.4.3 Shale Gas   /   35
2.5 Composition   /   35
2.6 Properties   /   37
    2.6.1 Density   /   37
    2.6.2 Heat of Combustion (Energy Content)   /   40
    2.6.3 Volatility, Flammability, and Explosive Properties   /   40
    2.6.4 Environmental Properties   /   43
2.7 Gas Processing   /   44
    2.7.1 Olamine Processes   /   45
    2.7.2 Physical Solvent Processes   /   47
    2.7.3 Metal Oxide Processes   /   49
    2.7.4 Methanol-Based Processes   /   51
    2.7.5 Carbonate-Washing and Water-Washing Processes   /   52

2.7.6 Sulfur Recovery Processes / 55
2.7.7 Hydrogenation and Hydrolysis Processes / 57
2.8 Uses / 57

Chapter 3. Fuels from Petroleum and Heavy Oil   63

3.1 History / 63
3.2 Petroleum Recovery / 66
3.3 Petroleum Refining / 67
     3.3.1 Dewatering and Desalting / 67
     3.3.2 Distillation / 68
     3.3.3 Thermal Processes / 70
     3.3.4 Catalytic Processes / 74
     3.3.5 Hydroprocesses / 77
     3.3.6 Reforming Processes / 80
     3.3.7 Isomerization Processes / 82
     3.3.8 Alkylation Processes / 83
     3.3.9 Polymerization Processes / 84
     3.3.10 Deasphalting / 85
     3.3.11 Dewaxing / 86
3.4 Heavy Oil Refining / 88
3.5 Petroleum Products and Fuels / 90
     3.5.1 Gaseous Fuels / 90
     3.5.2 Liquid Fuels / 92
     3.5.3 Solid Fuels / 100

Chapter 4. Fuels from Tar Sand Bitumen   103

4.1 Occurrence and Reserves / 104
     4.1.1 Canada / 105
     4.1.2 United States / 105
     4.1.3 Venezuela / 107
     4.1.4 Other Countries / 107
4.2 Structure and Properties of Tar Sand / 108
     4.2.1 Mineralogy / 109
     4.2.2 Properties / 109
4.3 Chemical and Physical Properties of Tar Sand Bitumen / 110
     4.3.1 Composition / 112
     4.3.2 Properties / 114
4.4 Mining Technology / 115
4.5 Bitumen Recovery / 116
     4.5.1 The Hot Water Process / 116
     4.5.2 In Situ Processes / 117
4.6 Upgrading, Refining, and Fuel Production / 120
     4.6.1 Primary Conversion / 122
     4.6.2 Secondary Upgrading / 124
     4.6.3 Other Processes / 124
     4.6.4 Hydrogen Production / 125
4.7 Synthetic Crude Oil / 125
4.8 The Future / 126

Chapter 5. Fuels from Coal   131

5.1 Occurrence and Reserves / 131
5.2 Formation and Types / 134

5.2.1 Coal Formation / 134
5.2.2 Coal Types / 135
5.3 Mining and Preparation / 137
    5.3.1 Surface Mining / 138
    5.3.2 Underground Mining / 139
    5.3.3 Mine Safety and Environment Effects / 139
    5.3.4 Coal Preparation / 141
5.4 Uses / 141
5.5 Gaseous Fuels / 145
    5.5.1 Gasifiers / 146
    5.5.2 Gaseous Products / 148
    5.5.3 Physicochemical Aspects / 150
    5.5.4 Gasification Processes / 153
    5.5.5 Underground Gasification / 156
    5.5.6 The Future / 157
5.6 Liquid Fuels / 157
    5.6.1 Physicochemical Aspects / 158
    5.6.2 Liquefaction Processes / 159
    5.6.3 Products / 161
    5.6.4 The Future / 162
5.7 Solid Fuels / 163
    5.7.1 Coke / 163
    5.7.2 Charcoal / 164
5.8 Environmental Aspects of Coal Utilization / 164
5.9 Clean Coal Technologies / 165

Chapter 6. Fuels from Oil Shale   169

6.1 Origin / 172
6.2 History / 173
6.3 Occurrence and Development / 174
    6.3.1 Australia / 175
    6.3.2 Brazil / 176
    6.3.3 Canada / 176
    6.3.4 China / 177
    6.3.5 Estonia / 177
    6.3.6 Germany / 178
    6.3.7 Israel / 178
    6.3.8 Jordan / 178
    6.3.9 Morocco / 179
    6.3.10 Russian Federation / 179
    6.3.11 Thailand / 179
    6.3.12 United States / 179
6.4 Liquid Fuels / 181
    6.4.1 Thermal Decomposition of Oil Shale / 181
    6.4.2 Mining and Retorting / 182
    6.4.3 In Situ Technologies / 184
6.5 Refining Shale Oil / 186
6.6 The Future / 192

Chapter 7. Fuels from Synthesis Gas   197

7.1 Gasification of Coal / 198
    7.1.1 Chemistry / 199
    7.1.2 Processes / 201
    7.1.3 Gasifiers / 202

7.2 Gasification of Petroleum Fractions / *203*
    7.2.1 Chemistry / *204*
    7.2.2 Processes / *206*
7.3 Steam-methane Reforming / *209*
7.4 Gasification of Other Feedstocks / *212*
7.5 Fischer-Tropsch / *213*

Chapter 8. Fuels from Biomass — 221

8.1 Biomass Fuels / *223*
    8.1.1 Feedstock Types / *224*
    8.1.2 Feedstock Properties / *227*
8.2 The Chemistry of Biomass / *228*
8.3 Processes / *231*
    8.3.1 Process Types / *231*
    8.3.2 Environmental Issues / *240*
8.4 Fuels from Biomass / *241*
    8.4.1 Gaseous Fuels / *242*
    8.4.2 Liquid Fuels / *244*
    8.4.3 Solid Fuels / *248*
    8.4.4 Biofuels from Synthesis Gas / *249*
8.5 Uses / *250*
8.6 A Biorefinery / *251*
    8.6.1 Bioconversion / *252*
    8.6.2 Thermal Conversion / *254*
    8.6.3 Greenhouse Gas Production / *257*
    8.6.4 Other Aspects / *257*
8.7 The Future / *259*

Chapter 9. Fuels from Crops — 265

9.1 Energy Crops / *266*
    9.1.1 Short Rotation Coppice / *267*
    9.1.2 Miscanthus / *267*
    9.1.3 Reed Canary Grass / *268*
    9.1.4 Cordgrass / *268*
    9.1.5 Switchgrass / *268*
    9.1.6 Reed Plants / *269*
    9.1.7 Jerusalem Artichoke / *269*
    9.1.8 Sorghum / *269*
    9.1.9 Hemp / *270*
9.2 Processes / *270*
    9.2.1 Direct Combustion / *270*
    9.2.2 Gasification / *272*
    9.2.3 Pyrolysis / *274*
    9.2.4 Anaerobic Digestion / *275*
9.3 Ethanol / *278*
    9.3.1 Ethanol Production / *279*
9.4 Other Alcohols / *281*
    9.4.1 Methanol / *281*
    9.4.2 Propanol and Butanol / *282*
9.5 Biodiesel / *282*
    9.5.1 Feedstocks for Biodiesel / *283*
    9.5.2 Transesterification (Alcoholysis) / *284*
    9.5.3 Catalytic Transesterification / *284*
    9.5.4 Noncatalytic Supercritical Methanol Transesterification / *286*

9.5.5 Effect of Reaction Parameters on Conversion Yield / 287
9.5.6 Properties / 288
9.5.7 Technical Standards / 289
9.5.8 Uses / 289
9.6 Hydrocarbons / 290

Chapter 10. Fuels from Wood — 297

10.1 History / 297
10.2 Composition and Properties / 301
10.3 Energy from Wood / 304
10.4 Fuels From Wood / 306
    10.4.1 Gaseous Fuels / 307
    10.4.2 Liquid Fuels / 310
    10.4.3 Solid Fuels / 313

Chapter 11. Fuels from Domestic and Industrial Waste — 323

11.1 Industrial and Domestic Waste / 325
11.2 Effects of Waste / 326
11.3 Fuels from Waste Incineration / 328
11.4 Gaseous Fuels / 332
    11.4.1 Chemistry / 333
    11.4.2 Reactors / 333
11.5 Liquid Fuels / 334
    11.5.1 Hydrocarbon Fuels / 335
    11.5.2 Biodiesel / 336
    11.5.3 Ethanol / 338
    11.5.4 Other Alcohols / 339
11.6 Solid Fuels / 340
    11.6.1 Briquette Manufacture / 341
    11.6.2 Bagasse Briquettes / 342
    11.6.3 Sawdust Briquettes / 342
    11.6.4 Urban Waste Briquettes / 342
11.7 The Future / 342

Chapter 12. Landfill Gas — 345

12.1 Landfill Classification / 346
12.2 Landfill Gas / 347
12.3 Biogas and Other Gases / 351
    12.3.1 Biogas / 351
    12.3.2 Renewable Natural Gas / 352
    12.3.3 Gober Gas / 352
12.4 Formation of Landfill Gas / 353
    12.4.1 Organic Reactions / 355
    12.4.2 Inorganic Reactions / 356
    12.4.3 Landfill Leachate / 356
    12.4.4 Landfill Gas Monitoring / 357
    12.4.5 Landfill Mining / 358
12.5 Gas Migration / 359
    12.5.1 Diffusion (Uniform Concentration) / 360
    12.5.2 Pressure / 360
    12.5.3 Permeability / 360
    12.5.4 Other Factors Affecting Gas Migration / 361

12.6 Bioreactors / *362*
12.7 Gasification of Landfill Waste / *363*
12.8 Power Generation / *364*

Appendix A. Definition and Properties of Fuels and Feedstocks from Different Sources
367

Appendix B. Comparison of the Properties of Gaseous Fuels from Different Sources
377

Appendix C. Comparison of the Properties of Liquid Fuels from Different Sources
381

Appendix D. Comparison of the Properties of Solid Fuels from Different Sources
389

Glossary   393
Index   403

PREFACE

Petroleum-based fuels are well-established products that have served industry and consumers for more than 100 years. For the foreseeable future automotive fuels will still be largely based on liquid hydrocarbons. The specifications of such fuels will however continue to be adjusted as they have been and are still being adjusted to meet changing demands from consumers. Traditional crude oil refining underwent increasing levels of sophistication to produce fuels of appropriate specifications. Increasing operating costs continuously put pressure on refining margins but it remains problematic to convert all refinery streams into products with acceptable specifications at a reasonable return.

However, the time is running out and petroleum, once considered inexhaustible, is now being depleted at a rapid rate. As the amount of available petroleum decreases, the need for alternate technologies to produce liquid fuels that could potentially help prolong the liquid fuels culture and mitigate the forthcoming effects of the shortage of transportation fuels that has been suggested to occur under the Hubbert peak oil theory (Hirsch, 2005).

To mitigate the influence of the *oil peak* and the subsequent depletion of supplies, unconventional (or nonpetroleum derived) fuels and synthetic fuels are becoming major issues in the consciousness of oil importing countries.

On the other hand, synthetic fuels, such as gasoline and diesel from other sources, are making headway into the fuel balance. For example, biodiesel from plant sources is similar to diesel, but has differences that include higher cetane rating (45–60 compared to 45–50 for petroleum-derived diesel) and it acts as a cleaning agent to get rid of dirt and deposits. As with alcohols and petrol engines, taking advantage of biodiesel's high cetane rating potentially overcomes the energy deficit compared to ordinary number 2 diesel.

For example, coal (coal-to-liquids), natural gas (gas-to-liquids), and oil shale (shale-to-liquids) have been touted for decades. At this time, the potential for liquid fuels from various types of biomass (Chap. 8) is also seeing prominence. Shortages of the supply of petroleum and the wish for various measures of energy independence are a growing part of the national psyche of many countries.

However, the production of liquid fuels from sources other than petroleum has a checkered history. The on-again-off-again efforts that are the result of political maneuvering has seen to it that the race to secure self-sufficiency by the production of nonconventional fuels has never got much further than the starting gate! This is due in no small part to the price fluctuations of crude oil (currently in excess of $90 per barrel) and the lack of foresight by various levels of government. It must be realized that for decades the price of petroleum has always been maintained at a level that was sufficiently low to discourage the establishment of a synthetic fuels industry. However, we are close to the time when the lack of preparedness for the production of nonconventional fuels may set any national government on its heels.

In the near term, the ability of conventional fuel sources and technologies to support the global demand for energy will depend on how efficiently the energy sector can match available energy resources with the end user and how efficiently and cost effectively the energy can be delivered. These factors are directly related to the continuing evolution of a truly global energy market.

In the long-term, one cannot create a sustainable energy future by treating energy as an independent topic. Rather, its role and interrelationship with other markets and other infrastructure demand further attention and consideration. Greater energy efficiency will depend on the developing world market's ability to integrate resources within a common structure.

The dynamics are now coming into place for the establishment of a synthetic fuels industry and it is up to various levels of government not only to promote the establishment of such an industry but to lead the way recognizing that it is not only supply and demand but the available and variable technology. For example, the technology of the tar sand industry and the oil shale industry is not the 1970s. The processes for recovery of the raw materials and the processing options have changed in an attempt to increase the efficiency of oil production. Various national events (for the United States) and international events (for other countries) have made it essential that we move ahead to develop fuels from nonconventional sources.

Voices are being raised for the establishment of an industry that produces and develops liquid fuels from nonconventional sources but there is still a long way to go. Incentives are still needed to develop such resources.

There is a cone of silence in many government capitals that covers the cries to develop nonconventional fuel sources. Hopefully, the silence will end within the near future, before it is too late.

In the context of the present book, the United States Energy Policy Act of 1992 (Section 301) defines alternative fuels as "methanol, denatured ethanol, and other alcohols; mixtures containing 85 percent or more (or such other percentage, but not less than 70 percent, as determined by the Secretary, by rule, to provide for requirements relating to cold start, safety, or vehicle functions) by volume of methanol, denatured ethanol, and other alcohols with gasoline or other fuels; natural gas; liquefied petroleum gas; hydrogen; coal-derived liquid fuels; fuels (other than alcohol) derived from biological materials; electricity (including electricity from solar energy); and any other fuel the Secretary determines, by rule, is substantially not petroleum and would yield substantial energy security benefits and substantial environmental benefits" (https://energy.navy.mil/publications/law_US/92epact/hr_0301.htm). It is this definition that is used to guide the contents of this book and show that sources that are *substantially* "not petroleum" are available as sources of fuels.

This book is written to assist the reader understand the options that are available for the production of synthetic fuel from nonconventional sources. For, the purposes of this book, nonconventional sources are those sources of gaseous, liquid, and solid fuels other than petroleum and heavy oil.

In addition, the book includes appendices that contain lists of the chemical and physical properties of the fuel sources and the fuels in order to assist the researcher understand the nature of the feedstocks as well as the nature of the products. If a product cannot be employed for its hoped-for-use, it is not a desirable product and must be changed accordingly. Such plans can only be made when the properties of the original product are understood.

James G. Speight, Ph.D., D.Sc.

CHAPTER 1
FUEL SOURCES

Fuel sources (gas, liquid, and solid) are those sources that can be used to roduce fuels (gas, liquid, and solid), which are combustible or energy-generating molecular species that can be harnessed to create mechanical energy.

Petroleum-based fuels are well-established products that have served industry and consumers for more than one hundred years. Over the past four decades, in spite of the *energy shocks* of the 1970s, there has been rapid escalation in fuel demand to the point that many countries, particularly the United States, are net importers of petroleum and petroleum products and this is projected to continue (Fig. 1.1). However, the time is running out and these fuel sources, once considered inexhaustible, are now being depleted at a rapid rate. In fact, there is little doubt that the supplies of crude oil are being depleted with each year that passes. However, it is not clear just how long it will take to reach the bottom of the well!

The impact of an oil deficiency can be overcome by serious planning for the world *beyond petroleum* (the slogan used by BP, formerly British Petroleum) but it is a trade off. The trade off is between having a plentiful supply of liquid fuels versus the higher cost (initially with a fall in production costs as technology advances) for the petroleum replacements. The flaw in this plan, of course, is its acceptance by the various levels of government in the oil consuming nations as the politicians think of re-election. And so, the matter falls into the hands of the consumers and requires recognition that the price of fuels will rise and may even continue to rise in the short-term. At least until serious options are mature and the relevant technologies being applied are on-stream.

Thus, as the amount of available petroleum decreases, the need for alternate technologies to produce liquid fuel grows (Table 1.1) (Green and Willhite, 1998). These fuels could potentially help prolong the liquid fuels culture and mitigate the forthcoming effects of the shortage of transportation fuels that has been suggested to occur under the Hubbert peak oil theory (Hirsch, 2005).

The Hubbert peak oil theory is based on the fundamental observation that the amount of oil under the ground is finite and proposes that for any given geographic area, from an individual oil field to the planet as a whole, the rate of petroleum production tends to follow a bell-shaped curve. The theory also proposed the means to show how to calculate the point of maximum production in advance based on discovery rates, production rates, and cumulative production. Early in the curve (pre-peak), the production rate increases due to the discovery rate and the addition of infrastructure. Late in the curve (post-peak), production declines due to resource depletion.

To mitigate the influence of the *oil peak* and the subsequent depletion of supplies, unconventional (or nonpetroleum-derived) fuels are of increasing interest in the consciousness of oil importing countries.

An *alternative fuel* or *synthetic fuel* is defined according to the context of its usage. In the context of substitutes for petroleum-based fuel, the term alternative fuel or synthetic implies any available fuel or energy source and may also refer to a fuel derived from a renewable energy sources. However, in the context of environmental sustainability, alternative fuel often implies an ecologically benign renewable fuel.

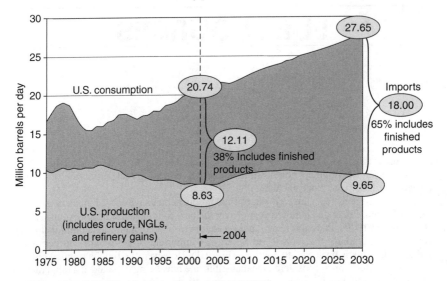

Source: U.S. DOE EIA "Annual energy outlook 2006."

FIGURE 1.1 Abbreviated history of consumption and production of petroleum and petroleum products in the United States.

For the purposes of this book, *alternate fuels* or *synthetic fuels* include liquid and gaseous fuels, as well as clean solid fuels, produced by the conversion of coal, oil shale or tar sands, and various forms of biomass. Such fuels are making headway into the fuel balance. For example, biodiesel from plant sources is similar to diesel, but has differences that include higher cetane rating (45–60 compared to 45–50 for petroleum-derived diesel) and it acts as a cleaning agent to get rid of dirt and deposits. As with alcohols and petrol engines, taking advantage of the high cetane number of biodiesel potentially overcomes the energy deficit compared to ordinary number 2 diesel.

In addition, coal (coal-to-liquids), natural gas (gas-to-liquids), and oil shale (shale-to-liquids) have been touted for decades. At this time, the potential for liquid fuels from various types of biomass is also seeing prominence (Chap. 8). Shortages of the supply of petroleum and the wish for various measures of energy independence are a growing part of the national psyche of many countries (Freeman, 2007).

However, the production of liquid fuels from sources other than petroleum has a checkered history. The on-again-off-again efforts that are the result of political maneuvering has seen to it that the race to secure self-sufficiency by the production of nonconventional fuels has never got much further than the starting gate! This is due in no small part to the price fluctuations of crude oil (i.e., gasoline) and the lack of foresight by various levels of government. It must be realized that for decades the price of petroleum has always been maintained at a level that was sufficiently low to discourage the establishment of a synthetic fuels industry. However, we are close to the time when the lack of preparedness for the production of nonconventional fuels may set any national government on its heels.

The dynamics are now coming into place for the establishment of a synthetic fuels industry and it is up to various levels of government not only to promote the establishment of such an industry but to lead the way recognizing that dynamics are not only supply and demand but the available and variable technology. For example, the technology of the tar sand industry and the oil shale industry is not the 1970s. The processes for recovery of the raw

TABLE 1.1 Screening Criteria for Enhanced Oil Recovery Methods

EOR method	°API	Viscosity [cp]	Composition	Oil saturation [%PV]	Formation type	Net thickness [m]	Permeability [md]	Depth [m]	T [°C]	Cost [USD/bbl]
N_2 (& flue gas)	>35/48	<0.4/0.2	High% C_1–C_7	>40/75	Sandstone/ carbonate	Thin unless dipping	—	>2000	—	—
Hydrocarbon	>23/41	<3/0.5	High% C_2–C_7	>30/80	Sandstone/ carbonate	Thin unless dipping	—	>1350	—	—
CO_2	>22/36	<10/1.5	High% C_5–C_{12}	>20/55	Sandstone/ carbonate	—	—	>600	120^4	2–8
Micellar/ polymer/ alkaline	>20/35	<35/13	Light, intermediate	>35/53	Sandstone	—	>10/450	<3000/1100	<95/25	8–12
Polymer flooding	>15/<40	<150/>10	—	>70/80	Sandstone	—	>10/800	<3000	<95/60	5–10
Combustion	>10/16	<5000/1200	—	>50/72	High porosity sand/ sandstone	>3	>50	<4000/1200	>40/55	3–6
Steam	>8/13.5	<200,000/4700	—	40/66	High porosity sand/ sandstone	>6	>200	<1500/500	—	3–6

PV = pore volume.

materials and the processing options have changed in an attempt to increase the efficiency of oil production. Various national events (for the United States) and international events (for other countries) have made it essential that we move ahead to develop fuels from nonconventional sources.

Voices are being raised for the establishment of an industry that produces and develops liquid fuels from nonconventional sources but there is still a long way to go. Incentives are still needed to develop such resources.

There is a cone of silence in many government capitals that covers the cries to develop nonconventional fuel sources. Hopefully, the silence will end in the near future, before it is too late.

1.1 CONVENTIONAL FUEL SOURCES

For the purposes of this book, petroleum is recognized as the prominent conventional fuel source. Thus, in the context of the definition of an *alternative fuel* or *synthetic fuel* that is defined in the context of substitutes for petroleum-based fuel the term *conventional fuel* implies any available fuel derived from petroleum.

Petroleum and the equivalent term *crude oil* cover a vast assortment of materials that consist of gaseous, liquid, and solid hydrocarbon-type chemical compounds that occur in sedimentary deposits throughout the world (Speight, 2007). When petroleum occurs in a reservoir that allows the crude material to be recovered by pumping operations as a free-flowing dark- to light-colored liquid, it is often referred to as *conventional* petroleum.

The U.S. Congress has defined tar sands as the several rock types that contain an extremely viscous hydrocarbon which is not recoverable in its natural state by conventional oil well production methods including currently used enhanced recovery techniques (US Congress, 1976). By inference, heavy oil, which can be recovered in its natural state by conventional oil-well production methods including currently used enhanced recovery techniques does not fall into the same category as tar sand bitumen and therefore is a type of petroleum. Thus, heavy oil is a type of petroleum that is different from conventional petroleum insofar the flow properties are reduced and a heavy oil is much more difficult to recover from the subsurface reservoir. These materials have a high viscosity (and low API [American Petroleum Institute] gravity) relative to the viscosity (and API gravity) of conventional petroleum and recovery of heavy oil usually requires thermal stimulation of the reservoir (Speight, 2007 and references cited therein).

The definition of heavy oil is usually based on the API gravity or viscosity but the definition is quite arbitrary. Although there have been attempts to rationalize the definition based upon viscosity, API gravity, and density (Speight, 2007) such definitions based on physical properties are inadequate and a more precise definition should involve some reference to the recovery method.

However, in a very general sense (and not in any sense meant to indicate classification of heavy oil), the term *heavy oil* is often applied to a petroleum that has a gravity greater than 20°API, and usually, but not always, a sulfur content higher than 2 percent w/w. Furthermore, in contrast to conventional crude oil, heavy oil is darker in color and may even be black. The term *heavy oil* has also been arbitrarily used to describe the heavy oil that requires thermal stimulation of recovery from the reservoir and (albeit incorrectly) the *bitumen* in bituminous sand (tar sand, oil sand) formations from which the highly viscous bituminous material is recovered by a mining operation (Chap. 4).

Petroleum varies widely in composition and variations of the composition of heavy oil add further complexity to the use of these feedstocks in the production of liquid fuels. The variations in composition are generally reflected in variations in the API gravity, viscosity, and potential for coke formation in thermal process, to mention only three of the affected properties. In addition to the organic constituents, there are also metal-containing constituents (notably those compounds that contain vanadium and nickel) which usually occur in the more viscous crude oil in amounts up to several hundred parts per million.

Physical methods of fractionation of petroleum and heavy oil can be achieved to produce four bulk generic fractions: saturates, aromatics, resins, and asphaltenes (Speight, 2001). Relative amounts of these fractions have often been equated to the behavior of petroleum or heavy-oil feedstocks during recovery and refining (Speight, 2007).

1.2 NONCONVENTIONAL FUEL SOURCES

Nonconventional fuel sources are sources of fuels (alternate or synthetic fuels) other than traditional petroleum (Tables 1.2–1.3) (Cooke, 2005).

Gaseous, liquid, or solid synthetic fuels are obtained by converting a carbonaceous material to another form. The most abundant naturally occurring materials suitable for this purpose are (a) tar sand, (b) coal, and (c) oil shale. The conversion of these raw materials to synthetic fuels can replace depleted, unavailable, or costly supplies of natural fuels.

Biomass is another carbonaceous material that can also be converted to synthetic fuels—the fermentation of grain to produce alcohol is the best-known example. Wood is also an abundant and accessible source of bio-energy and the procedures for the gasification of cellulosic materials have much in common with the conversion of coal to gas.

Currently, nonconventional oil production is less efficient and some types have a larger environmental impact relative to conventional oil production. Nonconventional types of production include: tar sand, coal, oil shale, and biofuels as well as liquid fuels from natural gas through processes such as the Fischer-Tropsch process. These nonconventional sources of oil may be increasingly relied upon as fuel for transportation as the price of conventional petroleum increases and supplies dwindle.

However, conventional sources of liquid fuels from petroleum are currently preferred because they provide a much higher ratio of extracted energy over energy used in extraction and refining processes. Technology, such as using steam injection in heavy oil reservoirs continues to serve as a means of extracting heavy oil while mining serves as the only commercial production of tar sand bitumen.

TABLE 1.2 Differences between Conventional and Nonconventional Oil (see also Table 1.3)

Conventional oil	Nonconventional oil
Mobile, low viscosity liquid oil	Tar sand—immobile in the natural state, viscous, near solid
Many wells flow on their own, or otherwise are produced by pumping	Must be mined, or heated-decomposed and then pumped
Wells often produce water that must be disposed of	Requires up to 3 bbl of water for every barrel of oil produced
Nearing "peak oil"	Largely untapped resource
High discovery cost; relatively low production cost	Locations known; relatively high production cost
Production quickly ramps up, peaks, and declines	Steady production for next 100 years
Reserves primarily outside North America	Extensive reserves in United States, Canada, Venezuela
Often influences foreign policy	Allows energy independence

TABLE 1.3 Advantages and Disadvantages of Developing Heavy Oil and Tar Sand Resources

ADVANTAGES:	• The Western Hemisphere alone contains more than 1 trillion barrels of recoverable unconventional oil. This is at least 35 years worth of oil at current global consumption rates.
	• Most tar sand and oil shale reserves are found in the United States and Canada with enough producible oil to meet 25 percent of the U.S. current oil demand for 400 years.
	• The Green River Basin in Colorado, Wyoming, and Utah has oil shale deposits that contain up to 800 billion barrels of recoverable oil—3 times the size of the oil reserves in Saudi Arabia.
	• Oil shale deposits in the Western United States can ultimately support production of up to 10 million barrels per day.
	• Most nonconventional oil production will come from stable countries (e.g., Canada) that do not belong to a cartel.
	• Nonconventional oil can reduce the U.S. dependence on foreign oil.
	• Nonconventional oil from tar sand and oil shale is generally compatible with existing pipeline and refinery infrastructure.
	• Tar sand and oil shale can be an *energy bridge* to the *beyond oil* era.
	• Will force rejection of the Hubbert Peak and the peak oil theory.
DISADVANTAGES:	• Production cannot be ramped up as quickly as conventional oil production.
	• Production of tar sand bitumen requires high volumes of expensive natural gas.
	• Tar sand operations emit large amounts of carbon dioxide.
	• No swing suppliers that can turn the taps on and off in response to global market shocks.
	• The U.S. Strategic Petroleum Reserve (SPR) may need to maintain larger stockpiles of oil.
	• High environmental impact of tar sands strip mining.
	• Environmental impact of oil shale development not fully assessed.
	• Environmental issues could be a constraint to future development.

1.2.1 Tar Sand Bitumen

Tar sand bitumen is another source of liquid fuels that is distinctly separate from conventional petroleum (US Congress, 1976).

Tar sand, also called *oil sand* (in Canada), or the more geologically correct term *bituminous sand* is commonly used to describe a sandstone reservoir that is impregnated with a heavy, viscous bituminous material. Tar sand is actually a mixture of sand, water, and bitumen but many of the tar sand deposits in countries other than Canada lack the water layer that is believed to facilitate the hot water recovery process. The heavy bituminous material has a high viscosity under reservoir conditions and cannot be retrieved through a well by conventional production techniques.

Geologically, the term *tar sand* is commonly used to describe a sandstone reservoir that is impregnated with bitumen, a naturally occurring material that is solid or near solid and is substantially immobile under reservoir conditions. The bitumen cannot be retrieved through a well by conventional production techniques, including currently used enhanced recovery techniques. In fact, tar sand is defined (FE-76-4) in the United States as:

> The several rock types that contain an extremely viscous hydrocarbon which is not recoverable in its natural state by conventional oil well production methods including currently used enhanced recovery techniques. The hydrocarbon-bearing rocks are variously known as bitumen-rocks oil, impregnated rocks, tar sands, and rock asphalt.

In addition to this definition, there are several tests that must be carried out to determine whether or not, in the first instance, a resource is a tar sand deposit (Speight, 2001 and references cited therein). Most of all, a core taken from a tar sand deposit, the bitumen isolated therefrom, are certainly not identifiable by the preliminary inspections (sight and touch) alone.

In the United States, the final determinant is whether or not the material contained therein can be recovered by primary, secondary, or tertiary (enhanced) recovery methods (US Congress 1976).

The relevant position of tar sand bitumen in nature is best illustrated by comparing its position relevant to petroleum and heavy oil. Thus, petroleum is referred to generically as a *fossil energy resource* that is derived from organic sediment (Fig. 1.2) and is further classified as a hydrocarbon resource.

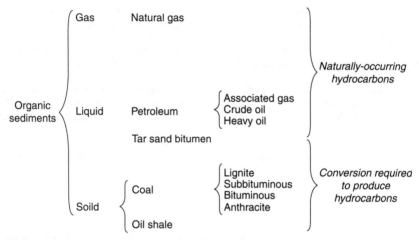

FIGURE 1.2 Informal classification of organic sediments by their ability to produce hydrocarbons.

The inclusion of tar sand bitumen, coal, and oil shale kerogen in this subdivision of organic sediments is automatic since these two natural resources (coal and oil shale kerogen) will produce hydrocarbons by thermal decomposition (high-temperature processing). Thus, if either coal and/or oil shale kerogen is to be included in the term hydrocarbon resources, it is more appropriate that they be classed as *hydrocarbon-producing resources*.

It is incorrect to refer to tar sand bitumen as *tar* or *pitch*. In many parts the name *bitumen* is used as the name for *road asphalt*. Although the word tar is somewhat descriptive of the black bituminous material, it is best to avoid its use with respect to natural materials. More correctly, the name tar is usually applied to the heavy product remaining after the destructive distillation of coal or other organic matter. *Pitch* is the distillation residue of the various types of tar.

Physical methods of fractionation of tar sand bitumen can also produce the four generic fractions: saturates, aromatics, resins, and asphaltenes. However, for tar sand bitumen, the fractionation produces shows that bitumen contains high proportions of the molecularly complex *asphaltene constituents* and *resin constituents*, even in amounts up to 50 percent w/w (or higher) of the bitumen with much lower proportions of saturates and aromatics than petroleum or heavy oil. In addition, the presence of ash-forming metallic constituents, including organo-metallic compounds such as, vanadium and nickel, is also a distinguishing feature of bitumen.

Currently, the only commercial production of bitumen from tar sand deposits occurs in northeastern Alberta (Canada) where mining operations are currently used to recover the tar sand. After mining, the tar sands are transported to an extraction plant, where a hot-water process separates the bitumen from sand, water, and minerals. The separation takes place in separation cells. Hot water is added to the sand, and the resulting slurry is piped to the extraction plant where it is agitated. The combination of hot water and agitation releases bitumen from the oil sand and causes tiny air bubbles to attach to the bitumen droplets, which float to the top of the separation vessel, where the bitumen can be skimmed off. Further processing removes residual water and solids. The bitumen is then transported and converted to *synthetic crude oil* by thermal processes. Approximately 2 tons of tar sands are required to produce 1 bbl of oil.

Both mining and processing of tar sands involve a variety of environmental impacts, such as global warming and greenhouse gas emissions, disturbance of mined land and impacts on wildlife and air and water quality. The development of a commercial tar sands industry in the United States would also have significant social and economic impacts on local communities. Of special concern in the relatively arid western United States is the large amount of water required for tar sands processing. Currently, tar sands extraction and processing require several barrels of water for each barrel of oil produced, though some of the water can be recycled.

To some observers, this proves the viability of the entire process while to others the energy requirements for the production of the synthetic crude oil make it marginally feasible for a significant percentage of world oil production to be extracted from tar sand.

Nevertheless synthetic crude oil is produced that has given Canada a measure of self sufficiency (at a cost) that is currently moving toward 1,500,000 bbl of synthetic crude oil per day.

1.2.2 Coal

Coal is a fossil fuel formed as an organic sediment (Fig. 1.2) in swamp ecosystems where plant remains were saved by water and mud from oxidization and biodegradation. Coal is a combustible black or brownish-black organic rock and is composed primarily of carbon along with assorted other elements, including sulfur. It is extracted from the ground by coal mining, either underground mining or open-pit mining (surface mining).

As geologic processes apply pressure to peat over time, it is transformed successively into: (a) lignite—also referred to as brown coal and is the lowest rank of coal and used almost exclusively as fuel for steam-electric power generation; (b) sub-bituminous coal—whose properties range from those of lignite to those of bituminous coal and are used primarily as fuel for steam-electric power generation; (c) bituminous coal—a dense coal, usually black, sometimes dark brown, often with well-defined bands of bright and dull material, used primarily as fuel in steam-electric power generation, with substantial quantities also used for heat and power applications in manufacturing and to make coke; and (d) anthracite—the highest rank; a harder, glossy, black coal used primarily for residential and commercial space heating.

The crude oil price has been sharply rising in the twenty-first century and there are indications that the high crude oil price is here to stay, rather than a temporary phenomenon. Even after considering the changes in various economic factors involving energy industries, production of transportation fuels or fuel oils via coal liquefaction is certainly an outstanding option for the sustainable future. Further, the products of coal liquefaction can be refined and formulated to possess the properties of conventional transportation fuels, as such requiring neither change in distribution nor lifestyle changes for consumers.

There are inherent technological advantages with the conversion of coal to liquid products since *coal liquefaction* can produce clean liquid fuels that can be sold as transportation fuels such as gasoline and diesel. There are three principal routes by which liquid fuels can be produced from solid coal: (a) direct conversion to liquids, (b) direct hydrogenation of coal, and (c) indirect conversion to liquids.

The direct liquefaction of coal by the Bergius process (liquefaction by hydrogenation) is also available. Several other direct liquefaction processes have been developed (Speight, 1994).

The Bergius process has not been used outside Germany, where such processes were operated both during World War I and World War II. Several other direct liquefaction processes have been developed, among these being the SRC-I and SRC-II (Solvent Refined Coal) processes developed by Gulf Oil and implemented as pilot plants in the United States in the 1960s and 1970s (Speight, 1994 and references cited therein).

The direct hydrogenation of coal was explored by the Nuclear Utility Services (NUS) Corporation in 1976 and involved the thermal conversion of dried, pulverized coal mixed with roughly 1 percent by weight molybdenum catalyst. The process yielded a limited amount of propane and butane, a synthetic naphtha (the precursor to gasoline), small amounts of NH_3, and significant amounts of carbon dioxide (Lowe et al., 1976).

Another process to manufacture liquid hydrocarbons from coal is low temperature carbonization (LTC) (Karrick process). Coal is coked at temperatures between 450 and 700°C compared to 800 to 1000°C for metallurgic coke. The lower temperatures optimize the production of coal tar that is richer in lighter hydrocarbons than high-temperature coal tar. The coal tar is then further processed into fuels.

The current process objectives of coal liquefaction by direct methods are mainly focused on easing operating process severity, minimizing hydrogen requirement, and making the product liquid more environmentally acceptable. Due to the recent trends of high and fluctuating petroleum price in the world market, the relative process economics of coal liquefaction is also much more favorable. Considering the vast amount of coal reserve throughout the world and its global distribution of major deposits, this alternative is even more attractive and also very practical.

Coal can also be converted into liquid fuels by *indirect liquefaction* which involves gasification of coal to mixtures of carbon monoxide and hydrogen (*synthesis gas*) followed by application of the Fischer-Tropsch process (Chap. 7) in which the synthesis gas is converted to hydrocarbons under catalytic conditions of temperature and pressure.

Fischer-Tropsch synthesis for production of synfuels is an established technology. Fossil fuels or biomass are converted into syngas via steam reforming, autothermal reforming, or gasification.

The Fischer-Tropsch process of *indirect* synthesis of liquid hydrocarbons was used in Germany for many years and is currently used by Sasol in South Africa. In the process, coal is gasified to produce synthesis gas (syngas; a balanced purified mixture of carbon monoxide and hydrogen) and the syngas condensed using Fischer-Tropsch catalysts to produce low-boiling hydrocarbons which are further processed into gasoline and diesel. Syngas can also be converted to methanol, which can be used as a fuel, fuel additive, or further processed into gasoline via the Mobil M-Gas process.

South Africa has established this technology on a large scale during the past 50 years, and Sasol is operating a coal-to-liquids plant. The product mix consists of 80 percent diesel and 20 percent naphtha. China has expressed interest in an advanced version of the Sasol conversion process. A number of other, small-scale projects have already started operation or are under construction (Coaltrans, 2003).

In the process, methanol is first made from methane (natural gas) in a series of three reactions:

Steam reforming: $CH_4 + H_2O \rightarrow CO + 3H_2$ $\Delta rH = +206$ kJ/mol

Water shift reaction: $CO + H_2O \rightarrow CO_2 + H_2$ $\Delta rH = +206$ kJ/mol

Methanol synthesis: $2H_2 + CO \rightarrow CH_3OH$ $\Delta rH = -92$ kJ/mol

Overall: $CO_2 + CO + 5 H_2 \rightarrow 2 CH_3OH + H_2O +$ heat

The methanol is then converted to gasoline by a dehydration step to produce dimethyl ether:

$$2CH_3OH \rightarrow CH_3OCH_3 + H_2O$$

This is then further dehydrated over a zeolite catalyst, ZSM-5, to give gasoline.

$$n+2CH_3OH \rightarrow CH_3(CH_2)_nCH_3 + n+2H_2O$$

Many of the methods for the production of fuels from coal (as well as the conversion of coal to syngas) release carbon dioxide in the conversion process, far more than is released in the production of liquid fuels from petroleum. If these methods were adopted to replace declining petroleum supplies, carbon dioxide emissions would be greatly increased on a global scale. Hence, carbon dioxide sequestration has been proposed to avoid releasing it into the atmosphere, though no pilot projects have confirmed the feasibility of this approach on a wide scale. Sequestration, however, may well add to the costs of synthetic fuels.

Fischer-Tropsch synthesis from natural gas is another established technology. Further expansion is planned for the year 2010, the bulk of this capacity being located in the Middle East (Qatar) (Chemical Market Reporter, 2004; IEA, 2004). The conversion efficiency is about 55 percent, with a theoretical maximum of about 78 percent. Due to the energy loss, this process makes only economic sense for cheap stranded gas. As the cost for liquefied natural gas transportation declines and demand increases, the importance or need for such options may also decline.

1.2.3 Oil Shale

Oil shale is a fine-grained sedimentary rock containing relatively large amounts of organic matter (called *kerogen*), an organic sediment (Fig. 1.2) from which significant amounts of shale oil and combustible gas can be extracted by destructive distillation (Longwell, 1990; Scouten, 1990; Lee, 1991; Bartis et al., 2005).

Oil shale, or the kerogen contained therein, does not have definite geological definition or a specific chemical formula. Different type of oil shale vary by the chemical consist, type of kerogen, age, and depositional history, including the organisms from which they were derived. Based upon environment of deposition, oil shale could be divided into three groups which are of terrestrial (land) origin, lacustrine (lake) origin, and marine (sea) origin.

The term *oil shale* is a misnomer. It does not contain oil nor is it commonly shale. The organic material is chiefly kerogen, and the *shale* is usually a relatively hard rock, called marl. Properly processed, kerogen can be converted into a substance somewhat similar to petroleum. However, the kerogen in oil shale has not gone through the *oil window* by which petroleum is produced and to be converted into a liquid hydrocarbon product, it must be heated to a high temperature. By this process the organic material is converted into a liquid, which must be further processed to produce oil which is said to be better than the lowest grade of oil produced from conventional oil deposits, but of a lower quality than the upper grades of conventional oil.

Oil shale occurs in many parts of the world ranging from deposits of little or no economic value to those that occupy thousands of square miles and contain many billions of barrels of potentially extractable shale oil. Total world resources of oil shale are conservatively estimated at 2.6 trillion barrels of oil equivalent. With the continuing decline of petroleum supplies, accompanied by increasing costs of petroleum-based products, oil shale presents opportunities for supplying some of the fossil energy needs of the world in the years ahead.

In the United States there are two principal types of oil shale: (a) Green River shale from the Green River Formation in Colorado, Utah, and Wyoming, and (b) the Devonian-Mississippian black shale of the eastern and midwestern states. The Green River shale is considerably richer in organic material, occurs in thicker seams, and therefore is more likely to be exploited for synthetic fuel manufacture.

In the Green River oil shale, the kerogen is not bound to a particular type of rock such as shale and the largest concentrations of kerogen are found in sedimentary nonreservoir rocks such as marlstone (a mix of carbonates, silicates, and clay). In contrast the black shale of the eastern and midwestern states is true shale, insofar as it is composed predominantly of the illite.

The organic content of oil shale is much higher than those of normal and ordinary rocks, and typically range from 1 to 5 percent w/w (lean shale) to 15 to 20 percent w/w (rich shale). This natural resource is widely scattered in the entire world, and occurrences are scientifically closely linked to the history and geologic evolution of the earth. Due to its abundance and wide distribution throughout the world, its utilization has a long history, both documented and undocumented. It is also obvious that the shale must have been relatively easy sources for domestic energy requirements for the ancient world, mainly due to the ease of handling and transportation; solid fuels were more convenient in the earlier human history and the examples are plentiful, including wood and coal.

There are two conventional approaches to oil shale processing. In one, the shale is fractured in situ and heated to obtain gases and liquids by wells. The second is by mining, transporting, and heating the shale to about 450°C, adding hydrogen to the resulting product, and disposing of and stabilizing the waste. Both processes use considerable amounts of water. The total energy and water requirements together with environmental and monetary costs (to produce shale oil in significant quantities) have so far made production uneconomic. During and following the oil crisis of the 1970s, major oil companies, working on some of the richest oil shale deposits in the world in western United States, spent several billion dollars in various unsuccessful attempts to commercially extract shale oil.

The amount of shale oil that can be recovered from a given deposit depends upon many factors. Some deposits or portions thereof, such as large areas of the Devonian black shale in eastern United States, may be too deeply buried to economically mine in the foreseeable future. Surface land uses may greatly restrict the availability of some oil shale deposits for development, especially those in the industrial western countries. The bottom line in developing a large oil shale industry will be governed by the price of petroleum. When the price of shale oil is comparable to that of crude oil because of diminishing resources of crude, then shale oil may find a place in the world fossil energy mix.

In order to extract hydrocarbons (or, oil in "loose" terms), the oil shale is typically subjected to a thermal treatment, scientifically categorized as "destructive distillation." A collective scientific term for hydrocarbons in oil shale is called *kerogen*, an ill-defined macromolecule which, when heated, undergoes both physical and chemical change. Physical changes involve phase changes, softening, expansion, and oozing through pores, while chemical changes are typically involving bond cleavages mostly on carbon–carbon bonds that result in smaller and simpler molecules. The chemical change is often termed as pyrolysis or thermal decomposition. The pyrolysis reaction is endothermic in nature, requiring heat, and produces lighter molecules thereby increasing the pressure.

In addition to the kerogen pyrolysis reaction, carbonate decomposition reactions are also included here as principal chemical reactions, due to their abundant existence and also to their reaction temperature ranges that overlap the kerogen pyrolysis temperature range. Other mineral matters of oil shale that are worthy of note are alumina, nahcolite, and dawsonite. Some of the processes are designed to recover these mineral matters for economic benefit to the overall process. The pyrolysis reaction is quite active at a temperature above 400°C, where most of the commercial retorting processes are operated. Most of the ex situ processes utilize the spent (processed) shale as a char source to supply the process heat, thus accomplishing higher energy efficiency for the process. The typical temperature required to carry out such pyrolysis reaction is in the range of 450 to 520°C. In order to make the efficiency of oil extraction higher, oil shale rocks need to be ground to finer particle sizes, thus alleviating mass transfer resistance and at the same time facilitating smoother flow for cracked hydrocarbons to escape out of the rock matrix. Due to the poor porosity or permeability of oil shale rock, the rock matrix often goes through stress fracture during pyrolysis operation, typically noticed as crackling.

Major drawbacks of this type of process involve (a) "mining first" operation, which is costly, (b) transportation or conveying the mined shale to retorting facilities, (c) size reduction such as rubblizing, grinding, or milling, and (d) returning the spent shale back to the environment. In the current energy market of the globe, any major transportation of unprocessed (raw) shale would be economically unfavorable, unless the raw shale contain very high levels of oil contents or liquid fuel price from conventional petroleum source is substantially higher. In ex situ operations, heavy reliance on earth-moving equipment (EME) and rock-handling equipment, such as rock pump and heavy-duty hammer mill, is noticeable. Furthermore, it is likely to be constrained by the underlying market principle that the recoverable value including any tax credits or incentives from oil shale has to be favorable after considering all the cost factors including mining, transportation, and processing. Often, the mass percentage of oil content of oil shale or the volume of recoverable oil from unit mass of oil shale is used as a measuring parameter. The latter is called Fischer assay, which is based on the ASTM standard under a prescribed condition of retorting. However, this value should not be considered as the maximum recoverable oil content for the shale or the oil content itself in the shale.

Several of the ex situ retorting processes have been commercially tested on large scales and also proven effective for designed objectives. Some of the successfully demonstrated processes include: (a) Gas Combustion retort process, (b) TOSCO (The Oil Shale Corporation) process, (c) Union Oil retorting process, (d) Lurgi-Ruhrgas (LR) process, (e) Superior Oil's multi-mineral process, and (f) Petróleo Brasileiro (Petrobrás) process. Partial combustion of residual char provides the thermal energy for heating the shale via direct contact, thus achieving energy efficiency. TOSCO uses heated ceramic balls to provide the thermal energy for heating the shale by direct contact, and also successfully implements multi-levels of heat recovery and energy integration strategy. The Union Oil retorting process is unique and innovative with utilizing well-designed rock pumps and adopting a number of designs for heating shale in the retort. The LR process produces hydrocarbons from oil shale by bringing raw shale in contact with hot fine-grained solid heat carrier, which can be just spent shale. The Petrobrás process was operated for about 10 years in southern Brazil, treating over 3,500,000 tons of Irati (Permian age) oil shale to produce more than 1,500,000 bbl of shale oil and 20,000 tons of sulfur.

In situ retorting of oil shale does not involve any mining operation, except starter holes and implementation digging. Therefore, in situ retorting does not require any transportation of shale out of the oil shale field. In situ retorting is often called subsurface retorting. The advantages of in situ retorting processes include: (a) no need for mining, (b) no need for oil shale transportation, and (c) cost and labor effectiveness. However, difficulties are in the domain of (a) process control and reliability, (b) environmental and ecological impact before and after the processing, (c) long-term ground water contamination, and (d) process efficiency. Sites for in situ processing are put back to normal vegetated areas or to the original forms of the environment as close as possible, upon completion.

There are several advantages associated with oil shale commercialization and they are (a) worldwide abundance and distribution, (b) surface mining or in situ processing possibilities, (c) source for high-quality crude products, (d) source for aliphatic liquid fuels, and last but by no means least (e) oil shale is a politically less sensitive fossil fuel resource.

The quality of oil shale can be very simply represented by its oil content in the shale. To compare the oil contents as recoverable amounts of hydrocarbon from a wide variety of oil shale, a standardized method of oil content determination is needed. Fischer assay is most generally used for this purpose and it has definite merits based on its simplicity and use of a common apparatus (Fischer assay).

The mostly aliphatic nature of the shale oil is very attractive from the environmental and processing standpoints, since aromatics in liquid fuel are generally viewed negatively due to the high potential for evaporative and fugitive emission that introduces a high level of volatile organic compounds (VOCs) into the atmosphere. In fact, there are several issues that relate to the environmental constraints in exploiting oil shale formations:

1. Water availability and salinity change in water.
2. Availability and development of other energy and mineral resources.
3. Ecology and preservation of natural character of the region.
4. Sources of fugitive dust from such operations as crushing, sizing, transfer conveying, vehicular traffic, and wind erosion.
5. Gaseous emissions such as H_2S, NH_3, CO, SO_2, NO_x, and trace metals.
6. Nature of the region and the population density.
7. Outdoor recreation in the oil shale region.
8. Environmental permits and regulations.

Oil shale can be ignited and burst into fire, if conditions are met. Depending upon the shale types and their hydrocarbon contents, the self-ignition temperature (SIT) of dry shale in the atmosphere varies widely from as low as 135 to 420°C. The finer the particle, the stronger is the possibility of catching fire spontaneously. However, it is generally too expensive to grind oil shale to fine meshes for processing. This threshold value is not generally set for all types of oil shale or processes; however, it is estimated to be about 1 to 3 mm as a minimum. Oozing oils from raw or spent shale can complicate the safety matters by exposing not only potentially hazardous air pollutants (HAPs) to the environment, but also highly combustible matters in contact with air. This can be especially true with spent shale transportation, if the residual hydrocarbons are not burnt off for heat recovery for the process. Reburial or disposal of spent shale potentially renders an ecological and environmental concern. Since spent shale is the shale that has gone through a thermal treatment process, it is more likely to become a source for leaching of minerals and organics, that may be harmful to the ecologic constituents, and contaminating the ground waterway.

For much of the twentieth century, the Naval Petroleum and Oil Shale Reserves served as a contingency source of fuel for the U.S. military, as the government owned both. However, in the later part of the twentieth century, the Naval Petroleum and Oil Shale Reserves were determined to no longer serve the national defense purpose envisioned earlier in the century. Privatization of these properties followed and commercial exploitation was prompted, as evidenced by the Elk Hill field (California) operated by Occidental Petroleum

Corporation. Following the sale of the Elk Hills by the U.S. Department of Energy, two of the Naval Oil Shale Reserves, both in Colorado are being transferred to Department of Interior's Bureau of Land Management, thereby offering commercial mining leasing for gas and petroleum production. Considering the environmental success of Elk Hills field project, it is foreseeable that many of the oil shale fields in the world can be more seriously explored for commercial exploitation for petroleum (shale oil crude) and gas (natural gas and LPG) generation.

Environmental technology has also been constantly enhanced for better management of underground waterways and surface water treatment, thus alleviating the burden of environmental constraints from oil shale industries. The most significant merits of shale oil are in its relative ease of producing high-quality liquid fuel via relatively simple processing of oil shale rock. High-quality liquid fuels, especially from alternative sources, are going to be more and more strongly demanded due to their minimal impact on lifestyle changes. Other factors, that may have significant effects on boosting the interest in oil shale commercialization, may include the ever-increasing petroleum market price, the fluctuating supply of petroleum resources, the steeply rising cost of natural gas, the nondwindling appetite of petroleum-like liquid fuels, and the public fear of petroleum depletion.

It may not be a long stretch in prediction for the world energy market in the future that oil shale will be a very valuable commodity as an alternative source for clean liquid fuel and the resource is not easily depletable. Transportation of oil shale will then become a very important issue, since the fuel cost structure is more likely to justify the long distance transportation of oil shale. However, until that time, oil shale will always remain as a contingency plan, as an invaluable alternative source for petroleum-like liquid fuel. There is always the advantage and temptation of using oil shale since it minimizes the lifestyle change in energy consumption pattern even without relying on the petroleum resource.

1.2.4 Natural Gas

Natural gas is a gaseous fossil fuel consisting primarily of methane but including significant quantities of ethane, butane, propane, carbon dioxide, nitrogen, helium, and hydrogen sulfide. It is found in natural gas fields (*unassociated natural gas, non-associated natural gas*), oil fields (*associated natural gas*) and in coal seams or beds (hence *coalbed methane*).

Natural gas is often informally referred as simply gas and before it can be used as a fuel, it must undergo extensive processing (refining) to remove almost all materials other than methane. The by-products of that processing include ethane, propane, butanes, pentanes and higher molecular weight hydrocarbons, elemental sulfur, and sometimes helium and nitrogen.

Natural gas is colorless in its pure form and is a combustible mixture of hydrocarbon gases and while the major constituents is methane, ethane, propane, butane and pentane are also present but the composition of natural gas varies widely.

Natural gas can also be used to produce alternative liquid fuels and the process is often referred to as gas-to-liquid (GTL) (Table 1.4). The term *alternative fuel* includes methanol, ethanol, and other alcohols, mixtures containing methanol, and other alcohols with gasoline or other fuels, biodiesel, fuels (other than alcohol) derived from biologic materials, and any other fuel that is substantially not a petroleum product.

The production of liquid fuels from sources other than petroleum broadly covers liquid fuels that are produced from tar sand (oil sand) bitumen, coal, oil shale, and natural gas. Synthetic liquid fuels have characteristics approaching those of liquid fuels generated from petroleum but differ because the constituents of synthetic liquid fuels do not occur naturally in the source material used for the production. Thus, the creation of liquids to be used as fuels from sources other than natural crude petroleum broadly defines synthetic liquid fuels. For much of the twentieth century, the synthetic fuels emphasis was on liquid

TABLE 1.4 Advantages and Disadvantages of Gas-to-Liquids (GTL) Technology

ADVANTAGES:	• Allows owners of remote gas reserves a way to bring their gas to market.
	• Tighter air quality standards will create high demand for low sulfur diesel, like that produced by GTL.
	• Diesel fuel is ultra-low sulfur free and has a higher cetane number than diesel from crude oil.
	• Allows stranded to be turned into useful/valuable products.
	• Products are compatible with existing tankers, pipelines, and storage facilities.
	• Engines running on GTL fuels pollute less.
	• Greater global use of GTL-made gasoline and diesel could slow down oil demand.
DISADVANTAGES:	• If gas or products need to be imported, does not reduce dependence on foreign energy.
	• Conversion plants are expensive to build.
	• Many large gas reserves are found in politically unstable areas.
	• Products will have to be transported from distant production centers—adds to cost.

products derived from coal upgrading or by extraction or hydrogenation of organic matter in coke liquids, coal tars, tar sands, or bitumen deposits. More recently, the potential for natural gas as a source of liquid fuels has been recognized and attention is now on the development of natural gas as a source of liquid fuels.

Projected shortages of petroleum make it clear that, for the remainder of the twenty-first century, alternative sources of liquid fuels are necessary. Such sources (e.g., natural gas) are available but the exploitation technologies are in general not as mature as for petroleum. The feasibility of the upgrading of natural gas to valuable chemicals, especially liquid fuels, has been known for years. However, the high cost of the steam reforming and the partial oxidation processes, used for the conversion of natural gas to syngas has hampered the widespread exploitation of natural gas. Other sources include tar sand (also called oil sand or bituminous sand) (Berkowitz and Speight, 1975; Speight, 1990) and coal (Speight, 1994) that are also viable sources of liquid fuels.

The potential of natural gas, which typically has 85 to 95 percent methane, has been recognized as a plentiful and clean alternative feedstock to crude oil. Currently, the rate of discovery of proven natural gas reserves is increasing faster than the rate of natural gas production. Many of the large natural gas deposits are located in areas where abundant crude oil resources lie such as in the Middle East. However, huge reserves of natural gas are also found in many other regions of the world, providing oil-deficient countries access to a plentiful energy source. The gas is frequently located in remote areas far from centers of consumption, and pipeline costs can account for as much as one-third of the total natural gas cost. Thus tremendous strategic and economic incentives exist for gas conversion to liquids; especially if this can be accomplished on site or at a point close to the wellhead where shipping costs becomes a minor issue.

However, despite reduced prominence, coal technology continues to be a viable option for the production of liquid fuels in the future. World petroleum production is expected ultimately to level off and then decline and despite apparent surpluses of natural gas, production is expected to suffer a similar decline. Coal gasification to syngas is utilized to synthesize liquid fuels in much the same manner as natural gas steam reforming technology. But the important aspect is to use the natural gas reserves when they are available and to maximize the use of these reserves by conversion of natural gas to liquid fuels.

1.2.5 Gas Hydrates

Gas hydrates are crystalline solids in which a hydrocarbon, usually methane is trapped in a lattice of ice. They occur in the pore spaces of sediments, and may form cements, nodes, or layers. Gas hydrates are found in naturally-occurring deposits under ocean sediments or within continental sedimentary rock formations. The worldwide amounts of carbon bound in gas hydrates is conservatively estimated to total twice the amount of carbon to be found in all known fossil fuels on Earth.

Methane trapped in marine sediments as a hydrate represents such an immense carbon reservoir that it must be considered a dominant factor in estimating unconventional energy resources; the role of methane as a *greenhouse gas* also must be carefully assessed. Hydrates store immense amounts of methane, with major implications for energy resources and climate, but the natural controls on hydrates and their impacts on the environment are very poorly understood.

Gas hydrates occur abundantly in nature, both in Arctic regions and in marine sediments. Gas hydrate is a crystalline solid consisting of gas molecules, usually methane, each surrounded by a cage of water molecules. It looks very much like ice. Methane hydrate is stable in ocean floor sediments at water depths greater than 300 m, and where it occurs, it is known to cement loose sediments in a surface layer several hundred meters thick.

This estimate is made with minimal information from U.S. Geological Survey (USGS) and other studies. Extraction of methane from hydrates could provide an enormous energy and petroleum feedstock resource. Additionally, conventional gas resources appear to be trapped beneath methane hydrate layers in ocean sediments. The immense volumes of gas and the richness of the deposits may make methane hydrates a strong candidate for development as an energy resource.

Because the gas is held in a crystal structure, gas molecules are more densely packed than in conventional or other unconventional gas traps. Gas-hydrate-cemented strata also act as seals for trapped free gas. These traps provide potential resources, but they can also represent hazards to drilling, and therefore must be well understood. Production of gas from hydrate-sealed traps may be an easy way to extract hydrate gas because the reduction of pressure caused by production can initiate a breakdown of hydrates and a recharging of the trap with gas.

Seafloor slopes of 5° and less should be stable on the Atlantic continental margin, yet many landslide scars are present. The depth of the top of these scars is near the top of the hydrate zone, and seismic profiles indicate less hydrate in the sediment beneath slide scars. Evidence available suggests a link between hydrate instability and occurrence of landslides on the continental margin. A likely mechanism for initiation of land sliding involves a breakdown of hydrates at the base of the hydrate layer. The effect would be a change from a semi-cemented zone to one that is gas-charged and has little strength, thus facilitating sliding. The cause of the breakdown might be a reduction in pressure on the hydrates due to a sea-level drop, such as occurred during glacial periods when ocean water became isolated on land in great ice sheets.

1.2.6 Biomass

Biomass, in the present context, refers to living and recently dead biologic material which can be used as fuel or for industrial production. For example, manure, garden waste, and crop residues are all sources of biomass. Biomass is a *renewable* energy source, unlike other resources such as petroleum, natural gas, tar sand, coal, and oil shale. But like coal and petroleum, biomass is a form of stored solar energy.

Biofuel is derived from *biomass* (recently living organisms or their metabolic byproducts) (Hudson, 2004) and has the potential to produce fuels that are more environmentally benign than petroleum-based fuels (American Coalition for Ethanol, 2004). In addition, ethanol, a crop-based fuel alcohol (Chaps. 8 and 9), adds oxygen to gasoline thereby helping to improve vehicle performance and reduce air pollution. Biodiesel, an alternative or additive to petroleum diesel, is a nontoxic, renewable resource created from soybean or other oil crops (Pacific Biodiesel, 2004). Agricultural products specifically grown for use as biofuels include crops (Chap. 9) such as corn, soybeans, flaxseed, rapeseed, wheat, sugar beet, sugar cane, palm oil, and *Jatropha* (Chap. 10). Biodegradable outputs from industry, agriculture, forestry, and households can be used as well; examples include straw, timber, manure, rice husks, sewage, biodegradable waste, and food leftovers. They are converted to biogas through anaerobic digestion. Biomass used as fuel often consists of underutilized types, like chaff and animal waste. The quality of timber or grassy biomass does not have a direct impact on its value as an energy source.

Unlike other forms of renewable energy, biofuels do not reduce the amount of greenhouse gases in the atmosphere. The combustion of biofuels produces carbon dioxide and other greenhouse gases. The carbon in biofuels is often taken to have been recently extracted from atmospheric carbon dioxide by plants as they have grown. The potential for biofuels to be considered *carbon neutral* depends upon the carbon that is emitted being reused by plants for further growth. Clearly however, cutting down trees in forests that have grown for hundreds or thousands of years for use as a biofuel, without the replacement of this biomass would not have a carbon-neutral effect.

It is generally believed that a way to reduce the amount of carbon dioxide released into the atmosphere is to use biofuels to replace nonrenewable sources of energy. Dried compressed peat is also sometimes considered a biofuel. However, it does not meet the criteria of being a renewable biofuel. Though more recent than petroleum or coal, on the time scale of human industrialization, peat is a fossil fuel and burning it does contribute to atmospheric carbon dioxide.

The production of biofuels to replace oil and natural gas is in active development, focusing on the use of cheap organic matter (usually cellulose, agricultural and sewage waste) in the efficient production of liquid and gas biofuels which yield high net energy gain. One advantage of biofuel over most other fuel types is that it is biodegradable, and so relatively harmless to the environment if spilled.

Direct biofuels are biofuels that can be used in existing unmodified petroleum engines. Because engine technology changes all the time, direct biofuel can be hard to define; a fuel that works well in one unmodified engine may not work in another. In general, newer engines are more sensitive to fuel than older engines, but new engines are also likely to be designed with some amount of biofuel in mind.

Straight vegetable oil can be used in many older diesel engines (equipped with indirect injection systems), but only in the warmest climates. Usually it is turned into biodiesel instead. No engine manufacturer explicitly allows any use of vegetable oil in their engines.

Biodiesel can be a direct biofuel. In some countries manufacturers cover many of their diesel engines under warranty for 100 percent biodiesel use. Many people have run thousands of miles on biodiesel without problem, and many studies have been made on 100 percent biodiesel.

Butanol is often claimed as a direct replacement for gasoline. It is not in wide spread production at this time, and engine manufacturers have not made statements about its use. While on paper (and a few laboratory tests) it appears that butanol has sufficiently similar characteristics with gasoline such that it should work without problem in any gasoline engine, no widespread experience exists.

Ethanol is the most common biofuel, and over the years many engines have been designed to run on it. Many of these could not run on regular gasoline, so it is debatable whether ethanol is a replacement in them. In the late 1990s, engines started appearing that by design can use either fuel. Ethanol is a direct replacement in these engines, but it is debatable if these engines are unmodified, or factory modified for ethanol.

In reality, small amounts of biofuel are often blended with traditional fuels. The biofuel portion of these fuels is a direct replacement for the fuel they offset, but the total offset is small. For biodiesel, 5 percent or 20 percent are commonly approved by various engine manufacturers.

Using waste biomass to produce energy can reduce the use of fossil fuels, reduce greenhouse gas emissions, and reduce pollution and waste management problems. A recent publication by the European Union highlighted the potential for waste-derived bioenergy to contribute to the reduction of global warming. The report concluded that 140 million barrels of oil equivalent will be available from biomass by 2020, 46 percent from bio-wastes: municipal solid waste (MSW), agricultural residues, farm waste, and other biodegradable waste streams (European Environment Agency, 2006; Marshall, 2007).

Landfill sites generate gases as the waste buried in them undergoes anaerobic digestion. These gases are known collectively as landfill gas (LFG); this can be burned and is a source of renewable energy. Landfill gas (Chap. 12) contains approximately 50 percent methane (the same gas that is found in some types of natural gas) and can be burned either directly for heat or to generate electricity for public consumption. If LFG is not harvested, it escapes into the atmosphere; this is not desirable because methane is a greenhouse gas (GHG), and is more harmful than carbon dioxide. Methane has a global warming potential of 23 relative to carbon dioxide (i.e., 1 ton of methane produces the same GHG effect as 23 tons of carbon dioxide).

Anaerobic digestion can be used as a distinct waste management strategy to reduce the amount of waste sent to landfill and generate methane, or biogas. Any form of biomass can be used in anaerobic digestion and will break down to produce methane, which can be harvested and burned to generate heat, power, or to power certain automobiles.

1.2.7 Bioalcohol

Alcohol fuels are usually of biologic rather than petroleum sources. When obtained from biologic sources, they are referred to bioalcohols (e.g., bioethanol). It is important to note that there is no chemical difference between biologically produced alcohols and those obtained from other sources. However, ethanol that is derived from petroleum should not be considered safe for consumption as this alcohol contains about 5 percent methanol and may cause blindness or death. This mixture may also not be purified by simple distillation, as it forms an azeotropic mixture.

Bioalcohols are still in developmental and research stages. Use of optimized crops with higher yields of energy (Chaps. 8 and 9), elimination of pesticides and fertilizers based on petroleum, and a more rigorous accounting process will help improve the feasibility of bioalcohols as fuels.

Alcohols are a useful type of liquid fuel because they combust rapidly and are often cheap to produce. However, their acceptance is hampered by the fact that their production often requires as much or even more fossil fuel than they replaced since they are typically not primary sources of energy; however, they are a convenient way to store the energy for transportation. No type of fuel production is 100 percent energy-efficient, thus some energy is always lost in the conversion. This energy can be supplied by the original source, or from other sources like fossil fuel reserves, solar radiation (either through photosynthesis or photovoltaic panels), or hydro, wind, or nuclear energy.

Methanol is the lowest molecular weight and simplest alcohol, produced from the natural gas component methane. It is also called *methyl alcohol* or *wood alcohol*, the latter because it was formerly produced from the distillation of wood.

Ethanol, also known as *grain alcohol* or *ethyl alcohol,* is most commonly used in alcoholic beverages. However, it may also be used as a fuel, most often in combination with gasoline. For the most part, it is used in a gasoline-to-ethanol ratio of 9:1 to reduce the negative environmental effects of gasoline. Ethanol can be readily produced by fermentation of simple sugars that are converted from starch crops. Feedstocks for such fermentation ethanol include corn, barley, potato, rice, and wheat. This type of ethanol may be called *grain ethanol*, whereas ethanol produced from cellulose biomass such as trees and grasses is called *bioethanol* or *biomass ethanol*. Both grain ethanol and bioethanol are produced via biochemical processes, while *chemical ethanol* is synthesized by chemical synthesis routes that do not involve fermentation.

There is increasing interest in the use of a blend of 85 percent fuel ethanol with 15 percent gasoline. This fuel blend called E85 has a higher fuel octane than premium gasoline allowing in properly optimized engines increase both power and fuel economy over gasoline.

Butanol, also known as *butyl alcohol*, may be used as a fuel with the normal combustion engine, typically as a product of the ferment of biomass with the bacterium *Clostridium acetobutylicum.*

The advantages of butanol are the high octane rating (over 100) and high energy content, only about 10 percent lower than gasoline, and subsequently about 50 percent more energy-dense than ethanol, 100 percent more so than methanol. The major disadvantage of butanol is the high flashpoint (35°C, 95°F).

1.3 SYNTHETIC FUELS

The production of synthetic fuels is essentially a hydrogen-addition process. Fuels such as gasoline and natural gas have an atomic hydrogen/carbon ratio on the order of 2.0 while the fuel sources have lower atomic hydrogen/carbon ratio (on the order of 1.0 to 1.5). The source of hydrogen can be intramolecular in which a carbonaceous low-hydrogen residue (e.g., coke) is produced or intermolecular in which hydrogen is added from an external source.

On the one hand, that is, intramolecular hydrogenation, pyrolysis of the feedstock in the absence of any added agent produces volatile (high-hydrogen) products and a nonvolatile (low-hydrogen) coke. In pyrolysis the carbon content is reduced by heating the raw hydrocarbon until it thermally decomposes to yield solid carbon, together with gases and liquids having higher fractions of hydrogen than the original material.

On the other hand, that is, hydrogenation from an external source, the hydrogenation is either direct or indirect. Direct hydrogenation involves exposing the raw material to hydrogen at high pressure. Indirect hydrogenation involves reaction of the feedstock with steam, and the hydrogen is generated within the system.

Thus, gaseous or liquid synthetic fuels are obtained by converting a carbonaceous material to a gaseous or liquid form, respectively. In the United States and many other countries, the most abundant naturally occurring materials suitable for this purpose are coal and oil shale. Tar sands are also suitable, and large deposits are located in Canada. The conversion of these raw materials is carried out to produce synthetic fuels to replace depleted, unavailable, or costly supplies of natural fuels. However, the conversion may also be undertaken to remove sulfur or nitrogen that would otherwise be burned, giving rise to undesirable air pollutants. Another reason for conversion is to increase the calorific value of the original raw fuel by removing unwanted constituents such as ash, and thereby to produce a fuel which is cheaper to transport and handle.

Although most of the emphasis for the production of synthetic fuels is on synthetic fuels from coal, oil shale, and tar sands, biomass can also be converted to synthetic fuels and the fermentation of grain to produce alcohol is a well known example. However, in many countries, grain is an expensive product which is generally thought to be more useful for its food value. Wood is an abundant and accessible source of bioenergy but it is not known whether its use to produce synthetic fuels is economic. The procedures for the gasification of cellulose-containing materials have much in common with the conversion of coal to gas. Most of the conversion principles to be discussed are, however, applicable to the spectrum of carbonaceous or cellulosic materials which occur naturally, are grown, or are waste.

Synthetic fuel or *synfuel* is any liquid fuel obtained from any of the aforementioned fuel sources (i.e., tar sand, coal, oil shale, natural gas, natural gas hydrates, and biomass) as well as biologic alcohol and through the agency of the Fischer-Tropsch synthesis. For the purposes of this text, the term *synthetic fuel* also includes liquid fuels derived from crops, wood, waste plastics, and landfill materials. In a similar manner, *synthetic gaseous fuel* (*syngas*) and *synthetic solid fuel* can also (but less often) refer to gaseous fuels and solid fuels produced from the same sources.

In fact, *synthesis gas* (*syngas*) is the name given to a gas mixture that contains varying amounts of carbon monoxide and hydrogen generated by the gasification of a carbon containing fuel to a gaseous product with a heating value. Examples include steam reforming of natural gas or liquid hydrocarbons to produce hydrogen, the gasification of coal and in some types of waste-to-energy gasification facilities. The name comes from their use as intermediates in creating *synthetic natural gas* (SNG) and for producing ammonia or methanol. Syngas is also used as an intermediate in producing synthetic petroleum for use as a fuel or lubricant via Fischer-Tropsch synthesis and previously the Mobil methanol to gasoline process.

Syngas consists primarily of carbon monoxide, carbon dioxide, and hydrogen, and has less than half the energy density of natural gas. It is combustible and often used as a fuel source or as an intermediate for the production of other chemicals. Syngas for use as a fuel is most often produced by gasification of coal or municipal waste:

$$C + O_2 \rightarrow CO_2$$

$$CO_2 + C \rightarrow 2CO$$

$$C + H_2O \rightarrow CO + H_2$$

When used as an intermediate in the large-scale, industrial synthesis of hydrogen and ammonia, it is also produced from natural gas (via the steam reforming reaction):

$$CH_4 + H_2O \rightarrow CO + 3H_2$$

The syngas produced in large waste-to-energy gasification facilities is used as fuel to generate electricity.

Coal gasification processes (Chap. 5) are reasonably efficient and were used for many years to manufacture illuminating gas (coal gas) for gas lighting, before electric lighting became widely available.

When syngas contains a significant amount of nitrogen, the nitrogen must be removed. Cryogenic processing has great difficulty in recovering pure carbon monoxide when relatively large volumes of nitrogen are present, as carbon monoxide and nitrogen have very similar boiling points (i.e., $-191.5°C$ and $-195.79°C$, respectively). Instead there is technology that selectively removes carbon monoxide by complexation/decomplexation of carbon monoxide in a proprietary solvent containing cuprous aluminum chloride ($CuAlCl_4$) dissolved in an organic liquid such as toluene. The purified carbon monoxide can have purity greater than 99 percent. The reject gas from the process can contain carbon dioxide, nitrogen, methane, ethane, and hydrogen. The reject gas can be further processed on a pressure swing absorption system to remove hydrogen and the hydrogen and carbon dioxide recombined in the proper ratio for methanol production, Fischer-Tropsch diesel, and so on.

The process of producing synfuels is often referred to as *coal-to-liquids* (CTL), *gas-to-liquids* (GTL), or *biomass-to-liquids* (BTL), depending on the initial feedstock.

Gas-to-liquids is a process for conversion of natural gas into longer-chain hydrocarbons. Thus, methane-rich gases are converted into liquid fuels either via direct conversion or via syngas as an intermediate, for example, using the Fischer-Tropsch process. Using such processes, refineries can convert some of their gaseous waste products into valuable fuel oils, which can be sold as or blended only with diesel fuel. The process may also be used for the economic extraction of gas deposits in locations where it is not economic to build a pipeline. This process will be increasingly significant as crude oil resources are depleted, while natural gas supplies are projected to last into the twenty-second century.

The best known synthesis process is the Fischer-Tropsch synthesis which was used on a large scale in Germany during World War II. Other processes include the Bergius process, the Mobil process, and the Karrick process. An intermediate step in the production of synthetic fuel is often syngas, a stoichiometric mixture of carbon monoxide and hydrogen, which is sometimes directly used as an industrial fuel.

The leading company in the commercialization of synthetic fuel is Sasol, a company based in South Africa. Sasol currently operates the world's only commercial coal-to-liquids facility at Secunda, with a capacity of 150,000 bbl/day.

The U. S. Department of Energy has projected that domestic consumption of synthetic fuel made from coal and natural gas could rise to almost 4.0 million barrels per day in 2030 based on a price of $57 per barrel of high sulfur crude (Annual Energy Outlook 2006, Table 14, page 52). However, depending on price scenarios, synthetic fuels require a relatively high price of crude oil as well as a *relatively low price* for production (i.e., crude oil price is not the only parameter), in order to be competitive with petroleum-based fuels without subsidies. However, synthetic fuels do offer the potential to supplement or replace petroleum-based fuels if oil prices continue to rise.

Several factors do make synthetic fuels attractive relative to conventional petroleum-based fuels:

1. The raw material (coal and tar sand) is available in quantities sufficient to meet current demand for centuries.
2. Many of the raw materials can produce gasoline, diesel, or kerosene directly without the need for additional refining steps such as reforming or cracking.
3. In some cases, there is no need to convert vehicle engines to use a different fuel.
4. The distribution network is already in place.

However, with higher costs of production and higher risks, companies may be well inclined to seek tax credits for the production of synthetic fuels from nonconventional sources. In allowing these credits, governments will be well advised to consider the value of a measure of energy independence. Canada asked this question in the 1970s with respect to the development of the Athabasca oil sands (tar sands) located in north-eastern Alberta. As a result, production is now almost 1,000,000 bbl/day of synthetic crude oil which makes a considerable difference to Canada's imports of nondomestic crude oil.

1.4 PRODUCTION OF SYNTHETIC FUELS

1.4.1 Thermal Decomposition

The production of fuels from alternate fuel sources usually (but not always) involves a degree of thermal conversion. In a very general sense, thermal decomposition is often used to mean liquid production by thermal decomposition but gaseous and solid product may also be produced.

For example, *cracking* (*pyrolysis*) refers to the decomposition of organic matter by heat in the absence of air. Thermal decomposition is frequently used to mean the same, although it generally connotes the breakdown of inorganic compounds. The petroleum industry tends to use the words *cracking* and *coking* for the thermal decomposition of petroleum constituents.

When coal, oil shale, or tar sands are thermally decomposed, hydrogen-rich volatile matter is distilled and a carbon-rich solid residue is left behind. The carbon and mineral matter remaining behind is the residual char. In this regard, the term *carbonization* is sometimes used as a synonym for coal pyrolysis. However, carbonization has as its aim the production of a solid char, whereas in synthetic fuel production greatest interest centers on liquid and gaseous hydrocarbons.

Thermal decomposition is one method to produce liquid fuels from coal, and it is the principal method used to convert oil shale and tar sand bitumen to liquid fuels. Moreover, as gasification and liquefaction are carried out at elevated temperatures pyrolysis may be considered a first stage in any conversion process.

Of most interest in the production of synthetic fuels is the prediction of the rate and amount of volatile yield and product distribution for a given raw material and pyrolysis parameters. Among the important chemical variables are the elemental composition and the functional compositions of the organic and inorganic matter, as well as the composition of the ambient gas in which the pyrolysis takes place. Among the more important basic physical variables are the final temperature, the time and rate of heating, the particle size distribution, the type and duration of any quenching, and the pressure. An indication of the uncertainty existing in this field is that at present there is no agreement on whether yield, that is, the loss in mass of the raw material from pyrolysis, is changed with heating rate.

The best understood pyrolysis processes are the cracking and coking of petroleum, (Chap. 3). However, the predictive capability for producing any fuel from an alternate fuel source is very speculative, especially since the properties of that fuel source (even petroleum) can vary with the origin.

First and foremost, assuming all process parameters are equal, the *composition of the raw material* is important in determining the yield of distillable products. The principal material property defining the yield is the atomic hydrogen-to-carbon ratio (derived from the elemental analysis). On the other hand, the composition of the volatile products evolved during thermal decomposition is largely determined by the raw organic material.

The *reaction temperature* affects both the amount and composition of the volatile yields. When a fuel produces volatile products, the residence time of the products within the hot zone and the temperature of the hot zone can markedly affect the distortion of the final products. Secondary and tertiary products will be formed from the primary products and to use the delayed coking process as an analogy (Chap. 3), the products may undergo several thermal alternations before stabilization in the final molecular form.

Pressure will also affect the yield of distillable products since there is also a relationship between pressure and residence time. Generally, higher pressures favor cracking reactions and produce higher yields of lower molecular weight hydrocarbon gases, whereas lower pressure will lead to larger tar and oil fractions.

Thermal decomposition in *hydrogen atmosphere*, (*hydrocracking* or *hydropyrolysis*) can increase the yield of distillable products because the attendant hydrogenation diminishes the tendency for the formation of higher molecular weigh products (tar and coke).

For the solid feedstocks, *particle size* is known to influence product yield. Larger particles heat up more slowly, so the average particle temperatures will be lower, and hence volatile yields may be expected to be less. In addition, the time taken for the thermal products to diffuse out of the larger particle (i.e., longer residence time in the hot zone) also contributes to product distribution. However, if the particle size is sufficiently small, the feedstock will be heated relatively uniformly, and, with rapid diffusion of the products from the hot zone, a different product slate can be anticipated.

1.4.2 Gasification

Gasification is the conversion of a solid or liquid into a gas at high temperature in a controlled amount of oxygen. In a broad sense it includes evaporation by heating, although the term is generally reserved for processes involving chemical change. For example, the term *coal gasification* refers to the overall process of converting coal to a product gas, including the initial pyrolysis and subsequent gas thermal upgrading steps. The resulting gas mixture (*synthesis gas*, *syngas*) is a fuel.

Gasification is a very efficient method for extracting energy from many different types of organic materials, and also has applications in waste disposal. The syngas combusts cleanly into water vapor and carbon dioxide. Alternatively, syngas may be converted efficiently to methane via the Sabatier process or gasoline/diesel-like synthetic fuels via the Fischer-Tropsch process (Sec. 1.4.3).

The Sabatier process involves the reaction of hydrogen with carbon dioxide at elevated temperatures and pressures in the presence of a nickel- or ruthenium-containing catalyst to produce methane and water:

$$CO_2 + 4H_2 \rightarrow CH_4 + 2H_2O$$

Usually, the nickel or ruthenium is supported on alumina.

Inorganic components of the input material, such as metals and minerals, are trapped in the char and may or may not be environmentally safe because of the potential for the inorganic constituents to leach (caused by rain, melting snow, or acid rain) into the surrounding environment.

The advantage of gasification is that using the syngas is more efficient than direct combustion of the original fuel; more of the energy contained in the fuel is extracted. The syngas may be burned directly in internal combustion engines, used to produce methanol and hydrogen or converted via the Fischer-Tropsch process (Sec. 1.4.3) into synthetic fuel. Gasification can also begin with materials that are not otherwise useful fuels, such as biomass or organic waste.

Gasification of coal and petroleum is currently used on a wide scale to generate electricity. However, almost any type of organic material can be used as the raw material for gasification, such as biomass, wood, or even waste plastic.

However, gasification relies on chemical processes at elevated temperatures (>700°C), which distinguishes it from biologic processes such as anaerobic digestion that produce biogas.

Regardless of the final fuel form, gasification itself and subsequent processing neither emits nor traps greenhouse gases such as carbon dioxide. Combustion of synthesis or derived fuels does of course emit carbon dioxide. However, biomass gasification is perceived to play a role in a renewable energy economy because biomass production removes carbon dioxide from the atmosphere. While other biofuel technologies such as biogas and biodiesel are also carbon neutral, gasification (a) uses a wider variety of feedstocks, (b) can be used to produce a wider variety of products, and (c) is an efficient method of extracting energy from biomass.

Therefore, it is not surprising that biomass gasification is one of the most technically and economically convincing energy possibilities for the production of alternate fuels.

1.4.3 The Fischer-Tropsch Process

The *Fischer-Tropsch process* is a catalyzed chemical reaction in which carbon monoxide and hydrogen are converted into liquid hydrocarbons of various forms. Typical catalysts used are based on iron and cobalt. The principal purpose of this process is to produce a synthetic petroleum substitute for use as synthetic lubrication oil or as synthetic fuel. The process is currently used in South Africa to produce liquid fuels from syngas (produced from coal by gasification).

Chemically, the Fischer-Tropsch process is represented as the conversion of carbon monoxide and hydrogen to hydrocarbons and water:

$$nCO + (2n+1)H_2 \rightarrow C_nH_{(2n+2)} + nH_2O$$

The initial reactants in the above reaction (i.e., CO and H_2) can be produced by other reactions such as the partial combustion of a hydrocarbon:

$$C_nH_{(2n+2)} + \tfrac{1}{2} nO_2 \rightarrow (n+1)H_2 + nCO$$

Or by the gasification of coal or biomass:

$$C + H_2O \rightarrow H_2 + CO$$

The energy needed for this endothermic reaction of coal or biomass and steam is usually provided by (exothermic) combustion with air or oxygen. This leads to the following reaction:

$$2C + O_2 \rightarrow CO$$

The mixture of carbon monoxide and hydrogen is called *syngas*. The resulting hydrocarbon products are refined to produce the desired synthetic fuel.

The carbon dioxide and carbon monoxide is generated by partial oxidation of coal and wood-based fuels. The utility of the process is primarily in its role in producing fluid hydrocarbons from a solid feedstock, such as coal or solid carbon-containing wastes of various types. Nonoxidative pyrolysis of the solid material produces syngas which can be used directly as a fuel without being taken through Fischer-Tropsch transformations. If a liquid fuel, lubricant, or wax is required, the Fischer-Tropsch process can be applied successfully in the manufacture.

Part of the issue with the Fischer-Tropsch process is that it produces a mixture of hydrocarbons—many of which are not useful as fuel. However, the use of molecularly-specific catalysts to convert these undesirable hydrocarbons into specific liquid fuels is established. These catalysts work by rearranging the carbon atoms, transforming six-carbon atom hydrocarbons, for example, into two- and ten-carbon atom hydrocarbons.

The liquid fuels produced in this way have several potential advantages not the least of which is the absence of the odiferous, particle-producing aromatics. The liquid fuels formed by the Fischer-Tropsch process burn much cleaner and are environmentally more acceptable.

With respect to coal, there are several processes for the conversion of coal to gas. One in particular, the Karrick process, is a low temperature carbonization process in which coal is heated at 360 to 749°C (680 to 1380°F) in the absence of air to produce oil and gas. For example, Karrick processing of 1 ton (2000 lb) of coal yields up to 1 bbl of coal tar (12 percent by weight, rich in lower molecular weight hydrocarbons suitable for processing into fuels), 3000 ft^3 of fuel gas (a mixture of hydrogen, carbon monoxide, methane, and other volatile hydrocarbons), and 1500 lb of solid smokeless char or semicoke. Smokeless char can be used for utility boilers and cooking coal in steel smelters, yields more heat than raw coal, and can be converted to water gas which, in turn, can be converted to hydrocarbon fuel by the Fischer-Tropsch process.

1.4.4 Bioprocesses

A *bioprocess* is any process that uses complete living cells or organisms or their components (e.g., bacteria, enzymes) to effect a desired physical and/or chemical change in the feedstock. Transport of energy and mass is fundamental to many biological and environmental processes.

Modern bioprocess technology used this principle and is actually an extension of older methods for developing useful products by taking advantage of natural biologic activities. Although more sophisticated, modern bioprocess technology is based on the same principle; combining living matter (whole organisms or enzymes) with nutrients under the conditions necessary to make the desired end product. Bioprocesses have become widely used in several fields of commercial biotechnology, such as production of enzymes (used, e.g., in food processing and waste management) and antibiotics.

Since bioprocesses use living material, they offer several advantages over conventional chemical methods of production. Bioprocesses usually require lower temperature, pressure, and pH (the measure of acidity) and can use renewable resources (biomass) as raw materials. In addition, greater quantities can be produced with less energy consumption.

In most bioprocesses, enzymes are used to catalyze the biochemic reactions of whole microorganisms or their cellular components. The biologic catalyst causes the reactions to occur but is not changed. After a series of such reactions which take place in large vessels (*fermenters* or *fermentation tanks*), the initial raw materials are chemically changed to form the desired end product. Nevertheless, there are challenges to the use of bioprocesses in the production of synthetic fuels.

First, the conditions under which the reactions occur must be rigidly maintained. Temperature, pressure, pH, oxygen content, and flow rate are some of the process parameters that must be kept at very specific levels. With the development of automated and computerized equipment, it is becoming much easier to accurately monitor reaction conditions and thus increase production efficiency.

Second, the reactions can result in the formation of many unwanted by-products. The presence of contaminating waste material often poses a twofold problem related to (a) the means to recover (or separate) the end product in a way that leaves as little residue as possible in the catalytic system, and (b) the means by which the desired product can be isolated in pure form.

1.5 REFERENCES

American Coalition for Ethanol: "Benefits of Ethanol," 2004, http://www.ethanol.org.

Bartis, J. T., T. LaTourrette, L. Dixon, D. J. Peterson, and G. Cecchine: "Oil Shale Development in the United States Prospects and Policy Issues," Report No. MG-414-NETL, RAND Corporation, Santa Monica, Calif., 2005.

Chemical Market Reporter: "GTL Could Become Major Chemicals Feedstock," January 12, 2004.

Coaltrans: "Coal Liquefaction Enters New Phase," May–June, 2003, pp. 24–25.

Cooke, R. R.: "Industry Analysis," *Oil, Jihad and Destiny: Will Declining Oil Production Plunge Our Planet into a Depression?* Opportunity Analysis, Saratoga, Calif., 2005, chap. 2, pp. 5–31.

European Environment Agency: "How Much Bioenergy Can Europe Produce Without Harming the Environment?" EEA Report No. 7, 2006.

Freeman, S. D.: *Winning Our Energy Independence: An Insider Show How,* Gibbs Smith, Salt Lake City, Utah, 2007.

Gielen, D. and F. Unander: "Alternative fuels: An Energy Technology Perspective," Report No. EET/2005/01. International Energy Agency, Paris, France, 2005.

Green, D. W. and G. P. Willhite: *Enhanced Oil Recovery, Textbook Vol. 6,* Society of Petroleum Engineers, Richardson, Tex., 1998.

IEA: "World Energy Outlook," IEA/OECD, Paris, France, 2004.

Lee, S.: *Oil Shale Technology,* CRC Press, Boca Raton, Fla., 1991.

Marshall, A. T.: "Bioenergy from Waste: A Growing Source of Power," *Waste Management World Magazine,* April 2007, pp. 34–37.

Hirsch, R. L., R. Bezdek, and R. Wendling: "Peaking of World Oil Production: Impacts, Mitigation, and Risk Management. United States Department of Energy, National Energy Technology Laboratory," Study by Science Applications International Corporation. 2005.

Hudson, J.: "Biotech Reduces the Cost of Ethanol Production 12-fold," *Comstock's Business,* June 2004, pp. 86–87.

Longwell, J. P. (Chairman): "Fuels to Drive Our Future," Committee on Production Technologies for Liquid Transportation Fuels, Energy Engineering Board Commission on Engineering and Technical Systems, National Research Council, National Academic Press, Washington DC, 1990.

Lowe, P. A., W. C. Schroeder, and A. L. Liccardi: "Technical Economies, Synfuels and Coal Energy Symposium, Solid-Phase Catalytic Coal Liquefaction Process," Proceedings, The American Society of Mechanical Engineers. 1976.

Pacific Biodiesel: "Fuel Properties," 2004, http://www.biodiesel.com/theFuel.htm.

Scouten, C.: In *Fuel Science and Technology Handbook*, J. G. Speight (ed.), Marcel Dekker Inc., New York, 1990.

Speight, J. G.: *The Chemistry and Technology of Coal,* 2nd ed., Marcel Dekker Inc., New York, 1994.

Speight, J. G.: *Handbook of Petroleum Analysis,* John Wiley & Sons Inc., New York, 2001.

Speight, J. G.: *The Chemistry and Technology of Petroleum,* 4th ed., CRC Press-Taylor and Francis Group, Boca Raton, Fla., 2007.

US Congress. "Public Law FEA-76-4," Congress of the United States of America, Washington, DC, 1976.

CHAPTER 2
NATURAL GAS

Natural gas (also called *marsh gas* or *swamp gas* in older texts and more recently *landfill gas*) is a naturally occurring gaseous fossil fuel that is found in oil fields and natural gas fields, and in coalbeds.

For clarification, *natural gas* is not the same as *town gas*, which is manufactured from coal and the terms *coal gas*, *manufactured gas*, *producer gas*, and *syngas* (*synthetic natural gas*, SNG) are also used for gas produced from coal. Depending on the production process, gas from coal is a mixture of hydrogen, carbon monoxide, methane, and volatile hydrocarbons in varying amounts with small amounts of carbon dioxide and nitrogen as impurities.

Prior to the development of resources, virtually all fuel and lighting gas was manufactured from coal and the history of *natural gas cleaning* has its roots in *town gas cleaning* (Chap. 5). The by-product coal tar produced during the manufacture of gas from coal was an important feedstock for the chemical industry. The development of manufactured gas paralleled that of the industrial revolution and urbanization.

2.1 HISTORY

The uses of natural gas did not necessarily parallel its discovery. In fact, the discovery of natural gas dates from ancient times in the Middle East. During recorded historical time, there was little or no understanding of what natural gas was; it posed somewhat of a mystery to man. Sometimes, such things as lightning strikes would ignite natural gas that was escaping from under the earth's crust. This would create a fire coming from the earth, burning the natural gas as it seeped out from underground. These fires puzzled most early civilizations, and were the root of much myth and superstition. One of the most famous of these types of flames was found in ancient Greece, on Mount Parnassus approximately 1000 B.C. A goat herdsman came across what looked like a *burning spring*, a flame rising from a fissure in the rock. The Greeks, believing it to be of divine origin, built a temple on the flame. This temple housed a priestess who was known as the Oracle of Delphi, giving out prophecies she claimed were inspired by the flame.

These types of springs became prominent in the religions of India, Greece, and Persia. Unable to explain where these fires came from, they were often regarded as divine, or supernatural. However the energy value of natural gas was not recognized until approximately 900 B.C. in China and the Chinese drilled the first known natural gas well in 211 B.C. The Chinese formed crude pipelines out of bamboo shoots to transport the gas, where it was used to boil sea water, for separating the salt and making it drinkable.

Natural gas was discovered and identified in America as early as 1626, when French explorers discovered natives igniting gases that were seeping into and around Lake Erie. However, Britain was the first country to commercialize the use of natural gas. Around 1785, natural gas produced from coal was used to light houses, as well as streetlights.

Manufactured natural gas of this type (as opposed to naturally occurring gas) was first brought to the United States in 1816, when it was used to light the streets of Baltimore, Maryland. This manufactured gas was much less efficient, and less environment friendly, than modern natural gas that comes from underground.

In 1821 in Fredonia, United States, residents observed gas bubbles rising to the surface from a creek. William Hart, considered as *America's father of natural gas*, dug there the first natural gas well in North America (Speight, 1993, Chap. 1 and references cited therein; Speight, 2007b). The American natural gas industry got its beginnings in this area. In 1859, Colonel Edwin Drake (a former railroad conductor who adopted the title "Colonel" to impress the townspeople) dug the first well. Drake hit oil and natural gas at 69 ft below the surface of the earth.

More recently, natural gas was discovered as a consequence of prospecting for crude oil. Natural gas was often an unwelcome by-product, as natural gas reservoirs were tapped in the drilling process and workers were forced to stop drilling to let the gas vent freely into the air. Now, and particularly after the crude oil shortages of the 1970s, natural gas has become an important source of energy in the world.

Throughout the nineteenth century, natural gas was used almost exclusively as source of light and its use remained localized because of lack of transport structures, making difficult to transport large quantities of natural gas through long distances. There was an important change in 1890 with the invention of leak proof pipeline coupling but transportation of natural gas to long distance customers did not become practical until the 1920s as a result of technological advances in pipelines. Moreover, it was only after World War II that the use of natural gas grew rapidly because of the development of pipeline networks and storage systems.

2.2 FORMATION AND OCCURRENCE

Natural gas is found occurring with petroleum (*associated gas*) or alone (*nonassociated gas*) in reservoirs and in coalbeds (*coalbed methane*) (Mokhatab et al., 2006 and references cited therein; Speight, 2007a and references cited therein).

Natural gas often occurs in conjunction with crude oil, although natural gas reservoirs that contain condensate (higher molecular weight hydrocarbons up to about C_{10}) but no crude oil are equally well known.

Like crude oil, natural gas has been generated over geologic time by the decay of animal remains and plant remains (*organic debris*) that has occurred over millions of years. Over time, the mud and soil that covered the organic debris changed to rock and trapped the debris beneath the newly formed rock sediments. Pressure and, to some extent, heat (the geothermal gradient) changed some of the organic material into coal, some into oil (petroleum), and some into natural gas. Whether or not the debris formed coal, petroleum, or gas depended upon the nature of the debris and the localized conditions under which the changes occurred.

Although the geothermal gradient varies from place to place, it is generally on the order of 25 to 30°C/km (15°F/1000 ft or 120°C/1000 ft, i.e., 0.015°C/ft of depth or 0.012°C/ft of depth), that is, approximately 1°C for every 100 ft below the surface. Thus, with increasing depth of the reservoir, there is a tendency for crude oil to become lighter insofar as it contains increasing amounts of low molecular weight hydrocarbons and decreasing amounts of the higher molecular weight constituents.

However, there is considerable discussion about the heat to which the organic precursors have been subjected. Cracking temperatures ($\geq 300°C$, $\geq 572°F$) are not by any means certain as having played a role in natural gas formation. Maturation of the organic debris through temperature effects occurred over geologic time (millennia) and shortening the time to laboratory time and increasing the temperature to above and beyond the cracking temperature (at which the chemistry changes) does not offer conclusive proof of high temperatures (Speight, 2007a).

Nevertheless, at some point during or after the maturation process, the gas and crude oil migrate from the source rock either upward or sideways or in both directions (subject to the structure of the accompanying and overlying geological formations). Eventually, the gas and crude oil became trapped in reservoirs that may be many miles from the source rock. It is rare that the source rock and the reservoir were one and the same. Thus, a natural gas field may have a series of layers of crude oil/gas and gas reservoirs in the subsurface. In some instances, the natural gas and crude oil parted company leading to the occurrence of reservoirs containing only gas (nonassociated gas).

Reservoirs generally comprise a geologic formation that is made up of layers of porous, sedimentary rock, such as sandstone, in which the gas can collect. However, for retention of the gas each trap must have an impermeable base rock and an impermeable cap rock to prevent further movement of the gas. Such formations, known as *reservoirs* or *traps* (i.e., naturally occurring storage areas) vary in size and can retain varying amounts of gas.

There are a number of different types of these formations, but the most common is, characteristically, a folded rock formation such as an anticline as occurs in many petroleum reservoirs (Fig. 2.1), that traps and holds natural gas. On the other hand, a reservoir may be formed by a geologic fault that occurs when the normal sedimentary layers sort of split vertically, so that impermeable rock shifts down to trap natural gas in the more permeable limestone or sandstone layers. Essentially, the geologic formation which layers impermeable rock over more porous, oil and gas rich sediment has the potential to form a reservoir.

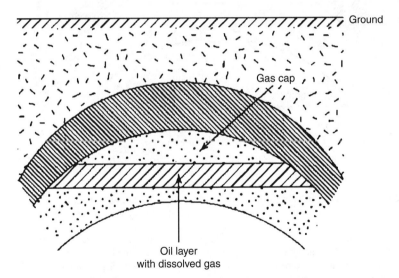

FIGURE 2.1 An anticlinal reservoir containing oil and (associated) gas.

Reservoirs vary in size from a few hundred meters to tens of kilometers across in plain, and tens to hundreds of meters thick, with the gas trapped against an impermeable layer similar to crude oil traps (Speight, 2007a). Some reservoirs may be only hundreds of feet below the surface. Other reservoirs are thousands, even tens of thousands of feet underground. In the United States, several reservoirs have been discovered at depths greater than 30,000 ft. Many offshore wells are drilled in thousands of feet of water and penetrate tens of thousands of feet into the sediments below the sea floor.

Natural gas reservoirs, like crude oil reservoirs, exist in many forms such as the dome (syncline-anticline) structure (Fig. 2.2), with water below, or a dome of gas with a crude oil rim and water below the oil. When the water is in direct contact with the gas, pressure effects may dictate that a considerable portion of the gas (20 percent or more) is dissolved in the crude oil as well as in the water. As gas is produced (or recovered) from the reservoir, the reservoir pressure declines allowing the dissolved gas to enter the gas phase. In addition, and because of the variability of reservoir structure, gas does not always flow equally to wells placed throughout the length, breadth, and depth of the reservoir and at equal pressure. Recovery wells must be distributed throughout the reservoir to recover as much of the gas as efficiently as possible.

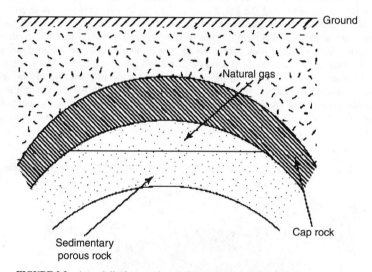

FIGURE 2.2 An anticlinal reservoir containing unassociated natural gas.

As the gas pressure in the reservoir declines, the reservoir energy (i.e., reservoir pressure) declines, and the gas requires stimulation for continued production. Furthermore, reduction in the gas pressure may allow compaction of the reservoir rock by the weight of rock above eventually resulting in subsidence of the surface above the reservoir. This can be gradual process or a sudden catastrophic process depending on the structures of the geologic formation above the reservoir.

A reservoir containing *wet* gas with a large amount of valuable *natural gas liquids* (any hydrocarbons other than methane such as ethane, propane, and butane) and even light crude oil and condensate has to be treated carefully. When the reservoir pressure drops below the critical point for the mixture, the liquids may condense out and remain in the reservoir. Thus it is necessary to implement a *cycling* process in which the wet gas is produced to the surface and the natural gas liquids are condensed as a separate stream and the gas is compressed and injected back into the reservoir to maintain the pressure.

Once brought from underground (recovered from the reservoir), natural gas is refined to remove impurities like water, other gases, sand, and other compounds. Some hydrocarbons are removed and sold separately, including propane and butane. Other impurities are also removed, like hydrogen sulfide (the refining of which can produce sulfur, which is then also sold separately). After refining, the clean natural gas is transmitted through a network of pipelines that deliver natural gas to its point of use.

In terms of reserves, natural gas is produced on all continents except Antarctica (BP, 2006). The world's largest producer is Russia. The United States, Canada, and the Netherlands are also important producers. The proven reserves of natural gas are of the order of in excess of 3600 trillion cubic feet (1 Tcf = 1×10^{12}). Approximately 300 Tcf exist in the United States and Canada. It should also be remembered that the total gas resource base (like any fossil fuel or mineral resource base) is dictated by economics. Therefore, when resource data are quoted some attention must be given to the cost of recovering those resources. Most important, the economics must also include a cost factor that reflects the willingness to secure total, or a specific degree of, energy independence.

Natural gas is produced in many countries around the world and most of those countries produce both oil and natural gas; a few produce only natural gas. The ten largest natural gas-producing countries are: United States, Canada, Russia, United Kingdom, Algeria, Netherlands, Iran, Indonesia, Norway, and Uzbekistan (EIA, 2004; 2006). Because transportation costs add to the cost of natural gas, in most countries natural gas is consumed within the country or exported to a neighboring country by pipeline. Technology for liquefying natural gas so that it can be transported in tankers is improving (Mokhatab et al., 2006). As technology continues to expand the options for gas transportation, demand for natural gas is expected to grow.

A common misconception about natural gas is that resources are being depleted at an alarming rate and the supplies are quickly running out. In fact, there is a vast amount of natural gas estimated still to be retrieved from a variety of reservoirs. However, many proponents of the rapid-depletion theory believe that price spikes indicate that natural gas resources are depleted beyond the point of no return. However, price spikes of any commodity are not always caused by waning resources but can be the outcome of other forces at work in the marketplace.

2.3 CONVENTIONAL GAS

2.3.1 Associated Gas

Associated or dissolved natural gas occurs either as free gas in a petroleum reservoir or as gas in solution in the petroleum. Gas that occurs as a solution in the petroleum is *dissolved* gas whereas the gas that exists in contact with the petroleum (*gas cap*) is *associated* gas.

Crude oil cannot be produced without producing some of its associated gas, which comes out of solution as the pressure is reduced on the way to and on the surface. Properly designed well completions and reservoir management are used to minimize the production of associated gas so as to retain the maximum energy in the reservoir and thus increase ultimate recovery. Crude oil in the reservoir with minimal or no dissolved associated gas is rare and *dead crude oil* is often difficult to produce as there is little energy to drive it.

After the production fluids are brought to the surface, they are separated at a tank battery at or near the production lease into a hydrocarbon liquid stream (crude oil or *gas condensate*), a produced water stream (brine or salty water), and a gaseous stream. The gaseous stream is traditionally very rich (*rich gas*) in *natural gas liquids* (NGLs). Natural gas liquids include ethane, propane, butanes, and pentane (C_5H_{12}) and higher molecular weight hydrocarbons. The higher molecular weight hydrocarbon product, which may also contain some pentane, is commonly referred to as *natural gasoline*.

Rich gas has a high heating value and a high hydrocarbon dew point. However, the terms *rich gas* and *lean gas*, as used in the gas processing industry, are not precise indicators of gas quality but only indicate the relative amount of natural gas liquids in the gas stream. When referring to natural gas liquids in the natural gas stream, the term *gallons per thousand cubic feet* of gas is used as a measure of hydrocarbon richness.

Thus, in the case of associated gas, crude oil may be assisted up the wellbore by gas lift (Speight, 1993, 2007b). Thus, gas is compressed into the annulus of the well and then injected by means of a gas lift valve near the bottom of the well into the crude oil column in the tubing. At the top of the well the crude oil and gas mixture passes into a separation plant that drops the pressure down to nearly atmospheric in two stages. The crude oil and water exits the bottom of the lower pressure separator, from where it is pumped to tanks for separation of the crude oil and water. The gas produced in the separators is recompressed and the gas that comes out of solution with the produced crude oil (surplus gas) is then treated to separate out the natural gas liquids that are treated in a gas plant to provide propane and butane or a mixture of the two (liquefied petroleum gas, LPG). The higher boiling residue, after the propane and butane are removed, is condensate, which is mixed with the crude oil or exported as a separate product.

The gas itself is then *dry* and, after compression, is suitable to be injected into the natural gas system where it substitutes for natural gas from the nonassociated gas reservoir. Pretreated associated gas from other fields enters the system at this stage (Manning and Thompson, 1991). Another use for the gas is as fuel for the gas turbines on site. This gas is treated in a fuel gas plant to ensure it is clean and at the correct pressure. The startup fuel gas supply will be from the main gas system, but facilities exist to collect and treat low-pressure gas from the various other plants as a more economical fuel source.

2.3.2 Nonassociated Gas

Nonassociated gas (sometimes called *gas well gas*) is produced from geologic formations that typically do not contain much, if any, crude oil, or higher boiling hydrocarbons (*gas liquids*) than methane. However, nonassociated gas can contain non–hydrocarbon gases such as carbon dioxide and hydrogen sulfide.

The nonassociated gas recovery system is somewhat simpler than the associated gas recovery system. The gas flows up the well under its own energy, through the wellhead control valves and along the flow line to the treatment plant. Treatment requires the temperature of the gas to be reduced to a point dependent upon the pressure in the pipeline so that all liquids that would exist at pipeline temperature and pressure condense and are removed.

The water in the gas must also be dealt with to stop the formation of *gas hydrates* (Sec. 2.4.2) that may block the pipes. One method is to inject ethylene glycol (*glycol*) that combines with the water and is later recovered by a glycol plant. The treated gas then flows from the top of the treatment vessel and into the pipeline. The water is treated in a glycol plant to recover the glycol. Any natural gas liquids are pumped as additional feedstock to the liquefied petroleum gas plant.

2.3.3 Liquefied Natural Gas

When natural gas is cooled to a temperature of approximately $-160°C$ ($-260°F$) at atmospheric pressure, it condenses to a liquid (liquefied natural gas, LNG). One volume of this liquid takes up about one six-hundredth the volume of natural gas. Liquefied natural gas weighs less than one-half that of water, actually about 45 percent as much. Liquefied natural gas is odorless, colorless, noncorrosive, and nontoxic. When vaporized it burns only in concentrations of 5 to 15 percent when mixed with air (Sec. 2.4). Neither liquefied natural gas, nor its vapor, can explode in an unconfined environment. Since liquefied natural gas takes less volume and weight, it presents more convenient options for storage and transportation.

2.4 UNCONVENTIONAL GAS

The boundary between conventional gas and unconventional gas resources is not well defined, because they result from a continuum of geologic conditions. Coal seam gas, more frequently called coalbed methane, is frequently referred to as unconventional gas. Tight shale gas and gas hydrates are also placed into the category of *unconventional gas*.

2.4.1 Coalbed Methane

Coalbed methane and gas hydrates (Sec. 2.4.2) are two relatively new resources that is recognized as having plentiful supplies of methane and other lower boiling hydrocarbons (Berecz and Balla-Achs, 1983; Sloan, 1997; Gudmundsson et al., 1998; Max, 2000; Sloan, 2000).

Coalbed methane (CBM) is the generic term given to methane gas held in underground coal seams and released or produced when the water pressure within the seam is reduced by pumping from either vertical or inclined to horizontal surface holes. The methane is predominantly formed during the coalification process whereby organic matter is slowly transformed into coal by increasing temperature and pressure as the organic matter is buried deeper and deeper by additional deposits of organic and inorganic matter over long periods of geologic time. This is referred to as *thermogenic coalbed methane*.

Alternatively, and more often (but not limited to) in lower rank and thermally immature coals, recent bacterial processes (involving naturally occurring bacteria associated with meteoric water recharge at outcrop or sub-crop) can dominate the generation of coalbed methane. This is referred to as late stage *biogenic coalbed methane*.

During the coalification process, a range of chemical reactions take place which produce substantial quantities of gas. While much of this gas escapes into the overlying or underlying rock, a large amount is retained within the forming coal seams. However, unlike conventional natural gas reservoirs, where gas is trapped in the pore or void spaces of a rock such as sandstone, methane formed and trapped in coal is actually adsorbed onto the coal grain surfaces or micropores and held in place by reservoir (water) pressure. Therefore because the micropore surface area is very large, coal can potentially hold significantly more methane per unit volume than most sandstone reservoirs.

The amount of methane stored in coal is closely related to the rank and depth of the coal; the higher the coal rank and the deeper the coal seam is presently buried (causing pressure on coal) the greater its capacity to produce and retain methane. Because coal has a very large internal surface area of over 1 billion square feet per ton of coal, it can hold on average three times as much gas in place as the same volume of a conventional sandstone reservoir at equal depth and pressure. In order to allow the "absorbed" gas to be released from the coal it is often necessary to lower the pressure on the coal. This generally involves removing the water contained in the coalbed. After the gas is released from the internal surfaces of the coal it moves through the coal's internal matrix until it reaches natural fracture networks in the coal known as *cleats*. The gas then flows through these cleats or fractures until it reaches the well bore.

Gas derived from coal is generally pure and requires little or no processing because it is solely methane and not mixed with heavier hydrocarbons, such as ethane, which is often present in conventional natural gas. Coalbed methane has a slightly higher energy value than some natural gases. Coal seam gas well productivity depends mostly on reservoir pressure and water saturation.

To recover coalbed methane, multi-well patterns are necessary to dewater the coal and to establish a favorable pressure gradient. Since the gas is adsorbed on the surface of the coal and trapped by reservoir pressure, initially there is low gas production and high water production. Therefore, an additional expense relates to the disposal of coalbed water, which may be saline, acidic, or alkaline. As production continues, water production declines and gas production increases, before eventually beginning a long decline. In general, however, coal seam gas recovery rates have been low and unpredictable. Average per-well conventional gas production in a mature gas-rich basin is about five times higher than average per-well coal seam gas production. Thus, several times as many wells have to be drilled in coal seams than in conventional gas accumulations to achieve similar gas recovery levels.

2.4.2 Gas Hydrates

In addition to coalbed methane (Sec. 2.4.1), another relatively new, and possibly large source of methane that can be expected to extend the availability of natural gas, is *methane hydrate* (also called *gas hydrate* or *methane ice*) (Berecz and Balla-Achs, 1983; Sloan, 1997; Gudmundsson et al., 1998; Max, 2000; Sloan, 2000). Their production technologies have only recently been developed and these sources are now becoming economically competitive.

A *gas hydrate* is a molecule consisting of an ice lattice or cage in which low molecular weight hydrocarbon molecules, such as methane, are embedded.

The two major conditions that promote hydrate formation are thus: (a) high gas pressure and low gas temperature, and (b) the gas at or below its water dew point with free water present.

Gas hydrates are common constituents of the shallow marine geosphere and occur both in deep sedimentary structures, and as outcrops on the ocean floor. Methane hydrates are believed to form by migration of gas from depth along geologic faults, followed by precipitation, or crystallization, on contact of the rising gas stream with cold sea water.

At high pressures methane hydrates remain stable at temperatures up to 18°C and the typical methane hydrate contain one molecule of methane for every six molecules of water that forms the ice cage, but this ratio is dependent on the number of methane molecules that fit into the various cage structures of the water lattice. One liter of solid methane hydrate can contain up to 168 L of methane gas.

Methane hydrates are restricted to the shallow lithosphere (i.e., less than 2000 m depth). Furthermore, necessary conditions are found only either in polar continental sedimentary rocks where surface temperatures are less than 0°C; or in oceanic sediment at water depths greater than 300 m where the bottom water temperature is around 2°C (35°F). Continental deposits have been located in Siberia and Alaska in sandstone and siltstone beds at less than 800 m depth.

The methane in gas hydrates is dominantly generated by bacterial degradation of organic matter in low oxygen environments. Organic matter in the uppermost few centimeters of sediments is first attacked by aerobic bacteria, generating carbon dioxide, which escapes from the sediments into the water column. In this region of aerobic bacterial activity sulfates are reduced to sulfides. If the sedimentation rate is low (<1 cm per 1000 years), the organic carbon content is low (<1 percent), and oxygen is abundant, aerobic bacteria use up all the organic matter in the sediments. But where sedimentation rates and the organic carbon content are high, the pore waters in the sediments are anoxic at depths of only a few centimeters, and methane is produced by anaerobic bacteria.

The presence of clathrates at a given site can often be determined by observation of a *bottom simulating reflector* (BSR), which is a seismic reflection at the sediment to clathrate stability zone interface caused by the different density between normal sediments and sediments laced with clathrates.

The size of the oceanic methane hydrate reservoir is not well defined and estimates of its size have varied considerably over a wide range. However, improvements in understanding the nature of the gas hydrate resource have revealed that hydrates only form in a narrow range of depths (such as in the area of continental shelves) and typically are found at low concentrations (0.9–1.5 percent by volume) at sites where they do occur. Recent estimates constrained by direct sampling suggest the global inventory lies between 1×10^{15} and 5×10^{15} m^3 of gas.

The potential for hydrate development is high and commercial development of natural gas hydrates could cause a shifting in energy supply geopolitics because of the amount of natural gas sequestered in hydrate form. If only a small part of the gas can be produced, especially adjacent to countries that are presently energy importers, there will be major economic and geopolitical consequences that would dramatically change the present energy supply system.

2.4.3 Shale Gas

Large continuous gas accumulations are sometimes present in low permeability shale, (tight) sandstones, siltstones, sandy carbonates, limestone, dolomite, and chalk. Such gas deposits are commonly classified as unconventional because their reservoir characteristics differ from conventional reservoirs and they require stimulation to be produced economically.

The tight gas is contained in lenticular or blanket reservoirs that are relatively impermeable, which occur downdip from water-saturated rocks and cut across lithologic boundaries. They often contain a large amount of in-place gas, but exhibit low recovery rates. Gas can be economically recovered from the better quality continuous tight reservoirs by creating downhole fractures with explosives or hydraulic pumping. The nearly vertical fractures provide a pressure sink and channel for the gas, creating a larger collecting area so that the gas recovery is at a faster rate. Sometimes massive hydraulic fracturing is required, using half a million gallons of gelled fluid and a million pounds of sand to keep the fractures open after the fluid has been drained away.

In the United States, unconventional gas accumulations account for about 2 Tcf of gas production per year, some 10 percent of total gas output. In the rest of the world, however, gas is predominantly recovered from conventional accumulations.

2.5 COMPOSITION

Natural gas is a combustible mixture of hydrocarbon gases (Mokhatab et al., 2006). The principle constituents of natural gas, in varying amounts depending upon the source (Table 2.1) (Speight, 1993, 2007b) are methane (CH_4), ethane (C_2H_6), propane (C_3H_8), butanes (C_4H_{10}), pentanes (C_5H_{12}), hexane (C_6H_{14}), heptane (C_7H_{16}), and sometimes trace amounts of octane (C_8H_{18}), and higher molecular weight hydrocarbons. Some aromatics

TABLE 2.1 Range of Composition of Natural Gas

Constituents	Formula	Amount present
Methane	CH_4	70–90%
Ethane	C_2H_6	
Propane	C_3H_8	0–20%
Butane	C_4H_{10}	
Pentane and higher hydrocarbons	C_5H_{12}	0–10%
Carbon dioxide	CO_2	0–8%
Oxygen	O_2	0–0.2%
Nitrogen	N_2	0–5%
Hydrogen sulfide, carbonyl sulfide	H_2S, COS	0–5%
Rare gases: argon, helium, neon, xenon	A, He, Ne, Xe	Trace

[BTX—benzene (C_6H_6), toluene ($C_6H_5CH_3$), and the xylene ($CH_3C_6H_4CH_3$)] can also be present, raising safety issues due to their toxicity. The non–hydrocarbon gas portion of the natural gas contains nitrogen (N_2), carbon dioxide (CO_2), helium (He), hydrogen sulfide (H_2S), water vapor (H_2O), and other sulfur compounds such as carbonyl sulfide (COS) and mercaptans (e.g., CH_3SH) and trace amounts of other gases. Carbon dioxide and hydrogen sulfide are commonly referred to as *acid gases* since they form corrosive compounds in the presence of water. Nitrogen, helium, and carbon dioxide are also referred to as *diluents* since none of these burn, and thus they have no heating value. Mercury can also be present either as a metal in vapor phase or as an organo-metallic compound in liquid fractions. Concentration levels are generally very small, but even at very small concentration levels, mercury can be detrimental due its toxicity and its corrosive properties (reaction with aluminum alloys).

However, in its purest form, the natural gas that is delivered to the consumer is almost pure methane and the remaining hydrocarbons and non-hydrocarbons have been removed through refining. The non-hydrocarbon constituents include, but are not limited to, carbon dioxide (CO_2), hydrogen sulfide (H_2S), nitrogen (N_2), and helium (He). Because natural gas is colorless, shapeless, odorless, and tasteless in its pure form, it is not possible to see or smell natural gas. Therefore, an odorant (a *mercaptan* also called a *thiol* that is a sulfur-containing compound having the general formula R-SH) is added to natural gas for safety reasons so that it can be smelled if there is a gas leak. A mercaptan is a chemical odorant that smells a little like rotten eggs or skunk spray.

There are several general definitions that have been applied to natural gas that are based on composition. For example, *lean gas* is gas in which methane is the major constituent. On the other hand, *wet gas* contains considerable amounts of the higher molecular weight hydrocarbons. To further define the terms *dry* and *wet* in quantitative measures, the term *dry* natural gas indicates that there is less than 0.1 gal (1 gal, U.S. = 264.2 m³) of gasoline vapor (higher molecular weight paraffins) per 1000 ft³ (1 ft³ = 0.028 m³). The term *wet natural gas* indicates that there are such paraffins present in the gas, in fact more than 0.1 gal/1000 ft³. Natural gas is considered *dry* when it is almost pure methane, having had most of the other commonly associated hydrocarbons removed. When other hydrocarbons are present, the natural gas is *wet*.

Sour gas contains hydrogen sulfide whereas *sweet gas* contains very little, if any, hydrogen sulfide. *Residue gas* is natural gas from which the higher molecular weight hydrocarbons have been extracted and *casing head gas* is derived from petroleum but is separated at the separation facility at the wellhead.

2.6 PROPERTIES

The properties of unrefined natural gas are variable because the composition of natural gas is never constant. Therefore, the properties and behavior of unrefined and refined natural gas (Table 2.2), as determined by a series of standard test methods (Table 2.3), are best understood by investigating the properties and behavior of the constituents.

TABLE 2.2 General Properties of Unrefined (Left Hand Data) and Refined (Right Hand Data) Natural Gas

Relative molar mass	20–16
Carbon content (weight %)	73–75
Hydrogen content (weight %)	27–25
Oxygen content (weight %)	0.4–0
Hydrogen-to-hydrogen atomic ratio	3.5–4.0
Density relative to air @15°C	1.5–0.6
Boiling temperature (°C/1 atm)	−162
Autoignition temperature (°C)	540–560
Octane number	120–130
Methane number	69–99
Vapor flammability limits (volume %)	5–15
Flammability limits	0.7–2.1
Lower heating/calorific value (Btu)	900
Methane concentration (volume %)	100–80
Ethane concentration (volume %)	5–0
Nitrogen concentration (volume %)	15–0
Carbon dioxide concentration (volume %)	5–0
Sulfur concentration (ppm, mass)	5–0

http://www.visionengineer.com/env/alt_ng_prop.php.

Thus, assuming that the natural gas has been cleaned (i.e., any constituents such as carbon dioxide and hydrogen sulfide have been removed and the only constituents remaining are hydrocarbons), the properties and behavior of natural gas becomes a study of the properties and behavior of the relevant hydrocarbons (see for example, Speight, 2005a).

The composition of natural gas varies depending on the field, the formation, or the reservoir from which it is extracted and that are an artifact of its formation. The different hydrocarbons that form natural gas can be separated using their different physical properties as weight, boiling point, or vapor pressure. Depending on its content of higher molecular weight hydrocarbon components, natural gas can be considered as rich (5 or 6 gal or more of recoverable hydrocarbon components per cubic feet) or lean (less than 1 gal of recoverable hydrocarbon components per cubic feet).

2.6.1 Density

Density is a physical property of matter, is a measure of the relative *heaviness* of hydrocarbons and other chemicals at a constant volume, and each constituents of natural gas has a unique density associated with it. Another term, *specific gravity* is commonly used is relation to the properties of hydrocarbons. The *specific gravity* of a substance is a comparison of its density to that of water. For most chemical compounds (i.e., those that are solid or liquid), the density is measured relative to water (1.00). For gases, the density is more likely to be compared to the density of air (also given the number 1.00 but this is arbitrary

TABLE 2.3 Standard Test Methods for Natural Gas (ASTM, 2007)

D1070-03 Standard Test Methods for Relative Density of Gaseous Fuels
D1071-83(2003) Standard Test Methods for Volumetric Measurement of Gaseous Fuel Samples
D1072-90(1999) Standard Test Method for Total Sulfur in Fuel Gases
D1142-95(2000) Standard Test Method for Water Vapor Content of Gaseous Fuels by Measurement of Dew-Point Temperature
D1826-94(2003) Standard Test Method for Calorific (Heating) Value of Gases in Natural Gas Range by Continuous Recording Calorimeter
D1945-03 Standard Test Method for Analysis of Natural Gas by Gas Chromatography
D1946-90(2000) Standard Practice for Analysis of Reformed Gas by Gas Chromatography
D1988-91(2000) Standard Test Method for Mercaptans in Natural Gas Using Length-of-Stain Detector Tubes
D3588-98(2003) Standard Practice for Calculating Heat Value, Compressibility Factor, and Relative Density of Gaseous Fuels
D3956-93(2003) Standard Specification for Methane Thermophysical Property Tables
D3984-93(2003) Standard Specification for Ethane Thermophysical Property Tables
D4084-94(1999) Standard Test Method for Analysis of Hydrogen Sulfide in Gaseous Fuels (Lead Acetate Reaction Rate Method)
D4150-03 Standard Terminology Relating to Gaseous Fuels
D4362-93(2003) Standard Specification for Propane Thermophysical Property Tables
D4468-85(2000) Standard Test Method for Total Sulfur in Gaseous Fuels by Hydrogenolysis and Rateometric Colorimetry
D4650-93(2003) Standard Specification for Normal Butane Thermophysical Property Tables
D4651-93(2003) Standard Specification for Isobutane Thermophysical Property Tables
D4784-93(2003) Standard for LNG Density Calculation Models
D4810-88(1999) Standard Test Method for Hydrogen Sulfide in Natural Gas Using Length-of-Stain Detector Tubes
D4888-88(1999) Standard Test Method for Water Vapor in Natural Gas Using Length-of-Stain Detector Tubes
D4891-89(2001) Standard Test Method for Heating Value of Gases in Natural Gas Range by Stoichiometric Combustion
D4984-89(1999) Standard Test Method for Carbon Dioxide in Natural Gas Using Length-of-Stain Detector Tubes
D5287-97(2002) Standard Practice for Automatic Sampling of Gaseous Fuels
D5454-93(1999) Standard Test Method for Water Vapor Content of Gaseous Fuels Using Electronic Moisture Analyzers
D5503-94(2003) Standard Practice for Natural Gas Sample-Handling and Conditioning Systems for Pipeline Instrumentation
D5504-01 Standard Test Method for Determination of Sulfur Compounds in Natural Gas and Gaseous Fuels by Gas Chromatography and Chemiluminescence
D5954-98 Standard Test Method for Mercury Sampling and Measurement in Natural Gas by Atomic Absorption Spectroscopy
D6228-98(2003) Standard Test Method for Determination of Sulfur Compounds in Natural Gas and Gaseous Fuels by Gas Chromatography and Flame Photometric Detection
D6273-98(2003) Standard Test Methods for Natural Gas Odor Intensity
D6350-98(2003) Standard Test Method for Mercury Sampling and Analysis in Natural Gas by Atomic Fluorescence Spectroscopy

and bears no relationship to the density of water). As a comparison, the density of liquefied natural gas is approximately 0.41 to 0.5 kg/L, depending on temperature, pressure, and composition; in comparison the density of water is 1.0 kg/L.

Density values (including those of natural gas hydrocarbons; Fig. 2.3) are given at room temperature unless otherwise indicated by a superscript figure. For example, 2.487[15] indicates a density of 2.487 g/cm at 15°C.

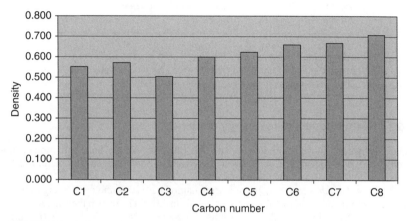

FIGURE 2.3 Carbon number and density of natural gas hydrocarbons (up to octane, C_8H_{18}).

The density of any gas compared to the density of air is the *vapor density* and is a very important characteristic of natural gas constituents (Fig. 2.4). Furthermore, the statement is often made that *natural gas is lighter than air*. This statement often arises because of the continued insistence by engineers and scientists that the properties of a mixture are determined by the mathematical average of the properties of the individual constituents of the mixture. Such mathematical bravado and inconsistency of thought is *detrimental to safety* and needs to be qualified.

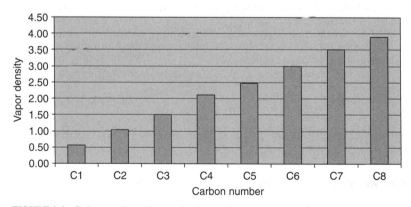

FIGURE 2.4 Carbon number and vapor density (relative to air = 1.0) of natural gas hydrocarbons (up to octane, C_8H_{18}).

Relative to air, methane is less dense (Table 2.4) but the other hydrocarbon constituents of unrefined natural gas (i.e., ethane, propane, butane, etc.) are denser than air (Fig. 2.4). Therefore, should a natural gas leak occur in field operations, especially where the natural gas contains constituents other than methane, only methane dissipates readily into the air and the higher molecular weight hydrocarbon constituents pose a considerable risk if the ambient conditions are such that they accumulate at ground level.

TABLE 2.4 Boiling Point and Density of Methane Relative to Air and Water

Boiling point (760 mm Hg)	−161.5°C (−258.7°F)
Gas specific gravity	0.55–0.64 (air = 1.00)
Specific gravity of liquefied natural gas	0.42–0.46 (water = 1.00)
Gas density (varies slightly)	0.0438 lb/scf

2.6.2 Heat of Combustion (Energy Content)

All of the hydrocarbon constituents of natural gas are combustible, but nonflammable non-hydrocarbon components (carbon dioxide, hydrogen sulfide, nitrogen, and helium) detract slightly from the heating value of natural gas.

The *heat of combustion* (*energy content*) of natural gas is the amount of energy obtained from the burning of a volume of natural gas and is measured in British thermal units (Btu). The value of natural gas is calculated by its Btu content. One Btu is the quantity of heat required to raise the temperature of one pound of water of 1°F at atmospheric pressure. A cubic foot of natural gas has an energy content of approximately 1031 Btu, but the range of values is between 500 and 1500 Btu depending upon the composition of the gas.

Thus, the energy content of natural gas is variable because natural gas has variations in the amount and types of energy gases (methane, ethane, propane, butane) it contains: the more noncombustible gases in the natural gas, the lower the energy (Btu). In addition, the volume mass of energy gases which are present in a natural gas accumulation also influences the Btu value of natural gas. The more carbon atoms in a hydrocarbon gas, the higher its Btu value. It is necessary to conduct the Btu analysis of natural gas at each stage of the supply chain. Gas chromatographic process analyzers are used in order to conduct fractional analysis of the natural gas streams, separating natural gas into identifiable components. The components and their concentrations are converted into a gross heating value in Btu-cubic foot.

In the United States, at retail, natural gas is often sold in units of therms (th) (1 therm = 100,000 Btu). Wholesale transactions are generally done in decatherms (Dth), or in thousand decatherms (MDth), or in million decatherms (MMDth). A million decatherms is roughly a billion cubic feet of natural gas.

The gross heats of combustion of crude oil and its products are given with fair accuracy by the equation:

$$Q = 12{,}400 - 2100d^2$$

where d is the 60/60°F specific gravity. Deviation from the formula is generally less than 1 percent.

2.6.3 Volatility, Flammability, and Explosive Properties

The *boiling point* (boiling temperature) of a substance is the temperature at which the vapor pressure of the substance is equal to atmospheric pressure.

At the boiling point, a substance changes its state from liquid to gas. A stricter definition of boiling point is the temperature at which the liquid and vapor (gas) phases of a substance can exist in equilibrium. When heat is applied to a liquid, the temperature of the liquid rises until the *vapor pressure* of the liquid equals the pressure of the surrounding atmosphere (gases). At this point there is no further rise in temperature, and the additional heat energy supplied is absorbed as *latent heat* of vaporization to transform the liquid into gas. This transformation occurs not only at the surface of the liquid (as in the case of *evaporation*) but also throughout the volume of the liquid, where bubbles of gas are formed. The boiling point of a liquid is lowered if the pressure of the surrounding atmosphere (gases) is decreased. On the other hand, if the pressure of the surrounding atmosphere (gases) is increased, the boiling point is raised. For this reason, it is customary when the boiling point of a substance is given to include the pressure at which it is observed, if that pressure is other than standard, that is, 760 mm Hg or 1 atm (STP, Standard Temperature and Pressure).

The boiling points of petroleum fractions are rarely, if ever, distinct temperatures. It is, in fact, more correct to refer to the boiling ranges of the various fractions; the same is true of natural gas. To determine these ranges, the material in question is tested in various methods of distillation, either at atmospheric pressure or at reduced pressure. Thus, the boiling points of the hydrocarbon constituents of natural gas increase with molecular weight and the initial boiling point of natural gas corresponds to the boiling point of the most volatile constituents (i.e., methane) (Fig. 2.5).

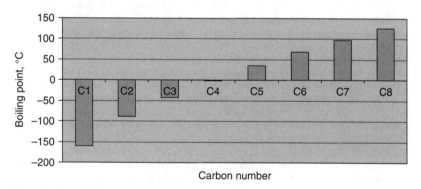

FIGURE 2.5 Carbon number and boiling points of natural gas hydrocarbons (up to octane, C_8H_{18}).

Purified natural gas is neither corrosive nor toxic, its ignition temperature is high, and it has a narrow flammability range, making it an apparently safe fossil fuel compared to other fuel sources. In addition, purified natural gas (i.e., methane) having a specific gravity (0.60) lower than that of air (1.00) rises if escaping and dissipates from the site of any leak.

However, methane is highly flammable, burns easily and almost completely. Therefore, natural gas can also be hazardous to life and property through an explosion. When natural gas is confined, such as within a house or in a coal mine, concentration of the gas can reach explosive mixtures that, if ignited, results in blasts that could destroy buildings.

The *flash point* (FP) of petroleum or a petroleum product, including natural gas, is the temperature to which the product must be heated under specified conditions to give off sufficient vapor to form a mixture with air that can be ignited momentarily by a specified flame (ASTM D56, D92, and D93). As with other properties, the flash point is dependent on the composition of the gas and the presence of other hydrocarbon constituents (Fig. 2.6).

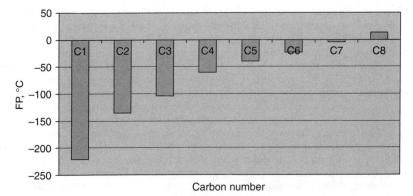

FIGURE 2.6 Carbon number and flash point of natural gas hydrocarbons (up to octane, C_8H_{18}).

The *fire point* is the temperature to which the gas must be heated under the prescribed conditions of the method to burn continuously when the mixture of vapor and air is ignited by a specified flame (ASTM D92).

From the viewpoint of safety, information about the flash point is of most significance at or slightly above the maximum temperatures (30 to 60°C, 86 to 140°F)) that may be encountered in storage, transportation, and use of liquid petroleum products, in either closed or open containers. In this temperature range the relative fire and explosion hazard can be estimated from the flash point. For products with flash point below 40°C (104°F) special precautions are necessary for safe handling. Flash points above 60°C (140°F) gradually lose their safety significance until they become indirect measures of some other quality.

The flash point of a petroleum product is also used to detect contamination. A substantially lower flash point than expected for a product is a reliable indicator that a product has become contaminated with a more volatile product, such as gasoline. The flash point is also an aid in establishing the identity of a particular petroleum product.

A further aspect of volatility that receives considerable attention is the vapor pressure of petroleum and its constituent fractions. The *vapor pressure* is the force exerted on the walls of a closed container by the vaporized portion of a liquid. Conversely, it is the force that must be exerted on the liquid to prevent it from vaporizing further (ASTM D323). The vapor pressure increases with temperature for any given gasoline, liquefied petroleum gas, or other product. The temperature at which the vapor pressure of a liquid, either a pure compound of a mixture of many compounds, equals 1 atmosphere (14.7 psi, absolute) is designated as the boiling point of the liquid.

The flammable range is expressed by the *lower explosive limit* (LEL) and the *upper explosive limit* (UEL). The LEL is the concentration of natural gas in the air below which the propagation of a flame will not occur on contact with an ignition source. The LEL for natural gas is 5 percent by volume in air and, in most cases, the smell of gas would be detected well before combustion conditions are met. The UEL is the concentration of natural gas in the air above which the propagation of a flame will not occur on contact with an ignition source. The natural gas UEL is 15 percent by volume in air.

Explosions caused by natural gas leaks occur a few times each year. Individual homes, small businesses, and boats are most frequently affected when an internal leak builds up gas inside the structure. Frequently, the blast will be enough to significantly damage a building but leave it standing. Occasionally, the gas can collect in high enough quantities to cause a deadly explosion, disintegrating one or more buildings in the process.

TABLE 2.5 General Properties of the Constituents of Natural Gas up to and Including n-Octane (C_8H_{18}) as well as Toluene, Ethyl Benzene, and Xylene.

Constituent	Molecular weight	Specific Gravity	Vapor density air = 1	Boiling point °C	Ignition temperature °C	Flash point °C
Methane	16	0.553	0.56	−160	537	−221
Ethane	30	0.572	1.04	−89	515	−135
Propane	44	0.504	1.50	−42	468	−104
Butane	58	0.601	2.11	−1	405	−60
Pentane	72	0.626	2.48	36	260	−40
Hexane	86	0.659	3.00	69	225	−23
Benzene	78	0.879	2.80	80	560	−11
Heptane	100	0.668	3.50	98	215	−4
Octane	114	0.707	3.90	126	220	13
Toluene	92	0.867	3.20	161	533	4
Ethyl benzene	106	0.867	3.70	136	432	15
Xylene	106	0.861	3.70	138	464	17

In any form, a minute amount of odorant that has an obvious smell is added to the otherwise colorless and odorless gas, so that leaks can be detected before a fire or explosion occurs. Odorants are considered nontoxic in the extremely low concentrations occurring in natural gas delivered to the end user.

2.6.4 Environmental Properties

The environmental issues regarding the use of natural gas are discussed in detail in Chap. 8 but a brief mention of such properties is also warranted here. However, in order to fully evaluate the environmental effects of natural gas, the general properties of the constituents (Table 2.5) must also be considered in addition to the effects of the combustion properties.

Currently, natural gas represents approximately one quarter of the energy consumed in the United States with increases in use projected for the next decade. These increases are expected because emissions of greenhouse gases are much lower with the consumption of natural gas relative to other fossil fuel consumption (Table 2.6). For example, natural gas, when burned, emits lower quantities of greenhouse gases and criteria pollutants per unit of energy produced, than other fossil fuels. This occurs in part because natural gas is fully

TABLE 2.6 Fossil Fuel Emission Levels—Pounds per Billion Btu of Energy Input

Pollutant	Natural gas	Petroleum	Coal
Carbon dioxide	117,000	164,000	208,000
Carbon monoxide	40	33	208
Nitrogen oxides	92	448	457
Sulfur dioxide	1	1,122	2,591
Particulates	7	84	2,744
Mercury	0.000	0.007	0.016

Source: EIA: *Natural Gas Issues and Trends*, Energy Information Administration, Washington, D.C., 1998.

combusted more easily and in part because natural gas contains fewer impurities than any other fossil fuel. However, the major constituent of natural gas, methane, also contributes directly to the greenhouse effect through venting or leaking of natural gas into the atmosphere (Speight, 2005b).

Purified natural gas (methane) is the cleanest of all the fossil fuels. The main products of the combustion of natural gas are carbon dioxide and water vapor. Coal and petroleum release higher levels of harmful emissions, including a higher ratio of carbon emissions, nitrogen oxides (NO_x), and sulfur dioxide (SO_2). Coal and fuel oil also release ash particles into the environment; substances that do not burn but instead are carried into the atmosphere and contribute to pollution. The combustion of purified natural gas, on the other hand, releases very small amounts of sulfur dioxide and nitrogen oxides, virtually no ash or particulate matter, and lower levels of carbon dioxide, carbon monoxide, and other reactive hydrocarbons.

Natural gas has no known toxic or chronic physiologic effects (i.e., it is not poisonous) but it is dangerous insofar as an atmosphere rich in natural gas will result in death to humans and animals. Exposure to a moderate concentration of natural gas may result in a headache or similar symptoms due to oxygen deprivation but it is likely that the smell (through the presence of the odorant) would be detected well in advance of concentrations being high enough for this to occur.

In fact, in the natural gas and refining industries (Speight, 2005b), as in other industries, air emissions include point and nonpoint sources. Point sources are emissions that exit stacks and flares and, thus, can be monitored and treated. Nonpoint sources are *fugitive emissions* that are difficult to locate and capture. Fugitive emissions occur throughout refineries and arise from, for example, the thousands of valves, pipe connections, seals in pumps and compressors, storage tanks, pressure relief valves, and flanged joints.

While individual leaks are typically small, the sum of all fugitive leaks at a gas processing plant can be one of its largest emission sources. These leaks can release methane and volatile constituents of natural gas into the air. Companies can minimize fugitive emissions by designing facilities with the fewest possible components and connections and avoiding components known to cause significant fugitive emissions. When companies quantify fugitive emissions, this provides them with important information they can then use to design the most effective leak repair program for their company. Directed inspection and maintenance programs are designed to identify the source of these leaks and to prioritize and plan their repair in a timely fashion. A reliable and effective directed inspection and maintenance plan for an individual facility will be composed of a number of components, including methods of leak detection, a definition of what constitutes a leak, set schedules and targeted devices for leak surveys, and allowable repair time.

A directed inspection and maintenance program begins with a baseline survey to identify and quantify leaks. Quantification of the leaks is critical because this information is used to determine which leaks are serious enough to justify their repair costs. Repairs are then made only to the leaking components that are cost-effective to fix. Subsequent surveys are then scheduled and designed based on information collected from previous surveys, permitting operators to concentrate on the components that are more likely to leak. Some natural gas companies have demonstrated that directed inspection and maintenance programs can profitably eliminate as much as 95 percent of gas losses from equipment leaks.

2.7 GAS PROCESSING

Gas processing (gas refining) usually involves several processes to remove: (a) oil; (b) water; (c) elements such as sulfur, helium, and carbon dioxide; and (d) natural gas liquids (Chap. 6). In addition, it is often necessary to install scrubbers and heaters at or near the wellhead that serve primarily to remove sand and other large-particle impurities. The heaters ensure that the temperature of the natural gas does not drop too low and form a hydrate (Chap. 1) with the water vapor content of the gas stream.

Many chemical processes are available for processing or refining natural gas. However, there are many variables in the choice of process of the choice of reining sequence that dictate the choice of process or processes to be employed. In this choice, several factors must be considered: (a) the types and concentrations of contaminants in the gas; (b) the degree of contaminant removal desired; (c) the selectivity of acid gas removal required; (d) the temperature, pressure, volume, and composition of the gas to be processed; (e) the carbon dioxide–hydrogen sulfide ratio in the gas; and (f) the desirability of sulfur recovery due to process economics or environmental issues.

In addition to hydrogen sulfide and carbon dioxide, gas may contain other contaminants, such as mercaptans (also called *thiols*, R-SH) and carbonyl sulfide (COS). The presence of these impurities may eliminate some of the sweetening processes since some processes remove large amounts of acid gas but not to a sufficiently low concentration. On the other hand, there are those processes that are not designed to remove (or are incapable of removing) large amounts of acid gases. However, these processes are also capable of removing the acid gas impurities to very low levels when the acid gases are there in low to medium concentrations in the gas.

Process selectivity indicates the preference with which the process removes one acid gas component relative to (or in preference to) another. For example, some processes remove both hydrogen sulfide and carbon dioxide; other processes are designed to remove hydrogen sulfide only. It is very important to consider the process selectivity for, say, hydrogen sulfide removal compared to carbon dioxide removal that ensures minimal concentrations of these components in the product, thus the need for consideration of the carbon dioxide to hydrogen sulfide in the gas stream.

To include a description of all of the possible process for gas cleaning is beyond the scope of this book. Therefore, the focus of this chapter is a selection of the processes that are an integral part within the concept of production of a product (methane) for sale to the consumer.

2.7.1 Olamine Processes

The most commonly used technique for gas processing is to first direct the gas flow through a tower containing an amine (olamine) solution. Amines absorb sulfur compounds from natural gas and can be reused repeatedly. After desulfurization, the gas flow is directed to the next section, which contains a series of filter tubes. As the velocity of the stream reduces in the unit, primary separation of remaining contaminants occurs due to gravity. Separation of smaller particles occurs as gas flows through the tubes, where they combine into larger particles which flow to the lower section of the unit. Further, as the gas stream continues through the series of tubes, a centrifugal force is generated which further removes any remaining water and small solid particulate matter.

The processes that have been developed to accomplish gas purification vary from a simple one—wash operation—to complex multi-step recycling systems (Speight, 1993; Mokhatab et al., 2006; Speight, 2007b). In many cases, the process complexities arise because of the need for recovery of the materials used to remove the contaminants or even recovery of the contaminants in the original, or altered, form (Kohl and Riesenfeld, 1985; Newman, 1985).

As currently practiced, acid gas removal processes involve the chemical reaction of the acid gases with a solid oxide (such as iron oxide) or selective absorption of the contaminants into a liquid (such as ethanolamine) that is passed countercurrent to the gas. Then the absorbent is stripped of the gas components (regeneration) and recycled to the absorber. The process design will vary and, in practice, may employ multiple absorption columns and multiple regeneration columns.

Amine (olamine) washing of natural gas involves chemical reaction of the amine with any acid gases with the liberation of an appreciable amount of heat and it is necessary to compensate for the absorption of heat. Amine derivatives such as monoethanolamine or ethanolamine (MEA), diethanolamine (DEA), triethanolamine (TEA), methyldiethanolamine (MDEA), diisopropanolamine (DIPA), and diglycolamine (DGA) have been used in commercial applications (Kohl and Riesenfeld, 1985; Speight, 1993; Polasek and Bullin, 1994; Mokhatab et al., 2006; Speight, 2007b).

Processes that use MDEA became popular with the natural gas industry because of its high selectivity for hydrogen sulfide over carbon dioxide. This high selectivity allows for a reduced solvent circulation rate, as well as a richer hydrogen sulfide feed to the sulfur recovery unit. The reaction of MDEA with hydrogen sulfide is almost instantaneous. However, the reaction of MDEA with carbon dioxide is much slower; the reaction rate of MDEA with carbon dioxide is slower than that of carbon dioxide with MEA.

Depending upon the application, special solutions such as mixtures of amines; amines with physical solvents such as sulfolane and piperazine; and amines that have been partially neutralized with an acid such as phosphoric acid may also be used (Bullin, 2003).

The proper selection of the amine can have a major impact on the performance and cost of a sweetening unit. However, many factors must be considered when selecting an amine for a sweetening application (Polasek and Bullin, 1994). Considerations for evaluating an amine type in gas treating systems are numerous. It is important to consider all aspects of the amine chemistry and type since the omission of a single issue may lead to operational issues. While studying each issue, it is important to understand the fundamentals of each amine solution.

$$2(RNH_2) + H_2S \leftrightarrow (RNH_3)_2S$$

$$(RNH_3)_2S + H_2S \leftrightarrow 2(RNH_3)HS$$

$$2(RNH_2) + CO_2 \leftrightarrow RNHCOONH_3R$$

These reactions are reversible by changing the system temperature. Ethanolamine also reacts with carbonyl sulfide and carbon disulfide to form heat-stable salts that cannot be regenerated.

Diethanolamine is a weaker base than ethanolamine and therefore the diethanolamine system does not typically suffer the same corrosion problems but does reacts with hydrogen sulfide and carbon dioxide:

$$2R_2NH + H_2S \leftrightarrow (R_2NH_2)_2S$$

$$(R_2NH_2)_2S + H_2S \leftrightarrow 2(R_2NH_2)HS$$

$$2R_2NH + CO_2 \leftrightarrow R_2NCOONH_2R_2$$

Diethanolamine also removes carbonyl sulfide and carbon disulfide partially as its regenerable compound with carbonyl sulfide and carbon disulfide without much solution losses.

The general process flow diagram for an amine sweetening plant varies little, regardless of the aqueous amine solution used as the sweetening agent (Fig. 2.7). The sour gas containing hydrogen sulfide and/or carbon dioxide will nearly always enter the plant through an inlet separator (scrubber) to remove any free liquids and/or entrained solids. The sour gas then enters the bottom of the absorber column and flows upward through the absorber in

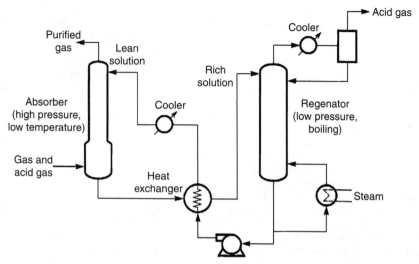

FIGURE 2.7 The amine (olamine) process for gas sweetening.

intimate counter-current contact with the aqueous amine solution, where the amine absorbs acid gas constituents from the gas stream. Sweetened gas leaving the top of the absorber passes through an outlet separator and then flows to a dehydration unit (and compression unit, if necessary) before being considered ready for sale.

In many units the rich amine solution is sent from the bottom of the absorber to a flash tank to recover hydrocarbons that may have dissolved or condensed in the amine solution in the absorber. The rich solvent is then preheated before entering the top of the stripper column. The amine–amine heat exchanger serves as heat conservation device and lowers total heat requirements for the process. A part of the absorbed acid gases will be flashed from the heated rich solution on the top tray of the stripper. The remainder of the rich solution flows downward through the stripper in counter-current contact with vapor generated in the reboiler. The reboiler vapor (primarily steam) strips the acid gases from the rich solution. The acid gases and the steam leave the top of the stripper and pass overhead through a condenser, where the major portion of the steam is condensed and cooled. The acid gases are separated in the separator and sent to the flare or to processing. The condensed steam is returned to the top of the stripper as reflux.

Lean amine solution from the bottom of the stripper column is pumped through an amine–amine heat exchanger and then through a cooler before being introduced to the top of the absorber column. The amine cooler serves to lower the lean amine temperature to the 100°F range. Higher temperatures of the lean amine solution will result in excessive amine losses through vaporization and also lower acid gas carrying capacity in the solution because of temperature effects.

2.7.2 Physical Solvent Processes

At present two of the most widely used physical solvent processes for gas cleaning are *Selexol* and *Rectisol* processes. The Selexol process solvent is the dimethyl ether of polyethylene glycol, while the Rectisol solvent is methanol.

The principal benefits of physical solvents are: (a) high selectivity for hydrogen sulfide over carbonyl sulfide and carbon dioxide, (b) high loadings at high acid gas partial pressures, (c) solvent stability, and (d) low heat requirements because most of the solvent can be regenerated by a simple pressure letdown.

The performance of a physical solvent can be easily predicted. The solubility of a compound in the solvent is directly proportional to its partial pressure in the gas phase, hence, the improvement in the performance of physical solvent processes with increasing gas pressure.

The Selexol process has been used since the late 1960s. The process solvent is a mixture of dimethyl ethers of polyethylene glycol [$CH_3(CH_2CH_2O)_nCH_3$] where n is between 3 and 9. The solvent is chemically and thermally stable, and has a low vapor pressure that limits its losses to the treated gas. The solvent has a high solubility for carbon dioxide, hydrogen sulfide, and carbonyl sulfide. It also has appreciable selectivity for hydrogen sulfide over carbon dioxide. The process can be configured in various ways, depending on the requirements for the level of hydrogen sulfide/carbon dioxide selectivity, the depth of sulfur removal, the need for bulk carbon dioxide removal, and whether the gas needs to be dehydrated (Fig. 2.8). The gas stream from the low-pressure flash is combined with the acid gas from the regenerator. This combined gas stream is then sent to a sulfur recovery unit. However, the hydrogen sulfide content could be too low for use in a conventional Claus plant.

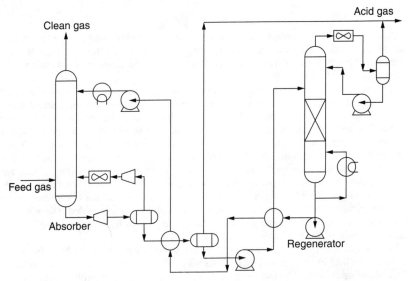

FIGURE 2.8 The Selexol process.

The Rectisol process is the most widely used physical solvent gas-treating process in the world. The process uses chilled methanol (methyl alcohol, CH_3OH) at a temperature of about −40 to −80°F. The selectivity (by methanol) for hydrogen sulfide over carbon dioxide at these temperatures is about 6:1, a little lower than that of the Selexol process at its usual operating temperature. However, the solubility of hydrogen sulfide and carbonyl sulfide in methanol, at typical process operating temperatures, are higher than in Selexol and allow for very deep sulfur removal. The high selectivity for hydrogen sulfide over carbon dioxide, combined with the ability to remove carbonyl sulfide, is the primary advantage of the process.

The need to refrigerate the solvent is the main disadvantage of the process, resulting in high capital and operating costs.

The Sulfinol process, developed in the early 1960s, is a combination process that uses a mixture of amines and a physical solvent. The solvent consists of an aqueous amine and sulfolane. In operation, this process is in many respects identical to the familiar amine method and its equipment components similar to those found in amine units. The main difference is that while the conventional amine process employs a fairly diluted concentration of amine in water, removing the acid gas by chemical reaction, the Sulfinol system uses a mixture of highly concentrated amine and a physical solvent removing the acid gases by physical and chemical reactions. The concentrations of the amine and the physical solvent vary with the type of feed gas in each application. Common Sulfinol mixtures are in the range of 40 percent amine (also called DIPA), 40 percent sulfolane (an organic solvent), and 20 percent water.

2.7.3 Metal Oxide Processes

Hydrogen sulfide scavengers have been used for many years. An example is the *iron sponge process* (*dry box process*) that is the oldest and still the most widely used batch process for sweetening of natural gas and natural gas liquids (Duckworth and Geddes, 1965; Anerousis and Whitman, 1984; and Zapffe, 1963). The process was implemented during the nineteenth century and has been in use in Europe and the United States for over 100 years. Hydrogen sulfide scavengers are appropriate for use at the low concentrations of hydrogen sulfide where conventional chemical absorption and physical solvents are not economical. During recent years, hydrogen sulfide scavenger technology has been expanded with many new materials coming on the market and others being discontinued. Overall, the simplicity of the process, low capital costs, and relatively low chemical (iron oxide) cost continue to make the process an ideal solution for hydrogen sulfide removal.

The *sponge* consists of wood shavings impregnated with a hydrated form of iron oxide. The wood shavings serve as a carrier for the active iron oxide powder. Hydrogen sulfide is removed by reacting with iron oxide to form ferric sulfide. The process is usually best applied to gases containing low to medium concentrations [300 ppm (parts per million)] of hydrogen sulfide or mercaptans. This process tends to be highly selective and does not normally remove significant quantities of carbon dioxide. As a result, the hydrogen sulfide stream from the process is usually high purity. The use of iron sponge process for sweetening sour gas is based on adsorption of the acid gases on the surface of the solid sweetening agent followed by chemical reaction of ferric oxide (Fe_2O_3) with hydrogen sulfide:

$$2Fe_2O_3 + 6H_2S \rightarrow 2Fe_2S_3 + 6H_2O$$

The reaction requires the presence of slightly alkaline water and a temperature below 43°C (110°F) and bed alkalinity should be checked regularly, usually on a daily basis. A pH level on the order of 8 to 10 should be maintained through the injection of caustic soda with the water. If the gas does not contain sufficient water vapor, water may need to be injected into the inlet gas stream.

The ferric sulfide produced by the reaction of hydrogen sulfide with ferric oxide can be oxidized with air to produce sulfur and regenerate the ferric oxide:

$$2Fe_2S_3 + 3O_2 \rightarrow 2Fe_2O_3 + 6S$$

$$S_2 + 2O_2 \rightarrow 2SO_2$$

The regeneration step, that is, the reaction with oxygen is exothermic and air must be introduced slowly so the heat of reaction can be dissipated. If air is introduced quickly the heat of reaction may ignite the bed. Some of the elemental sulfur produced in the regeneration step remains in the bed. After several cycles this sulfur will cake over the ferric oxide, decreasing the reactivity of the bed. Typically, after 10 cycles the bed must be removed and a new bed introduced into the vessel.

In some designs the iron sponge may be operated with continuous regeneration by injecting a small amount of air into the sour gas feed. The air regenerates ferric sulfide while hydrogen sulfide is removed by ferric oxide. This process is not as effective at regenerating the bed as the batch process and requires a higher-pressure air stream (Arnold and Stewart, 1999).

In the process (Fig. 2.9), the sour gas should pass down through the bed. In the case where continuous regeneration is to be utilized a small concentration of air is added to the sour gas before it is processed. This air serves to continuously regenerate the iron oxide, which has reacted with hydrogen sulfide, which serves to extend the on-stream life of a given tower but probably serves to decrease the total amount of sulfur that a given weight of bed will remove. The number of vessels containing iron oxide can vary from one to four. In a two-vessel process, one of the vessels would be on-stream removing hydrogen sulfide from the sour gas while the second vessel would either be in the regeneration cycle or having the iron sponge bed replaced.

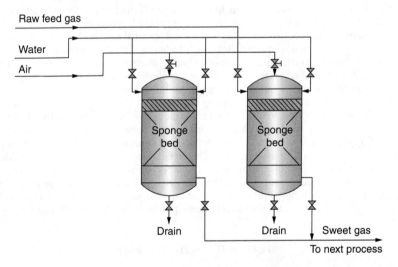

FIGURE 2.9 Typical iron oxide process flow sheet. Maddox, R. N. *Gas and Liquid Sweetening*, 2nd ed., Campbell Petroleum Series, Norman, Okla., 1974.

Generally, the iron oxide process is suitable only for small to moderate quantities of hydrogen sulfide. Approximately 90 percent of the hydrogen sulfide can be removed per bed, but bed clogging by elemental sulfur occurs and the bed must be discarded and the use of several beds in series is not usually economical. Removal of larger amounts of hydrogen sulfide from gas streams requires a continuous process, such as the Ferrox process or the Stretford process. The *Ferrox process* is based on the same chemistry as the iron oxide process except that it is fluid and continuous. The *Stretford process* employs a solution containing vanadium salts and anthraquinone disulfonic acid (Maddox, 1974).

The natural gas should be wet when passing through an iron sponge bed as drying of the bed will cause the iron sponge to lose its capacity for reactivity. If the gas is not already water-saturated or if the influent stream has a temperature greater than 50°C (approximately 120°F), water with soda ash is sprayed into the top of the contactor to maintain the desired moisture and alkaline conditions during operation.

Slurry processes were developed as alternatives to the iron sponge process. Slurries of iron oxide have been used to selectively absorb hydrogen sulfide (Fox, 1981; Samuels, 1988). The chemical cost for these processes is higher than that for iron sponge process but this is partially offset by the ease and lower cost with which the contact tower can be cleaned out and recharged. Obtaining approval to dispose of the spent chemicals, even if they are nonhazardous, is time consuming.

2.7.4 Methanol-Based Processes

Methanol is probably one of the most versatile solvents in the natural gas processing industry. Historically, methanol was the first commercial organic physical solvent and has been used for hydrate inhibition, dehydration, gas sweetening, and liquids recovery (Kohl and Nielsen, 1997). Most of these applications involve low temperature where methanol's physical properties are advantageous compared with other solvents that exhibit high viscosity problems or even solids formation. Operation at low temperatures tends to suppress methanol's most significant disadvantage, high solvent loss. Furthermore, methanol is relatively inexpensive and easy to produce making the solvent a very attractive alternate for gas processing applications.

The use of methanol has been further exploited in the development of the Rectisol process either alone or as toluene-methanol mixtures to more selectively remove hydrogen sulfide and slip carbon dioxide to the overhead product (Ranke and Mohr, 1985). Toluene has an additional advantage insofar as carbonyl sulfide is more soluble in toluene than in methanol. The Rectisol process was primarily developed to remove both carbon dioxide and hydrogen sulfide (along with other sulfur-containing species) from gas streams resulting from the partial oxidation of coal, oil, and petroleum residua. The ability of methanol to absorb these unwanted components made it the natural solvent of choice. Unfortunately, at cold temperatures, methanol also has a high affinity for hydrocarbon constituents of the gas streams. For example, propane is more soluble in methanol than is carbon dioxide. There are two versions of the Rectisol process (Hochgesand, 1970)—the two-stage and the once-through. The first step of the two-stage process is desulfurization before shift conversion; the concentrations of hydrogen sulfide and carbon dioxide are about 1 and 5 percent by volume, respectively. Regeneration of the methanol following the desulfurization of the feed gas produces high sulfur feed for sulfur recovery. The once-through process is only applicable for high pressure partial oxidation products. The once-through process is also applicable when the hydrogen sulfide to carbon dioxide content is unfavorable, in the neighborhood of 1:50 (Esteban et al., 2000).

Recently, a process using methanol has been developed in which the simultaneous capability to dehydrate, to remove acid gas, and to control hydrocarbon dew point (Rojey and Larue, 1988; Rojey et al., 1990). The IFPEXOL-1 is used for water removal and hydrocarbon dew point control; the IFPEXOL-2 process is used for acid gas removal. The novel concept behind the IFPEXOL-l process is to use a portion of the water-saturated inlet feed to recover the methanol from the aqueous portion of the low temperature separator. That approach has solved a major problem with methanol injection in large facilities, the methanol recovery via distillation. Beyond that very simple discovery, the cold section of the process is remarkably similar to a basic methanol injection process. Modifications to the process include water washing the hydrocarbon liquid from the low temperature separator to enhance the methanol recovery. The IFPEXOL-2 process for acid gas removal is very similar to an amine-type process except for the operating temperatures. The absorber operates below −20°F to minimize methanol losses, and the regenerator operates at about 90 psi. Cooling is required on the regenerator condenser to recover the methanol. This process usually follows the IFPEXOL-1 process so excessive hydrocarbon absorption is not as great a problem (Minkkinen and Jonchere, 1997).

2.7.5 Carbonate-Washing and Water-Washing Processes

Carbonate washing is a mild alkali process for emission control by the removal of acid gases (such as carbon dioxide and hydrogen sulfide) from gas streams (Speight, 1993, 2007b) and uses the principle that the rate of absorption of carbon dioxide by potassium carbonate increases with temperature. It has been demonstrated that the process works best near the temperature of reversibility of the reactions:

$$K_2CO_3 + CO_2 + H_2O \rightarrow 2KHCO_3$$

$$K_2CO_3 + H_2S \rightarrow KHS + KHCO_3$$

Water washing, in terms of the outcome, is analogous to washing with potassium carbonate (Kohl and Riesenfeld, 1985), and it is also possible to carry out the desorption step by pressure reduction. The absorption is purely physical and there is also a relatively high absorption of hydrocarbons, which are liberated at the same time as the acid gases.

The process using *potassium phosphate* is known as phosphate desulphurization, and it is used in the same way as the Girbotol process to remove acid gases from liquid hydrocarbons as well as from gas streams. The treatment solution is a water solution of tripotassium phosphate (K_3PO_4), which is circulated through an absorber tower and a reactivator tower in much the same way as the ethanolamine is circulated in the Girbotol process; the solution is regenerated thermally.

Other processes include the *Alkazid process* (Fig. 2.10), which removes hydrogen sulfide and carbon dioxide using concentrated aqueous solutions of amino acids. The hot

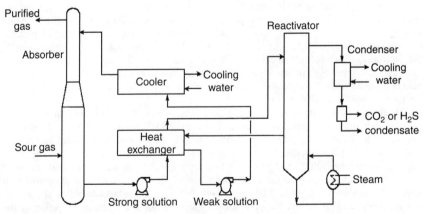

FIGURE 2.10 The Alkazid process flow diagram. Speight, J. G.: *Gas Processing: Environmental Aspects and Methods*, Butterworth Heinemann, Oxford, England, 1993.

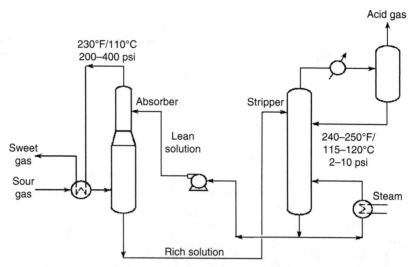

FIGURE 2.11 The Hot Potassium Carbonate process flow diagram. Speight, J. G.: *Gas Processing: Environmental Aspects and Methods*, Butterworth Heinemann, Oxford, England, 1993.

potassium carbonate process (Fig. 2.11) decreases the acid content of natural and refinery gas from as much as 50 percent to as low as 0.5 percent and operates in a unit similar to that used for amine treating. The *Giammarco-Vetrocoke process* is used for hydrogen sulfide and/or carbon dioxide removal (Fig. 2.12). In the hydrogen sulfide removal section, the reagent consists of sodium or potassium carbonates containing a mixture of arsenites and arsenates; the carbon dioxide removal section utilizes hot aqueous alkali carbonate solution activated by arsenic trioxide or selenous acid or tellurous acid.

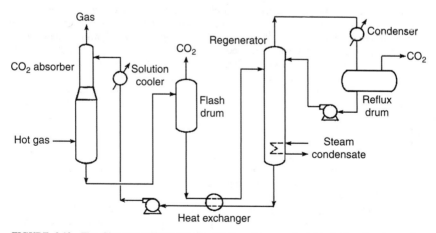

FIGURE 2.12 The Giammarco-Vetrocoke process flow diagram. Speight, J. G.: *Gas Processing: Environmental Aspects and Methods*, Butterworth Heinemann, Oxford, England, 1993.

Molecular sieves are highly selective for the removal of hydrogen sulfide (as well as other sulfur compounds) from gas streams and over continuously high absorption efficiency. They are also an effective means of water removal and thus offer a process for the simultaneous dehydration and desulphurization of gas. Gas that has excessively high water content may require upstream dehydration, however. The molecular sieve process (Fig. 2.13) is similar to the iron oxide process. Regeneration of the bed is achieved by passing heated clean gas over the bed.

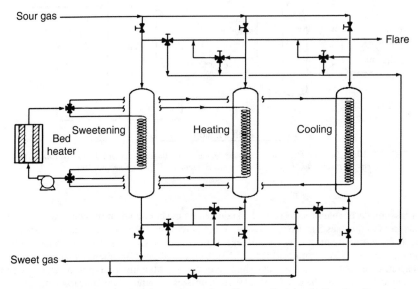

FIGURE 2.13 The molecular sieve process flow diagram. Speight, J. G.: *Gas Processing: Environmental Aspects and Methods*, Butterworth Heinemann, Oxford, England, 1993.

The molecular sieves are susceptible to poisoning by such chemicals as glycols and require thorough gas-cleaning methods before the adsorption step. Alternatively, the sieve can be offered some degree of protection by the use of *guard beds* in which a less expensive catalyst is placed in the gas stream before contact of the gas with the sieve, thereby protecting the catalyst from poisoning. This concept is analogous to the use of guard beds or attrition catalysts in the petroleum industry (Speight, 1993, 2007b).

Until recently, the use of membranes for gas separation has been limited to carbon dioxide removal (Alderton, 1993). Improvements in membrane technology have now made membranes competitive in other applications in the natural gas area. New membrane materials and configurations exhibit superior performance and offer improved stability against contaminants found in natural gas. The new membranes are targeted at three separations: nitrogen, carbon dioxide/hydrogen sulfide, and natural gas liquids (Baker et al., 2002). The process uses a two-step membrane system design; the methane-selective membranes do not need to be operated at low temperatures, and capital and operating costs are within economically acceptable limits.

New membranes have been developed (Lokhandwala and Jacobs, 2000) for the gas industry. For example, the membranes allow permeation of condensable vapors, such as C_{3+} hydrocarbons, aromatics, and water vapor, while rejecting the noncondensable gases, such as methane, ethane, nitrogen, and hydrogen. During the past 15 years, more than 50 systems have been installed in the chemical process industry worldwide. The main applications are nitrogen removal and recovery of natural gas liquids (Lokhandwala, 2000; Hall and Lokhandwala, 2004).

2.7.6 Sulfur Recovery Processes

The side stream from acid gas treating units consists mainly of hydrogen sulfide and/or carbon dioxide. Carbon dioxide is usually vented to the atmosphere but sometimes is recovered for carbon dioxide floods. Hydrogen sulfide could be routed to an incinerator or flare, which would convert the hydrogen sulfide to sulfur dioxide. The release of hydrogen sulfide to the atmosphere may be limited by environmental regulations. There are many specific restrictions on these limits, and the allowable limits are revised periodically. In any case, environmental regulations severely restrict the amount of hydrogen sulfide that can be vented or flared in the regeneration cycle.

Most sulfur recovery processes use chemical reactions to oxidize hydrogen sulfide and produce elemental sulfur. These processes are generally based either on the reaction of hydrogen sulfide and oxygen or hydrogen sulfide and sulfur dioxide. Both reactions yield water and elemental sulfur. These processes are licensed and involve specialized catalysts and/or solvents. These processes can be used directly on the produced gas stream. Where large flow rates are encountered, it is more common to contact the produced gas stream with a chemical or physical solvent and use a direct conversion process on the acid gas liberated in the regeneration step.

Currently, the Claus sulfur recovery process (Fig. 2.14) is the most widely used technology for recovering elemental sulfur from sour gas. Conventional three-stage Claus plants can approach 98 percent sulfur recovery efficiency. However, since environmental regulations have become stricter, sulfur recovery plants are required to recover sulfur with over 99.8 percent efficiency. To meet these stricter regulations, the Claus process underwent various modifications and add-ons.

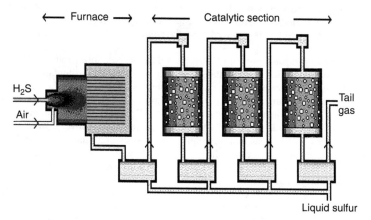

FIGURE 2.14 The Claus process. Maddox, R. N.: *Gas and Liquid Sweetening*, 2nd ed., Campbell Petroleum Series, Norman, Okla., 1974. and http://www.nelliott.demon.co.uk/company/claus.html.

The chemistry of the Claus process involves partial oxidation of hydrogen sulfide to sulfur dioxide and the catalytically promoted reaction of hydrogen sulfide and sulfur dioxide to produce elemental sulfur. The reactions are staged and are as follows:

Thermal stage:

$$H_2S + 3/2\ O_2 \rightarrow SO_2 + H_2O$$

Thermal and catalytic stage:

$$SO_2 + 2H_2S \rightarrow 3S + 2H_2O$$

The efficiency of sulfur recovery depends upon such things as feed composition, age of the catalyst, and number of reactor stages. Typical sulfur recovery efficiencies for Claus plants are 90 to 96 percent for a two-stage plant and 95 to 98 percent for a three-stage plant. Because of equilibrium limitations and other sulfur losses, overall sulfur recovery efficiency in a Claus unit usually does not exceed 98 percent.

The off-gas leaving a Claus plant is referred to as tail gas, and, in the past was burned to convert the unreacted hydrogen sulfide to sulfur dioxide, before discharge to the atmosphere, which has a much higher toxic limit. However, the increasing standards of efficiency required by the pressure from environmental protection has led to the development of a large number of Claus tail gas clean-up units, based on different concepts, in order to remove the last remaining sulfur species (Gall and Gadelle, 2003).

The oxygen-blown Claus process was originally developed to increase capacity at existing conventional Claus plants and to increase flame temperatures of gases having low hydrogen sulfide content. The process has also been used to provide the capacity and operating flexibility for sulfur plants where the feed gas is variable in flow and composition such as often found in refineries.

Liquid *redox* sulfur recovery processes are liquid-phase oxidation processes which use a dilute aqueous solution of iron or vanadium to remove hydrogen sulfide selectively by chemical absorption from sour gas streams. These processes can be used on relatively small or dilute hydrogen sulfide stream to recover sulfur from the acid gas stream or, in some cases, they can be used in place of an acid gas removal process. The mildly alkaline lean liquid scrubs the hydrogen sulfide from the inlet gas stream, and the catalyst oxidizes the hydrogen sulfide to elemental sulfur. The reduced catalyst is regenerated by contact with air in the oxidizer(s). Sulfur is removed from the solution by flotation or settling, depending on the process.

The *wet oxidation processes* are based on reduction-oxidation (redox) chemistry to oxidize the hydrogen sulfide to elemental sulfur in an alkaline solution containing an oxygen carrier. Vanadium and iron are the two oxygen carriers that are used. The best example of a process using the vanadium carrier is the Stretford process. The most prominent examples of the processes using iron as a carrier are the LO-CAT process and the SulFerox process. Both processes are capable of up to 99 percent or more sulfur recovery. However, using the processes for Claus tail gas treating requires hydrolysis of all the sulfur dioxide in the tail gas to hydrogen sulfide because the sulfur dioxide will react with the buffering base potassium hydroxide (KOH) and form potassium sulfate (K_2SO_4) which will consume the buffering solution and quickly saturate it.

Tail gas-treating process involves the removal of the remaining sulfur compounds from gases remaining after sulfur recovery. Tail gas from a typical Claus process, whether a conventional Claus or one of the extended versions of the process, usually contains small but varying quantities of carbonyl sulfide, carbon disulfide, hydrogen sulfide, and sulfur dioxide as well as sulfur vapor. In addition, there may be hydrogen, carbon monoxide, and carbon dioxide in the tail gas.

2.7.7 Hydrogenation and Hydrolysis Processes

The reduction of carbonyl sulfide, carbon disulfide, sulfur dioxide, and sulfur vapor in Claus tail gas to hydrogen sulfide is necessary when sulfur recovery of more than 99.9 percent is required. Usually the sulfur recovery level is set by the allowable emissions of sulfur from the tail gas incinerator. In addition, the reduction of carbonyl sulfide is done on raw synthesis gas when the downstream acid gas removal process is unable to remove carbonyl sulfide to a sufficient extent to meet sulfur emissions regulations from combustion of the cleaned fuel gas. These sulfur compounds are reduced to hydrogen sulfide by hydrogenation or by hydrolysis, at a raised temperature, over a catalytic bed.

In these processes, elemental sulfur and sulfur dioxide are reduced mainly via hydrogenation, while carbonyl sulfide and carbon disulfide are mainly hydrolyzed to hydrogen sulfide. Sulfur and sulfur dioxide are virtually completely converted to hydrogen sulfide when an excess of hydrogen is present.

The SCOT (Shell Claus off-gas treating) process was developed in the early 1970s and consists of a combination of a catalytic hydrogenation/hydrolysis step and an amine scrubbing unit. The hydrogenation/hydrolysis of the sulfur compounds in the tail gases from the Claus unit has already been covered above. The early SCOT units consisted of a hydrogenation/hydrolysis reactor and a conventional amine unit (Fig. 2.7). The Claus tail gas, after being reduced in the reactor, is cooled in a quench column and scrubbed by a Sulfinol solution. The clean tail gas goes to a Claus incinerator and the acid gas rich solution is regenerated in a stripping column. The acid gas off the top of the stripper is recycled back to the Claus plant for further conversion of the hydrogen sulfide. The absorber is operated at near atmospheric pressure and the amine solvent is not highly loaded with acid gases. Because the solution is not highly loaded, unlike high pressure operation, there is no need for an intermediate flash vessel and the loaded solution goes directly to a stripper.

2.8 USES

Natural gas is approximately currently one quarter of the energy resources of the world with use projected to increase over the next two decades (Fig. 2.15). However, to understand the use of natural gas, it is necessary to review the history of natural gas over the past 2000 years as well as the use of natural gas in the United States (Table 2.7).

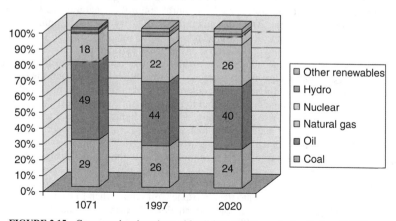

FIGURE 2.15 Current and projected use of fossil fuel resources and other fuels until 2020.

TABLE 2.7 Abbreviated Timeline for the Use of Natural Gas

1620	French missionaries recorded that Indians in what is now New York state ignited gases in the shallows of Lake Erie and in the streams flowing into the lake. It was in this same area (Fredonia, NY) that the natural gas industry in America began.
1803	Gas lighting system patented in London by Frederick Winsor.
1812	First gas company founded in London.
1815	Metering for households, invented in 1815 by Samuel Clegg, and put into general use during the 1840s.
1816	First US gas company (using manufactured gas) founded in Baltimore.
1817	The lighting of the first gas lamp on the corner of Market and Lemon streets in Baltimore, MD, on February 7th marks the effective birth of the gas industry in the United States.
1821	First natural gas from the wellhead used in Fredonia, NY for house lighting.
1826	World's first gas cooker was devised in England by James Sharp, but it was not until 1851 that such equipment came into use in America.
1840	The first industrial use of natural gas in the United States is recorded near Centerville, PA, when gas is used to evaporate brine to make salt.
1850	Fifty or more U.S. cities were burning public utility gas.
1859	Edwin L. Drake dug the first well and hit oil and natural gas near Titusville, PA. An iron 2-in diameter gas pipeline was built, running 5 1/2 miles from the well to Titusville proving that natural gas could be brought safely from its underground source to be used for practical purposes.
1880	Thomas Edison postulated replacing gas lighting by electric lighting.
1885	Carl Auer von Welsbach in Germany developed a practical gas mantle.
1904	Gas is used for the first time to power central heating and to provide a large-scale supply of hot water in London.
1915	Depleted reservoirs are used for the first time to store gas.
1938	In the United States, The Natural Gas Act of 1938 established federal authority over interstate pipelines, including the authority to set rates.
1951	Natural gas (coalbed methane) is produced from coal, while it is still underground in the coal seam, at a colliery at Newman Spinney, England.
1959	LNG is produced for the first time on an industrial scale in LA. It will be transported to Britain for the first by the vessel Methane Pioneer.

After the discovery by the Chinese more than 2000 years ago that the energy in natural gas could be harnessed and used as a heat source, the use of natural gas has grown. As already noted above, the American natural gas industry got its beginnings in the mid–nineteenth century and most in gas industry observers characterize the "Drake" well (*q.v.*, above) as beginning of the natural gas industry in America.

During most of the nineteenth century, natural gas was used almost exclusively as a source of light. Without a pipeline infrastructure, it was difficult to transport the gas very far, or into homes to be used for heating or cooking. Most of the natural gas produced in this era was manufactured from coal, as opposed to transport from a well. Near the end of the nineteenth century, with the rise of electricity, natural gas lights were converted to electric lights. This led producers of natural gas to look for new uses for their product.

One of the first lengthy pipelines was constructed in 1891. This pipeline was 120 miles long, and carried natural gas from wells in central Indiana to the city of Chicago. However, this early pipeline was very rudimentary, and was not very efficient at transporting natural gas. Without any way to transport it effectively, natural gas discovered pre–World War II was usually just allowed to vent into the atmosphere, or burnt, when found alongside coal and oil, or simply left in the ground when found alone. It wasn't until the 1920s that any significant effort was put into building a pipeline infrastructure. However, it wasn't until after the World War II that welding techniques, pipe rolling, and metallurgic advances allowed for the construction of reliable pipelines. This post-war pipeline construction boom lasted well into the 1960, and allowed for the construction of thousands of miles of pipeline in America.

Once the transportation of natural gas was possible, new uses for natural gas were discovered. These included using natural gas to heat homes and operate appliances such as water heaters and oven ranges. Industry began to use natural gas in manufacturing and processing plants. Also, natural gas was used to heat boilers to generate electricity. The transportation infrastructure had made natural gas easy to obtain, and it was becoming an increasingly popular form of energy.

There are so many different applications for this fossil fuel that it is hard to provide an exhaustive list of all of its uses, and new uses are continuously being discovered. Natural gas has many applications: for domestic use, industrial use, and for transportation. In addition, natural gas is also a raw material for many common products such as paints, fertilizer, plastics, antifreeze, dyes, photographic film, medicines, and explosives.

2.9 NATURAL GAS AND THE ENVIRONMENT

Natural gas burns more cleanly than other fossil fuels. It has fewer emissions of sulfur, carbon, and nitrogen than coal or oil, and it has almost no ash particles left after burning. Being a clean fuel is one reason that the use of natural gas, especially for electricity generation, has grown so much and is expected to grow even more in the future.

As with other fuels, there are also environmental concerns with the use of natural gas. Burning natural gas produces carbon dioxide, which is the most important greenhouse gas. Methane (the major constituent of natural gas) is also a greenhouse gas. Small amounts of methane can sometimes leak into the atmosphere from wells, storage tanks, and pipelines.

Exploring and drilling for natural gas will always have some impact on land and marine habitats. But new technologies have greatly reduced the number and size of areas disturbed by drilling (sometimes called *footprints*). Satellites, global positioning systems, remote sensing devices, and 3-D and 4-D seismic technologies, make it possible to discover natural gas reserves while drilling fewer wells. Plus, the application of horizontal- and directional-drilling methods make it possible for a single well to produce gas from larger areas.

2.10 REFERENCES

Alderton, P. D.: "Natural Gas Treating Using Membranes," Proceedings, 2nd GPA Technical Meeting, GCC Chapter, Bahrain, Oct. 27, 1993.

Anerousis, J. P. and S. K. Whitman: "An Updated Examination of Gas Sweetening by the Iron Sponge Process," SPE 13280, Proceedings, SPE Annual Technical Conference & Exhibition, Houston, Tex., 1984.

Arnold, K. and M. Stewart: *Surface Production Operations, Volume 2: Design of Gas-Handling systems and Facilities*, 2nd ed, Gulf Professional Publishing, Houston, Tex., 1999.

ASTM: *Annual Book of Standards,* American Society for Testing and Materials (ASTM), West Conshohocken, PA. 2007.

Baker, R. W., K. A. Lokhandwala, J. G. Wijmans, and A. R. Da Costa: "Two-Step Process for Nitrogen Removal from Natural Gas," United States Patent 6425267, 2002.

Berecz, E. and M. Balla-Achs: *Gas Hydrates,* Elsevier, Amsterdam, 1983.

Bullin, J. A: "Why Not Optimize Your Amine Sweetening Unit," Proceedings, GPA Europe Annual Conference, Heidelberg, Germany, Sept. 25–27, 2003.

BP: "Statistical Review of World Energy 2005," British Petroleum, 2006, http://www.bp.com/centres/energy2005.

Duckworth, G. L., and J. H. Geddes: *Oil & Gas Journal,* 63(37), 1965, pp. 94–96.

EIA: *Natural Gas Issues and Trends,* Energy Information Administration, Washington, D.C., 1998.

EIA: *International Energy Annual,* Energy Information Administration, Washington, D.C., Mar., 2004.

EIA: "Annual Energy Outlook 2006 with Projections to 2030," Report #:DOE/EIA-0383, International Energy Annual, Energy Information Administration, Washington, D.C., Feb., 2006.

Esteban, A., V. Hernandez, and K. Lunsford: "Exploit the Benefits of Methanol," Proceedings, 79th GPA Annual Convention, Atlanta, GA., March, 2000.

Fox, I.: "Process for Scavenging Hydrogen Sulfide from Hydrocarbon Gases," United States Patent, 4246274, 1981.

Gall, A. L., and D. Gadelle: "Technical and Commercial Evaluation of Processes for Claus Tail Gas Treatment," Proceedings, GPA Europe Technical Meeting, Paris, France, Feb., 2003.

Gollmar, H. A.: in *Chemistry of Coal Utilization,* H. H. Lowry (ed.), John Wiley & Sons, Inc., New York, 1945, pp. 947–1007.

Gudmundsson, J. S., V. Andersson, O. I. Levik, and M. Parlaktuna: "Hydrate Concept for Capturing Associated Gas," Proceedings, SPE European Petroleum Conference, The Hague, Netherlands, Oct. 20–22, 1998.

Hall, P., and K. A. Lokhandwala: "Advances in Membrane Materials Provide New Gas Processing Solutions," Proceedings, GPA Annual Convention, New Orleans, LA. Mar., 2004.

Hochgesand, G.: *Ind. Eng. Chem.,* 62(7), 1970, pp. 37–43.

Kohl, A. L. and R. B. Nielsen: *Gas Purification,* Gulf Publishing Company, Houston, Tex., 1997.

Kohl, A. L. and F. C. Riesenfeld: *Gas Purification,* 4th ed., Gulf Publishing Company, Houston, Tex., 1985.

Lokhandwala, K. A.: "Fuel Gas Conditioning Process," United States Patent 6,53965, 2000.

Lokhandwala, K. A. and M. L. Jacobs: "New Membrane Application in Gas Processing," Proceedings, GPA Annual Convention, Atlanta, GA., Mar., 2000.

Maddox, R. N.: *Gas and Liquid Sweetening,* 2nd ed., Campbell Petroleum Series, Norman, Okla., 1974.

Manning, F.S., and R. E. Thompson: *Oil Field Processing of Petroleum, Volume 1: Natural Gas,* Pennwell Publishing Company, Tulsa, Okla., 1991.

Max, M. D. (ed.): *Natural Gas in Oceanic and Permafrost Environments,* Kluwer Academic Publishers, Dordrecht, Netherlands, 2000.

Minkkinen, A. and J. P. Jonchere: "Methanol Simplifies Gas Processing," Proceedings, 5th GPA-GCC Chapter Technical Conference, Bahrain. May 6, 1997.

Mokhatab, S., W. A. Poe, and J. G. Speight: *Handbook of Natural Gas Transmission and Processing,* Elsevier, Amsterdam, Netherlands, 2006.

Newman, S. A. (ed.): *Acid and Sour Gas Treating Processes,* Gulf Publishing Company, Houston, Tex., 1985.

Polasek, J., and J. A. Bullin: "Selecting Amines for Sweetening Units," Proceedings, GPA Regional Meeting, Tulsa, Okla., 1994.

Ranke, G., and V. H. Mohr: "The Rectisol Wash New Developments in Acid Gas Removal from Synthesis Gas," in *Acid and Sour Gas Treating Processes,* S. A. Newman (ed.), Gulf Publishing Company, Houston, Tex., 1985.

Rojey, A., and J. Larue: "Integrated Process for the Treatment of a Methane-Containing Wet Gas in Order to Remove Water Therefrom," United States Patent 4775395, 1988.

Rojey, A., A. Procci, and J. Larue: "Process and Apparatus for Dehydration, De-acidification, and Separation of Condensate from a Natural Gas," United States Patent 4979966, 1990.

Samuels, A.: "Gas Sweetener Associates," Technical Manual, 3-88, Metairie, LA. 1988.

Sloan, E. D.: *Clathrates of Hydrates of Natural Gas,* Marcel Dekker Inc., New York. 1997.

Sloan, E. D.: "Clathrates Hydrates: The Other Common Water Phase," *Ind. Eng. Chem. Res.*, 39, 2000, pp. 3123–3129.

Speight, J. G.: *Gas Processing: Environmental Aspects and Methods,* Butterworth Heinemann, Oxford, England, 1993.

Speight, J.G. (ed.): *Lange's Handbook of Chemistry,* 16th ed., McGraw-Hill, New York, 2005a.

Speight, J. G.: *Environmental Analysis and Technology for the Refining Industry,* John Wiley & Sons Inc., Hoboken, N.J., 2005b.

Speight, J. G.: *The Chemistry and Technology of Petroleum,* 4th ed., CRC-Taylor and Francis Group, Boca Raton, Fla., 2007a.

Speight, J. G.: *Natural Gas: A Basic Handbook,* Gulf Publishing Company, Houston Tex., 2007b.

Zapffe, F.: *Oil & Gas Journal,* 61(33), 1963, pp. 103–104.

CHAPTER 3
FUELS FROM PETROLEUM AND HEAVY OIL

3.1 HISTORY

A petroleum refinery is a group of manufacturing plants (Fig. 3.1) which are used to separate petroleum into fractions and the subsequent treating of these fractions to yield marketable products, particularly fuels (Kobe and McKetta, 1958; Nelson, 1958; Gruse and Stevens, 1960; Bland and Davidson, 1967; Hobson and Pohl, 1973). The configuration of refineries may vary from refinery to refinery. Some refineries may be more oriented toward the production of gasoline (large reforming and/or catalytic cracking) whereas the configuration of other refineries may be more oriented toward the production of middle distillates such as jet fuel and gas oil.

In general, crude oil, once refined, yields three basic groupings of products that are produced when it is broken down into cuts or fractions (Table 3.1). The gas and gasoline cuts form the lower boiling products and are usually more valuable than the higher boiling fractions and provide gas (liquefied petroleum gas), naphtha, aviation fuel, motor fuel, and feedstocks, for the petrochemical industry. Naphtha, a precursor to gasoline and solvents, is extracted from both the light and middle range of distillate cuts and is also used as a feedstock for the petrochemical industry. The middle distillates refer to products from the middle boiling range of petroleum and include kerosene, diesel fuel, distillate fuel oil, and light gas oil. Waxy distillate and lower boiling lubricating oils are sometimes included in the middle distillates. The remainder of the crude oil includes the higher boiling lubricating oils, gas oil, and residuum (the nonvolatile fraction of the crude oil). The residuum can also produce heavy lubricating oils and waxes but is more often used for asphalt production. The complexity of petroleum is emphasized insofar as the actual proportions of light, medium, and heavy fractions vary significantly from one crude oil to another.

The refining industry has been the subject of the four major forces that affect most industries and which have hastened the development of new petroleum-refining processes: (a) the demand for products such as gasoline, diesel, fuel oil, and jet fuel; (b) feedstock supply, specifically the changing quality of crude oil and geopolitics between different countries and the emergence of alternate feed supplies such as bitumen from tar sand, natural gas, and coal; (c) environment regulations that include more stringent regulations in relation to sulfur in gasoline and diesel; and (d) technology development such as new catalysts and processes.

The general trend throughout refining has been to produce more products from each barrel of petroleum and to process those products in different ways to meet the product specifications for use in modern engines. Overall, the demand for gasoline has rapidly expanded and demand has also developed for gas oils and fuels for domestic central heating, and fuel oil for power generation, as well as for light distillates and other inputs, derived from crude oil, for the petrochemical industries.

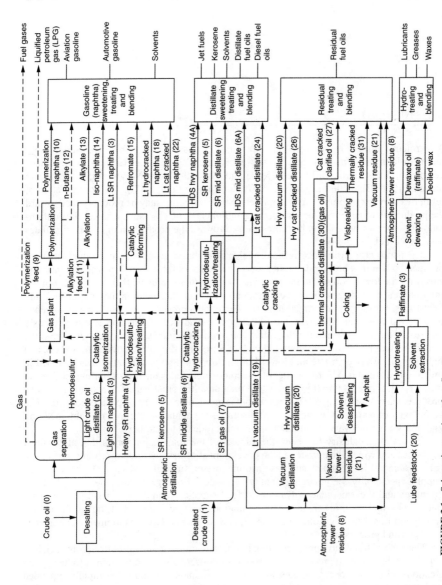

FIGURE 3.1 Schematic overview of a refinery.

TABLE 3.1 Crude Petroleum Is a Mixture of Compounds That Can Be Separated into Different Generic Boiling Fractions

	Boiling range*	
Fraction	°C	°F
Light naphtha	−1–150	30–300
Gasoline	−1–180	30–355
Heavy naphtha	150–205	300–400
Kerosene	205–260	400–500
Light gas oil	260–315	400–600
Heavy gas oil	315–425	600–800
Lubricating oil	>400	>750
Vacuum gas oil	425–600	800–1100
Residuum	>510	>950

*For convenience, boiling ranges are converted to the nearest 5°C.

To convert crude oil into desired products in an economically feasible and environmentally acceptable manner. Refinery process for crude oil are generally divided into three categories: (a) separation processes, of which distillation is the prime example; (b) conversion processes, of which coking and catalytic cracking are prime example; and (c) finishing processes, of which hydrotreating to remove sulfur is a prime example.

The simplest refinery configuration is the *topping refinery*, which is designed to prepare feedstocks for petrochemical manufacture or for production of industrial fuels in remote oil-production areas. The topping refinery consists of tankage, a distillation unit, recovery facilities for gases and light hydrocarbons, and the necessary utility systems (steam, power, and water-treatment plants). Topping refineries produce large quantities of unfinished oils and are highly dependent on local markets, but the addition of hydrotreating and reforming units to this basic configuration results in a more flexible *hydroskimming refinery*, which can also produce desulfurized distillate fuels and high-octane gasoline. These refineries may produce up to half of their output as residual fuel oil, and face increasing market loss as the demand for low-sulfur (even no-sulfur) fuel oil increases.

The most versatile refinery configuration today for fuel production is the *conversion refinery*. A conversion refinery incorporates all the basic units found in both the topping and hydroskimming refineries, but it also features gas oil conversion plants such as catalytic cracking and hydrocracking units, olefin conversion plants such as alkylation or polymerization units, and, frequently, coking units for sharply reducing or eliminating the production of residual fuels. Modern conversion refineries may produce two-thirds of their output as unleaded gasoline, with the balance distributed between liquefied petroleum gas, jet fuel, diesel fuel, and a small quantity of coke. Many such refineries also incorporate solvent extraction processes for manufacturing lubricants and petrochemical units with which to recover propylene, benzene, toluene, and xylene for further processing into polymers.

Finally, the yields and quality of refined petroleum products produced by any given oil refinery depends on the mixture of crude oil used as feedstock and the configuration of the refinery facilities. Light/sweet crude oil is generally more expensive and has inherent great yields of higher value low-boiling products such as naphtha, gasoline, jet fuel, kerosene, and diesel fuel. Heavy sour crude oil is generally less expensive and produces greater yields of lower value high-boiling products that must be converted into low-boiling products.

This chapter presents an overview of petroleum refining in order for the reader to place each process in the correct context of fuels production.

3.2 PETROLEUM RECOVERY

Petroleum is recovered from reservoir by various methods, the most advantageous of which is the use of the inherent reservoir energy that allows the petroleum to flow to the surface (Speight, 2007). However, the production rates from reservoirs depend on a number of factors, such as reservoir pressure, rock type, and permeability, fluid saturations and properties, extent of fracturing, number of wells, and their locations.

Primary oil recovery depends on the natural energy contained in the reservoir to drive the oil through the complex pore network to producing wells. The driving energy may come from liquid expansion and evolution of gas dissolved in the oil as reservoir pressure is lowered during production, expansion of free gas in a gas "cap," influx of natural water from an aquifer, or combinations of these effects. The recovery efficiency for primary production is generally low when liquid expansion and solution gas evolution are the driving mechanisms. Higher recoveries are associated with reservoirs having water or gas cap drives and from reservoirs where gravity effectively promotes drainage of the oil from the pores. Eventually, the natural drive energy is dissipated. When this occurs, energy must be supplied to the reservoir to produce additional oil.

Secondary oil recovery involves introducing energy into a reservoir by injecting gas or water under pressure. The injected fluids maintain reservoir pressure and displace a portion of the remaining crude oil to production wells.

Waterflooding is the principal secondary recovery method and currently accounts for almost half of the U.S. daily oil production. Limited use is made of gas injection because of the value of the natural gas. However, when gravity drainage is effective, pressure maintenance by gas injection can be very efficient. Certain reservoir systems, such as those with very viscous oils and low permeability or geologically complex reservoirs, respond poorly to conventional secondary recovery techniques. In these reservoirs improved geologic understanding and use of enhanced oil recovery (EOR) operations should be employed as early as possible.

Conventional primary and secondary recovery processes, at existing levels of field development, will ultimately produce about one-third of the original oil in place (OOIP) in discovered reservoirs. For individual reservoirs the recovery ranges from the extremes of less than 5 percent to as much as 80 percent of the OOIP. The range chiefly reflects the degree of reservoir complexity or heterogeneity. The more complex the reservoir, the lower the achievable recovery.

Of the remaining two-thirds of original oil in place in domestic reservoirs, a portion of this oil can be recovered through advanced secondary recovery methods involving improved sweep efficiency in poorly swept zones of the reservoirs. For these reservoirs, well placement and completion techniques need to be pursued consistent with the degree of reservoir heterogeneity. Such improved secondary oil recovery can be accomplished using advanced geologic models of complex reservoirs.

The balance of the remaining two-thirds of unrecovered oil is oil that is or will be residual to efficient sweep by secondary recovery processes. Portions of this residual oil can be recovered by tertiary or enhanced oil recovery methods. The intent of enhanced oil recovery methods is to increase ultimate oil production beyond that achieved by primary and secondary recovery methods by increasing the volume of rock contacted by the injected fluids (improving the sweep efficiency), reducing the residual oil remaining in the swept zones (increasing the displacement efficiency), or by reducing the viscosity of heavier oils.

Current enhanced oil recovery methods can be broadly grouped into three categories: (a) thermal methods, (b) miscible methods, and (c) chemical methods (Speight, 2007 and references cited therein). These processes differ considerably in complexity, the physical mechanisms responsible for oil recovery, and maturity of the technology derived from field applications. Enhanced oil recovery methods using microorganisms or electrical heating have been proposed; their current state of development is not complete.

Thermal recovery methods include cyclic steam injection, steam flooding, and in situ combustion. The thermal methods are used to reduce the viscosity of the oil and provide pressure so that the oil will flow more easily to the production wells.

The steam processes are the most evolved enhanced oil recovery methods in terms of field experience. For example, steam-based processes are most often applied in reservoirs containing heavy oil in place of, rather than following, primary or secondary recovery methods.

In situ combustion, an alternate thermal process, has been field tested under a wide variety of reservoir conditions, but few projects have proved economic and advanced to commercial scale. In the current economic climate, in situ combustion methods are once again being given serious consideration for the recovery of heavy oil and bitumen from tar sand formations.

Miscible methods use carbon dioxide, nitrogen, or hydrocarbons as miscible solvents to flood the reservoir and can produce 10 to 15 percent of the original oil in place. The solvents mix with the oil without an interface and are very effective in displacing the oil from the reservoir. Their greatest potential is enhancing the recovery of low-viscosity oils. Unfortunately, these methods do not always achieve high sweep efficiency.

The chemical methods include polymer-flooding, surfactant-(micellar/polymer, microemulsion)flooding, and alkaline-flooding processes. These methods take advantage of physical attributes of chemicals injected along with a displacing water driver to improve recovery. Polymer flooding is conceptually simple and inexpensive, but it produces only small amounts of incremental oil. It improves waterflooding by using polymers to thicken the water to increase its viscosity to near that of the reservoir oil so that displacement is more uniform and a greater portion of the reservoir is contacted.

Surfactant flooding is complex and requires detailed laboratory testing to support field project design, but it can produce as much as 50 to 60 percent of residual oil. Surfactants are injected into the reservoir to reduce the interfacial tension between the residual oil and flood water to release the oil from the reservoir rock. The surfactant causes the oil droplets to coalesce into an oil bank that can be pushed to production wells. Improvements in displacement efficiency clearly have been shown; however, sweep efficiency is a serious issue in applying this method.

Once the oil is recovered it is transported to the refinery where conversion to various products takes place.

3.3 PETROLEUM REFINING

3.3.1 Dewatering and Desalting

Petroleum is recovered from the reservoir mixed with a variety of substances: gases, water, and dirt (minerals). Thus, refining actually commences with the production of fluids from the well or reservoir and is followed by pretreatment operations that are applied to the crude oil either at the refinery or prior to transportation. Pipeline operators, for instance, are insistent upon the quality of the fluids put into the pipelines; therefore, any crude oil to be shipped by pipeline or, for that matter, by any other form of transportation must meet rigid specifications in regard to water and salt content. In some instances, sulfur content, nitrogen content, and viscosity may also be specified.

Field separation, which occurs at a field site near the recovery operation, is the first attempt to remove the gases, water, and dirt that accompany crude oil coming from the ground. The separator may be no more than a large vessel that gives a quieting zone for gravity separation into three phases: gases, crude oil, and water containing entrained dirt.

Desalting is a water-washing operation performed at the production field and at the refinery site for additional crude oil cleanup (Fig. 3.2). If the petroleum from the separators contains water and dirt, water washing can remove much of the water-soluble minerals and entrained solids. If these crude oil contaminants are not removed, they can cause operating problems during refinery processing, such as equipment plugging and corrosion as well as catalyst deactivation.

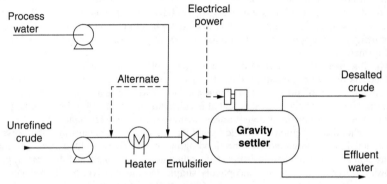

FIGURE 3.2 An electrostatic desalting unit.

3.3.2 Distillation

Distillation was the first method by which petroleum was refined. In the early stages of refinery development, when illuminating and lubricating oils were the main products, distillation was the major and often only refinery process. At that time gasoline was a minor, but more often unwanted, product. As the demand for gasoline increased, conversion processes were developed because distillation could no longer supply the necessary quantities of this volatile product.

It is possible to obtain fuels ranging from gaseous materials taken off at the top of the distillation column to a nonvolatile residue or reduced crude (*bottoms*), with correspondingly lighter materials at intermediate points. The reduced crude may then be processed by vacuum or steam distillation in order to separate the high-boiling lubricating oil fractions without the danger of decomposition, which occurs at high (>350°C, >662°F) temperatures. Atmospheric distillation may be terminated with a lower boiling fraction (*cut*) if it is felt that vacuum or steam distillation will yield a better-quality product, or if the process appears to be economically more favorable. Not all crude oils yield the same distillation products and the nature of the crude oil dictates the processes that may be required for refining.

Atmospheric Distillation. The distillation unit is a collection of distillation units but, in contrast to the early battery units, a tower is used in the modern-day refinery (Fig. 3.3) and brings about an efficient degree of fractionation (separation).

The feed to a distillation tower is heated by flow through pipes arranged within a large furnace. The heating unit is known as a pipe still heater or pipe still furnace, and the heating unit and the fractional distillation tower make up the essential parts of a distillation unit or pipe still. The pipe still furnace heats the feed to a predetermined temperature—usually a

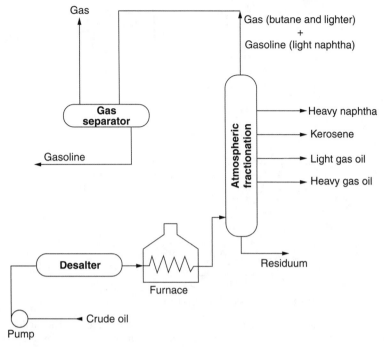

FIGURE 3.3 An atmospheric distillation unit.

temperature at which a predetermined portion of the feed will change into vapor. The vapor is held under pressure in the pipe in the furnace until it discharges as a foaming stream into the fractional distillation tower. Here the unvaporized or liquid portion of the feed descends to the bottom of the tower to be pumped away as a bottom nonvolatile product, while the vapors pass up the tower to be fractionated into gas oils, kerosene, and naphtha.

Pipe still furnaces vary greatly and, in contrast to the early units where capacity was usually 200 to 500 bbl/day, can accommodate 25,000 bbl, or more of crude petroleum per day. The walls and ceiling are insulated with firebrick and the interior of the furnace is partially divided into two sections: a smaller convection section where the oil first enters the furnace and a larger section (fitted with heaters) and where the oil reaches its highest temperature.

All of the primary fractions from a distillation unit are equilibrium mixtures and contain some proportion of the lighter constituent's characteristic of a lower boiling fraction. The primary fractions are *stripped* of these constituents (*stabilized*) before storage or further processing.

Vacuum Distillation. Vacuum distillation as applied to the petroleum-refining industry evolved because of the need to separate the less volatile products, such as lubricating oils, from the petroleum without subjecting these high-boiling products to cracking conditions. The boiling point of the heaviest cut obtainable at atmospheric pressure is limited by the temperature (ca 350°C; ca 662°F) at which the residue starts to decompose (*crack*). When the feedstock is required for the manufacture of lubricating oils, further fractionation without cracking is desirable and this can be achieved by distillation under vacuum conditions.

Operating conditions for vacuum distillation (Fig. 3.4) are usually 50 to 100 mmHg (atmospheric pressure = 760 mmHg). In order to minimize large fluctuations in pressure in the

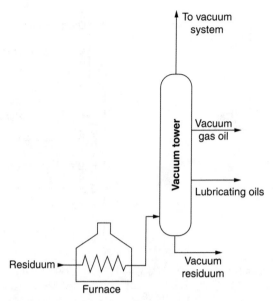

FIGURE 3.4 A vacuum distillation unit.

vacuum tower, the units are necessarily of a larger diameter than the atmospheric units. Some vacuum distillation units have diameters on the order of 45 ft (14 m). By this means, a heavy gas oil may be obtained as an overhead product at temperatures of about 150°C (302°F), and lubricating oil cuts may be obtained at temperatures of 250 to 350°C (482–662°F), feed and residue temperatures being kept below the temperature of 350°C (662°F), above which cracking will occur. The partial pressure of the hydrocarbons is effectively reduced still further by the injection of steam. The steam added to the column, principally for the stripping of asphalt in the base of the column, is superheated in the convection section of the heater.

The fractions obtained by vacuum distillation of the reduced crude (atmospheric residuum) from an atmospheric distillation unit depend on whether or not the unit is designed to produce lubricating or vacuum gas oils. In the former case, the fractions include (a) heavy gas oil, which is an overhead product and is used as catalytic cracking stock or, after suitable treatment, a light lubricating oil; (b) lubricating oil (usually three fractions—light, intermediate, and heavy), which is obtained as a side-stream product; and (c) asphalt (or residuum), which is the bottom product and may be used directly as, or to produce, asphalt and which may also be blended with gas oils to produce a heavy fuel oil.

3.3.3 Thermal Processes

Cracking distillation (thermal decomposition with simultaneous removal of distillate) was recognized as a means of producing the valuable lighter product (kerosene) from heavier nonvolatile materials. In the early days of the process (1870–1900) the technique was very simple—a batch of crude oil was heated until most of the kerosene had been distilled from it and the overhead material had become dark in color. At this point distillation was discontinued and the heavy oils were held in the hot zone, during which time some of the high-molecular-weight components were decomposed to produce low-molecular-weight products. After a suitable time, distillation was continued to yield light oil (kerosene) instead of the heavy oil that would otherwise have been produced.

Thermal Cracking. One of the earliest conversion processes used in the petroleum industry is the thermal decomposition of higher boiling materials into lower boiling products. The heavier oils produced by cracking are light and heavy gas oils as well as a residual oil which could also be used as heavy fuel oil. Gas oils from catalytic cracking were suitable for domestic and industrial fuel oils or as diesel fuels when blended with straight-run gas oils. The gas oils produced by cracking were also a further important source of gasoline. In a once-through cracking operation all of the cracked material is separated into products and may be used as such. However, the gas oils produced by cracking (cracked gas oils) are more resistant to cracking (more refractory) than gas oils produced by distillation (straight-run gas oils) but could still be cracked to produce more gasoline. This was achieved using a later innovation (post-1940) involving a recycle operation in which the cracked gas oil was combined with fresh feed for another trip through the cracking unit. The extent to which recycling was carried out affected the yield of gasoline from the process.

The majority of the thermal cracking processes use temperatures of 455 to 540°C (851–1004°F) and pressures of 100 to 1000 psi; the Dubbs process may be taken as a typical application of an early thermal cracking operation. The feedstock (reduced crude) is preheated by direct exchange with the cracking products in the fractionating columns. Cracked gasoline and heating oil are removed from the upper section of the column. Light and heavy distillate fractions are removed from the lower section and are pumped to separate heaters. Higher temperatures are used to crack the more refractory light distillate fraction. The streams from the heaters are combined and sent to a soaking chamber where additional time is provided to complete the cracking reactions. The cracked products are then separated in a low-pressure flash chamber where a heavy fuel oil is removed as bottoms. The remaining cracked products are sent to the fractionating columns.

Visbreaking. Visbreaking (viscosity breaking) is essentially a process of the post-1940 era and was initially introduced as a mild thermal cracking operation that could be used to reduce the viscosity of residua to allow the products to meet fuel oil specifications. Alternatively, the visbroken residua could be blended with lighter product oils to produce fuel oils of acceptable viscosity. By reducing the viscosity of the residuum, visbreaking reduces the amount of light heating oil that is required for blending to meet the fuel oil specifications. In addition to the major product, fuel oil, material in the gas oil and gasoline boiling range is produced. The gas oil may be used as additional feed for catalytic cracking units, or as heating oil.

In a typical visbreaking operation (Fig. 3.5), a crude oil residuum is passed through a furnace where it is heated to a temperature of 480°C (896°F) under an outlet pressure of about 100 psi. The heating coils in the furnace are arranged to provide a soaking section of low heat density, where the charge remains until the visbreaking reactions are completed and the cracked products are then passed into a flash-distillation chamber. The overhead material from this chamber is then fractionated to produce a low-quality gasoline as an overhead product and light gas oil as bottom. The liquid products from the flash chamber are cooled with a gas oil flux and then sent to a vacuum fractionator. This yields a heavy gas oil distillate and a residual tar of reduced viscosity.

Coking. Coking is a thermal process for the continuous conversion of heavy, low-grade oils into lighter products. Unlike visbreaking, coking involved compete thermal conversion of the feedstock into volatile products and coke (Table 3.2). The feedstock is typically a residuum and the products are gases, naphtha, fuel oil, gas oil, and coke. The gas oil may be the major product of a coking operation and serves primarily as a feedstock for catalytic cracking units. The coke obtained is usually used as fuel but specialty uses, such as electrode manufacture, production of chemicals and metallurgic coke are also possible and increases the value of the coke. For these uses, the coke may require treatment to remove sulfur and metal impurities.

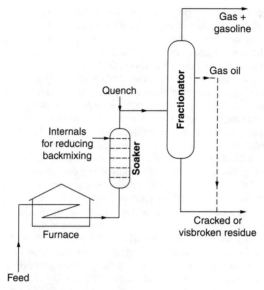

FIGURE 3.5 A soaker visbreaker.

TABLE 3.2 Comparison of Visbreaking with Delayed Coking and Fluid Coking

---

Visbreaking
Purpose: to reduce viscosity of fuel oil to acceptable levels
Conversion is not a prime purpose
Mild [470–495°C (878–923°F)] heating at pressures of 50–200 psi
  Reactions quenched before going to completion
  Low conversion (10%) to products boiling less than 220°C (428°F)
Heated coil or drum (soaker)

Delayed Coking
Purpose: to produce maximum yields of distillate products
Moderate [480–515°C (896–959°F)] heating at pressures of 90 psi
  Reactions allowed to proceed to completion
  Complete conversion of the feedstock
Soak drums (845–900°F) used in pairs (one on stream and one off stream being decoked)
Coked until drum solid
Coke removed hydraulically from off-stream drum
Coke yield: 20–40% by weight (dependent upon feedstock)
Yield of distillate boiling below 220°C (428°F): ca 30% (but feedstock dependent)

Fluid Coking
Purpose: to produce maximum yields of distillate products
Severe [480–565°C (896–1049°F)] heating at pressures of 10 psi
  Reactions allowed to proceed to completion
  Complete conversion of the feedstock
Oil contacts refractory coke
Bed fluidized with steam; heat dissipated throughout the fluid bed
Higher yields of light ends ($<C_5$) than delayed coking
Less coke made than delayed coking (for one particular feedstock)

After a gap of several years, the recovery of heavy oils either through secondary recovery techniques from oil sand formations caused a renewal of interest in these feedstocks in the 1960s and, henceforth, for coking operations. Furthermore, the increasing attention paid to reducing atmospheric pollution has also served to direct some attention to coking, since the process not only concentrates pollutants such as feedstock sulfur in the coke, but also can usually yield volatile products that can be conveniently desulfurized.

Delayed coking is a semi-continuous process (Fig. 3.6) in which the heated charge is transferred to large soaking (or coking) drums, which provide the long residence time needed to allow the cracking reactions to proceed to completion. The feed to these units is normally an atmospheric residuum although cracked residua are also used.

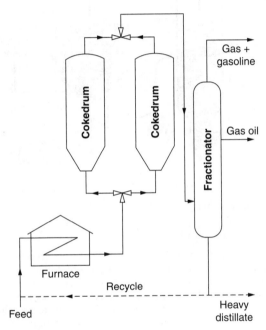

FIGURE 3.6 A delayed coker.

The feedstock is introduced into the product fractionator where it is heated and lighter fractions are removed as a side streams. The fractionator bottoms, including a recycle stream of heavy product, are then heated in a furnace whose outlet temperature varies from 480 to 515°C (896–959°F). The heated feedstock enters one of a pair of coking drums where the cracking reactions continue. The cracked products leave as overheads, and coke deposits form on the inner surface of the drum. To give continuous operation, two drums are used; while one is on stream, the other is being cleaned. The temperature in the coke drum ranges from 415 to 450°C (779–842°F) with pressures from 15 to 90 psi.

Overhead products go to the fractionator, where naphtha and heating oil fractions are recovered. The nonvolatile material is combined with preheated fresh feed and returned to the reactor. The coke drum is usually on stream for about 24 hours before becoming filled with porous coke after which the coke is removed hydraulically. Normally, 24 hours are required to complete the cleaning operation and to prepare the coke drum for subsequent use on stream.

Fluid coking is a continuous process (Fig. 3.7) which uses the fluidized solids technique to convert atmospheric and vacuum residua to more valuable products. The residuum is coked by

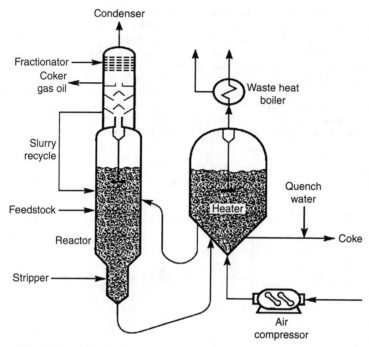

FIGURE 3.7 A fluid coker.

being sprayed into a fluidized bed of hot, fine coke particles, which permits the coking reactions to be conducted at higher temperatures and shorter contact times than can be employed in delayed coking. Moreover, these conditions result in decreased yields of coke; greater quantities of more valuable liquid product are recovered in the fluid-coking process.

Fluid coking uses two vessels, a reactor and a burner; coke particles are circulated between these to transfer heat (generated by burning a portion of the coke) to the reactor. The reactor holds a bed of fluidized coke particles, and steam is introduced at the bottom of the reactor to fluidize the bed.

Flexicoking (Fig. 3.8) is also a continuous process that is a direct descendent of fluid coking. The unit uses the same configuration as the fluid coker but has a gasification section in which excess coke can be gasified to produce refinery fuel gas. The flexicoking process was designed during the late 1960s and the 1970s as a means by which excess coke-make could be reduced in view of the gradual incursion of the heavier feedstocks in refinery operations. Such feedstocks are notorious for producing high yields of coke (>15 percent by weight) in thermal and catalytic operations.

3.3.4 Catalytic Processes

Catalytic cracking (Table 3.3) has a number of advantages over thermal cracking: (a) the gasoline produced has a higher octane number; (b) the catalytically cracked gasoline consists largely of isoparaffins and aromatics, which have high octane numbers and greater chemical stability than monoolefins and diolefins which are present in much greater quantities in

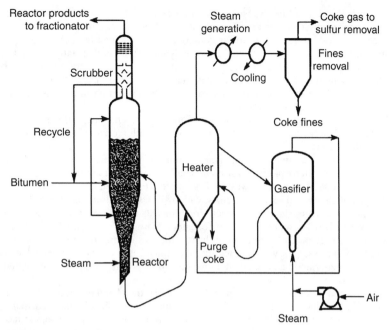

FIGURE 3.8 Flexicoking process.

TABLE 3.3 Summary of Catalytic Cracking Processes

Conditions
Solid acidic catalyst (silica-alumina, zeolite, etc.)
Temperature: 480–540°C (896–1004°F) (solid/vapor contact)
Pressure: 10–20 psi
Provisions needed for continuous catalyst replacement with heavier feedstocks (residua)
Catalyst may be regenerated or replaced

Feedstocks
Gas oils and residua
Residua pretreated to remove salts (metals)
Residua pretreated to remove high molecular weight (asphaltic constituents)

Products
Lower molecular weight than feedstock
Some gases (feedstock and process parameters dependent)
Isoparaffins in product
Coke deposited on catalyst

Variations
Fixed bed
Moving bed
Fluidized bed

thermally cracked gasoline. Substantial quantities of olefinic gases suitable for polymer gasoline manufacture and smaller quantities of methane, ethane, and ethylene are produced by catalytic cracking. Sulfur compounds are changed in such a way that the sulfur content of catalytically cracked gasoline is lower than in thermally cracked gasoline. Catalytic cracking produces less heavy residual or tar and more of the useful gas oils than does thermal cracking. The process has considerable flexibility, permitting the manufacture of both motor and aviation gasoline and a variation in the gas oil yield to meet changes in the fuel oil market.

The several processes currently employed in catalytic cracking differ mainly in the method of catalyst handling, although there is overlap with regard to catalyst type and the nature of the products.

The catalyst, which may be an activated natural or synthetic material, is employed in bead, pellet, or microspherical form and can be used as a fixed bed, moving bed, or fluid bed. The fixed bed process was the first process to be used commercially and uses a static bed of catalyst in several reactors, which allows a continuous flow of feedstock to be maintained. Thus, the cycle of operations consists of (a) flow of feedstock through the catalyst bed, (b) discontinuance of feedstock flow and removal of coke from the catalyst by burning, and (c) insertion of the reactor on stream. The moving bed process uses a reaction vessel (in which cracking takes place) and a kiln (in which the spent catalyst is regenerated) and catalyst movement between the vessels is provided by various means.

The fluid bed process (Fig. 3.9) differs from the fixed bed and moving bed processes, insofar as the powdered catalyst is circulated essentially as a fluid with the feedstock. The several fluid catalytic cracking processes in use differ primarily in mechanical design. Side-by-side reactor-regenerator construction along with unitary vessel construction (the reactor either above or below the regenerator) is the two main mechanical variations.

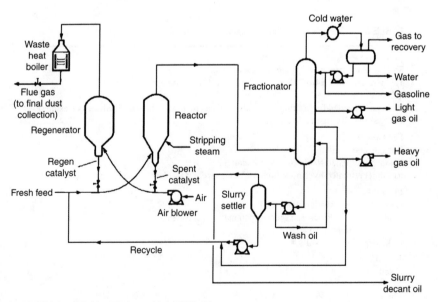

FIGURE 3.9 A fluid catalytic cracking (FCC) unit.

Natural clays have long been known to exert a catalytic influence on the cracking of oils, but it was not until about 1936 that the process using silica-alumina catalysts was developed sufficiently for commercial use. Since then, catalytic cracking has progressively supplanted thermal cracking as the most advantageous means of converting distillate oils into gasoline. The main reason for the wide adoption of catalytic cracking is the fact that a better yield of higher octane gasoline can be obtained than by any known thermal operation. At the same time the gas produced consists mostly of propane and butane with less methane and ethane. The production of heavy oils and tars, higher in molecular weight than the charge material, is also minimized, and both the gasoline and the uncracked "cycle oil" are more saturated than the products of thermal cracking.

Cracking crude oil fractions to produce fuels occurs over many types of catalytic materials, but high yields of desirable products are obtained with hydrated aluminum silicates. These may be either activated (acid-treated) natural clays of the bentonite type of synthesized silica-alumina or silica-magnesia preparations. Their activity to yield essentially the same products may be enhanced to some extent by the incorporation of small amounts of other materials such as the oxides of zirconium, boron (which has a tendency to volatilize away on use), and thorium. Natural and synthetic catalysts can be used as pellets or beads and also in the form of powder; in either case replacements are necessary because of attrition and gradual loss of efficiency. It is essential that they be stable to withstand the physical impact of loading and thermal shocks, and that they withstand the action of carbon dioxide, air, nitrogen compounds, and steam. They also should be resistant to sulfur and nitrogen compounds and synthetic catalysts, or certain selected clays, appear to be better in this regard than average untreated natural catalysts.

The catalysts are porous and highly adsorptive and their performance is affected markedly by the method of preparation. Two chemically identical catalysts having pores of different size and distribution may have different activity, selectivity, temperature coefficients of reaction rates, and responses to poisons. The intrinsic chemistry and catalytic action of a surface may be independent of pore size but small pores produce different effects because of the manner in which hydrocarbon vapors are transported in and out of the pore systems.

3.3.5 Hydroprocesses

Hydroprocesses use the principle that the presence of hydrogen during a thermal reaction of a petroleum feedstock will terminate many of the coke-forming reactions and enhance the yields of the lower boiling components such as gasoline, kerosene, and jet fuel (Table 3.4).

Hydrogenation processes for the conversion of petroleum fractions and petroleum products may be classified as *destructive* and *nondestructive*. Destructive hydrogenation (hydrogenolysis or hydrocracking) is characterized by the conversion of the higher molecular weight constituents in a feedstock to lower boiling products. Such treatment requires severe processing conditions and the use of high hydrogen pressures to minimize polymerization and condensation reactions that lead to coke formation.

Nondestructive or simple hydrogenation is generally used for the purpose of improving product quality without appreciable alteration of the boiling range. Mild processing conditions are employed so that only the more unstable materials are attacked. Nitrogen, sulfur, and oxygen compounds undergo reaction with the hydrogen to remove ammonia, hydrogen sulfide, and water, respectively. Unstable compounds which might lead to the formation of gums, or insoluble materials, are converted to more stable compounds.

TABLE 3.4 Summary of Hydrocracking Process Operations

Conditions
Solid acid catalyst (silica-alumina with rare earth metals, various other options)
Temperature: 260–450°C (500–842°F) (solid/liquid contact)
Pressure: 1,000–6,000 psi hydrogen
Frequent catalysts renewal for heavier feedstocks
Gas oil: catalyst life up to 3 years
Heavy oil/tar sand bitumen: catalyst life less than 1 year

Feedstocks
Refractory (aromatic) streams
Coker oils
Cycle oils
Gas oils
Residua (as a full hydrocracking or hydrotreating option)
In some cases, asphaltic constituents (S, N, and metals) removed by deasphalting

Products
Lower molecular weight paraffins
Some methane, ethane, propane, and butane
Hydrocarbon distillates (full range depending on the feedstock)
Residual tar (recycle)
Contaminants (asphaltic constituents) deposited on the catalyst as coke or metals

Variations
Fixed bed (suitable for liquid feedstocks)
Ebullating bed (suitable for heavy feedstocks)

Hydrotreating. Hydrotreating (Fig. 3.10) is carried out by charging the feed to the reactor, together with hydrogen in the presence of catalysts such as tungsten-nickel sulfide, cobalt-molybdenum-alumina, nickel oxide-silica-alumina, and platinum-alumina. Most processes employ cobalt-molybdenum catalysts which generally contain about 10 percent of molybdenum oxide and less than 1 percent of cobalt oxide supported on alumina. The temperatures employed are in the range of 260 to 345°C (500–653°F), while the hydrogen pressures are about 500 to 1000 psi.

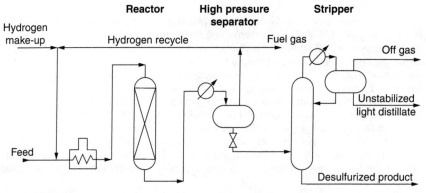

FIGURE 3.10 A distillate hydrotreater for hydrodesulfurization.

Hydrocracking. *Hydrocracking* is similar to catalytic cracking, with hydrogenation superimposed and with the reactions taking place either simultaneously or sequentially. Hydrocracking was initially used to upgrade low-value distillate feedstocks, such as cycle oils (high aromatic products from a catalytic cracker which usually are not recycled to extinction for economic reasons), thermal and coker gas oils, and heavy-cracked and straight-run naphtha. These feedstocks are difficult to process by either catalytic cracking or reforming, since they are characterized usually by a high polycyclic aromatic content and/or by high concentrations of the two principal catalyst poisons—sulfur and nitrogen compounds.

A comparison of hydrocracking with hydrotreating is useful in assessing the parts played by these two processes in refinery operations. Hydrotreating of distillates may be defined simply as the removal of nitrogen—sulfur and oxygen-containing compounds by selective hydrogenation. The hydrotreating catalysts are usually cobalt plus molybdenum or nickel plus molybdenum (in the sulfide) form impregnated on an alumina base. The hydrotreated operating conditions are such that appreciable hydrogenation of aromatics will not occur—1000 to 2000 psi hydrogen and about 370°C (698°F). The desulfurization reactions are usually accompanied by small amounts of hydrogenation and hydrocracking.

The commercial processes for treating, or finishing, petroleum fractions with hydrogen all operate in essentially the same manner as single-stage or two-stage processes (Fig. 3.11). The feedstock is heated and passed with hydrogen gas through a tower or reactor filled with catalyst pellets. The reactor is maintained at a temperature of 260 to 425°C (500–797°F) at

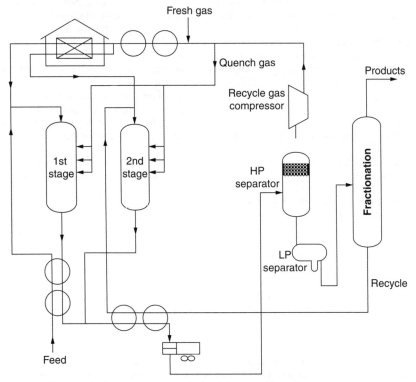

FIGURE 3.11 A single-stage or two-stage (optional) hydrocracking unit.

pressures from 100 to 1000 psi, depending on the particular process, the nature of the feedstock, and the degree of hydrogenation required. After leaving the reactor, excess hydrogen is separated from the treated product and recycled through the reactor after removal of hydrogen sulfide. The liquid product is passed into a stripping tower where steam removes dissolved hydrogen and hydrogen sulfide and, after cooling the product is taken to product storage or, in the case of feedstock preparation, pumped to the next processing unit.

3.3.6 Reforming Processes

When the demand for higher-octane gasoline developed during the early 1930s, attention was directed to ways and means of improving the octane number of fractions within the boiling range of gasoline. Straight-run (distilled) gasoline frequently had very low octane numbers, and any process that would improve the octane numbers would aid in meeting the demand for higher octane number gasoline. Such a process (called thermal reforming) was developed and used widely, but to a much lesser extent than thermal cracking. Thermal reforming was a natural development from older thermal cracking processes; cracking converts heavier oils into gasoline whereas reforming converts (reforms) gasoline into higher octane gasoline. The equipment for thermal reforming is essentially the same as for thermal cracking, but higher temperatures are used.

Thermal Reforming. In the thermal reforming process a feedstock such as 205°C (401°F) end-point naphtha or a straight-run gasoline is heated to 510 to 595°C (950–1103°F) in a furnace, much the same as a cracking furnace, with pressures from 400 to 1000 psi (27–68 atm). As the heated naphtha leaves the furnace, it is cooled or quenched by the addition of cold naphtha. The material then enters a fractional distillation tower where any heavy products are separated. The remainder of the reformed material leaves the top of the tower to be separated into gases and reformate. The higher octane number of the reformate is due primarily to the cracking of longer chain paraffins into higher octane olefins.

The products of thermal reforming are gases, gasoline, and residual oil or tar, the latter being formed in very small amounts (about 1 percent). The amount and quality of the gasoline, known as reformate, is very dependent on the temperature. A general rule is: the higher the reforming temperature, the higher the octane number, but the lower the yield of reformate.

Thermal reforming is less effective and less economic than catalytic processes and has been largely supplanted. As it used to be practiced, a single-pass operation was employed at temperatures in the range 540 to 760°C (1004–1400°F) and pressures of about 500 to 1000 psi (34–68 atm). The degree of octane number improvement depended on the extent of conversion but was not directly proportional to the extent of crack per pass. However at very high conversions, the production of coke and gas became prohibitively high. The gases produced were generally olefinic and the process required either a separate gas polymerization operation or one in which C3 to C4 gases were added back to the reforming system.

More recent modifications of the thermal reforming process due to the inclusion of hydrocarbon gases with the feedstock are known as gas reversion and polyforming. Gaseous olefins gases produced by cracking and reforming can be converted into liquids boiling in the gasoline range by heating them under high pressure. Since the resulting liquids (polymers) have high octane numbers, they increase the overall quantity and quality of gasoline produced in a refinery.

Catalytic Reforming. Like thermal reforming, catalytic reforming converts low-octane gasoline into high-octane gasoline (reformate). When thermal reforming could produce reformate with research octane numbers of 65 to 80 depending on the yield, catalytic reforming produces reformate with octane numbers on the order of 90 to 95. Catalytic reforming is conducted in the presence of hydrogen over hydrogenation-dehydrogenation catalysts, which may be supported on alumina or silica-alumina. Depending on the catalyst, a definite sequence of reactions takes place, involving structural changes in the feedstock. This more modern concept actually rendered thermal reforming somewhat obsolescent.

The commercial processes available for use can be broadly classified as the moving bed, fluid bed, and fixed bed types. The fluid bed and moving bed processes used mixed nonprecious metal oxide catalysts in units equipped with separate regeneration facilities. Fixed bed processes use predominantly platinum-containing catalysts in units equipped for cycle, occasional, or no regeneration.

Catalytic reformer feeds are saturated (i.e., not olefinic) materials; in the majority of cases that feed may be a straight-run naphtha but other by-product low-octane naphtha (e.g., coker naphtha) can be processed after treatment to remove olefins and other contaminants. Hydrocracker naphtha that contains substantial quantities of naphthenes is also a suitable feed.

Dehydrogenation is a main chemical reaction in catalytic reforming and hydrogen gas is consequently produced in large quantities. The hydrogen is recycled though the reactors where the reforming takes place to provide the atmosphere necessary for the chemical reactions and also prevents the carbon from being deposited on the catalyst, thus extending its operating life. An excess of hydrogen above whatever is consumed in the process is produced, and, as a result, catalytic reforming processes are unique in that they are the only petroleum refinery processes to produce hydrogen as a by-product.

Catalytic reforming usually is carried out by feeding a naphtha (after pretreating with hydrogen if necessary) and hydrogen mixture to a furnace where the mixture is heated to the desired temperature, 450 to 520°C (842–968°F), and then passed through fixed bed catalytic reactors at hydrogen pressures of 100 to 1000 psi (7–68 atm) (Fig. 3.12). Normally several reactors are used in series with heaters located between adjoining reactors in order to compensate for the endothermic reactions taking place. Sometimes as many as four or five reactors are kept on stream in series while one or more is being regenerated.

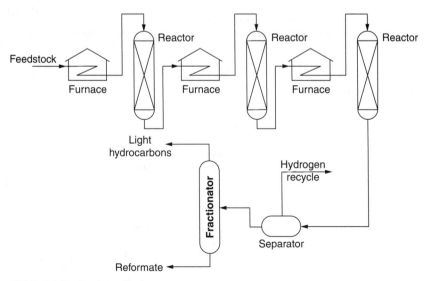

FIGURE 3.12 Catalytic reforming.

The composition of a reforming catalyst is dictated by the composition of the feedstock and the desired reformate. The catalysts used are principally molybdena-alumina, chromia-alumina, or platinum on a silica-alumina or alumina base. The nonplatinum catalysts are widely used in regenerative process for feeds containing, for example, sulfur, which poisons platinum catalysts, although pretreatment processes (e.g., hydrodesulfurization) may permit platinum catalysts to be employed.

The purpose of platinum on the catalyst is to promote dehydrogenation and hydrogenation reactions, that is, the production of aromatics, participation in hydrocracking, and rapid hydrogenation of carbon-forming precursors. For the catalyst to have an activity for isomerization of both paraffins and naphthenes—the initial cracking step of hydrocracking—and to participate in paraffin dehydrocyclization, it must have an acid activity. The balance between these two activities is most important in a reforming catalyst. In fact, in the production of aromatics from cyclic saturated materials (naphthenes), it is important that hydrocracking be minimized to avoid loss of the desired product and, thus, the catalytic activity must be moderated relative to the case of gasoline production from a paraffinic feed, where dehydrocyclization and hydrocracking play an important part.

3.3.7 Isomerization Processes

Catalytic reforming processes provide high-octane constituents in the heavier gasoline fraction but the normal paraffin components of the lighter gasoline fraction, especially butanes, pentanes, and hexanes, have poor octane ratings. The conversion of these normal paraffins to their isomers (isomerization) yields gasoline components of high octane rating in this lower boiling range. Conversion is obtained in the presence of a catalyst (aluminum chloride activated with hydrochloric acid), and it is essential to inhibit side reactions such as cracking and olefin formation.

Isomerization processes are to provide additional feedstock for alkylation units or high-octane fractions for gasoline blending (Table 3.5). Straight-chain paraffins (*n*-butane, *n*-pentane, *n*-hexane) are converted to respective isocompounds by continuous catalytic

TABLE 3.5 Component Streams for Gasoline

		Boiling range	
Stream	Producing process	°C	°F
Paraffinic			
Butane	Distillation	0	32
Conversion			
Isopentane	Distillation	27	81
	Conversion		
	Isomerization		
Alkylate	Alkylation	40–150	104–302
Isomerate	Isomerization	40–70	104–158
Naphtha	Distillation	30–100	86–212
Hydrocrackate	Hydrocracking	40–200	104–392
Olefinic			
Catalytic naphtha	Catalytic cracking	40–200	104–392
Cracked naphtha	Steam cracking	40–200	104–392
Polymer	Polymerization	60–200	140–392
Aromatic			
Catalytic reformate	Catalytic reforming	40–200	104–392

(aluminum chloride, noble metals) processes. Natural gasoline or light straight-run gasoline can provide feed by first fractionating as a preparatory step. High volumetric yields (>95 percent) and 40 to 60 percent conversion per pass are characteristic of the isomerization reaction.

Aluminum chloride was the first catalyst used to isomerize butane, pentane, and hexane. Since then, supported metal catalysts have been developed for use in high-temperature processes which operate in the range 370 to 480°C (698–896°F) and 300 to 750 psi (20–51 atm) (Fig. 3.13), while aluminum chloride plus hydrogen chloride are universally used for the low-temperature processes. Nonregenerable aluminum chloride catalyst is employed with various carriers in a fixed-bed or liquid contactor. Platinum or other metal catalyst processes utilized fixed bed operation and can be regenerable or nonregenerable. The reaction conditions vary widely depending on the particular process and feedstock, [40–480°C (104–896°F)] and 150 to 1000 psi (10–68 atm).

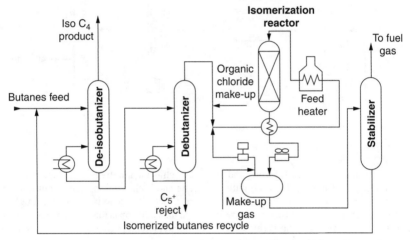

FIGURE 3.13 A butane isomerization unit.

3.3.8 Alkylation Processes

The combination of olefins with paraffins to form higher isoparaffins is termed alkylation. Since olefins are reactive (unstable) and are responsible for exhaust pollutants, their conversion to high-octane isoparaffins is desirable when possible. In refinery practice, only isobutane is alkylated, by reaction with isobutene or normal butene and isooctane is the product. Although alkylation is possible without catalysts, commercial processes use aluminum chloride, sulfuric acid, or hydrogen fluoride as catalysts, when the reactions can take place at low temperatures, minimizing undesirable side reactions, such as polymerization of olefins.

Alkylate is composed of a mixture of isoparaffins which have octane numbers that vary with the olefins from which they were made. Butylenes produce the highest octane numbers, propylene the lowest and pentylenes the intermediate values. All alkylates, however, have high octane numbers (>87) which makes them particularly valuable.

The alkylation reaction as now practiced is the union, through the agency of a catalyst, of an olefin (ethylene, propylene, butylene, and amylene) with isobutane to yield high-octane branched-chain hydrocarbons in the gasoline boiling range. Olefin feedstock is derived from the gas produced in a catalytic cracker, while isobutane is recovered by refinery gases or produced by catalytic butane isomerization. To accomplish this, either ethylene or propylene is combined with isobutane at 50 to 280°C (122–536°F) and 300 to 1000 psi (20–68 atm) in the presence of metal halide catalysts such as aluminum chloride. Conditions are less stringent in catalytic alkylation; olefins (propylene, butylene, or pentylene) are combined with isobutane in the presence of an acid catalyst (sulfuric acid or hydrofluoric acid) at low temperatures and pressures [1–40°C (34–104°F) and 14.8–150 psi (1–10 atm)] (Fig. 3.14).

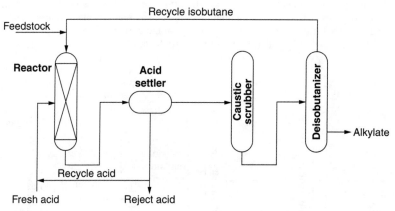

FIGURE 3.14 An alkylation unit (sulfuric acid catalyst).

Sulfuric acid, hydrogen fluoride, and aluminum chloride are the general catalysts used commercially. Sulfuric acid is used with propylene and higher boiling feeds, but not with ethylene, because it reacts to form ethyl hydrogen sulfate. The acid is pumped through the reactor and forms an air emulsion with reactants, and the emulsion is maintained at 50 percent acid. The rate of deactivation varies with the feed and isobutane charge rate. Butene feeds cause less acid consumption than the propylene feeds.

Aluminum chloride is not currently used as an alkylation catalyst but when employed, hydrogen chloride is used as a promoter and water is injected to activate the catalyst as an aluminum chloride/hydrocarbon complex. Hydrogen fluoride is used for alkylation of higher boiling olefins and the advantage of hydrogen fluoride is that it is more readily separated and recovered from the resulting product.

3.3.9 Polymerization Processes

Polymerization is the process by which olefin gases are converted to liquid products which may be suitable for gasoline (polymer gasoline) or other liquid fuels. The feedstock usually consists of propylene and butylenes from cracking processes or may even be selective olefins for dimer, trimer, or tetramer production.

Polymerization may be accomplished thermally or in the presence of a catalyst at lower temperatures. Thermal polymerization is regarded as not being as effective as catalytic polymerization but has the advantage that it can be used to "polymerize" saturated materials that cannot be induced to react by catalysts. The process consists of vapor-phase cracking of, for example, propane and butane followed by prolonged periods at the high temperature [510–595°C (950–1103°F)] for the reactions to proceed to near completion.

Olefins can also be conveniently polymerized by means of an acid catalyst (Fig. 3.15). Thus, the treated, olefin-rich feed stream is contacted with a catalyst (sulfuric acid, copper pyrophosphate, phosphoric acid) at 150 to 220°C (302–428°F) and 150 to 1200 psi (10–81 atm), depending on feedstock and product requirement.

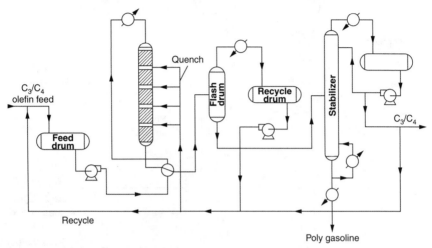

FIGURE 3.15 A polymerization unit.

Phosphates are the principal catalysts used in polymerization units; the commercially used catalysts are liquid phosphoric acid, phosphoric acid on kieselguhr, copper pyrophosphate pellets, and phosphoric acid film on quartz. The latter is the least active, but the most used and easiest one to regenerate simply by washing and recoating; the serious disadvantage is that tar must occasionally be burned off the support. The process using liquid phosphoric acid catalyst is far more responsible to attempt to raise production by increasing temperature than the other processes.

3.3.10 Deasphalting

Solvent deasphalting processes are a major part of refinery operations (Bland and Davidson, 1967; Hobson and Pohl, 1973; Gary and Handwerk, 1994; Speight and Ozum, 2002; Speight, 2007) and are not often appreciated for the tasks for which they are used. In the solvent deasphalting processes, an alkane is injected into the feedstock to disrupt the dispersion of components and causes the polar constituents to precipitate. Propane (or sometimes propane/butane mixtures) is extensively used for deasphalting and produces a deasphalted oil (DAO) and propane deasphalter asphalt (PDA or PD tar) (Dunning and Moore, 1957). Propane has unique solvent properties; at lower temperatures [38–60°C (100–140°F)], paraffins are very soluble in propane and at higher temperatures [about 93°C (199°F)] all hydrocarbons are almost insoluble in propane.

A *solvent deasphalting* unit (Fig. 3.16) processes the residuum from the vacuum distillation unit and produces deasphalted oil (DAO), used as feedstock for a fluid catalytic cracking unit, and the asphaltic residue (deasphalter tar, deasphalter bottoms) which, as a residual fraction, can only be used to produce asphalt or as a blend stock or visbreaker feedstock for low-grade fuel oil. Solvent deasphalting processes have not realized their maximum

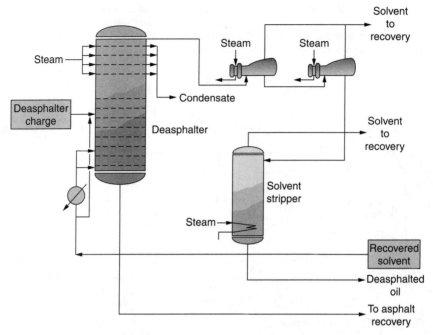

FIGURE 3.16 Propane deasphalting.

potential. With ongoing improvements in energy efficiency, such processes would display its effects in a combination with other processes. Solvent deasphalting allows removal of sulfur and nitrogen compounds as well as metallic constituents by balancing yield with the desired feedstock properties.

3.3.11 Dewaxing

Paraffinic crude oils often contain microcrystalline or paraffin waxes. The crude oil may be treated with a solvent such as methyl ethyl ketone (MEK) to remove this wax before it is processed. This is not a common practice, however and *solvent dewaxing processes* are designed to remove wax from lubricating oils to give the product good fluidity characteristics at low temperatures (e.g., low pour points) rather than from the whole crude oil. The mechanism of solvent dewaxing involves either the separation of wax as a solid that crystallizes from the oil solution at low temperature or the separation of wax as a liquid that is extracted at temperatures above the melting point of the wax through preferential selectivity of the solvent. However, the former mechanism is the usual basis for commercial dewaxing processes.

In the *solvent dewaxing process* (Fig. 3.17) the feedstock is mixed with one to four times its volume of a ketone (Scholten, 1992). The mixture is then heated until the oil is in solution and the solution is chilled at a slow, controlled rate in double-pipe, scraped-surface exchangers. Cold solvent, such as filtrate from the filters, passes through the 2-in annular space between the inner and outer pipes and chills the waxy oil solution flowing through the inner 6-in pipe.

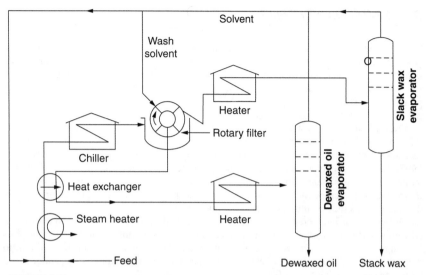

FIGURE 3.17 A solvent dewaxing unit.

To prevent wax from depositing on the walls of the inner pipe, blades, or scrapers extending the length of the pipe and fastened to a central rotating shaft scrape off the wax. Slow chilling reduces the temperature of the waxy oil solution to 2°C (35°F), and then faster chilling reduces the temperature to the approximate pour point required in the dewaxed oil. The waxy mixture is pumped to a filter case into which the bottom half of the drum of a rotary vacuum filter dips. The drum (8 ft in diameter, 14 ft in length), covered with filter cloth, rotates continuously in the filter case. Vacuum within the drum sucks the solvent and the oil dissolved in the solvent through the filter cloth and into the drum. Wax crystals collect on the outside of the drum to form a wax cake, and as the drum rotates, the cake is brought above the surface of the liquid in the filter case and under sprays of ketone that wash oil out of the cake and into the drum. A knife-edge scrapes off the wax, and the cake falls into the conveyor and is moved from the filter by the rotating scroll.

The recovered wax is actually a mixture of wax crystals with a little ketone and oil, and the filtrate consists of the dewaxed oil dissolved in a large amount of ketone. Ketone is removed from both by distillation, but before the wax is distilled, it is deoiled, mixed with more cold ketone, and pumped to a pair of rotary filters in series, where further washing with cold ketone produces a wax cake that contains very little oil. The deoiled wax is melted in heat exchangers and pumped to a distillation tower operated under vacuum, where a large part of the ketone is evaporated or flashed from the wax. The rest of the ketone is removed by heating the wax and passing it into a fractional distillation tower operated at atmospheric pressure and then into a stripper where steam removes the last traces of ketone.

An almost identical system of distillation is used to separate the filtrate into dewaxed oil and ketone. The ketone from both the filtrate and wax slurry is reused. Clay treatment or hydrotreating finishes the dewaxed oil as previously described. The wax (*slack wax*) even though it contains essentially no oil as compared to 50 percent in the slack wax obtained by cold pressing, is the raw material for either sweating or wax recrystallization, which subdivides the wax into a number of wax fractions with different melting points.

3.4 HEAVY OIL REFINING

The demand for petroleum and petroleum products has shown a sharp growth in recent years (Speight, 2007) this could well be the last century for petroleum refining, as we know it. The demand for transportation fuels and fuel oil is forecast to continue to show a steady growth in the future. The simplest means to cover the demand growth in low-boiling products is to increase the imports of light crude oils and low-boiling petroleum products, but these steps may be limited in the future.

Over the past three decades, crude oils available to refineries have generally decreased in API gravity. There is, nevertheless, a major focus in refineries on the ways in which heavy feedstocks might be converted into low-boiling high-value products (Khan and Patmore, 1997). Simultaneously, the changing crude oil properties are reflected in changes such as an increase in asphaltene constituents, an increase in sulfur, metal, and nitrogen contents. Pretreatment processes for removing such constituents or at least negating their effect in thermal process they would also play an important role.

Difficult-to-refine feedstocks, such as heavy oil and residua, are characterized by low API gravity (high density) and high viscosity, high initial boiling point, high carbon residue, high nitrogen content, high sulfur content, and high metals content. In addition, to these properties, the heavy feedstocks also have an increased molecular weight and reduced hydrogen content (Speight, 2007).

The limitations of processing these heavy feedstocks depend to a large extent on the tendency for coke formation and the deposition of metals and coke on the catalyst. However, the essential step required of refineries is the upgrading of heavy feedstocks, particularly residua (McKetta, 1992; Dickenson et al., 1997). In fact, the increasing supply of heavy crude oils is a matter of serious concern for the petroleum industry. In order to satisfy the changing pattern of product demand, significant investments in refining conversion processes will be necessary to profitably utilize these heavy crude oils. The most efficient and economic solution to this problem will depend to a large extent on individual country and company situations. However, the most promising technologies will likely involve the conversion of vacuum bottom residual oils, asphalt from deasphalting processes, and super-heavy crude oils into useful low-boiling and middle distillate products.

Upgrading heavy oil upgrading and residua began with the introduction of desulfurization processes (Speight, 1984, 2000). In the early days, the goal was desulfurization but, in later years, the processes were adapted to a 10 to 30 percent partial conversion operation, as intended to achieve desulfurization and obtain low-boiling fractions simultaneously, by increasing severity in operating conditions. Refinery evolution has seen the introduction of a variety of residuum cracking processes based on *thermal cracking*, *catalytic cracking*, and *hydroconversion*. Those processes are different from one another in cracking method, cracked product patterns, and product properties, and will be employed in refineries according to their respective features. Thus, refining heavy feedstocks has become a major issue in modern refinery practice and several process configurations have evolved to accommodate the heavy feedstocks (RAROP, 1991; Shih and Oballa, 1991; Khan and Patmore, 1997).

Technologies for upgrading heavy feedstocks can be broadly divided into *carbon rejection* and *hydrogen addition* processes. *Carbon rejection* redistributes hydrogen among the various components, resulting in fractions with increased H/C atomic ratios and fractions with lower H/C atomic ratios. On the other hand, *hydrogen addition* processes involve reaction of heavy crude oils with an external source of hydrogen and result in an overall increase in H/C ratio. Within these broad ranges, all more common upgrading technologies can be subdivided as follows:

1. *Carbon rejection*: Delayed coking, fluid coking, flexicoking, residuum fluid catalytic cracking, and heavy oil cracking.
2. *Hydrogen addition*: LC-Fining, H-Oil.

Carbon rejection processes operate at moderate to high temperatures and low pressures and suffer from a lower liquid yield of transportation fuels than hydrogen addition processes, because a large fraction of the feedstock is rejected as solid coke high in sulfur and nitrogen (and gaseous product). The liquids are generally of poor quality and must be hydrotreated before they can be used as reformer or fluid catalytic-cracking (FCC) feeds to make transportation fuels.

In the delayed coking process, heavy oil is heated to above 480°C (896°F) and fed to a vessel where thermal cracking and polymerization occur. A typical product slate would be 10 percent gas, 30 percent coke, and only 60 percent liquids, the coke percentage increasing at the expense of liquid products as feeds become heavier. Since sulfur is concentrated in the coke, the coke market is limited to buyers that can control, or are not restricted by, emissions of sulfur oxides (SO_x).

In the fluid-coking process, heavy oil is fed to a reactor containing a 480 to 540°C (896–1004°F) bed of fluidized coke particles, where it cracks to produce lighter liquids, gases, and more coke. The coke is circulated to a burner vessel where a portion of the coke is burned to supply the heat required for the endothermic coking reactions. A portion of the remaining coke is returned to the reactor as fluidizing medium, and the balance is withdrawn as product. The net coke yield is only about 65 percent of that produced by delayed coking, but the liquids are of worse quality and the flue gas from the burner requires SO_x control.

The Flexicoking process is an extension of fluid coking. All but a small fraction of the coke is gasified to low-Btu gas (120 Btu/standard cubic feet) by addition of steam and air in a separate fluidized reactor. The heat required for both the gasification and thermal cracking is generated in this gasifier. A small amount of net coke (about 1 percent of feed) is withdrawn to purge the system of metals and ash. The liquid yield and properties are similar to those from fluid coking. The need for a coke market is eliminated or markedly reduced. The low-Btu gas can be burned in refinery furnaces and boilers or probably could also be used in cogeneration units to generate power and steam; but it must be used near the refinery since its heating value is too low to justify transportation. Unlike with fluid coking, SO_x is not an issue since sulfur is liberated in a reducing atmosphere (carbon monoxide and molecular hydrogen) inside the gasifier but hydrogen sulfide removal is required.

The Resid FCC process is an extension of gas oil FCC technology. In the process, resid (usually above 650°F boiling point, not vacuum resid) is fed to a 480 to 540°C (896–1004°F) fluidized bed of cracking catalyst. It is converted to predominantly gasoline-range boiling materials, and the carbon residue in the feed is deposited on the catalyst. The catalyst activity is then restored by burning the deposited coke in the regenerator. This also supplies the heat required to crack the feed in the next contacting cycle. The sulfur emissions are typically controlled by additives that bind the sulfur to the catalyst for later reduction to hydrogen sulfide in the fluid catalytic cracking reactor. The hydrogen sulfide is later processed to sulfur for sale as low-value by-product.

Ebullating bed processes (catalytic hydrocracking processes) such as the LC-Fining (developed by the Lummus Company) and the H-Oil process (developed by Hydrocarbon Research, Inc.) can be used to demetallize, desulfurize, and hydrocrack any heavy oil. Both processes involve high-pressure catalytic hydrogenation but runs at higher temperatures than fixed bed resid desulfurization processes [about 426–441°C (799–825°F)]. The feed passes upflow, expanding the catalyst bed with the ebullation and producing a backmixed isothermal system. Reactors are very large relative to fixed bed and are frequently staged to overcome the kinetic penalties associated with back mixing. A big advantage is the ability to add and withdraw catalyst while the unit is onstream, which allows the processing of oils with high metal concentrations than is practical with conventional fixed beds. Also, the ebullating bed eliminates coke plugging problems and allows high-temperature operation and high (70–90 percent) conversion of the vacuum resid. The disadvantages include hydrogen consumption that may be 20 to 100 percent higher than that for fixed bed resid desulfurization process, and loss of liquid and hydrogen to high gas yields. The distillate products require further hydrotreating and conversion to produce transportation fuels.

Thus, the options for refiners processing heavy high sulfur will be a combination of upgrading schemes and by-product utilization. Other heavy oil upgrading options include: (a) deep cut vacuum distillation, (b) solvent deasphalting prior to conversion, and (c) hydrogenation prior to conversion.

For the present, using a schematic refinery operation (Fig. 3.1), new processes for the conversion of residua and heavy oils will probably be used in concert with visbreaking with some degree of hydroprocessing as a primary conversion step. Other processes may replace or augment the deasphalting units in many refineries. Depending on the properties, an option for heavy oil, like tar sand bitumen, is to subject the feedstock to either delayed coking or fluid coking as the *primary upgrading* step with some prior distillation or topping (Speight, 2007). After primary upgrading, the product streams are hydrotreated and combined to form a *synthetic crude oil* that is shipped to a conventional refinery for further processing to liquid fuels.

The product qualities resulting from the various heavy oil upgrading technologies are quite variable and are strongly dependent on feed type, process type, and processing conditions. However, producing fuels of acceptable properties is possible (in all cases) with existing petroleum processing technology, although the economics vary with a given refinery.

However, there is not one single heavy-oil-upgrading solution that will fit all refineries. Heavy feedstock properties, existing refinery configuration, and desired product slate all can have a significant effect on the final configuration. Furthermore, a proper evaluation however is not a simple undertaking for an existing refinery. The evaluation starts with an accurate understanding of the nature of the feedstock; along with corresponding conversion chemistry need to be assessed. Once the options have been defined, development of the optimal configuration for refining the incoming feedstocks can be designed.

3.5 PETROLEUM PRODUCTS AND FUELS

Petroleum products and *fuels*, in contrast to *petrochemicals*, are bulk fractions that are derived from petroleum and have commercial value as a bulk product (Speight, 2007). In the strictest sense, petrochemicals are also petroleum products but they are individual chemicals that are used as the basic building blocks of the chemical industry.

The constant demand for fuels is the main driving force behind the petroleum industry. Other products, such as lubricating oils, waxes, and asphalt, have also added to the popularity of petroleum as a national resource. Indeed, fuel products derived from petroleum supply more than half of the world's total supply of energy. Gasoline, kerosene, and diesel oil provide fuel for automobiles, tractors, trucks, aircraft, and ships. Fuel oil and natural gas are used to heat homes and commercial buildings, as well as to generate electricity. Petroleum products are the basic materials used for the manufacture of synthetic fibers for clothing and in plastics, paints, fertilizers, insecticides, soaps, and synthetic rubber. The uses of petroleum as a source of raw material in manufacturing are central to the functioning of modern industry.

3.5.1 Gaseous Fuels

Natural Gas. Natural gas, which is predominantly methane, occurs in underground reservoirs separately or in association with crude oil (Speight, 2007). The principal types of gaseous fuels are oil (distillation) gas, reformed natural gas, and reformed propane or liquefied petroleum gas (LPG).

The principal constituent of natural gas is methane (CH_4). Other constituents are paraffinic hydrocarbons such as ethane (CH_3CH_3), propane ($CH_3CH_2CH_3$), and the butanes [$CH_3CH_2CH_2CH_3$ and/or $(CH_3)_3CH$]. Many natural gases contain nitrogen (N_2) as well as carbon dioxide (CO_2) and hydrogen sulfide (H_2S). Trace quantities of argon, hydrogen, and helium may also be present. Generally, the hydrocarbons having a higher molecular weight than methane, carbon dioxide, and hydrogen sulfide are removed from natural gas prior to its use as a fuel. Gases produced in a refinery contain methane, ethane, ethylene, propylene, hydrogen, carbon monoxide, carbon dioxide, and nitrogen, with low concentrations of water vapor, oxygen, and other gases.

Liquefied Petroleum Gas. Liquefied petroleum gas (*LPG*) is the term applied to certain specific hydrocarbons and their mixtures, which exist in the gaseous state under atmospheric ambient conditions but can be converted to the liquid state under conditions of moderate pressure at ambient temperature. These are the light hydrocarbons fraction of the paraffin series, derived from refinery processes, crude oil stabilization plants and natural gas processing plants comprising propane ($CH_3CH_2CH_3$), butane ($CH_3CH_2CH_2CH_3$), isobutane [$CH_3CH(CH_3)CH_3$] and to a lesser extent propylene ($CH_3CH=CH_2$), or butylene ($CH_3CH_2CH=CH_2$). The most common commercial products are propane, butane, or some mixture of the two and are generally extracted from natural gas or crude petroleum. Propylene and butylenes result from cracking other hydrocarbons in a petroleum refinery and are two important chemical feedstocks.

Mixed gas is a gas prepared by adding natural gas or liquefied petroleum gas to a manufactured gas, giving a product of better utility and higher heat content or Btu value.

The compositions of natural, manufactured, and mixed gases can vary so widely, no single set of specifications could cover all situations. The requirements are usually based on performances in burners and equipment, on minimum heat content, and on maximum sulfur content. Gas utilities in most states come under the supervision of state commissions or regulatory bodies and the utilities must provide a gas that is acceptable to all types of consumers and that will give satisfactory performance in all kinds of consuming equipment. However, there are specifications for liquefied petroleum gas (ASTM D1835) which depend upon the required volatility.

Since natural gas as delivered to pipelines has practically no odor, the addition of an odorant is required by most regulations in order that the presence of the gas can be detected readily in case of accidents and leaks. This odorization is provided by the addition of trace amounts of some organic sulfur compounds to the gas before it reaches the consumer. The standard requirement is that a user will be able to detect the presence of the gas by odor when the concentration reaches 1 percent of gas in air. Since the lower limit of flammability of natural gas is approximately 5 percent, this 1 percent requirement is essentially equivalent to one-fifth the lower limit of flammability. The combustion of these trace amounts of odorant does not create any serious problems of sulfur content or toxicity.

The different methods for gas analysis include absorption, distillation, combustion, mass spectroscopy, infrared spectroscopy, and gas chromatography (ASTM D2163, ASTM D2650, and ASTM D4424). Absorption methods involve absorbing individual constituents one at a time in suitable solvents and recording of contraction in volume measured. Distillation methods depend on the separation of constituents by fractional distillation and measurement of the volumes distilled. In combustion methods, certain combustible elements are caused to burn to carbon dioxide and water, and the volume changes are used to calculate composition. Infrared spectroscopy is useful in particular applications. For the most accurate analyses, mass spectroscopy, and gas chromatography are the preferred methods.

The specific gravity of product gases, including liquefied petroleum gas, may be determined conveniently by a number of methods and a variety of instruments (ASTM D1070, ASTM D4891).

The *heat value* of gases is generally determined at constant pressure in a flow calorimeter in which the heat released by the combustion of a definite quantity of gas is absorbed by a measured quantity of water or air. A continuous recording calorimeter is available for measuring heat values of natural gases (ASTM D1826).

The lower and upper limits of *flammability* of organic compounds indicate the percentage of combustible gas in air below and above which flame will not propagate. When flame is initiated in mixtures having compositions within these limits, it will propagate and therefore the mixtures are flammable. Knowledge of flammable limits and their use in establishing safe practices in handling gaseous fuels is important, for example, when purging equipment used in gas service, in controlling factory or mine atmospheres, or in handling liquefied gases.

Synthesis Gas. On the other hand, synthesis gas (syngas) is the name given to a gas mixture that is generated by the gasification of a carbon-containing fuel (e.g., petroleum coke, qv) to a gaseous product that contains varying amounts of carbon monoxide and hydrogen. The name *synthesis gas* originates from their use as intermediates in creating synthetic natural gas (SNG) and for producing ammonia and/or methanol. Syngas is also used as an intermediate in producing synthetic fuels via the Fischer-Tropsch reaction.

In the strictest sense, synthesis gas consists primarily of carbon monoxide and hydrogen, although carbon dioxide and nitrogen may also be present. The chemistry of syntehsis gas production is relatively simple but the reactions are often much more complex than indicated by simple chemical equations:

$$[C]_{petroleum\ coke} + O_2 \rightarrow CO_2$$

$$[C]_{petroleum\ coke} + CO_2 + C \rightarrow 2CO$$

$$[C]_{petroleum\ coke} + H_2O \rightarrow CO + H_2$$

Synthesis gas is combustible and often used as a fuel source or as an intermediate for the production of other chemicals. When used as an intermediate in the large-scale, industrial synthesis of hydrogen and ammonia, it is also produced from natural gas (via the steam-reforming reaction):

$$CH_4 + H_2O \rightarrow CO + 3H_2$$

Synthesis gas is also manufactured from waste and form coal but these feedtocks and processes are not discussed in the current context.

3.5.2 Liquid Fuels

Gasoline. *Gasoline*, also called *gas* (United States and Canada), or *petrol* (Great Britain) or *benzine* (Europe) is mixture of hydrocarbons that usually boil below 180°C (356°F) or, at most, below 200°C (392°F).

Gasoline is manufactured to meet specifications and regulations and not to achieve a specific distribution of hydrocarbons by class and size. However, chemical composition often defines properties. For example, volatility is defined by the individual hydrocarbon constituents and the lowest boiling constituent(s) defines the volatility as determined by certain test methods.

Automotive gasoline typically contains about almost 200 (if not several hundred) hydrocarbon compounds. The relative concentrations of the compounds vary considerably depending on the source of crude oil, refinery process, and product specifications. Typical hydrocarbon chain lengths range from C_4 through C_{12} with a general hydrocarbon distribution consisting of alkanes (4–8 percent), alkenes (2–5 percent), isoalkanes (25–40 percent), cycloalkanes (3–7 percent), cycloalkenes (1–4 percent), and aromatics (20–50 percent). However, these proportions vary greatly.

The majority of the members of the paraffin, olefin, and aromatic series (of which there are about 500) boiling below 200°C (392°F) have been found in the gasoline fraction of petroleum. However, it appears that the distribution of the individual members of straight-run gasoline (i.e., distilled from petroleum without thermal alteration) is not even.

Highly branched paraffins, which are particularly valuable constituents of gasoline(s), are not usually the principal paraffinic constituents of straight-run gasoline. The more predominant paraffinic constituents are usually the normal (straight-chain) isomers, which may dominate the branched isomer(s) by a factor of two or more. This is presumed to indicate the tendency to produce long uninterrupted carbon chains during petroleum maturation rather than those in which branching occurs. However, this trend is somewhat different for the cyclic constituents of gasoline, that is, cycloparaffins (naphthenes) and aromatics. In these cases, the preference appears to be for several short side chains rather than one long substituent.

Gasoline can vary widely in composition: even those with the same octane number may be quite different, not only in the physical makeup but also in the molecular structure of the constituents. For example, the Pennsylvania petroleum is high in paraffins (normal and branched), but the California and Gulf Coast crude oils are high in cycloparaffins. Low-boiling distillates with high content of aromatic constituents (above 20 percent) can be obtained from some Gulf Coast and West Texas crude oils, as well as from crude oils from the Far East. The variation in aromatics content as well as the variation in the content of normal paraffins, branched paraffins, cyclopentanes, and cyclohexanes involve characteristics of any one individual crude oil and may in some instances be used for crude oil identification. Furthermore, straight-run gasoline generally shows a decrease in paraffin content with an increase in molecular weight, but the cycloparaffins (naphthenes) and aromatics increase with increasing molecular weight. Indeed, the hydrocarbon type variation may also vary markedly from process to process.

The reduction of the lead content of gasoline and the introduction of reformulated gasoline has been very successful in reducing automobile emissions Further improvements in fuel quality have been proposed for the years 2000 and beyond. These projections are accompanied by a noticeable and measurable decrease in crude oil quality and the reformulated gasoline will help meet environment regulations for emissions for liquid fuels.

Gasoline was at first produced by distillation, simply separating the volatile, more valuable fractions of crude petroleum. Later processes, designed to raise the yield of gasoline from crude oil, decomposed higher molecular weight constituents into lower molecular weight products by processes known as *cracking*. And like typical gasoline, several processes produce the blending stocks for gasoline (Fig. 3.18).

Thermal cracking and catalytic cracking, once used to supplement the gasoline supplies produced by distillation, are now the major processes used to produce gasoline. In addition, other methods used to improve the quality of gasoline and increase its supply include *polymerization, alkylation, isomerization*, and *reforming*.

Polymerization is the conversion gaseous olefins, such as propylene and butylene into larger molecules in the gasoline range. *Alkylation* is a process combining an olefin and paraffin such as isobutane. *Isomerization* is the conversion of straight-chain hydrocarbons to branched-chain hydrocarbons. *Reforming* is the use of either heat or a catalyst to rearrange the molecular structure.

Despite the variations in the composition of the gasoline produced by the various available processes, this material is rarely if ever suitable for use as such. It is at this stage of a refinery operation that blending becomes important (Speight, 2007).

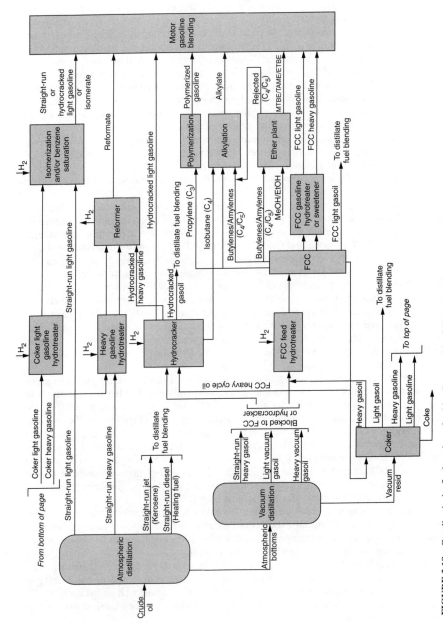

FIGURE 3.18 Gasoline is the final product after blending several refinery streams.

Despite the diversity of the processes within a modern petroleum refinery, no single stream meets all the requirements of gasoline. Thus, the final step in gasoline manufacture is *blending* the various streams into a finished product (Fig. 3.18). It is not uncommon for the finished gasoline to be made up of six or more streams and several factors make this flexibility critical: (a) the requirements of the gasoline specification (ASTM D 4814) and the regulatory requirements and (b) performance specifications that are subject to local climatic conditions and regulations.

Aviation gasoline is form of motor gasoline that has been especially prepared for use for aviation piston engines and is composed of paraffins and isoparaffins (50–60 percent), moderate amounts of naphthenes (20–30 percent), small amounts of aromatics (10 percent), and usually no olefins, whereas motor gasoline may contain up to 30 percent olefins and up to 40 percent aromatics. It has an octane number suited to the engine, a freezing point of −60°C (−76°F), and a distillation range usually within the limits of 30 to 180°C (86–356°F) compared to −1 to 200°C (30 to 390°F) for automobile gasoline.

The narrower boiling range of aviation gasoline ensures better distribution of the vaporized fuel through the more complicated induction systems of aircraft engines. Aircraft operate at altitudes at which the prevailing pressure is less than the pressure at the surface of the earth (pressure at 17,500 ft is 7.5 psi compared to 14.7 psi at the surface of the earth). Thus, the vapor pressure of aviation gasoline must be limited to reduce boiling in the tanks, fuel lines, and carburetors. Thus, the aviation gasoline does not usually contain the gaseous hydrocarbons (butanes) that give automobile gasoline the higher vapor pressures.

Under conditions of use in aircraft, olefins have a tendency to form gum, cause preignition, and have relatively poor antiknock characteristics under lean mixture (cruising) conditions; for these reasons olefins are detrimental to aviation gasoline. Aromatics have excellent antiknock characteristics under rich mixture (takeoff) conditions, but are much like the olefins under lean mixture conditions; hence the proportion of aromatics in aviation gasoline is limited. Some naphthenes with suitable boiling temperatures are excellent aviation gasoline components but are not segregated as such in refinery operations. They are usually natural components of the straight-run naphtha (aviation base stocks) used in blending aviation gasoline. The lower boiling paraffins (pentane and hexane), and both the high-boiling and low-boiling isoparaffins (isopentane to isooctane) are excellent aviation gasoline components. These hydrocarbons have high heat contents per pound and are chemically stable, and the isoparaffins have high octane numbers under both lean and rich mixture conditions.

Gasoline performance and hence quality of an automobile gasoline is determined by its resistance to knock, for example, *detonation* or *ping* during service. The antiknock quality of the fuel limits the power and economy that an engine using that fuel can produce: the higher the antiknock quality of the fuel, the more the power and efficiency of the engine. Thus, the performance ability of gasoline is measured by the *octane number*.

Octane numbers are obtained by the two test procedures, those obtained by the first method are called *motor octane numbers* (indicative of high-speed performance) (ASTM D-2700 and ASTM D-2723). Those obtained by the second method are called *research octane numbers* (indicative of normal road performance) (ASTM D-2699 and ASTM D-2722). Octane numbers quoted are usually, unless stated otherwise, research octane numbers.

In the test methods used to determine the antiknock properties of gasoline, comparisons are made with blends of two pure hydrocarbons, *n*-heptane and isooctane (2,2,4-trimethylpentane). Isooctane has an octane number of 100 and is high in its resistance to knocking; *n*-heptane is quite low (with an octane number of 0) in its resistance to knocking.

Extensive studies of the octane numbers of individual hydrocarbons have brought to light some general rules. For example, normal paraffins have the least desirable knocking characteristics, and these become progressively worse as the molecular weight increases.

Isoparaffins have higher octane numbers than the corresponding normal isomers, and the octane number increases as the degree of branching of the chain is increased. Olefins have markedly higher octane numbers than the related paraffins; naphthenes are usually better than the corresponding normal paraffins but rarely have very high octane numbers; and aromatics usually have quite high octane numbers.

Blends of n-heptane and isooctane thus serve as a reference system for gasoline and provide a wide range of quality used as an antiknock scale. The exact blend, which matches identically the antiknock resistance of the fuel under test, is found, and the percentage of isooctane in that blend is termed the *octane number* of the gasoline. For example, gasoline with a knocking ability which matches that of a blend of 90 percent isooctane and 10 percent n-heptane has an octane number of 90.

With an accurate and reliable means of measuring octane numbers, it was possible to determine the cracking conditions—temperature, cracking time, and pressure—that caused increases in the antiknock characteristics of cracked gasoline. In general it was found that higher cracking temperatures and lower pressures produced higher octane gasoline, but unfortunately more gas, cracked residua, and coke were formed at the expense of the volume of cracked gasoline.

To produce higher-octane gasoline, cracking coil temperatures were pushed up to 510°C (950°F), and pressures dropped from 1000 to 350 psi. This was the limit of thermal cracking units, for at temperatures over 510°C (950°F) coke formed so rapidly in the cracking coil that the unit became inoperative after only a short time on-stream. Hence it was at this stage that the nature of the gasoline-producing process was reexamined, leading to the development of other processes, such as reforming, polymerization, and alkylation for the production of gasoline components having suitably high octane numbers.

During the manufacture and distribution of gasoline, it comes into contact with water and particulate matter and can become *contaminated* with such materials. Water is allowed to settle from the fuel in storage tanks and the water is regularly withdrawn and disposed of properly. Particulate matter is removed by filters installed in the distribution system. (ASTM D 4814, App. X6).

Oxygenates are carbon-, hydrogen-, and oxygen-containing combustible liquids that are added to gasoline to improve performance. The addition of oxygenates gasoline is not new since ethanol (ethyl alcohol or grain alcohol) has been added to gasoline for decades. Thus, *oxygenated gasoline* is a mixture of conventional hydrocarbon-based gasoline and one or more oxygenates. The current oxygenates belong to one of two classes of organic molecules: alcohols and ethers. The most widely used oxygenates in the United States are ethanol, methyl tertiary-butyl ether (MTBE), and tertiary-amyl methyl ether (TAME). Ethyl tertiary-butyl ether (ETBE) is another ether that could be used. Oxygenates may be used in areas of the United States where they are not required as long as concentration limits (as refined by environment regulations) are observed.

Of all the oxygenates, methyl tertiary-butyl ether is attractive for a variety of technical reasons. It has a low vapor pressure, can be blended with other fuels without phase separation, and has the desirable octane characteristics. If oxygenates achieve recognition as vehicle fuels, the biggest contributor will probably be methanol, the production of which is mostly from synthesis gas derived from methane.

The higher alcohols also offer some potential as motor fuels. These alcohols can be produced at temperatures below 300°C (572°F) using copper oxide-zinc oxide-alumina catalysts promoted with potassium. Isobutyl alcohol is of particular interest because of its high octane rating, which makes it desirable as a gasoline-blending agent. This alcohol can be reacted with methanol in the presence of a catalyst to produce methyl tertiary-butyl ether. Although it is currently cheaper to make isobutyl alcohol from isobutylene, it can be synthesized from syngas with alkali-promoted zinc oxide catalysts at temperatures above 400°C (752°F).

Kerosene and Related Fuels. Kerosene (*kerosine*), also called paraffin or paraffin oil, is a flammable pale-yellow or colorless oily liquid with a characteristic odor. It is obtained from petroleum and used for burning in lamps and domestic heaters or furnaces, as a fuel or fuel component for jet engines, and as a solvent for greases and insecticides.

Kerosene is intermediate in volatility between gasoline and gas oil. It is a medium oil distilling between 150 and 300°C (302 and 572°F). Kerosene has a flash point about 25°C (77°F) and is suitable for use as an illuminant when burned in a wide lamp. The term *kerosene* is also too often incorrectly applied to various fuel oils, but a fuel oil is actually any liquid or liquid petroleum product that produces heat when burned in a suitable container or that produces power when burned in an engine.

Jet fuel is a light petroleum distillate that is available in several forms suitable for use in various types of jet engines. The major jet fuels used by the military are JP-4, JP-5, JP-6, JP-7, and JP-8. Briefly, JP-4 is a wide-cut fuel developed for broad availability. JP-6 is a higher cut than JP-4 and is characterized by fewer impurities. JP-5 is specially blended kerosene, and JP-7 is high flash point special kerosene used in advanced supersonic aircraft. JP-8 is kerosene modeled on Jet A-l fuel (used in civilian aircraft). From what data are available, typical hydrocarbon chain lengths characterizing JP-4 range from C_4 to C_{16}. Aviation fuels consist primarily of straight and branched alkanes and cycloalkanes. Aromatic hydrocarbons are limited to 20 to 25 percent of the total mixture because they produce smoke when burned. A maximum of 5 percent alkenes is specified for JP-4. The approximate distribution by chemical class is: straight chain alkanes (32 percent), branched alkanes (31 percent), cycloalkanes (16 percent), and aromatic hydrocarbons (21 percent).

Gasoline-type jet fuel includes all light hydrocarbon oils for use in aviation turbine power units that distill between 100 and 250°C (212 and 482°F). It is obtained by blending kerosene and gasoline or naphtha in such a way that the aromatic content does not exceed 25 percent in volume. Additives can be included to improve fuel stability and combustibility. *Kerosene-type jet fuel* is a medium distillate product that is used for aviation turbine power units. It has the same distillation characteristics and flash point as kerosene [between 150 and 300°C (302 and 572°F)], but not generally above 250°C (482°F). In addition, it has particular specifications (such as freezing point) which are established by the International Air Transport Association (IATA).

Chemically, kerosene is a mixture of hydrocarbons; the chemical composition depends on its source, but it usually consists of about 10 different hydrocarbons, each containing from 10 to 16 carbon atoms per molecule; the constituents include *n*-dodecane (n-$C_{12}H_{26}$), alkyl benzenes, and naphthalene and its derivatives. Kerosene is less volatile than gasoline; it boils between about 140°C (284°F) and 320°C (608°F).

Although the kerosene constituents are predominantly saturated materials, there is evidence for the presence of substituted tetrahydronaphthalene. Dicycloparaffins also occur in substantial amounts in kerosene. Other hydrocarbons with both aromatic and cycloparaffin rings in the same molecule, such as substituted indan, also occur in kerosene. The predominant structure of the dinuclear aromatics appears to be that in which the aromatic rings are condensed, such as naphthalene whereas the *isolated* two-ring compounds, such as biphenyl, are only present in traces, if at all.

Kerosene is now largely produced by cracking the less volatile portion of crude oil at atmospheric pressure and elevated temperatures.

In the early days, the poorer quality kerosene was treated with large quantities of sulfuric acid to convert them to marketable products. However, this treatment resulted in high acid and kerosene losses, but the later development of the *Edeleanu* process overcame these problems.

Kerosene is a very stable product, and additives are not required to improve the quality. Apart from the removal of excessive quantities of aromatics by the Edeleanu process, kerosene fractions may need only a lye wash or a doctor treatment if hydrogen sulfide is present to remove mercaptans.

Kerosene has been used as a fuel oil since the beginning of the petroleum-refining industry. As such, low proportions of aromatic and unsaturated hydrocarbons are desirable to maintain the lowest possible level of smoke during burning. Although some aromatics may occur within the boiling range assigned to kerosene, excessive amounts can be removed by extraction; that kerosene is not usually prepared from cracked products almost certainly excludes the presence of unsaturated hydrocarbons.

The essential properties of kerosene are flash point, fire point, distillation range, burning, sulfur content, color, and cloud point. In the case of the flash point (ASTM D-56), the minimum flash temperature is generally placed above the prevailing ambient temperature; the fire point (ASTM D-92) determines the fire hazard associated with its handling and use.

The boiling range (ASTM D-86) is of less importance for kerosene than for gasoline, but it can be taken as an indication of the viscosity of the product, for which there is no requirement for kerosene. The ability of kerosene to burn steadily and cleanly over an extended period (ASTM D-187) is an important property and gives some indication of the purity or composition of the product.

The significance of the total sulfur content of a fuel oil varies greatly with the type of oil and the use to which it is put. Sulfur content is of great importance when the oil to be burned produces sulfur oxides that contaminate the surroundings. The color of kerosene is of little significance, but a product darker than usual may have resulted from contamination or aging, and in fact a color darker than specified (ASTM D-156) may be considered by some users as unsatisfactory. Finally, the cloud point of kerosene (ASTM D-2500) gives an indication of the temperature at which the wick may become coated with wax particles, thus lowering the burning qualities of the oil.

Fuel Oil. Fuel oil is classified in several ways but generally may be divided into two main types: *distillate fuel oil* and *residual fuel oil*. Distillate fuel oil is vaporized and condensed during a distillation process and thus have a definite boiling range and do not contain high-boiling constituents. A fuel oil that contains any amount of the residue from crude distillation of thermal cracking is a residual fuel oil. The terms *distillate fuel oil* and *residual fuel oil* are losing their significance, since fuel oil is now made for specific uses and may be either distillates or residuals or mixtures of the two. The terms *domestic fuel oil, diesel fuel oil*, and *heavy fuel oil* are more indicative of the uses of fuel oils.

Diesel fuel oil is also a distillate fuel oil that distils between 180 and 380°C (356 and 716°F). Several grades are available depending on uses: diesel oil for diesel compression ignition (cars, trucks, and marine engines) and light heating oil for industrial and commercial uses.

Heavy fuel oil comprises all residual fuel oils (including those obtained by blending). Heavy fuel oil constituents range from distillable constituents to residual (nondistillable) constituents that must be heated to 260°C (500°F) or more before they can be used. The kinematic viscosity is above 10 cSt at 80°C (176°F). The flash point is always above 50°C (122°F) and the density is always higher than 0.900. In general, heavy fuel oil usually contains cracked residua, reduced crude, or cracking coil heavy product which is mixed (cut back) to a specified viscosity with cracked gas oils and fractionator bottoms. For some industrial purposes in which flames or flue gases contact the product (ceramics, glass, heat treating, and open hearth furnaces) fuel oils must be blended to contain minimum sulfur contents, and hence low-sulfur residues are preferable for these fuels.

No. 1 fuel oil is a petroleum distillate that is one of the most widely used of the fuel oil types. It is used in atomizing burners that spray fuel into a combustion chamber where the tiny droplets burn while in suspension. It is also used as a carrier for pesticides, as a weed killer, as a mold release agent in the ceramic and pottery industry, and in the cleaning industry.

It is found in asphalt coatings, enamels, paints, thinners, and varnishes. No. 1 fuel oil is a light petroleum distillate (straight-run kerosene) consisting primarily of hydrocarbons in the range C_9 to C_{16}. No. 1 fuel oil is very similar in composition to diesel fuel; the primary difference is in the additives.

No. 2 fuel oil is a petroleum distillate that may be referred to as domestic or industrial. The domestic fuel oil is usually lower boiling and a straight-run product. It is used primarily for home heating. Industrial distillate is a cracked product or a blend of both. It is used in smelting furnaces, ceramic kilns, and packaged boilers. No. 2 fuel oil is characterized by hydrocarbon chain lengths in the C_{11} to C_{20} range. The composition consists of aliphatic hydrocarbons (straight-chain alkanes and cycloalkanes) (64 percent), unsaturated hydrocarbons (alkenes) (1–2 percent), and aromatic hydrocarbons (including alkyl benzenes and 2-ring, 3-ring aromatics) (35 percent), but contains only low amounts of the polycyclic aromatic hydrocarbons (<5 percent).

No. 6 fuel oil (also called *Bunker C oil* or *residual fuel oil*) is the residuum from crude oil after naphtha-gasoline, No. 1 fuel oil, and No. 2 fuel oil have been removed. No. 6 fuel oil can be blended directly to heavy fuel oil or made into asphalt. Residual fuel oil is more complex in composition and impurities than distillate fuels. Limited data are available on the composition of No. 6 fuel oil. Polycyclic aromatic hydrocarbons (including the alkylated derivatives) and metal-containing constituents are components of No. 6 fuel oil.

Stove oil, like kerosene, is always a straight-run fraction from suitable crude oils, whereas other fuel oils are usually blends of two or more fractions, one of which is usually cracked gas oil. The straight-run fractions available for blending into fuel oils are heavy naphtha, light and heavy gas oils, reduced crude, and pitch. Cracked fractions such as light and heavy gas oils from catalytic cracking, cracking coil tar, and fractionator bottoms from catalytic cracking may also be used as blends to meet the specifications of the different fuel oils.

Since the boiling ranges, sulfur contents, and other properties of even the same fraction vary from crude oil to crude oil and with the way the crude oil is processed, it is difficult to specify which fractions are blended to produce specific fuel oils. In general, however, furnace fuel oil is a blend of straight-run gas oil and cracked gas oil to produce a product boiling in the 175 to 345°C (347–653°F) range.

Diesel fuel oil is essentially the same as furnace fuel oil, but the proportion of cracked gas oil is usually less since the high aromatic content of the cracked gas oil reduces the cetane value of the diesel fuel. Under the broad definition of diesel fuel, many possible combinations of characteristics (such as volatility, ignition quality, viscosity, gravity, stability, and other properties) exist. To characterize diesel fuels and thereby establish a framework of definition and reference, various classifications are used in different countries. An example is ASTM D975 in the United States in which grades No. 1-D and 2-D are distillate fuels, the types most commonly used in high speed engines of the mobile type, in medium speed stationary engines, and in railroad engines. No. 4-D covers the class of more viscous distillates and, at times, blends of these distillates with residual fuel oils. No. 4-D fuels are applicable for use in low- and medium-speed engines employed in services involving sustained load and predominantly constant speed.

Cetane number is a measure of the tendency of a diesel fuel to knock in a diesel engine. The scale is based upon the ignition characteristics of two hydrocarbons *n*-hexadecane (cetane) and 2,3,4,5,6,7,8-heptamethylnonane. Cetane has a short delay period during ignition and is assigned a cetane number of 100; heptamethylnonane has a long delay period and has been assigned a cetane number of 15. Just as the octane number is meaningful for automobile fuels, the cetane number is a means of determining the ignition quality of diesel fuels and is equivalent to the percentage by volume of cetane in the blend with heptamethylnonane, which matches the ignition quality of the test fuel (ASTM D-613).

The manufacture of fuel oils at one time largely involved using what was left after removing desired products from crude petroleum. Now fuel oil manufacture is a complex matter of selecting and blending various petroleum fractions to meet definite specifications, and the production of a homogeneous, stable fuel oil requires experience backed by laboratory control.

3.5.3 Solid Fuels

Coke is the residue left by the destructive distillation of petroleum residua. That formed in catalytic cracking operations is usually nonrecoverable, as it is often employed as fuel for refinery processes.

The composition of petroleum coke varies with the source of the crude oil, but in general, large amounts of high-molecular-weight complex hydrocarbons (rich in carbon but correspondingly poor in hydrogen) make up a high proportion. The solubility of petroleum *coke* in carbon disulfide has been reported to be as high as 50 to 80 percent, but this is in fact a misnomer, since the coke is the insoluble, honeycomb material that is the end product of thermal processes.

Petroleum coke is employed for a number of purposes, but its chief use is in the manufacture of carbon electrodes for aluminum refining, which requires a high-purity carbon—low in ash and sulfur free; the volatile matter must be removed by calcining. In addition to its use as a metallurgical reducing agent, petroleum coke is employed in the manufacture of carbon brushes, silicon carbide abrasives, and structural carbon (e.g., pipes and Rashig rings), as well as calcium carbide manufacture from which acetylene is produced:

$$\text{Coke} \rightarrow \text{CaC}_2$$

$$\text{CaC}_2 + \text{H}_2\text{O} \rightarrow \text{HC} \equiv \text{CH}$$

Marketable coke is coke that is relatively pure carbon and can be sold for use as fuel or for the manufacture of dry cells, electrodes, and the likes. *Needle coke (acicular coke)* is a highly crystalline petroleum coke used in the production of electrodes for the steel and aluminum industries. *Catalyst coke* is coke that has deposited on the catalysts used in oil refining, such as those in a catalytic cracker. This coke is impure and is only used for fuel.

Coke may be used to make fuel gases such as *water gas* and *producer gas*. From which, in turn, synthesis gas can be manufactured leading to a variety of other liquid fuel products.

Water gas is a mixture of carbon monoxide and hydrogen, made by passing steam over red-hot coke. Producer gas is a mixture of carbon monoxide, hydrogen, and nitrogen and is manufactured by passing air over red-hot coke (or any carbon-based char)

3.6 REFERENCES

ASTM: *Annual Book of Standards,* American Society for Testing and Materials, West Conshohocken, Pa., 2007.

Bland, W. F. and R. L. Davidson: *Petroleum Processing Handbook,* McGraw-Hill, New York, 1967.

Dickenson, R. L., F. E. Biasca, B. L. Schulman, and H. E. Johnson: *Hydrocarbon Processing,* 76(2), 1997, p. 57.

Dunning, H. N. and J. W. Moore: "Propane Removes Asphalts from Crudes," *Petroleum Refiner,* 36(5), 1957, pp. 247–250.

Gary, J. H., and G. E. Handwerk: *Petroleum Refining: Technology and Economics,* 4th ed., Marcel Dekker Inc., New York, 2001.

Gruse, W. A. and D. R. Stevens: *Chemical Technology of Petroleum,* McGraw-Hill, New York, 1960.

Hobson, G. D. and W. Pohl: *Modern Petroleum Technology,* Applied Science Publishers, Barking, England, 1973.

Khan, M. R., and D. J. Patmore: In *Petroleum Chemistry and Refining,* J.G. Speight (ed.). Taylor & Francis, Washington, D.C., 1997, chap. 6.

Kobe, K. A. and J. J. McKetta: *Advances in Petroleum Chemistry and Refining,* Interscience, New York, 1958.

McKetta, J. J. (ed.): *Petroleum Processing Handbook,* Marcel Dekker Inc., New York, 1992.

Nelson, W. L.: *Petroleum Refinery Engineering,* McGraw-Hill, New York, 1958.

RAROP: *RAROP Heavy Oil Processing Handbook,* Research Association for Residual Oil Processing, T. Noguchi (Chairman), Ministry of Trade and International Industry (MITI), Tokyo, Japan, 1991.

Scholten G. G.: In *Petroleum Processing Handbook,* J. J. McKetta (ed.), Marcel Dekker Inc., New York, 1992, p. 565.

Shih, S. S. and M. C. Oballa (ed.): "Tar Sand Upgrading Technology," Symposium Series No. 282, American Institute for Chemical Engineers, New York, 1991.

Speight, J. G.: In *Catalysis on the Energy Scene,* S. Kaliaguine and A. Mahay (ed.), Elsevier, Amsterdam, 1984.

Speight, J. G.: *The Desulfurization of Heavy Oils and Residua,* 2d ed., Marcel Dekker Inc., New York, 2000.

Speight, J. G. and B. Ozum: *Petroleum Refining Processes,* Marcel Dekker Inc., New York, 2002.

Speight, J. G.: *The Chemistry and Technology of Petroleum,* 4th ed., CRC Press, Taylor & Francis Group, Boca Raton, Fla., 2007.

CHAPTER 4

FUELS FROM TAR SAND BITUMEN

Tar sand bitumen is another source of liquid fuels that is distinctly separate from conventional petroleum (US Congress, 1976; Speight, 2005, 2007).

Tar sand (also called *oil sand* in Canada) or the more geologically correct term *bituminous sand* is commonly used to describe a sandstone reservoir that is impregnated with a heavy, viscous bituminous material. Tar sand is actually a mixture of sand, water, and bitumen but many of the tar sand deposits in countries other than Canada lack the water layer that is believed to facilitate the hot water recovery process. The heavy bituminous material has a high viscosity under reservoir conditions and cannot be retrieved through a well by conventional production techniques.

Geologically, the term *tar sand* is commonly used to describe a sandstone reservoir that is impregnated with bitumen, a naturally occurring material that is solid or near solid and is substantially immobile under reservoir conditions. The bitumen cannot be retrieved through a well by conventional production techniques, including currently used enhanced recovery techniques. In fact, tar sand is defined (FE-76-4) in the United States as:

> The several rock types that contain an extremely viscous hydrocarbon which is not recoverable in its natural state by conventional oil well production methods including currently used enhanced recovery techniques. The hydrocarbon-bearing rocks are variously known as bitumen-rocks oil, impregnated rocks, tar sands, and rock asphalt.

In addition to this definition, there are several tests that must be carried out to determine whether or not, in the first instance, a resource is a tar sand deposit (Speight, 2007 and references cited therein). Most of all, a core taken from a tar sand deposit, the bitumen isolated therefrom, are certainly not identifiable by the preliminary inspections (sight and touch) alone.

In the United States, the final determinant is whether or not the material contained therein can be recovered by primary, secondary, or tertiary (enhanced) recovery methods (US Congress, 1976).

The relevant position of tar sand bitumen in nature is best illustrated by comparing its position relevant to petroleum and heavy oil. Thus, petroleum is referred to generically as a fossil energy resource and is further classified as a hydrocarbon resource and, for illustrative (or comparative) purposes in this report, coal and oil shale kerogen have also been included in this classification. However, the inclusion of coal and oil shale under the broad classification of hydrocarbon resources has required (incorrectly) that the term hydrocarbon be expanded to include these resources. It is essential to recognize that resources such as coal, oil shale kerogen, and tar sand bitumen contain large proportions of heteroatomic species. Heteroatomic species are those organic constituents that contain atoms other than carbon and hydrogen, for example, nitrogen, oxygen, sulfur, and metals (nickel and vanadium).

Use of the term *organic sediments* is more correct and is preferred. The inclusion of coal and oil shale kerogen in the category of hydrocarbon resources is due to the fact that these two natural resources will produce hydrocarbons by thermal decomposition (high-temperature processing). Therefore, if either coal and/or oil shale kerogen is to be included in the term hydrocarbon resources, it is more appropriate that they be classed as hydrocarbon-producing resources under the general classification of organic sediments. Thus, tar sand bitumen stands apart from petroleum and heavy oil not only from the method of recovery but also from the means by which hydrocarbons are produced.

It is incorrect to refer to tar sand bitumen as *tar* or *pitch*. In many parts of the world the term *bitumen* is used as the name for *road asphalt*. Although the word tar is somewhat descriptive of the black bituminous material, it is best to avoid its use with respect to natural materials. More correctly, the name tar is usually applied to the heavy product remaining after the destructive distillation of coal or other organic matter. *Pitch* is the distillation residue of the various types of tar.

Physical methods of fractionation of tar sand bitumen can also produce the four generic fractions: saturates, aromatics, resins, and asphaltenes. However, for tar sand bitumen, the fractionation produces shows that bitumen contains high proportions of *asphaltenes* and *resins*, even in amounts up to 50 percent w/w (or higher) of the bitumen with much lower proportions of saturates and aromatics than petroleum or heavy oil. In addition, the presence of ash-forming metallic constituents, including such organometallic compounds as those of vanadium and nickel, is also a distinguishing feature of bitumen.

4.1 OCCURRENCE AND RESERVES

Tar sand deposits are widely distributed throughout the world in a variety of countries.

The various tar sand deposits have been described as belonging to two types: (a) materials that are found in *stratigraphic* traps, and (b) deposits that are located in *structural* traps. There are, inevitably, gradations, and combinations of these two types of deposits and a broad pattern of deposit entrapment are believed to exist.

The distinction between a structural trap (the usually description of a petroleum reservoir) and a stratigraphic trap is often not clear. For example, an *anticlinal trap* may be related to an underlying buried limestone reef. Beds of sandstone may wedge out against an anticline because of depositional variations or intermittent erosion intervals. Salt domes, formed by flow of salt at substantial depths, also have created numerous traps that are both a structural trap and a stratigraphic trap.

In general terms, the entrapment characteristics for the very large tar sand deposits all involve a combination of stratigraphic and structural traps. Entrapment characteristics for the very large tar sands all involve a combination of stratigraphic-structural traps. There are no very large tar sand accumulations having more than 4 billion barrels in place either in purely structural or in purely stratigraphic traps. In a regional sense, structure is an important aspect since all of the very large deposits occur on gently sloping homoclines.

Furthermore, the tar sand deposits of the world have been described as belonging to two types. These are (a) in situ deposits resulting from breaching and exposure of an existing geologic trap and (b) migrated deposits resulting from accumulation of migrating material at outcrop. However, there are inevitable gradations and combinations of these two types of deposits. The deposits have been laid down over a variety of geologic periods and in different entrapments and a broad pattern of deposit entrapment is believed to exist since all deposits occur along the rim of major sedimentary basins and near the edge of Precambrian shields. The deposits either transgress an ancient relief at the edge of the shield (e.g., those in Canada) or lie directly on the ancient basement (e.g., as in Venezuela, West Africa, and Madagascar).

A feature of major significance in at least five of the major areas is the presence of a regional cap (usually a widespread transgressive marine shale). Formations of this type occur in the Colorado Group in western Canada, in the Freites formation in eastern Venezuela, or in the Jurassic formation in Melville Island. The cap plays an essential role in restraining vertical fluid escape from the basin thereby forcing any fluids laterally into the paleo-delta itself. Thus, the subsurface fluids were channeled into narrow outlets at the edge of the basin.

The potential reserves of hydrocarbon liquids (available through conversion of the bitumen to *synthetic crude oil*) that occur in tar sand deposits have been variously estimated on a world basis to be in excess of 3 trillion barrels. However, the issue is whether or not these reserves can be recovered and conversed to synthetic crude oil. Geographic and geologic feature may well put many of these resources beyond the capabilities of current recovery technology requiring new approaches to recovery and conversion.

4.1.1 Canada

In Canada, the Athabasca deposit along with the neighbouring Wabasca, and Peace River deposits have been estimated to contain approximately 2 trillion barrels of bitumen.

The town of McMurray, about 240 miles north east of Edmonton, Alberta lies at the eastern margin of the largest accumulation in the world that is, in effect, three major accumulations within the Lower Cretaceous deposits. The McMurray-Wabasca reservoirs are found toward the base of the formation and are characteristically cross-bedded coarse grit and gritty sandstone that are unconsolidated or cemented by tar; fine-to-medium grained sandstone and silt occur higher in the sequence. Bluesky-Gething and Grand Rapids reservoirs are composed of sub-angular quartz and well-rounded chert grains. The McMurray-Wabasca tar sand deposit dips at between 5 and 25 ft/mile (1.5 and 8 m/mile) to the southwest. The Bluesky-Gething sands overlie several unconformities between the Mississippian and Jurassic deposits.

In the context of the Athabasca deposit, inconsistencies arise presumably because of the lack of mobility of the bitumen at formation temperature [approximately 4°C (39°F)]. For example, the proportion of bitumen in the tar sand increases with depth within the formation. Furthermore, the proportion of the nonvolatile asphaltenes or the nonvolatile asphaltic fraction (asphaltenes plus resins) in the bitumen also increases with depth within the formation that leads to reduced yields of distillate from the bitumen obtained from deeper parts of the formation. In keeping with the concept of higher proportions of asphaltic fraction (asphaltenes plus resins), variations (horizontal and vertical) in bitumen properties have been noted previously, as have variations in sulfur, nitrogen, and metals content.

Obviously, the richer tar sand deposits occur toward the base of the formation. However, the bitumen is generally of poorer quality than the bitumen obtained from near the top of the deposit insofar as the proportions of nonvolatile coke-forming constituents (asphaltenes plus resins) are higher (with increased proportions of nitrogen, sulfur, and metals) near the base of the formation.

4.1.2 United States

The major tar sand deposits of the United States occur within and around the periphery of the Uinta Basin, Utah. These include the Sunnyside, Tar Sand Triangle, Peor (PR) Springs, Asphalt Ridge, and sundry other deposits (Table 4.1). Asphalt Ridge lies on the northeastern

TABLE 4.1 Estimation of Utah Tar Sand Reserves

Deposit	Known reserves (million bbl)	Additional probable reserves (billion bbl)
Sunnyside	4,400	1,700
Tar Sand Triangle	2,500	420
PR Spring	2,140	2,230
Asphalt Ridge	820	310
Circle Cliffs	590	1,140
Other	1,410	1,530
Total	11,8060	7,330

margin of the central part of the Uinta Basin at the contact of the tertiary beds with the underlying Cretaceous Mesaverde Group. The Mesaverde Group is divided into three formations, two of which, the Asphalt Ridge sandstone and the Rim Rock sandstone are beach deposits containing the viscous bitumen. The Rim Rock sandstone is thick and uniform with good reservoir characteristics and may even be suitable for thermal-recovery methods. The Duchesne River formation (Lower Oligocene) also contains bituminous material but the sands tend to be discontinuous.

The Sunnyside deposits extend over a greater area than Asphalt Ridge and are located on the southwest flank of the Uinta Basin. The tar sand accumulations occur in sandstone of the Wasatch and lower Green River formations (Eocene). The Wasatch sandstone contains bitumen impregnation but is lenticular and occupies broad channels cut into the underlying shale and limestone; the Green River beds are more uniform and laterally continuous. The source of the bitumen in the Asphalt Ridge and Sunnyside accumulation is considered to be the Eocene Green River shale.

The Peor (PR) Springs accumulation is about 60 miles (96.5 km) east of the Sunnyside deposit and occurs as lenticular sandstone (Eocene Wasatch formation). There are two main beds from 30 to 85 ft (9–26 m) thick with an estimated overburden thickness of 0 to 250 ft (0–76 m). The tilt of the southern flank of the Uinta Basin has left this deposit relatively undisturbed except for erosion, which has stripped it of its cover allowing the more volatile constituents to escape. In the central southeast area of Utah, some deposits of bitumen-impregnated sandstone occur in Jurassic rock, but the great volume of in-place bitumen occurs in rocks of Triassic and Permian age. The Tar Sand Triangle is considered to be a single, giant stratigraphic trap containing the bitumen.

The Californian deposits are concentrated in the coastal region west of the San Andreas Fault. The largest deposit is the Edna deposit, which is located midway between Los Angeles and San Francisco. It consists of conglomerate, sandstone, diatomaceous sandstone, and siliceous shale. The deposit occurs as a stratigraphic trap and outcrops in scattered areas on both flanks of a narrow syncline. The deposit extends over an area of about 7000 acres and occurs from outcrop to a 100-ft (30-m) depth. The accumulations are considered to have been derived from the underlying organic and petroliferous Monterey shale.

The Sisquoc deposit (Upper Pliocene) is the second largest in California and occurs in sandstone in which there are as many as eight individual tar sand units. The total thickness of the deposit is about 185 ft (56 m) occurring over an area of about 175 acres with an overburden thickness between 15 and 70 ft (4.6–21 m). The reservoir sands lie above the Monterey shale, which has been suggested to be the source of the bitumen.

The third California deposit at Santa Cruz is located approximately 56 miles (90 km) from San Francisco. The material occurs in sandstone of the Monterey and Vaqueros formations, which are older than both the Edna and Sisquoc reservoir rocks. The Santa Cruz tar sands are discontinuous and overlie the pre-Cretaceous basement.

South Texas holds the largest tar sand deposits. The tar sand deposits occur in the San Miguel tar belt (Upper Cretaceous) mostly in Maverick and Zavala counties as well as in the Anacadro limestone (Upper Cretaceous) of the Uvalde district. The Kentucky tar sand deposits are located at Asphalt, Davis-Dismal Creek, and Kyrock; they all occur in nonmarine Pennsylvanian or Mississippian sediments. The three deposits appear as stratigraphic traps and are thought to have received their bitumen or bitumen precursor from the Devonian Chattanooga shale. Tar sand deposits in New Mexico occur in the Triassic Santa Rosa sandstone, which is an irregularly bedded, fine- to medium-grained micaceous sandstone.

Finally, in the context of the tar sand deposits in the United States, the tar sand deposits in Missouri occur over an area estimated at 2000 square miles in Barton, Vernon, and Cass Counties and the sandstone bodies that contain the bitumen are middle Pennsylvanian in age. The individual bitumen-bearing sands are approximately 50 ft (15 m) in thickness except where they occur in channels which may actually be as much as 250 ft (76 m) thick. The two major reservoirs are the Warner sandstone and the Bluejacket sandstone that at one time were regarded as blanket sands covering large areas. However, recent investigations suggest that these sands can abruptly grade into barren shale or siltstone.

4.1.3 Venezuela

Tar sand deposits in Venezuela occur in the Officina/Tremblador tar belt that is believed to contain bitumen-impregnated sands of a similar extent to those of Alberta, Canada. The Officina formation overlaps the Tremblador (Cretaceous) formation and the organic material is typical bitumen having an API (American Petroleum Institute) gravity less than 10°. The Guanaco Asphalt Lake occurs in deposits that rest on a formation of mid-Pliocene age. This formation, the Las Piedras, is principally brackish sandstone to freshwater sandstone with associated lignite. The Las Piedras formation overlies a marine Upper Cretaceous group; the Guanaco Lake asphalt is closely associated with the Guanaco crude oil field that produces heavy crude oil from shale and fractured argillite of the Upper Cretaceous group.

4.1.4 Other Countries

The Bemolanga (Madagascar) deposit is the third largest tar sand deposit presently known and extends over some 150 square miles in western Madagascar with a recorded overburden from 0 to 100 ft (0–30 m). The average pay zone thickness is 100 ft (30 m) with a total bitumen in-place quoted at approximately 2 billion barrels. The deposit is of Triassic age and the sands are cross-bedded continental sediments; the coarser, porous sands are more richly impregnated. The origin of the deposit is not clear; the most preferred source is the underlying shale or in down-dip formations implying small migration.

The largest tar sand deposit in Europe is that at Selenizza Albania. This region also contains the Patos oil field throughout which there occurs extensive bitumen impregnation. This deposit occurs in middle-upper Miocene lenticular sands, characterised by a brackish water fauna. Succeeding Pliocene conglomerate beds, which are more generally marine, are also locally impregnated with heavy crude oil. The Selenizza and Patos fields occupy the crestal portions of a north-south trending anticline. Faulting also controls the vertical distribution of the accumulation. The Miocene rests on Eocene limestone and it is these that are thought by some to be the source of the tar.

The Trinidad Asphalt (Pitch) Lake situated on the Gulf of Paria, 12 miles west south west of San Fernando and 138 ft (42 m) above sea level, occupies a depression in the

Miocene sheet sandstone. It overlies an eroded anticline of Upper Cretaceous age with remnants of an early tertiary formation still preserved on the flanks.

The Trinidad bitumen is currently mined and sold as road asphalt. Estimates of the amount available vary and a very approximate estimate indicates that at current production rates (approximately 50,000 tons/year) there is believed to be sufficient to last 150 years. There are no current plans to use the Trinidad bitumen as a source of synthetic fuels.

The Rumanian deposits are located at Derna deposits and occur (along with Tataros and other deposits) in a triangular section east and northeast of Oradia between the Sebos Koros and Berrettyo rivers. The tar sand occurs in the upper part of the Pliocene formation and the asphalt is characterised by its penetrating odour. The reservoir rock is nonmarine, representing freshwater deposition during a period of regression.

Tar sands occur at Cheildag, Kobystan and outcrop in the south flank of the Cheildag anticline; there are approximately 24 million barrels of bitumen in place. Other deposits in the former U.S.S.R. occur in the Olenek anticline (northeast of Siberia) and it has been claimed that the extent of asphalt impregnation in the Permian sandstone is of the same order of magnitude (in area and volume) as that of the Athabasca deposits. Tar sands have also been reported from sands at Subovka and the Notanebi deposit (Miocene sandstone) is reputed to contain 20 percent bitumen by weight. On the other hand, the Kazakhstan occurrence, near the Shubar-Kuduk oil field, is a bituminous lake with a bitumen content that has been estimated to be of the order of 95 percent by weight of the deposit.

Tar sand occurrences also occur in the Southern Llanos of Colombia where drilling has presented indications of deposits generally described as heavy crude oil, natural asphalt, and bitumen. Most of these occurrences are recorded below 1500 ft (457 m). The tar sands at Burgan in Kuwait and at the Inciarte and Bolivar coastal fields of the Maracaibo Basin are of unknown dimensions. Those at Inciarte have been exploited and all are directly or closely associated with large oil fields. The tar sands of the Bolivar coastal fields are above the oil zones in Miocene beds and are in a lithologic environment similar to that of the Officina-Tremblador tar belt. The small Miocene asphalt deposits in the Leyte Islands (Philippines) are extreme samples of stratigraphic entrapment and resemble some of the Californian deposits. Those of the Mefang Basin in Thailand are in Pliocene beds that overlie Triassic deposits and their distribution is stratigraphically controlled. There is a small accumulation at Chumpi, near Lima (Peru), which occurs in tuffaceous sands and it is believed to be derived from strongly deformed Cretaceous limestone from which a petroleum-type was distilled as a result of volcanic activity. Finally, tar sand deposits have also been recorded in Spain, Portugal, Cuba, Argentina, Thailand, and Senegal but most are poorly defined and are considered to contain (in-place) less than 1 million barrel of bitumen.

4.2 STRUCTURE AND PROPERTIES OF TAR SAND

Tar sand is a mixture of sand and other rock materials that is composed of approximately 80 to 85 percent sand, clay, and other mineral matter, 5 to 10 percent by weight of water, and from 1 to 18 percent by weight of bitumen.

Bitumen is a thick, viscous carbonaceous material that, at room temperature, is in a near solid state and which is immobile in the deposit.

Prior to processing, the bitumen is separated from the sand, other mineral materials, and formation water before it is delivered to downstream upgraders or refineries. Shallow oil sands deposits, less than about 250 ft (76 m) to the top of the tar sands zone, are exploited using surface mining to recover ore-grade oil sands, which are then delivered to an extraction plant for separation of bitumen from the sand, other minerals, and connate water. Deep oil sands, greater than about 250 ft (76 m) to the top of the oil sands zone, are exploited using in situ recovery techniques, whereby the bitumen is separated from the sand in situ and produced to the surface through wells drilled into the oil sands reservoir.

In order to accomplish this, the tar sand properties are of extreme importance and are outline below.

4.2.1 Mineralogy

The mineralogy of tar sand deposits is also worthy of note as it does affect the potential for recovery of the bitumen. Usually, more than 99 percent by weight of the tar sand mineral is composed of quartz sand and clays. In the remaining 1 percent, more than 30 minerals have been identified, mostly calciferous or iron based. Particle size ranges from large grains (99.9 percent is finer than 1000 μm) down to 44 μm (325 mesh), the smallest size that can be determined by dry screening. The size between 44 and 2 μm is referred to as slit; sizes below 2 μm (equivalent spherical diameter) are clay.

The Canadian deposits are largely unconsolidated sands with a porosity ranging up to 45 percent and have good intrinsic permeability. However, the deposits in the United States, in Utah, range from predominantly low-porosity, low-permeability consolidated sand to, in a few instances, unconsolidated sands. In addition, the bitumen properties are not conducive to fluid flow under normal reservoir conditions in either the Canadian or United States deposits. Nevertheless, where the general nature of the deposits prohibits the application of a mining technique (as in many of the United States deposits), a nonmining technique may be the only feasible bitumen recovery option.

By definition, tar sand is a mixture of sand, water, and bitumen with the sand component occurring predominantly as quartz. The arrangement of the sand, water, and bitumen has been assumed to be an arrangement whereby each particle of the sand is water-wet and a film of bitumen surrounds the water-wetted grains. The balance of the void volume is filled with bitumen, connate water, or gas; fine material, such as clay, occurs within the water envelope.

One additional aspect of the character of Athabasca tar sands is that the sand grains are not uniform in character. Grain-to-grain contact is variable and such a phenomenon influences attempts to repack mined sand, as may be the case in studies involving bitumen removal from the sand in laboratory-type in situ studies. This phenomenon also plays a major role in the expansion of the sand during processing where the sand to be returned to the mine site might occupy 120 to 150 percent of the volume of the original as-mined material.

The tar sand mass can be considered a four-phase system composed of solid phase (siltstone and clay), liquid phase (from fresh to more saline water), gaseous phase (natural gases), and viscous phase (black and dense bitumen, about 8° API).

In normal sandstone, sand grains are in grain-to-grain contact but tar sand is thought to have no grain-to-grain contact due to the surrounding of individual grains by fines with a water envelope and/or a bitumen film. The remaining void space might be filled with water, bitumen, and gas in various proportions. The sand material in the formation is represented by quartz and clays (99 percent by weight), where fines content is approximately 30 percent by weight; the clay content and clay size are important factors which affect the bitumen content.

4.2.2 Properties

Tar sand properties that are of general interest are bulk density, porosity, and permeability. Porosity is, by definition, the ratio of the aggregate volume of the interstices between the particles to the total volume and is expressed as a percentage. High-grade tar sand usually has porosity in the range from 30 to 35 percent that is somewhat higher than the porosity (5–25 percent) of most reservoir sandstone. The higher porosity of the tar sand has been attributed to the relative lack of mineral cement (chemically precipitated material that binds adjacent particles together and gives strength to the sand, which in most sandstone occupies a considerable amount of what was void space in the original sediment).

Permeability is a measure of the ability of a sediment or rock to transmit fluids. It is, to a major extent, controlled by the size and shapes of the pores as well as the channels (throats) between the pores; the smaller the channel, the more difficult it is to transmit the reservoir fluid (water, bitumen). Fine-grained sediments invariably have a lower permeability than coarse-grained sediments, even if the porosity is equivalent. It is not surprising that the permeability of the bitumen-free sand from the Alberta deposits is quite high. On the other hand, the bitumen in the deposits, immobile at formation temperatures [approximately 4°C (39°F)] and pressures, actually precludes any significant movement of fluids through the sands under unaltered formation conditions.

For the Canadian tar sands, bitumen contents from 8 to 14 percent by weight may be considered as normal (or average). Bitumen contents above or below this range have been ascribed to factors that influence impregnation of the sand with the bitumen (or the bitumen precursor). There are also instances where bitumen contents in excess of 12 percent by weight have been ascribed to gravity settling during the formative stages of the bitumen. Bitumen immobility then prevents further migration of the bitumen itself or its constituents.

The bitumen content of the tar sand of the United States varies from 0 to as much as 22 percent by weight. There are, however, noted relationships between the bitumen, water, fines, and mineral contents for the Canadian tar sands. Similar relationships may also exist for the United States tar sands but an overall lack of study has prevented the uncovering of such data.

While conventional crude oil and heavy oil either flows naturally or is pumped from the ground, tar sand must be mined and the bitumen removed from the sand and water. Tar sand currently represents 40 percent of Alberta's total oil production and about one-third of all the oil produced by Canada. By 2005, oil sands production is expected to represent 50 percent of Canada's total crude oil output and 10 percent of North American production. Although tar sands occur in more than 70 countries, the two largest are Canada and Venezuela, with the bulk being found in four different regions of Alberta, Canada: areas of Athabasca, Wabasha, Cold Lake, and Peace River. The sum of these covers an area of nearly 77,000 km^2. In fact, the reserve that is deemed to be technologically retrievable today is estimated at 280 to 300 billion barrels, which is larger than the Saudi Arabia crude oil reserves. The total reserves for Alberta, including oil not recoverable using current technology, are estimated at 1700 to 2500 billion barrels.

4.3 CHEMICAL AND PHYSICAL PROPERTIES OF TAR SAND BITUMEN

Tar sand bitumen is a naturally occurring material that is frequently found filling pores and crevices of sandstone, limestone, or argillaceous sediments or deposits where the permeability is low. Bitumen is reddish brown to black in color and occurs as a semisolid or solid that can exist in nature with no mineral impurity or with mineral matter contents that exceed 50 percent by weight.

Tar sand bitumen is extremely susceptible to oxidation by aerial oxygen. The oxidation process can change the properties (such as viscosity) to such an extent that precautions need to be taken not only in the separation of the bitumen from the sand but also during storage (Wallace, 1988; Wallace et al., 1988a, 1988b).

Bitumen is a high-boiling resource with little, if any, of the constituents boiling below 350°C (662°F). In fact, the boiling range may be approximately equivalent to the boiling range of an atmospheric residuum that is produced as a refinery product (Tables 4.2 and 4.3).

TABLE 4.2 Distillation Data (Cumulative Percent by Weight Distilled) for Bitumen and Crude Oil

Cut point		Cumulative percent by weight distilled		
°C	°F	Athabasca	PR Spring	Leduc (Canada)
200	392	3	1	35
225	437	5	2	40
250	482	7	3	45
275	527	9	4	51
300	572	14	5	
325	617	26	7	
350	662	18	8	
375	707	22	10	
400	752	26	13	
425	797	29	16	
450	842	33	20	
475	887	37	23	
500	932	40	25	
525	977	43	29	
538	1000	45	35	
538+	1000+	55	65	

TABLE 4.3 Properties of Selected Atmospheric (>650) and Vacuum (>1050) Residua

Feedstock	Gravity (°API)	Sulfur (wt %)	Nitrogen (wt %)	Nickel (ppm)	Vanadium (ppm)	Asphaltene (heptane) (wt %)	Carbon residue (Conradson) (wt %)
Arabian Light, >650°F	17.7	3.0	0.2	10.0	26.0	1.8	7.5
Arabian Light, >1050°F	8.5	4.4	0.5	24.0	66.0	4.3	14.2
Arabian Heavy, > 650°F	11.9	4.4	0.3	27.0	103.0	8.0	14.0
Arabian Heavy, >1050°F	7.3	5.1	0.3	40.0	174.0	10.0	19.0
Alaska, North Slope, >650°F	15.2	1.6	0.4	18.0	30.0	2.0	8.5
Alaska, North Slope, >1050°F	8.2	2.2	0.6	47.0	82.0	4.0	18.0
Lloydminster (Canada), >650°F	10.3	4.1	0.3	65.0	141.0	14.0	12.1
Lloydminster (Canada), >1050°F	8.5	4.4	0.6	115.0	252.0	18.0	21.4
Kuwait, >650°F	13.9	4.4	0.3	14.0	50.0	2.4	12.2
Kuwait, >1050°F	5.5	5.5	0.4	32.0	102.0	7.1	23.1
Tia Juana, >650°F	17.3	1.8	0.3	25.0	185.0		9.3
Tia Juana, >1050°F	7.1	2.6	0.6	64.0	450.0		21.6
Taching, >650°F	27.3	0.2	0.2	5.0	1.0	4.4	3.8
Taching, >1050°F	21.5	0.3	0.4	9.0	2.0	7.6	7.9
Maya, >650°F	10.5	4.4	0.5	70.0	370.0	16.0	15.0

There are wide variations both in the bitumen saturation of tar sand (0–18 percent by weight of bitumen) even within a particular deposit, and the viscosity is particularly high. Of particular note is the variation of the density of Athabasca bitumen with temperature and the maximum density difference between bitumen and water occurs at 70 to 80°C (158–176°F), hence the choice of the operating temperature of the hot water bitumen-extraction process.

The character of bitumen can be assessed in terms of API gravity, viscosity, and sulfur content (Table 4.4). Properties such as these help the refinery operator to gain an understanding of the nature of the material that is to be processed. Thus, initial inspection of the feedstock (conventional examination of the physical properties) is necessary. From this, it is possible to make deductions about the most logical means of refining.

TABLE 4.4 Properties of Synthetic Crude Oil from Athabasca Bitumen

Property		Bitumen	Synthetic crude oil
Gravity, °API		8	32
Sulfur, wt %		4.8	0.2
Nitrogen, wt %		0.4	0.1
Viscosity, cP at 100°F		500,000	10
Distillation profile, by wt %			
°C	°F		
0	32	0	5
30	86	0	30
220	428	1	60
345	653	17	90
550	1022	45	100

In the context of the Athabasca deposit, compositional inconsistencies arise because of the lack of mobility of the bitumen at formation temperature [approximately 4°C (39°F)]. For example, the proportion of bitumen in the tar sand increases with depth within the formation. Furthermore, the proportion destructive distillate from the bitumen also decreases with depth within the formation that leads to reduced yields of distillate from the bitumen obtained from deeper parts of the formation. Variations (horizontal and vertical) in bitumen properties have been noted previously, as have variations in sulfur content, nitrogen content, and metals content.

Obviously, the richer tar sand deposits occur toward the base of the formation. However, the bitumen is generally of poorer quality than the bitumen obtained from near the top of the deposit insofar as the proportions of nonvolatile coke-forming constituents (asphaltenes plus resins) are higher (with increased proportions of nitrogen, sulfur, and metals) near the base of the formation.

4.3.1 Composition

Elemental (Ultimate) Composition. The elemental analysis of tar sand bitumen has been widely reported. However, the data suffer from the disadvantage that identification of the source is very general (i.e., Athabasca bitumen) or analysis is quoted for separated bitumen that may have been obtained by, say, the hot water separation or solvent extraction and may not therefore represent the total bitumen on the sand.

However, of the available data, the elemental composition of tar sand bitumen is generally constant and falls into the same narrow range as for petroleum. In addition, the ultimate composition of the Alberta bitumen does not appear to be influenced by the proportion of bitumen in the tar sand or by the particle size of the tar sand minerals.

Of the data that are available for bitumen, the proportions of the elements vary over fairly narrow limits:

Carbon	83.4–0.5%
Hydrogen	10.4–0.2%
Nitrogen	0.4–0.2%
Oxygen	1.0–0.2%
Sulfur	5.0–0.5%
Metals (Ni and V)	>1000 ppm

The major exception to these narrow limits is the oxygen content of heavy oil and especially bitumen, which can vary from as little as 0.2 percent to as high as 4.5 percent. This is not surprising, since when oxygen is estimated by difference the analysis is subject to the accumulation of all of the errors in the other elemental data. In addition, bitumen is susceptible to aerial oxygen and the oxygen content is very dependent upon the sample history. For example, oxidation occurs during separation of the bitumen from the sand as well as when the samples are not protected by a blanket of nitrogen gas. The end result of the oxidation process is an increase in viscosity. Therefore, bitumen cannot be defined using viscosity or any other property that is susceptible to changes during the sample history. Similarly, a sample that is identified as tar sand bitumen without any consideration of the sample history will lead to an erroneous diagnosis.

Chemical Composition. The precise chemical composition of bitumen feedstocks is, despite the large volume of work performed in this area, largely speculative. In very general terms (and as observed from elemental analyses), bitumen is a complex mixture of (a) hydrocarbons, (b) nitrogen compounds, (c) oxygen compounds, (d) sulfur compounds, and (e) metallic constituents. However, this general definition is not adequate to describe the composition of bitumen as it relates to its behavior.

Fractional Composition. Bitumen can be separated into a variety of fractions using a myriad of different techniques.

In general, the fractions produced by these different techniques are called by names of convenience, namely: saturates, aromatics, resins, and asphaltenes. And much of the focus has been on the asphaltene fraction because of its high sulfur content and high coke-forming propensity. The use of composition data to model feedstock behavior during refining is becoming increasingly important in refinery operations.

Bitumen composition varies depending on the locale within a deposit due to the immobility of the bitumen at formation conditions. However, whether or not this is a general phenomenon for all tar sand deposits is unknown. The available evidence is specific to the Athabasca deposit. For example, bitumen obtained from the northern locales of the Athabasca deposit (Bitumount, Mildred-Ruth Lakes) has a lower amount (approximately 16–20 percent by weight) of the nonvolatile asphaltene fraction than the bitumen obtained from southern deposits (Abasand, Hangingstone River; approximately 22–23 percent by weight of asphaltenes). In addition, other data indicate that there is also a marked variation of asphaltene content in the tar sand bitumen with depth in the particular deposit.

Oxidation of bitumen with common oxidising agents, such as acid and alkaline peroxide, acid dichromate, and alkaline permanganate, occurs. Oxidation bitumen in solution, by air, and in either the presence or absence of a metal salt, also occurs.

Thus, changes in the fractional composition can occur due to oxidation of the bitumen during separation, handling, or storage. Similar chemical reactions (oxidation) will also occur in heavy oil giving rise to the perception that the sample is bitumen.

4.3.2 Properties

The specific gravity of bitumen shows a fairly wide range of variation. The largest degree of variation is usually due to local conditions that affect material lying close to the faces, or exposures, occurring in surface tar sand deposits. There are also variations in the specific gravity of the bitumen found in beds that have not been exposed to weathering or other external factors.

A very important property of the Athabasca bitumen (which also accounts for the success of the hot water separation process) is the variation of bitumen density (specific gravity) of the bitumen with temperature. Over the temperature range 30 to 130°C (86–266°F) the bitumen is lighter than water; hence (with aeration) floating of the bitumen on the water is facilitated and the logic of the hot water process is applied.

The API gravity of known U.S. tar sand bitumen ranges downward from about 14° API (0.973 specific gravity) to approximately 2° API (1.093 specific gravity). Although only a general relationship exists between density (gravity) and viscosity, very low gravity bitumen generally has very high viscosity. For instance, bitumen with a gravity of 5° or 6° API can have viscosity up to 5 million centipoises. Elements related to API gravity are viscosity, thermal characteristics, pour point, hydrogen content, and hydrogen-carbon ratio.

It is also evident that not only are there variations in bitumen viscosity between the major Alberta deposits, but there is also considerable variation of bitumen viscosity within the Athabasca deposit and even within one location. These observations are, of course, in keeping with the relatively high proportions of asphaltenes in the denser, highly viscous samples, a trait that appears to vary not only horizontally but also vertically within a deposit.

The most significant property of bitumen is its immobility under the conditions of temperatures and pressure in the deposit. While viscosity may present an indication of the immobility of bitumen, the most pertinent representation of this property is the *pour point*. (ASTM D-97) which is the lowest temperature at which oil will pour or flow when it is chilled without disturbance under definite conditions. When used in consideration with reservoir temperature, the pour point gives an indication of the liquidity of the heavy oil or bitumen and, therefore, the ability of the heavy oil or bitumen to flow under reservoir conditions.

Thus, Athabasca bitumen with a pour point of 50 to 100°C (122–212°F) and a deposit temperature of 4 to 10°C (39–50°F) is a solid or near solid in the deposit and will exhibit little or no mobility under deposit conditions. Similar rationale can be applied to the Utah bitumen where pour points of 35 to 60°C (95–140°F) have been recorded for the bitumen with formation temperatures on the order of 10°C (50°F) also indicate a solid bitumen within the deposit and therefore immobility in the deposit. On the other hand, the California oils exhibit pour points on the order of 2 to 10°C (35–50°F) at a reservoir temperature of 35 to 38°C (95–100°F) indicates that the oil is in the liquid state in the reservoir and therefore mobile.

Irrespective of the differences between the various tar sand bitumen the factor that they all have in common is the near-solid or solid nature and therefore immobility of the bitumen in the deposit. Conversely, heavy oil in different reservoirs has the commonality of being in the liquid state and therefore mobility in the reservoir.

In the more localised context of the Athabasca deposit, inconsistencies arise because of the lack of mobility of the bitumen at formation temperature. For example, the proportion of bitumen in the tar sand increases with depth within the formation. Furthermore, the proportion of asphaltenes in the bitumen or asphaltic fraction (asphaltenes plus resins) also increases with depth within the formation that leads to reduced yields of distillate from the bitumen obtained from deeper parts of the formation. In keeping with the concept of higher proportions of asphaltic fraction (asphaltenes plus resins), variations (horizontal and vertical) in bitumen properties have been noted previously, as have variations in sulfur content, nitrogen content, and metals content.

Nondestructive distillation data (Table 4.2) show that tar sand bitumen is a high-boiling material. There is usually little or no gasoline (naphtha) fraction in bitumen and the majority of the distillate falls in the gas oil–lubrication distillate range [greater than 260°C (500°F)]. Usually, in excess of 50 percent by weight of tar sand bitumen is nondistillable under the conditions of the test. On the other hand, heavy oil has a considerable proportion of its constituents that are volatile below 260°C (500°F).

4.4 MINING TECHNOLOGY

Proposed methods for recovery of bitumen from tar sand deposits are based either on in situ processes or on mining combined with some further processing or operation on the tar sands in situ. The typical in situ recovery methods are not applicable to bitumen recovery because bitumen, in its immobile state, is extremely difficult to move to a production well. Extreme processes are required, usually in the form of a degree of thermal conversion that produces free-flowing product oil that will flow to the well and reduce the resistance of the bitumen to flow. Tar sand deposits are not amenable to injection technologies such as steam soak and steam flooding. In fact, the only successful commercial method of recovering bitumen from tar sand deposits occurs at the two plants in Alberta (Canada) and involves use of a mining technique.

The equipment employed at a *tar sand mine* is a combination of mining equipment and an on-site transportation system that may (currently) either be conveyor belts and/or large trucks. The mining operation itself differs in detail depending upon the equipment; bucketwheel excavators sit on benches; the draglines sit on the surface.

The Suncor (formerly Great Canadian Oil Sands Ltd.) mining and processing plant, located 20 miles north of Fort McMurray, Alberta, started production in 1967. The Syncrude Canada mining and processing plant, located 5 miles (8 km) away from the Suncor plant, started production in 1978. In both projects, about half of the terrain is covered with muskeg, an organic soil resembling peat moss, which ranges from a few inches to 23 ft (7 m) in depth. The total overburden varies from 23 to 130 ft (7–40 m) in thickness.

Mining the Athabasca tar sands presents two major issues: in-place tar sand requires very large cutting forces and is extremely abrasive to cutting edges, and both the equipment and pit layouts must be designed to operate during the long Canadian winters at temperatures as low as −50°C (−58°F).

There are two approaches to open-pit mining of tar sand. The first uses a few bucketwheel excavators and large draglines in conjunction with belt conveyors. In the second approach, a multiplicity of smaller mining units of conventional design is employed.

Over time, different techniques have been used for oil sands mining. Suncor started operations using bucketwheel excavators that discharged their loads onto conveyor belts. The initial Syncrude operation used large draglines to remove oil sands ore from the mineface and place it in windrows from which bucketwheel reclaimers loaded it onto conveyor belts for transportation to the extraction plant. Suncor and Syncrude have now retired their bucketwheel and dragline-based mining systems.

Large mining trucks and power shovels were introduced to replace these early mining systems. By the early 1990s, Syncrude was mining about one-third of its ore using trucks and shovels, while Suncor totally converted to a truck and shovel operation in 1993. Truck and shovel mining is considerably more flexible and less prone to interruption of service than the earlier systems used. In the current mining systems, trucks capable of hauling up to 420 tons of material are loaded by electric- and hydraulic-power shovels with bucket capacities up to 60 yd^3. The trucks transport the oil sands to ore preparation facilities where the ore is crushed and prepared for transport to the extraction plant (where bitumen is separated from the sand).

In their early operations, Suncor and Syncrude used long conveyor systems for ore transportation. These systems have been replaced by hydrotransport with the first commercial applications of this technology occurring in the early 1990s. For hydrotransport, the oil sands ore is mixed with heated water (and chemicals in some cases) at the ore preparation plant to create oil sands slurry that is pumped via pipeline to the extraction plant. Hydrotransport preconditions the ore for extraction of crude bitumen and improves energy efficiency and environmental performance compared to conveyor systems.

4.5 BITUMEN RECOVERY

In terms of bitumen separation and recovery, the hot water process is, to date, the only successful commercial process to be applied to bitumen recovery from mined tar sand in North America. Many process options have been tested with varying degrees of success, and one of these options may even supersede the hot water process.

4.5.1 The Hot Water Process

The hot water process utilises the linear variation of bitumen density and the nonlinear variation of water density with temperature so that the bitumen that is heavier than water at room temperature becomes lighter than water at 80°C (176°F). Surface active materials in the tar sand also contribute to the process. The essentials of the hot water process involve a *conditioning* and *separation*. Other ancillary steps are also used but will not be covered here.

In the *conditioning* step, also referred to as mixing or pulping, tar sand feed is heated and mixed with water to form a pulp of 60 to 85 percent by weight of solids at 80 to 90°C (176–194°F). First the lumps of tar sand as-mined are reduced in size by ablation, that is, successive layers of lump are warmed and sloughed off revealing cooler layers. The conditioned pulp is screened through a double-layer vibrating screen. Water is then added to the screened material (to achieve more beneficial pumping conditions) and the pulp enters the *separation* cell through a central feed well and distributor. The bulk of the sand settles in the cell and is removed from the bottom as tailings, but the majority of the bitumen floats to the surface and is removed as froth. A middling stream (mostly of water with suspended fines and some bitumen) is withdrawn from approximately midway up the side of the cell wall.

Froth from the hot water process may be mixed with a hydrocarbon diluent, for example, coker naphtha, and centrifuged. The Suncor process employs a two-stage centrifuging operation and each stage consists of multiple centrifuges of conventional design installed in parallel. The bitumen product contains 1 to 2 percent by weight of mineral (dry bitumen basis) and 5 to 15 percent by weight of water (wet diluted basis). Syncrude also utilises a centrifuge system with naphtha diluent.

As noted, tailings are a byproduct of the oil sands extraction process. For each ton of tar sand in-place has a volume of about 16 ft^3, which will generate about 22 ft^3 of tailings giving a volume gain on the order of 40 percent. If the mine produces about 200,000 tons of tar sand per day, the volume expansion represents a considerable solids disposal problem. Tailings from the process consist of about 49 to 50 percent by weight of sand, 1 percent by weight of bitumen, and about 50 percent by weight of water. After bitumen extraction, the tailings are pumped to a settling basin. Coarse tailings settle rapidly and can be restored to a dry surface for reclamation. Fine tailings, consisting of slow-settling clay particles and water, are more problematic.

The U.S. tar sands have received considerably less attention than the Canadian deposits. Nevertheless, approaches to recover the bitumen from U.S. tar sands have been made. An attempt has been made to develop the hot water process for the Utah sands. The process differs significantly from that used for the Canadian sands due to the oil-wet Utah sands contrasting to the water-wet Canadian sands. This necessitates disengagement by hot water digestion in a high shear force field under appropriate conditions of pulp density and alkalinity. The dispersed bitumen droplets can also be recovered by aeration and froth flotation.

4.5.2 In Situ Processes

The other aboveground method of separating bitumen from tar sand after the mining operation involves direct heating of the tar sand without previous separation of the bitumen. Thus, the bitumen is not recovered as such but is an upgraded product. Although several processes have been proposed to accomplish, the common theme is to heat the tar sand to thermally decompose the bitumen to produce a volatile product with the coke remaining on the sand.

In general, the viscous nature of the bitumen and its immobility in the deposits has precluded other forms of recovery. However, bitumen recovery from deep deposits is not economical by a mining method. Therefore the bitumen viscosity must be reduced in situ to increase the mobility of bitumen to flow to wellbores that bring the bitumen to the surface.

Bitumen viscosity can be reduced in situ by increasing reservoir temperature or by injecting solvents. Steam-based thermal recovery is the primary recovery method for heavy oil in the Cold Lake and Peace River areas. Various steam-based methods have been shown to be inefficient for bitumen but more recently a method known as steam-assisted gravity drainage (SAGD) has been applied to the Athabasca tar sand with success.

In the process, a pair of horizontal wells, separated vertically by about 15 to 20 ft is drilled at the bottom of a thick unconsolidated sandstone reservoir. Steam is injected into the upper well. The heat reduces the oil viscosity to values as low as 1 to 10 cP (depending on temperature and initial conditions) and develops a *steam chamber* that grows vertically and laterally. The steam and gases rise because of their low density, and the oil and condensed water are removed through the lower well. The gases produced during SAGD tend to be methane with some carbon dioxide and traces of hydrogen sulfide.

To a small degree, the noncondensable gases tend to remain high in the structure, filling the void space, and even acting as a partial insulating blanket that helps to reduce vertical heat losses as the chamber grows laterally. At the pore scales, and at larger scales as well, flow is through counter-current, gravity-driven flow, and a thin and continuous oil film is sustained, giving high recoveries.

Operating the production and injection wells at approximately the same pressure as the reservoir eliminates viscous fingering and coning processes, and also suppresses water influx or oil loss through permeable streaks. This keeps the steam chamber interface relatively sharp, and reduces heat losses considerably. Injection pressures are much lower than the fracture gradient, which means that the chances of breaking into a thief zone, an instability problem which plagues all high-pressure steam injection processes, such as cyclic steam soak, are essentially zero.

Thus, the SAGD process, as for all gravity-driven processes, is extremely stable because the process zone grows only by gravity segregation, and there are no pressure-driven instabilities such as channeling, coning, and fracturing. It is vital in the SAGD process to maintain a volume balance, replacing each unit volume withdrawn with a unit volume injected, to maintain the processes in the gravity dominated domain. If bottom water influx develops, this indicates that the pressure in the water is larger than the pressure in the steam chamber, and steps must be taken to balance the pressures. Because it is not possible to reduce the pressure in the water zone, the pressure in the steam chamber and production well region must be increased. This can be achieved by increasing the operating pressure of the steam chamber through the injection rate of steam or through reduction of the production rate from the lower well. After some time, the pressures will become more balanced and the water influx ceases.

Clearly, a low pressure gradient between the bottom water and the production well must be sustained. If pressure starts to build up in the steam chamber zone, then loss of hot water can take place as well. In such cases, the steam chamber pressure must be reduced and perhaps also the production rate increased slightly to balance the pressures. In all these cases, the system tends to return to a stable configuration because of the density differences between the phases.

SAGD seems to be relatively insensitive to shale streaks and similar horizontal barriers, even up to several meters thick (3–6 ft), that otherwise would restrict vertical flow rates. This occurs because as the rock is heated, differential thermal expansion causes the shale to be placed under a tensile stress, and vertical fractures are created, which serve as conduits for steam (up) and liquids (down). As high temperatures hit the shale, the kinetic energy in the water increases, and absorbed water on clay particles is liberated. Thus, instead of expanding thermally, dehydration (loss of water) occurs and this leads to volumetric shrinkage of the shale barriers. As the shale shrink, the lateral stress (fracture gradient) drops until the pore pressure exceeds the lateral stress, which causes vertical fractures to open. Thus, the combined processes of gravity segregation and shale thermal fracturing make SAGD so efficient that recovery ratios of 60 to 70 percent are probably achievable even in cases where there are many thin shale streaks, although there are limits on the thickness of shale bed that can be traversed in a reasonable time.

Heat losses and deceleration of lateral growth mean that there is an economic limit to the lateral growth of the steam chamber. This limit is thought to be a chamber width of four times (4×) the vertical zone thickness. For thinner zones, horizontal well pairs would therefore have to be placed close together, increasing costs as well as providing lower total resources per well pair. In summary, the zone thickness limit (net pay thickness) must be defined for all reservoirs.

The cost of heat is a major economic constraint on all thermal processes. Currently, steam is generated with natural gas, and when the cost of natural gas rises, operating costs rise considerably. Thermally, SAGD is about twice as efficient as cyclic steam stimulation, with steam-oil ratios that are now approaching 2 (instead of 4 for cyclic steam soak), for similar cases. Combined with the high recovery ratios possible, SAGD will likely displace pressure-driven thermal process in all cases where the reservoir is reasonably thick.

Finally, because of the lower pressures associated with SAGD, in comparison to high pressure processes such as cyclic steam soak and steam drive, greater wellbore stability should be another asset, reducing substantially the number of sheared wells that are common in cyclic steam soak projects.

The Canadian tar sand has several SAGD projects in progress, since this region is home of one of the largest deposits of bitumen in the world (Canada and Venezuela have the world's largest deposits). However, in spite of the success reported for the Athabasca deposit, the SAGD process is not entirely without drawbacks. The process requires amounts of makeup fresh water and large water recycling facilities as well as a high energy demand to create the steam. In addition, gravity drainage being the operative means of bitumen separation from the reservoir rock, the process requires comparatively thick and homogeneous reservoirs; to date the process has not been tested on a wide variety of reservoirs. Different processes are still being developed and these include steam-and-gas push (SAGP) and expanding-solvent-SAGD

(ES-SAGD) in which a hydrocarbon solvent (to facilitate complete or partial dissolution of the bitumen and help reduce its viscosity) is mixed with the steam.

Alternative processes that are being reviewed for bitumen recovery (or atleast a derivative of the bitumen) from tar sand deposits include *vapor-assisted extraction* (VAPEX), which is a new process in which the physics of the process are essentially the same as for SAGD and the configuration of wells is generally similar. The process involves the injection of vaporized solvents such as ethane or propane to create a vapor-chamber through which the oil flows due to gravity drainage. The process can be applied in paired horizontal wells, single horizontal wells, or a combination of vertical and horizontal wells. The key benefits are significantly lower energy costs, potential for in situ upgrading and application to thin reservoirs, with bottom water or reactive mineralogy.

Because of the slow diffusion of gases and liquids into viscous oils, this approach, used alone, perhaps will be suited only for less viscous oils although preliminary tests indicate that there are micromechanisms that act so that the VAPEX dilution process is not diffusion rate limited and the process may be suitable for the highly viscous tar sand bitumen.

Nevertheless, VAPEX can undoubtedly be used in conjunction with SAGD methods (qv). A key factor is the generation of a three-phase system with a continuous gas phase so that as much of the oil as possible can be contacted by the gaseous phases, generating the thin oil-film drainage mechanism. Vertical permeability barriers are a problem, and must be overcome through hydraulic fracturing to create vertical permeable channels, or undercut by the lateral growth of the chamber beyond the lateral extent of the limited barrier, or *baffle*.

However, as with all solvent-based processes, there is the potential for solvent losses in the reservoir. These can arise, for example, due to unknown fissures in the reservoir rock as well as clay lenses to which the solvent will adhere.

In situ combustion in heavy oil reservoirs and bitumen deposits has been notoriously difficult to control but make a comeback with a new concept. The THAI (toe-to-heel air injection) process is based on the geometry of horizontal wells that may solve the problems that have plagued conventional in situ combustion. The well geometry enforces a short flow path so that any instability issues associated with conventional combustion are reduced or even eliminated.

In situ conversion, or underground refining, is a promising new technology to tap the extensive reservoirs of heavy oil and deposits of bitumen. The new technology (United States Patent 6016867; United States Patent 6016868) features the injection of high-temperature, high-quality steam and hot hydrogen ore into a formation containing heavy hydrocarbons to initiate conversion of the heavy hydrocarbons into lighter hydrocarbons. In effect, the heavy hydrocarbons undergo partial underground refining that converts them into a synthetic crude oil (or *syncrude*). The heavier portion of the syncrude is treated to provide the fuel and hydrogen required by the process, and the lighter portion is marketed as a conventional crude oil.

Thus, below ground, superheated steam and hot hydrogen are injected into a heavy oil or bitumen formation, which simultaneously produces the heavy oil or bitumen and converts it in situ (i.e., within the formation) into syncrude. Above ground, the heavier fraction of the syncrude is separated and treated on-site to produce the fuel and hydrogen required by the process, while the lighter fraction is sent to a conventional refinery to be made into petroleum products (United States Patent 6016867; United States Patent 6016868).

The potential advantages of an in situ process for bitumen and heavy oil include (a) leaving the carbon-forming precursors in the ground, (b) leaving the heavy metals in the ground, (c) reducing sand-handling, and (d) bringing a partially upgraded product to the surface. The extent of the upgrading can, hopefully, be adjusted by adjusting the exposure of the bitumen of heavy oil to the underground thermal effects.

In the *modified in situ extraction* processes, combinations of in situ and mining techniques are used to access the reservoir. A portion of the reservoir rock must be removed to enable application of the in situ extraction technology. The most common method is to enter the reservoir through a large-diameter vertical shaft, excavate horizontal drifts from the bottom of the shaft, and drill injection and production wells horizontally from the drifts. Thermal extraction processes are then applied through the wells. When the horizontal wells are drilled at or near the base of the tar sand reservoir, the injected heat rises from the injection wells through the reservoir, and drainage of produced fluids to the production wells is assisted by gravity.

4.6 UPGRADING, REFINING, AND FUEL PRODUCTION

The limitations of processing these heavy oil and bitumen depend to a large extent on the amount of nonvolatile higher molecular weight constituents, which also contain the majority of the heteroatoms (i.e., nitrogen, oxygen, sulfur, and metals such as nickel and vanadium). These constituents are responsible for high yields of thermal and catalytic coke. The majority of the metal constituents in crude oils are present as organometallic complexes, such as porphyrins. The rest are found in organic or inorganic salts that are soluble in water or in crude. In recent years, attempts have been made to isolate and to study the vanadium present in petroleum porphyrins.

When catalytic processes are employed, complex molecules (such as those that are present in the nonvolatile fraction) or those formed during the process, are not sufficiently mobile (i.e., they are strongly adsorbed by the catalyst) to be saturated by hydrogenation. The chemistry of the thermal reactions of some of these constituents dictates that certain reactions, once initiated, cannot be reversed and proceed to completion. Coke is the eventual product. These deposits deactivate the catalyst sites and eventually interfere with the hydroprocess.

Technologies for upgrading heavy crude feedstocks, such as residua and tar sand bitumen, can be broadly divided into carbon rejection and hydrogen addition processes.

Carbon rejection processes are those processes in which hydrogen is redistributed among the various components, resulting in fractions with increased hydrogen/carbon atomic ratios (distillates) and fractions with lower hydrogen/carbon atomic ratios (coke). On the other hand, *hydrogen addition processes* involve the reaction of heavy crude oils with an external source of hydrogen and result in an overall increase in hydrogen/carbon ratio. Within these broad ranges, all upgrading technologies can be subdivided as follows:

1. *Carbon rejection*: Visbreaking, steam cracking, fluid catalytic cracking, coking, and flash pyrolysis.
2. *Hydrogen addition*: Catalytic hydroconversion (hydrocracking) using active hydrodesulfurization catalysts, fixed-bed catalytic hydroconversion, ebullated catalytic-bed hydroconversion, thermal slurry hydroconversion, hydrovisbreaking, hydropyrolysis, donor solvent processes, and supercritical water upgrading.
3. *Separation processes*: Distillation, deasphalting, and supercritical extraction.

Thermal-cracking processes offer attractive methods of conversion of heavy oil and bitumen because they enable low operating pressure, while involving high operating temperature, without requiring expensive catalysts. Currently, the widest operated heavy oil and bitumen upgrading or conversion processes are visbreaking and delayed coking. And, these are still attractive processes for refineries from an economic point of view.

The fluid catalytic cracking process using vacuum gas oil feedstock was introduced into refineries in the 1930s. In recent years, because of a trend for low-boiling products, most refineries perform the operation by partially blending residues into vacuum gas oil. However, conventional fluid catalytic cracking processes have limits in when applied to processing heavy oils and bitumen, so residue fluid catalytic cracking processes have lately been employed one after another. Because the residue fluid catalytic cracking process enables efficient gasoline production directly from residues, it will play the most important role as a residue cracking process, along with the residue hydroconversion process. Another role of the *residuum fluid catalytic cracking process* is to generate high-quality gasoline blending stock and petrochemical feedstock. Olefins (propene, butenes, and pentenes) serve as feed for alkylation processes, for polymer gasoline, as well as for additives for reformulated gasoline.

Residuum hydrotreating processes have two definite roles: (a) desulfurization to supply low-sulfur fuel oils and (b) pretreatment of feed residua for residuum fluid catalytic cracking processes. The main goal is to remove sulfur, metal, and asphaltene contents from residua and other heavy feedstocks to a desired level. The major goal of *residuum hydroconversion* is cracking of heavy oil (and to some extent bitumen) with desulfurization, metal removal, denitrogenation, and asphaltene conversion. Residuum hydroconversion process offers production of kerosene and gas oil, and production of feedstocks for *hydrocracking, fluid catalytic cracking*, and petrochemical applications.

Finally, in terms of upgrading tar sand bitumen, *solvent deasphalting processes* have not realized their maximum potential. With ongoing improvements in energy efficiency, such processes would display its effects in a combination with other processes. Solvent deasphalting allows removal of sulfur and nitrogen compounds as well as metallic constituents by balancing yield with the desired feedstock properties.

Upgrading residua that are similar in character to tar sand bitumen began with the introduction of desulfurization processes that were designed to reduce the sulfur content of residua as well as some heavy crude oils and products therefrom. In the early days, the goal was desulfurization but, in later years, the processes were adapted to a 10 to 30 percent partial conversion operation, as intended to achieve desulfurization and obtain low-boiling fractions simultaneously, by increasing severity in operating conditions. Refinery evolution has seen the introduction of a variety of heavy feedstock residuum cracking processes based on thermal cracking, catalytic cracking, and hydroconversion. Those processes are different from one another in cracking method, cracked product patterns, and product properties, and will be employed in refineries according to their respective features.

In general terms, the quality of tar sand bitumen is low compared to that of conventional crude oil and heavy oil. Upgrading and refining bitumen requires a different approach to that used for upgrading heavy oil. In addition, the distance that the bitumen must be shipped to the refinery and in what form as well as product quality must all be taken into account when designing a bitumen refinery.

The low proportion of volatile constituents in bitumen [i.e., those constituents boiling below 200°C (392°F)] initially precluded distillation as a refining step, are recognized by thermal means and are necessary to produce liquid fuel streams. A number of factors have influenced the development of facilities that are capable of converting bitumen to a synthetic crude oil. A visbreaking product would be a hydrocarbon liquid that was still high in sulfur and nitrogen with some degree of unsaturation. This latter property enhances gum formation with the accompanying risk of pipeline fouling and similar disposition problems in storage facilities and fuel oil burners. A high sulfur content in finished products is environmentally unacceptable. In addition, high levels of nitrogen cause problems in the downstream processes, such as in catalytic cracking where nitrogen levels in excess of 3000 ppm will cause rapid catalyst deactivation; metals (nickel and vanadium) cause similar problems.

However, high-boiling constituents [i.e., those boiling in the range 200–400°C, (392–752°F)] can be isolated by distillation but, in general terms, more than 40 percent by weight of tar sand bitumen boils above 540°C (1004°F). Thus, a product of acceptable quality could be obtained by distillation to an appropriate cut point but the majority of the bitumen would remain behind to be refined by whichever means would be appropriate, remembering, of course, the need to balance fuel requirements and coke production. It is therefore essential that any bitumen-upgrading program convert the nonvolatile residuum to a low-boiling, low-viscosity, low-molecular-weight, high hydrocarbon/carbon ratio oil.

Bitumen is hydrogen-deficient that is upgraded by carbon removal (coking) or hydrogen addition (hydrocracking). There are two methods by which bitumen conversion can be achieved: (a) by direct heating of mined tar sand and (b) by thermal decomposition of separated bitumen. The latter is the method used commercially but the former deserves mention here since there is the potential for commercialization.

An early process involved a coker for bitumen conversion and a burner to remove carbon from the sand. A later proposal suggested that the Lurgi process might have applicability to bitumen conversion. A more modern approach has also been developed which also cracks the bitumen constituents on the sand. The processor consists of a large, horizontal, rotating vessel that is arranged in a series of compartments. The two major compartments are a preheating zone and a reaction zone. Product yields and quality are reported to be high.

Direct coking of tar sand with a fluid-bed technique has also been tested. In this process, tar sand is fed to a coker or still, where the tar sand is heated to approximately 480°C (896°F) by contact with a fluid bed of clean sand from which the coke has been removed by burning. Volatile portions of the bitumen are distilled. Residual portions are thermally cracked, resulting in the deposition of a layer of coke around each sand grain. Coked solids are withdrawn down a standpipe, fluid with air, and transferred to a burner or regenerator [operating at approximately 800°C (1472°F)] where most of the coke is burned off the sand grains. The clean, hot sand is withdrawn through a standpipe. Part (20–40 percent) is rejected and the remainder is recirculated to the coker to provide the heat for the coking reaction. The products leave the coker as a vapor, which is condensed in a receiver. Reaction off-gases from the receiver are recirculated to fluidize the clean, hot sand which is returned to the coker.

4.6.1 Primary Conversion

The overall upgrading process by which bitumen is converted to liquid fuels is accomplished in two steps. Initially, at the time of opening of both the Suncor and Syncrude plants, the first step is the *primary conversion process* or *primary upgrading process* (coking) that involves cracking the bitumen to lighter products that are more easily processed downstream. The *secondary upgrading process* involves hydrogenation of the primary products and is the means by which sulfur and nitrogen are removed from the primary products. The synthetic crude oil can then be refined to gasoline, jet fuel, and homes heating oil by conventional means.

There are two coking processes that have been applied to the production of liquids from Athabasca bitumen. Delayed coking is practised at the Suncor plant, whereas Syncrude employs a fluid-coking process which produces less coke than the delayed coking in exchange for more liquids and gases.

Thus, *coking* became the process of choice for bitumen conversion and bitumen is currently converted commercially by *delayed coking* (Suncor) and by *fluid coking* (Syncrude). In each case the charge is converted to distillate oils, coke, and light gases. The coke fraction and product gases can be used for plant fuel. The coker distillate is a partially upgraded material in itself and is a suitable feed for hydrodesulfurization to produce a low-sulfur synthetic crude oil.

Delayed coking is a semibatch process in which feed bitumen is heated before being fed to coking drums that provide sufficient residence time for the cracking reactions to occur. In the Suncor operation, bitumen conversion to liquids is on the order of more than 75 percent with fluid coking giving a generally higher yield of liquids compared to delayed coking. The remainder appears as coke (approximately 15 percent by weight) and gases.

The Suncor plant (in operation since 1967) involves a delayed coking technique followed by hydrotreating of the distillates to produce synthetic crude oil that has properties that are substantially different from the original bitumen (Table 4.4). The selection of delayed coking over less severe thermal processes, such as visbreaking, was based (at the time of planning, from 1960 to 1964) on the high yields of residuum produced in these alternate processes. The yields of coke from the residuum would have exceeded the plant fuel requirements, especially if the distillate had to be shipped elsewhere for hydrogen treatment as well as a more favorable product distribution and properties. Alternate routes for the disposal of the excess coke would be needed.

Fluid coking is a continuous process employing two vessels with fluid coke. It provides a better yield of overhead products than delayed coking. Feed oil flows to the reactor vessel where cracking and formation of coke occur; coke is combusted in the burner. Fluid-transfer lines between these vessels provide the coke circulation necessary for heat balance. The proportion of coke burned is just sufficient to satisfy heat losses and provide the heat for the cracking reactions.

In the fluid-coking process, whole bitumen (or topped bitumen) is preheated and sprayed into the reactor where it is thermally cracked in the fluidized coke bed at temperatures typically between 510°C and 540°C (950°F and 1004°F) to produce light products and coke. The coke is deposited on the fluidized coke particles while the light products pass overhead to a scrubbing section in which any high-boiling products are condensed and recombined with the reactor fresh feed. The uncondensed scrubber overhead passes into a fractionator in which liquid products of suitable boiling ranges for downstream hydrotreating are withdrawn. Cracked reactor gases (containing butanes and lower molecular weight hydrocarbon gases) pass overhead to a gas recovery section. The propane material ultimately flows to the refinery gas system and the condensed butane and butenes may (subject to vapor pressure limitations) be combined with the synthetic crude. The heat necessary to vaporize the feed and to supply the heat of reaction is supplied by hot coke which is circulated back to the reactor from the coke heater. Excess coke that has formed from the fresh feed and deposited on hot circulating coke in the fluidized reactor bed is withdrawn (after steam stripping) from the bottom of the reactor.

Sulfur is distributed throughout the boiling range of the delayed coker distillate, as with distillates from direct coking. Nitrogen is more heavily concentrated in the higher boiling fractions but is present in most of the distillate fractions. Raw coker naphtha contains significant quantities of olefins and diolefins that must be saturated by downstream hydrotreating. The gas oil has a high aromatic content typical of coker gas oils.

In addition to Suncor and Syncrude, Shell Canada Ltd. has also commenced operations in the Athabasca tar sand through its *Athabasca Oil Sands Project*; the project is a joint venture that consists of the Albian Sands Muskeg River Mine, the Shell Scotford Upgrader, and the Corridor Pipeline. The joint venture currently consists of Shell Canada Ltd. (60 percent), Chevron Canada Ltd. (20 percent), and Western Oil Sands LP (20 percent). Recently, Western Oil Sands LP has been purchased by Marathon Oil Corp.

In April 2003, the project commenced fully integrated operations and by April 2004, production capacity began to consistently exceed the project's daily design rate and at the end of 2005 daily production averaged 160,000 per day. Currently, the design capacity of 155,000 bbl/day of bitumen is being increased to 180,000 to 200,000 bbl/day. By 2010, planned expansion are expected to further increase bitumen throughputs by approximately 90,000 bbl/day, raising total expected production to 270,000 and 290,000 bbl/day. At the end of the expansion, the production is projected to be on the order of 525,000 bbl/day of bitumen with a lease operating life of 22 years. Expended production of the in situ projects at Foster Creek, MacKay River, and Christina Lake will provide a further 200,000 bbl/day of bitumen at the time of completion.

The Scotford Upgrader is located next to Shell Canada's Scotford Refinery near Fort Saskatchewan, Alberta where bitumen from the Muskeg River Mine is upgraded into synthetic crude oil. A significant portion of the output is sold to the Scotford Refinery for further processing.

In addition to the coking options, the LC-Fining ebullated-bed hydroconversion process is used at both the Syncrude Mildred Lake upgrader and the Shell Scotford Upgrader. The H-Oil ebullated-bed hydroconversion process is used at the Husky Lloydminster Upgrader.

4.6.2 Secondary Upgrading

Catalytic hydrotreating is used for secondary upgrading to remove impurities and enhance the quality of the final synthetic crude oil product.

In a typical catalytic hydrotreating unit, the feedstock is mixed with hydrogen, preheated in a fired heater, and then charged under high pressure to a fixed-bed catalytic reactor. Hydrotreating converts sulfur and nitrogen compounds present in the feedstock to hydrogen sulfide and ammonia. Sour gases from the hydrotreater(s) are treated for use as plant fuel. Hydrocracking may also be employed at this stage to improve product yield and quality.

Thus the primary liquid product (synthetic crude oil) is hydrotreated (*secondary upgrading*) to remove sulfur and nitrogen (as hydrogen sulfide and ammonia, respectively) and to hydrogenate the unsaturated sites exposed by the conversion process. It may be necessary to employ separate hydrotreaters for light distillates and medium-to-heavy fractions; for example, the heavier fractions require higher hydrogen partial pressures and higher operating temperatures to achieve the desired degree of sulfur and nitrogen removal. Commercial applications have therefore been based on the separate treatment of two or three distillate fractions at the appropriate severity to achieve the required product quality and process efficiency.

Hydrotreating is generally carried out in down-flow reactors containing a fixed bed of cobalt-molybdate catalysts. The reactor effluents are stripped of the produced hydrogen sulfide and ammonia. Any light ends are sent to the fuel gas system and the liquid products are recombined to form synthetic crude oil.

Finishing and stabilisation (hydrodesulfurization and saturation) of the liquid products is achieved by hydrotreating the liquid streams, as two or three separate streams (Speight, 1999 and references cited therein). This is necessary because of the variation in conditions and catalysts necessary for treatment of a naphtha fraction relative to the conditions necessary for treatment of gas oil. It is more efficient to treat the liquid product streams separately and then to blend the finished liquids to a synthetic crude oil. In order to take advantage of optimum operating conditions for various distillate fractions, the Suncor coker distillate is treated as three separate fractions: naphtha, kerosene, and gas oil. In the operation used by Syncrude, the bitumen products are separated into two distinct fractions: naphtha and mixed gas oils. Each plant combines the hydrotreated fractions to form synthetic crude oil that is then shipped by pipeline to a refinery.

4.6.3 Other Processes

Other processes which have also received some attention for bitumen upgrading include partial upgrading (a form of thermal deasphalting), flexicoking, the Eureka process, and various hydrocracking processes.

A *partial coking* or *thermal deasphalting* process provides a minimal upgrading of bitumen. In partial coking, the hot water process froth is distilled at atmospheric pressure and minerals and water are removed. A dehydrated mineral-free bitumen product is obtained that contains most of the asphaltenes and coke precursors. The process has been carried out in batch equipment in laboratory tests over periods ranging from 30 minutes to 4 hours. Thermal cracking begins as the liquid temperature passes 340°C (644°F). The distillation is continued into the range 370 to 450°C (698–842°F). With slow heating [10°C (50°F) temperature rise per hour] the coke production rate is approximately 1 percent weight of feed per hour. As the coke forms about the entrained mineral particles, 1 to 4 percent weight of coke up to 50 percent v/v of the feed is recovered as distillate. After this treatment the residue may be filtered to yield an essentially ash-free production suitable for applications such as metallurgic coke or production of bituminous paints, for which the original mineral content would have disqualified it.

In the *flexicoking process,* a gasifier vessel is added to the system in order to gasify excess coke with a gas-air mixture to a low heating-value gas that can be desulfurized and used as a plant fuel. The *Eureka process* is a variant of delayed coking and uses steam stripping to enhance yield and produce a heavy pitch rather than coke byproduct.

4.6.4 Hydrogen Production

Current synthetic crude oil production operations meet the hydrogen requirements using steam-methane reforming with natural gas used for both feedstock and fuel. However, the increasing price of natural gas is increasing production costs and partial oxidation (gasification) of a bitumen stream or coal could well be the mode of hydrogen generation for future projects.

4.7 SYNTHETIC CRUDE OIL

Synthetic crude oil is not a naturally occurring material and is currently produced by upgrading tar sand bitumen. Synthetic crude usually requires further refining to produce gasoline and other types of petroleum products. Another product *dilbit* (bitumen diluted with condensate or with naphtha) has also found a market.

Synthetic crude oil is a blend of naphtha, distillate, and gas oil range materials, with no residuum [1050°F+ (565°C+) material]. Canadian synthetic crude oil first became available in 1967 when Suncor (then Great Canadian Oil Sands) started to market a blend produced by hydrotreating the naphtha, distillate, and gas oil generated in a delayed coking unit. The light, sweet synthetic crude marketed by Suncor today is called Suncor Oil Sands Blend A (OSA). Syncrude Canada Ltd. started production in 1978, marketing a fully-hydrotreated blend utilizing fluidized-bed coking technology as the primary upgrading step. This product is referred to as Syncrude sweet blend (SSB).

Suncor is planning to build a third bitumen upgrader for its oil sands operations and plans to construct new sulfur recovery plant in support of its existing upgrader capacity. The eventual aim is to raise the production capacity of Suncor's oil sands processing to more than 500,000 bbl/day over the next 5 to 7 years. The new plant, which would be built about half a kilometer from the existing upgrader site, would include cokers, hydrotreaters, and a 50-km bitumen pipeline to connect the upgrader with Suncor's mining operations.

Syncrude has moved away from fluid coking as the only upgrading option for tar sand bitumen. The process begins with diluted bitumen fed into the diluent recovery units where water is removed and naphtha is recovered to be recycled through extraction. The resulting dry bitumen is fed to the fluid coker, the LC-Finer (a hydrocracker), and the vacuum distillation unit (VDU) for further processing.

Light and heavy gas oils are distilled off in the vacuum distillation unit and sent to hydrotreaters. The remaining bitumen is sent to the LC-Finer and coking unit. Light and heavy gas oils are formed in the LC-Finer, through the use of a catalyst to add hydrogen, and sent to hydrotreaters. The remaining bitumen is sent to the coking unit. Naphtha, light gas oil, and heavy gas oil are produced in the coker, through chemical reaction using hot coke, and sent to hydrotreaters. As before, the end result is a synthetic crude oil but the makeup of the synthetic crude oil can be varied in response to the market demand.

In comparison to a conventional crude oil (Brent), the Syncrude sweet blend has lower sulfur and no resid (as shipped); it contains significantly less naphtha-range material and more middle distillate and vacuum gas oil (Table 4.5).

TABLE 4.5 Comparison of Brent Crude Oil and Syncrude Sweet Blend

	Brent	Sweet blend
Gravity, °API	38.6	31.8
Sulfur, wt %	0.3	0.1
C_4-minus	2.9	3.4
C_5–350°F	28.6	15.0
350–650°F	29.6	44.0
650–1050°F	29.8	37.6
1050°F and above	9.1	0.0

Generally, sweet synthetic crude oil makes up the majority of the synthetic crude oil market but sour synthetic crude oil is also available. Suncor markets a range of sour synthetic blends, each tailored to meet specific refinery processing capabilities. One such synthetic crude oil is a blend of hydrotreated coker naphtha with nonhydrotreated coker distillate and coker heavy gas oil.

Suncor's sour synthetic crude oil is a blend of hydrotreated coker naphtha with straight-run distillate and straight-run vacuum gas oil. Other blends can also be made available, each with its own processing characteristics. While these sour crude blends still contain no resid fraction, they are generally sold to medium and heavy sour crude refineries.

Husky Oil started up a heavy conventional crude upgrader in 1990 using a combination of ebullated-bed hydroprocessing and delayed coking technologies. Their sweet synthetic crude is traded as Husky Sweet Blend (HSB). The Athabasca Oils Sands Project (AOSP) started producing sweet synthetic crude in 2003 called Premium Albian Synthetic (PAS) using ebullated-bed hydroprocessing technology.

4.8 THE FUTURE

For future tar sand development, the Government of the Province of Alberta, Canada, announced a standard royalty formula for the oil sands industry, have embraced the principles and, to a large degree, put into the practice the fiscal recommendations of the National Task Force on Oil Sands

Strategies. The Canadian Government plan to extend the mining tax regulation to include in situ operations. More than $3.4 billion in new projects and expansions have been waiting for the resolution of fiscal terms to allow the industry to move forward with a number of the projects that are in the initial stages of development. It is anticipated that such a move will encourage further development of the Canadian tar sand resources.

There have been numerous forecasts of world production and demand for conventional crude, all covering varying periods of time, and even after considering the impact of the conservation ethic, the development of renewable resources, and the possibility of slower economic growth, nonconventional sources of liquid fuels could well be needed to make up for the future anticipated shortfalls in conventional supplies.

This certainly applies to North America, which has additional compelling reasons to develop viable alternative fossil fuel technologies. Those reasons include, of course, the security of supply and the need to quickly reduce the impact of energy costs on the balance of payments. There has been the hope that the developing technology in North America will eventually succeed in applying the new areas of nuclear and solar energy to the energy demands of the population. However, the optimism of the 1970s has been succeeded by the reality of the 1980s and it is now obvious that these energy sources will not be the answer to energy shortfalls for the remainder of the present century. Energy demands will most probably need to be met by the production of more liquid fuels from fossil fuel sources.

There are those who suggest that we are indeed faced with the inevitable decline of the *liquid fuel culture*. This may be so, but the potential for greater energy availability from alternative fossil fuel technologies is high. North America is rich in coal, oil shale, and oil sands—so rich that with the development of appropriate technologies, it could be self sufficient well into the twenty-first century.

There is very little doubt that unlocking energy from the tar sand is a complex and expensive proposition. With conventional production, the gamble is taken in the search and the expenses can be high with no guarantee of a commercial find. With oil sands, the oil is known to be there, but getting it out has been the problem and has required gambling on the massive use of untried technology. There is no real market for the bitumen extracted from the oil sands and the oil sand itself is too bulky to be shipped elsewhere with the prospect of any degree of economic return. It is therefore necessary that the extraction and upgrading plants be constructed in the immediate vicinity of the mining operation.

To develop the present concept of oil from the oil sands, it is necessary to combine three operations, each of which contributes significantly to the cost of the venture: (a) a mining operation capable of handling 2 million tons or more, of oil sand per day, (b) an extraction process to release the heavy oil from the sand, and (c) an upgrading plant to convert the heavy oil to a synthetic crude oil.

For Suncor, being the first of the potential oil sands developers carried with it a variety of disadvantages. The technical problems were complex and numerous with the result that Suncor (onstream: 1967) had accumulated a deficit of $67 million by the end of 1976, despite having reported a $12 million profit for that year. However, with hindsight it appears that such a situation is not without some advantages. The early start in the oil sands gave Suncor a relatively low capital cost per daily barrel for a nonconventional synthetic crude oil operation. Total capital costs were about $300 million that, at a production rate of 50,000 bbl/day, places the capital cost at about $6000 per daily barrel.

It is perhaps worthy of mention here that a conventional refinery of the Imperial Oil Strathcona-type (150 to 300 × 10^3 bbl/day) may have cost at that time $100 to $400 million and have an energy balance (i.e., energy output/energy input) in excess of 90 percent. A tar sand refinery of the Suncor-Syncrude type may have an energy balance of the order of 70 to 75 percent.

The second oil sands plant erected by the Syncrude faced much stiffer capital costs. In fact, it was the rapidly increasing capital costs that nearly killed the Syncrude project.

Originally estimated at less than $1 billion, capital needs began to escalate rapidly in the early 1970s. The cost was more than one of the four partners wanted to pay and the number of participants dropped to three. Since the company dropping out held one of the largest interests, the loss was keenly felt and for a while the project was in jeopardy. It was finally kept alive through the participation of the Canadian government and the governments of the Provinces of Ontario and Alberta. The Canadian government took a 15 percent interest in the project while the Province of Alberta took a 10 percent interest and the Province of Ontario a 5 percent interest. The balance remained with three of the original participants. Imperial Oil Ltd., Gulf Oil Canada Ltd., and Canada-Cities Service Ltd. After that initial setback, progress became rapid and the project (located a few miles north of the Suncor plant) was brought to completion (onstream: 1978). The latest estimate of the cost of the plant is in the neighborhood of $2.5 billion. At a design level of 120,000 to 130,000 bbl/day, the capital cost is in excess of $20,000 per daily barrel.

For both the Suncor and Syncrude plants, the investment is broken down to four broad areas: mining (28–34 percent), bitumen recovery (approximately 12 percent), bitumen upgrading (28–30 percent), and offsites, including the power plant (16–24 percent).

In an economic treatment of, in this case, oil production, there are invariably attempts made to derive the costs on the basis of mathematic formulas. The effects of inflation on the capital outlay for the construction of an oil sands plant prohibit such mathematic optimism. For this reason, there are no attempts made here to "standardize" plant construction costs except to note that, say, the percentage of the outlay required for specific parts of a plant must be anticipated to be approximately the same whether the plant costs $300 million or $15 billion. The economics of an in situ project will be somewhat different because of the ongoing nature of the project and the offset of costs by revenue.

In the United States, oil sand economics is still very much a matter for conjecture. The estimates published for current and proposed Canadian operations are, in a sense, not applicable to operations in the United States because of differences in the production techniques that may be required. As an example, one estimate in particular showed a construction cost of (in 1978) $145 million for a 10,000 bbl/day extraction plant and it was conjectured that such an extraction plant would have to operate and be maintained between $2.00 and $5.00 per barrel.

Any degree of maximizing the production of liquid fuels will require the development of heavy oil fields and oil sand deposits. Recent inflationary aspects of plant construction costs have brought to a slowdown what was considered, for example, to be a natural evolution of a succession of oil sands plants in Canada. The significantly higher capital requirements ($10–$15 million) for the construction of large recovery/upgrading plants will not decrease but the suggestion that commercial planning be directed toward smaller scale nominal size (i.e., Suncor-type plant, 45,000–60,000 bbl/day) may be seeing fruition. On the other hand, oil sand plants could be developed on the minimodular concept thereby relieving some of the high capital outlay required before the production of 1 bbl of oil. Along these lines, there has also been a similar suggestion that several mining and recovery units produce feedstock for a larger upgrading facility. Either of these modular concepts could be developed throughout the life of the lease by the sole owner or by a consortium.

Activities related to the development of the tar sand resources of the United States have declined substantially since the early 1980s in line with the decline in the price of crude oil. In 1981, when the price of crude oil was approximately $35 per barrel, some 35 tar sand field projects in the United States were either operating or in the late planning stages. By mid-1985, these numbers had declined seriously.

The projects in operation in the early 1980s included 34 in situ projects and 9 mining/extraction projects, and production totaled in excess of 10,000 bbl/day. Almost all of this production occurred in California by means of in situ steam operations but this was from reservoirs where the oil in-place had viscosity in the range of 10,000 to 25,000 cP (somewhat lower than the viscosity of tar sand bitumen).

Nevertheless, the projects did cover a wide range of reservoir conditions: porosity ranged from 15 to 37 percent and permeability was up to 6000 millidarcies. Bitumen saturation was up to 90 percent of the pore space (up to 22 percent by weight of the tar sand) and the API gravity ranged from 2° to 14° API and viscosity from 1 to 2 million centipoise.

Finally, the fact that most of the tar sand resource in the United States is too deep for economic development is reflected in the ratio of the numbers of in situ projects to mining/extraction projects (almost 4:1).

Obviously, there are many features to consider when development of tar sand resources is planned. It is more important to recognize that what are important features for one resource might be less important in the development of a second resource. Recognition of this facet of tar sand development is a major benefit that will aid in the production of liquid fuels in an economic and effective manner.

Construction is underway for the next phase of oil sands growth as part of the Voyageur Project. This is expected to give production capacity of 350,000 bbl/day by 2008 and includes a new pair of coke drums—the largest ever constructed (using ConocoPhillips' ThruPlus delayed coking technology)—and a sulfur recovery plant.

Suncor's Firebag in situ operations are located 40 km northwest of the original oil sands plant and will form a key part of the increased bitumen supply to the upgraders as the project progresses.

Key elements of the Voyageur Project are the construction of a third oil sands upgrader in Fort McMurray; expansion of the bitumen supply; and the continuation of third-party bitumen supplies.

Plans call for the new upgrader to be constructed approximately half a kilometer southwest of existing Suncor upgrader facilities.

The new facility will include cokers, hydrotreaters, utilities support, and a 50 km hot bitumen pipeline to connect the upgrader with the Suncor in situ operations. Preparations have begun on the site but construction is not expected to begin until 2007 while some final decisions are made.

The upgrader has been designed to produce light crude oil and production from the new facility is expected to be brought online in phases starting in 2010 with full capacity of approximately 550,000 bbl/day targeted in 2012. It is estimated that constructing the upgrader will cost C$5.9 billion.

Suncor has also identified the need for additional pipeline capacity from Fort McMurray to Edmonton and the company is pursuing various options to accommodate the additional volumes. The construction of the upgrader would employ approximately 4000 workers. Approximately 300 new permanent jobs at Suncor's oil sands facility are expected to be created when the upgrader is in full operation.

Suncor also intends to build and operate a petroleum coke gasifier that would reduce the company's reliance on natural gas. The gasifier is planned to process about 20 percent of the proposed upgrader's petroleum coke (a byproduct of the upgrading process) into synthetic gas. The synthetic gas would then be used to supply hydrogen and fuel. The gasifier will add an estimated $600 million to the total cost of this project.

Suncor's Steepbank and millennium mines currently produce 263,000 bbl/day and its Firebag in situ project produces 35,000 bbl/day. It intends to spend C$3.2 billion to expand its mining operations to 400,000 bbl/day and in situ production to 140,000 bbl/day by 2008. An estimate of the recoverable oil resources on Suncor leases is 9 billion barrels of crude oil.

At current rates of production, the Athabasca oil sands reserves as a whole could last over 400 years. New in situ methods have been developed to extract bitumen from deep deposits by injecting steam to heat the sands and reduce the bitumen viscosity so that it can be pumped out like conventional crude oil.

Also the use of light hydrocarbon injection to decrease the viscosity of the bitumen is being investigated, which would use much less gas since not so much steam would be required for recovery. The standard extraction process usually requires huge amounts of natural gas. Suncor is offsetting the gas demands of expanded extraction operations by building a new petroleum coke gasifier.

Suncor's in situ plans will use recycled water in a closed system for steam generation. No additional surface or ground water will be required and no tailings ponds will be created. In situ is expected to disturb only about 10 percent of the surface land in the development area.

4.9 REFERENCES

ASTM D97. Standard Test Method for Pour Point of Petroleum Products. Annual Book of Standards, American Society for Testing and Materials, West Conshohocken, Pennsylvania. 2007.

Speight, J. G.: *The Desulfurization of Heavy Oils and Residua,* 2d ed. Marcel Dekker Inc., New York, 1999.

Speight, J. G.: "Natural Bitumen (Tar Sands) and Heavy Oil," in *Coal, Oil Shale, Natural Bitumen, Heavy Oil and Peat, from Encyclopedia of Life Support Systems (EOLSS),* Developed under the Auspices of the UNESCO, EOLSS Publishers, Oxford, UK, 2005, http://www.eolss.net.

Speight, J. G.: *Chemistry and Technology of Petroleum,* 4th ed. CRC-Taylor and Francis Group, Boca Raton, Fla., 2007.

US Congress: Public Law FEA-76-4, United States Congress, Washington, D.C., 1976.

Wallace, D. (ed.): *A Review of Analytical Methods for Bitumens and Heavy Oils,* Alberta Oil Sands Technology and Research Authority, Edmonton, Alberta, Canada, 1988, pp. 93–6.

Wallace, D., J. Starr, K. P. "Thomas, and S. M. Dorrence: "Characterization of Oil Sand Resources," Report on the Activities Concerning Annex 1 of the US-Canada Cooperative Agreement on Tar Sand and Heavy Oil. Alberta Oil Sands Technology and Research Authority, Edmonton, Alberta, Canada, 1988a, app. C, pp. 3–4.

Wallace, D., J. Starr, K. P. Thomas, and S. M. Dorrence: "Characterization of Oil Sand Resources," Report on the Activities Concerning Annex 1 of the US-Canada Cooperative Agreement on Tar Sand and Heavy Oil. Alberta Oil Sands Technology and Research Authority, Edmonton, Alberta, Canada, 1988b, p. 12.

CHAPTER 5
FUELS FROM COAL

Coal is a fossil fuel formed in swamp ecosystems where plant remains were saved by water and mud from oxidization and biodegradation. Coal is a combustible organic sedimentary rock (composed primarily of carbon, hydrogen, and oxygen) formed from ancient vegetation and consolidated between other rock strata to form coal seams. The harder forms, such as anthracite coal, can be regarded as organic metamorphic rocks because of a higher degree of maturation.

Coal is composed primarily of carbon along with assorted other elements, including sulfur. It is the largest single source of fuel for the generation of electricity worldwide, as well as the largest source of carbon dioxide emissions, which have been implicated as the primary cause of global warming. Coal is extracted from the ground by coal mining, either underground mining or open-pit mining (surface mining).

Coal is the one fossil energy source that can play a substantial role as a transitional energy source as one moves from the petroleum- and natural-gas-based economic system to the future economic system based on nondepletable or renewable energy systems. Coal has been used as an energy source for thousands of years. It has many important uses, but most significantly in electricity generation, steel and cement manufacture, and industrial process heating. In the developing world, the use of coal in the household, for heating and cooking, is important. For coal to remain competitive with other sources of energy in the industrialized countries of the world, continuing technologic improvements in all aspects of coal extraction have been necessary. Coal is often the only alternative when low-cost, cleaner energy sources are inadequate to meet growing energy demand.

According to BP Statistical Review of World Energy 2006 figures (BP, 2006), global consumption of coal grew from 2282 Mtoe in 1995 to 2930 Mtoe in 2005, an annual growth rate of 2.6 percent. Coal accounts for about 28 percent (hard coal 25 percent, soft brown coal 3 percent) of global primary energy consumption, surpassed only by crude oil (BGR, 2007). Developing countries use about 55 percent of the world's coal today; this share is expected to grow to 65 percent over the next 15 years (Balat and Ayar, 2004). In year 2050, coal will account for more than 34 percent of the world's primary energy demand.

5.1 OCCURRENCE AND RESERVES

Coal is found as successive layers, or seams, sandwiched between strata of sandstone and shale. Compared to other fossil fuels, coal reserves are the largest ones and are more evenly distributed worldwide.

With current consumption trends, the reserves-to-production (R/P) ratio of world proven reserves of coal is higher than that of world proven reserves of oil and gas—155 years versus 40 and 65 years, respectively. Total recoverable reserves of coal around the world are estimated at 696 billion metric tons of carbon equivalent (Btce) or 909 billion tons. Geographic distribution of coal reserves reveals that the largest deposits are located in the

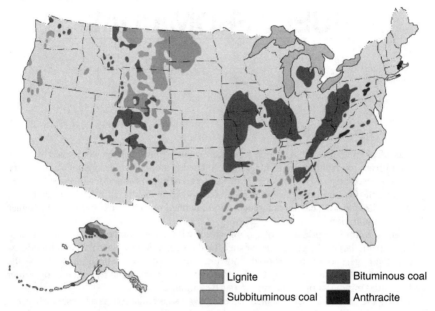

FIGURE 5.1 Coal fields of the United States.

United States (Fig. 5.1) (27.1 percent of the world reserves), FSU (former Soviet Unoin) (25.0 percent), China (12.6 percent), India (10.2 percent), Australia (8.6 percent), and South Africa (5.4 percent) (Fig. 5.2).

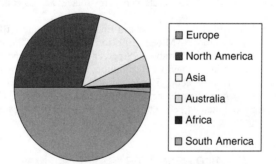

FIGURE 5.2 Global distribution of coal reserves.

Recoverable reserves are those quantities of coal which geologic and engineering information indicates with reasonable certainty can be extracted in the future under existing economic and operating conditions (Balat, 2007).

Global coal recoverable reserves are estimated to be on the order of 909 billion tonnes (equivalent to 1102.3 billion tons, where 1 tonne = 2240 lb and 1 ton = 2000 lb) that occur in the following regions: North America (28.0 percent), South and Central America (2.2 percent), Europe and Eurasia (31.6 percent), Africa (5.5 percent), and Asia Pacific

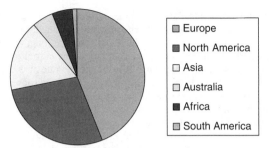

FIGURE 5.3 Global distribution of recoverable coal reserves.

(32.7 percent) (Fig. 5.3) there being a slight difference in the data when compared to global distraction of total coal resources (Fig. 5.2) (BP, 2007). Whichever way the data are considered, the amount of coal in the world is phenomenal.

However, for Europe the coal currently extracted within the EU (European Union) cannot meet the demand in the long-term, which is not even possible at present levels. The only European countries with important hard coal resources for economic extraction are Poland and the Czech Republic, but also those will be depleted before the end of this century at current production. Germany has only resources of subbituminous coal and lignite, which will likely be depleted in about 30 years at current rate of consumption (Spohn and Ellersdorfer, 2005).

Hard coal with a calorific value greater than 16,500 kJ/kg (>4.000 kcal/kg) is traded globally. The price is usually not significantly affected by transport costs. Soft brown coal with a calorific value less than 16,500 kJ/kg is mainly used locally by power plants near the coal deposits. Coal remains the most important fuel, now amounting to about 55 percent of the reserves of all nonrenewable fuels (Fig. 5.4), followed by oil with 26 percent (conventional oil 18.1 percent and nonconventional oil 7.4 percent) and natural gas with almost 15 percent, nuclear fuels account for about 4 percent (BGR, 2007).

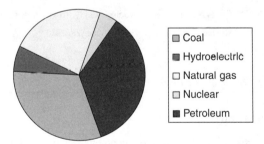

FIGURE 5.4 Coal as member of the energy-generating group.

The United States has the largest hard coal reserves (26 percent of global reserves), followed by Russia (12 percent), China (11 percent), India (10 percent), and Australia (9 percent). Soft brown coal reserves are 10 percent of global reserves. Australia has the largest soft brown coal reserves (19.2 percent of global reserves), followed by India (16.9 percent), the United States (16.1 percent), China (9.0 percent), Serbia and Montenegro (7.7 percent), Russia (5.0 percent), and Germany (3.2 percent) (BGR, 2007).

Thus, coal remains in adequate supply and at current rates of recovery and consumption, the world global coal reserves have been variously estimated to have an R/P ratio of at least 155 years. However, as with all estimates of resource longevity, coal longevity is subject to the assumed rate of consumption remaining at the current rate of consumption and, moreover, to technologic developments that dictate the rate at which the coal can be mined. And, moreover, coal is a fossil fuel and an *unclean* energy source that will only add to global warming. In fact, the next time electricity is advertised as a clean energy source, consider the means by which the majority of electricity is produced—almost 50 percent of the electricity generated in the United States is from coal (EIA, 2007).

Current projections are that the use of coal as an energy source will diminish by the year 2020 (Fig. 5.5) with natural gas use increasing as petroleum use also declines as a percent of the global energy production. However, the current author is of the opinion that coal use will increase as more liquids and gas (synthesis gas) are produced from coal and as environment technologies evolve and are capable of ensuring that coal is truly a clean and nonpolluting fuel.

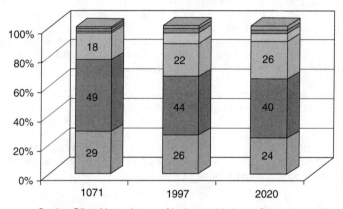

FIGURE 5.5 Estimated use of energy sources in the year 2020.

5.2 FORMATION AND TYPES

5.2.1 Coal Formation

The precursors to coal were plant remains (containing carbon, hydrogen, and oxygen) that were deposited in the Carboniferous period, between 345 and 280 million years ago (Fig. 5.6). As the plant remains became submerged under water, decomposition occurred in which oxygen and hydrogen were lost from the remains to leave a deposit with a high percentage of carbon. With the passage of time, layers of inorganic material such as sand and mud settled from the water and covered the deposits. The pressure of these overlying layers, as well as movements of the earth's crust acted to compress and harden the deposits, thus producing coal from the vegetal matter.

The plant material (vegetal matter) is composed mainly of carbon, hydrogen, oxygen, nitrogen, sulfur, and some inorganic mineral elements. When this material decays under

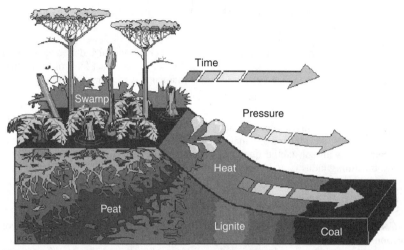

FIGURE 5.6 Formation of coal. (*Reproduced with permission. Copyright 2000. Kentucky Geological Survey, University of Kentucky*).

water, in the absence of oxygen, the carbon content increases. The initial product of this decomposition process is known as peat. The transformation of peat to lignite is the result of pressure exerted by sedimentary materials that accumulate over the peat deposits. Even greater pressures and heat from movements of the Earth's crust (as occurs during mountain building), and occasionally from igneous intrusion, cause the transformation of lignite to bituminous and anthracite coal.

5.2.2 Coal Types

Coal occurs in different forms or *types* (Fig. 5.7). Variations in the nature of the source material and local or regional variations in the coalification processes cause the vegetal matter to evolve differently. Thus, various classification systems exist to define the different types of coal.

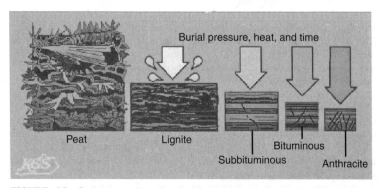

FIGURE 5.7 Coal types. (*Reproduced with permission. Copyright 2000. Kentucky Geological Survey, University of Kentucky*).

Thus, as geologic processes increase their effect over time, the coal precursors are transformed over time into:

1. Lignite—also referred to as brown coal, is the lowest rank of coal and used almost exclusively as fuel for steam-electric power generation. Jet is a compact form of lignite that is sometimes polished and has been used as an ornamental stone since the Iron Age.
2. Subbituminous coal—whose properties range from those of lignite to those of bituminous coal and are used primarily as fuel for steam-electric power generation.
3. Bituminous coal—a dense coal, usually black, sometimes dark brown, often with well-defined bands of bright and dull material, used primarily as fuel in steam-electric power generation, with substantial quantities also used for heat and power applications in manufacturing industry and to make coke.
4. Anthracite—the highest rank; a harder, glossy, black coal used primarily for residential- and commercial-space heating.

Coal classification systems are based on the degree to which coals have undergone coalification. Such varying degrees of coalification are generally called coal *ranks* (or *classes*). The determination of coal rank has a number of practical applications such as the definition of the coal properties. The properties include the amount of heat produced during combustion, the amount of gaseous products released upon heating, and the suitability of the coals for producing coke.

The *rank* of a coal indicates the progressive changes in carbon, volatile matter, and probably ash and sulfur that take place as coalification progresses from the lower rank lignite through the higher ranks of subbituminous, high-volatile bituminous, low-volatile bituminous, and anthracite. The rank of a coal should not be confused with its grade. A high rank (e.g., anthracite) represents coal from a deposit that has undergone the greatest degree of metamorphosis and contains very little mineral matter, ash, and moisture. On the other hand, any rank of coal, when cleaned of impurities through coal preparation will be of a higher grade.

The most commonly employed systems of classification are those based on analyses that can be performed relatively easily in the laboratory, as described by the American Society for Testing and Materials (ASTM) on the basis of fixed carbon content, volatile matter content, and calorific value. In addition to the major ranks (lignite, subbituminous, bituminous, and anthracite), each rank may be subdivided into coal groups, such as *high-volatile A bituminous coal*. Other designations, such as coking coal and steam coal, have been applied to coals, but they tend to differ from country to country.

The term *coal type* is also employed to distinguish between banded coals and nonbanded coals. Banded coals contain varying amounts of vitrinite and opaque material. They include bright coal, which contains more than 80 percent vitrinite, and splint coal, which contains more than 30 percent opaque matter. The nonbanded varieties include boghead coal, which has a high percentage of algal remains, and cannel coal with a high percentage of spores. The usage of all the above terms is quite subjective.

By analogy to the term mineral, which is applied to inorganic material, the term maceral is used to describe organic constituents present in coals. Maceral nomenclature has been applied differently by some European coal petrologists who studied polished blocks of coal using reflected-light microscopy (their terminology is based on morphology, botanical affinity, and mode of occurrence) and by some North American petrologists who studied very thin slices (thin sections) of coal using transmitted-light microscopy.

Three major maceral groups are generally recognized: *vitrinite*, *exinite*, and *inertinite*. The vitrinite group is the most abundant and is derived primarily from cell walls and woody tissues. Several varieties are recognized, for example, telinite (the brighter parts of vitrinite that make up cell walls) and collinite (clear vitrinite that occupies the spaces between cell walls).

Coal is also classified on the basis of its macroscopic appearance (generally referred to as coal rock type, lithotype, or *kohlentype*) and there are four main types: (a) vitamin (*Glanzkohle* or *charbon brillant*), which is dominated by vitrinite group macerals and appears glassy to the unaided eye, (b) clarain (*Glanzstreifenkohle* or *charbon semi-brillant*), which is composed of both vitrinite and exinite and has an appearance between that of vitrain and durain, (c) durain (*Mattkohle* or *charbon mat*), which is generally composed of fine-grained inertinites and exinites and has a dull, mat-like luster, and (d) fusion (*Faserkohle* or *charbon fibreux*), which is composed mainly of fusinite and resembles wood charcoal (because it soils the hands just as charcoal would).

Coal analysis may be presented in the form of *proximate* and *ultimate* analyses, whose analytic conditions are prescribed by organizations such as the ASTM. A typical proximate analysis includes the moisture content, ash yield (that can be converted to mineral matter content), volatile matter content, and fixed carbon content.

It is important to know the moisture and ash contents of a coal because they do not contribute to the heating value of a coal. In most cases ash becomes an undesirable residue and a source of pollution, but for some purpose (e.g., use as a chemical feedstock or for liquefaction) the presence of mineral matter may be desirable. Most of the heat value of a coal comes from its volatile matter, excluding moisture, and fixed carbon content. For most coals, it is necessary to measure the actual amount of heat released upon combustion, which is expressed in British thermal units (Btu) per pound.

Fixed carbon is the material, other than ash, that does not vaporize when heated in the absence of air. It is determined by subtracting the weight percent sum of the moisture, ash, and volatile matter—in weight percent from 100 percent.

Ultimate analyses are used to determine the carbon, hydrogen, sulfur, nitrogen, ash, oxygen, and moisture contents of a coal. For specific applications, other chemical analyses may be employed. These may involve, for example, identifying the forms of sulfur present; sulfur may occur in the form of sulfide minerals (pyrite and marcasite, FeS_2), sulfate minerals (gypsum, Na_2SO_4), or organically bound sulfur. In other cases the analyses may involve determining the trace elements present (e.g., mercury, chlorine), which may influence the suitability of a coal for a particular purpose or help to establish methods for reducing environmental pollution.

5.3 MINING AND PREPARATION

Early coal mining (i.e., the extraction of coal from the seam) was small-scale, the coal lying either on the surface, or very close to it. Typical methods for extraction included drift mining and bell pits. In Britain, some of the earliest drift mines (in the Forest of Dean) date from the medieval period. As well as drift mines, small scale shaft mining was used. This took the form of a bell pit, the extraction working outward from a central shaft, or a technique called room and pillar in which *rooms* of coal were extracted with pillars left to support the roofs. Both of these techniques however left considerable amount of usable coal behind.

Deep shaft mining started to develop in England in the late eighteenth century, although rapid expansion occurred throughout the nineteenth and early twentieth century. The counties of Durham and Northumberland were the leading coal producers and they were the sites of the first deep pits. Before 1800 a great deal of coal was left in places as support pillars and, as a result in the deep pits (300–1000 ft deep) of these two northern counties only about 40 percent of the coal could be extracted. The use of wood props to support the roof was an innovation first introduced around 1800. The critical factor was circulation of air and control of explosive gases.

138   CHAPTER FIVE

In the current context, coal mining depends on the following criteria: (a) seam thickness, (b) the overburden thickness, (c) the ease of removal of the overburden (surface mining), (d) the ease with which a shaft can be sunk to reach the coal seam (underground mining), (e) the amount of coal extracted relative to the amount that cannot be removed, and (f) the market demand for the coal.

There are two predominant types of mining methods that are employed for coal recovery. The first group consists of surface mining methods, in which the strata (overburden) overlying the coal seam are first removed after which the coal is extracted from the exposed seam (Fig. 5.8). Underground mining currently accounts for recovery of approximately 60 percent of the world recovery of coal.

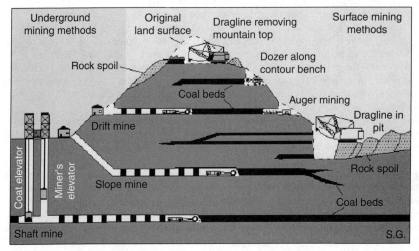

FIGURE 5.8 Coal mining. (*Reproduced with permission. Copyright 2000. Kentucky Geological Survey, University of Kentucky*).

5.3.1 Surface Mining

Surface mining is the application of coal-removal methods to reserves that are too shallow to be developed by other mining methods.

The characteristic that distinguishes *open-pit mining* is the thickness of the coal seam insofar as it is virtually impossible to backfill the immediate mined out area with the original overburden when extremely thick seams of coal are involved. Thus, the coal is removed either by taking the entire seam down to the seam basement (i.e., floor of the mine) or by benching (the staged mining of the coal seam).

Frequent use is made of a *drift mine* in which a horizontal seam of coal outcrops to the surface in the side of a hill or mountain, and the opening into the mine can be made directly into the coal seam. This type of mine is generally the easiest and most economic to open because excavation through rock is not necessary.

Another surface mine is a *slope mine* in which an inclined opening is used to trap the coal seam (or seams). A slope mine may follow the coal seam if the seam is inclined and outcrops, or the slope may be driven through rock strata overlying the coal to reach a seam that is below drainage. Coal transportation from a slope mine can be by conveyor or by track haulage (using a trolley locomotive if the grade is not severe) or by pulling mine cars up the slope using an electric hoist and steel rope if the grade is steep. The most common practice is to use a belt conveyor where grades do not exceed 18°.

On the other hand, *contour mining* prevails in mountainous and hilly terrain taking its name from the method in which the equipment follows the contours of the earth.

Auger mining is frequently employed in open-pit mines where the thickness of the overburden at the high-wall section of the mine is too great for further economic mining. This, however, should not detract from the overall concept and utility of auger mining as it is also applicable to underground operations. As the coal is discharged from the auger spiral, it is collected for transportation to the coal preparation plant or to the market. Additional auger lengths are added as the cutting head of the auger penetrates further under the high wall into the coal. Penetration continues until the cutting head drifts into the top or bottom, as determined by the cuttings returned, into a previous hole, or until the maximum torque or the auger is reached.

5.3.2 Underground Mining

The second method for the recovery of coal is underground (or deep) mining. This is a method in which the coal is extracted from a seam by means of a *shaft mine* enters earth by a vertical opening from the surface and descends to the coal seam and there overlying strata are not removed. In the mine, the coal is extracted from the seam by conventional mining, or by *continuous mining*, or by *longwall mining*, or by *shortwall mining*, or by *room and pillar mining*.

Conventional mining (also called *cyclic* mining) involves a sequence of operations in the order (a) supporting the roof, (b) cutting, (c) drilling, (d) blasting, (e) coal removal, and (f) loading. After the roof above the seam has been made safe by timbering or by roof bolting, one or more slots (a few inches wide and extending for several feet into the coal) are cut along the length of the coal face by a large, mobile cutting machine. The cut, or slot, provides a free face and facilitates the breaking up of the coal, which is usually blasted from the seam by explosives. These explosives (*permissible explosives*) produce an almost flame-free explosion and markedly reduce the amount of noxious fumes relative to the more conventional explosives. The coal may then be transported by rubber-tired electric vehicles (shuttle cars) or by chain (or belt) conveyor systems.

Continuous mining involved the use of a single machine (*continuous miner*) that breaks the coal mechanically and loads it for transport. Roof support is then installed, ventilation is advanced, and the coal face is ready for the next cycle. The method of secondary transportation is located immediately behind the continuous miner and requires installation of mobile belt conveyors.

The *longwall mining* system involves the use of a mechanical self-advancing roof in which large blocks of coal are completely extracted in a continuous operation. Hydraulic or self-advancing jacks (*chocks*) support the roof at the immediate face as the coal is removed. As the face advances, the strata are allowed to collapse behind the support units. Coal recovery is near that attainable with the conventional or continuous systems as well as efficient mining under extremely deep cover or overburden or when the roof is weak.

The *shortwall mining* system is a combination of the continuous mining and longwall mining concepts and offers good recovery of the in-place coal with a marked decrease in the costs for roof support.

Room and pillar mining is a means of developing a coal face and, at the same time, retaining supports for the roof. Thus, by means of this technique, rooms are developed from large tunnels driven into the solid coal with the intervening pillars of coal supporting the roof. The percentage of coal recovered from a seam depends on the number and size of protective pillars of coal thought necessary to support the roof safely and of the percentage of pillar recovery.

5.3.3 Mine Safety and Environment Effects

Mining operations are hazardous and each year a number of coal miners lose their lives or are seriously injured through the occurrence of roof falls, rock bursts, fires, and explosions.

The latter results when flammable gases (such as methane) trapped in the coal are released during mining operations and accidentally are ignited.

Provision of adequate ventilation is, amongst other aspects an essential safety feature of underground coal mining. In some mines, the average weight of air passing daily through the coal mines may be many times the total daily weight of coal produced. Not all of this air is required to enable miners to work in comfort. Most of it is required to dilute the harmful gases, frequently termed *damps* (German *dampf*, vapor), produced during mining operations.

The gas, which occurs naturally in the coal seams, is methane (CH_4, *firedamp*) that is a highly flammable gas and forms explosive mixtures with air (5–14 volume percent methane). The explosion can then cause the combustion of the ensuing coal dust thereby increasing the extent of the hazard. In order to render the gas harmless, it is necessary to circulate large volumes of air to maintain the proportion of methane below the critical levels. Long boreholes may be drilled in the strata ahead of the working face and the methane drawn out of the workings and piped to the surface (*methane drainage*).

Carbon monoxide (CO, *whitedamp*) is a particularly harmful gas; as little as 1 percent in the air inhaled can cause death. It is often found after explosions and occurs in the gases evolved by explosives.

Carbon dioxide (CO_2, *blackdamp*, *chokedamp*, or *stythe*) is found chiefly in old workings or badly ventilated headings.

Hydrogen sulfide (H_2S, *stinkdamp*) is one of the first gases to be produced when coal is heated out of contact with air. It occasionally occurs in small quantities along with the methane given off by outbursts and is sometimes present in the fumes resulting from blasting.

Afterdamp is the term applied to the mixture of gases found in a mine after an explosion or fire. The actual composition varies with the nature and amount of the materials consumed by the fire or with the extent to which firedamp or coal was involved in the explosion.

The continued inhalation of certain dusts is detrimental to health and may lead to reticulation of the lungs and eventually to fatal disease *pneumoconiosis* or *anthracosis* (*black lung disease*). Coal and silica dusts are particularly harmful and the methods that have been adopted to combat the dust hazard include the infusion of water under pressure into the coal before it is broken down; the spraying of water at all points where dust is likely to be formed; the installation of dust extraction units at strategic points; and the wearing of masks by miners operating drilling, cutting, and loading machinery.

The environmental aspects of coal mining are varied and range from aquifer disturbance in the subsurface to floral and faunal disturbance on the surface. In addition, the transportation of spoil or tipple to the surface from an underground mine where it is then deposited in piles or rows offers a new environment hazard. Toxic minerals and substances exposed during removal of the overburden include acidic materials, highly alkaline materials, and dilute concentrations of heavy metals. These materials can have an adverse effect on the indigenous wildlife by creating a hostile environment (often through poisoning the waterways and, in some cases, destruction of species. Thus, mine design should include plans to accommodate potentially harmful substances, which are generated by weathering of spoil piles.

Surface areas exposed during mining, as well as coal and rock waste (which were often dumped indiscriminately), weathered rapidly, producing abundant sediment and soluble chemical products such as sulfuric acid and iron sulfates. Nearby streams became clogged with sediment, iron oxides stained rocks, and *acid mine drainage* caused marked reductions in the numbers of plants and animals living in the vicinity. Potentially toxic elements, leached from the exposed coal and adjacent rocks, were released into the environment. Since the 1970s, however, stricter environment laws have significantly reduced the environment damage caused by coal mining.

Once the coal has been extracted it needs to be moved from the mine to the power plant or other place of use.

Over short distances coal is generally transported by conveyor or truck, whereas trains, barges, ships, or pipelines are used for long distances. Preventative measures are taken at every stage during transport and storage to reduce potential environment impacts. Dust can be controlled by using water sprays, compacting the coal, and enclosing the stockpiles. Sealed systems, either pneumatic or covered conveyors, can be used to move the coal from the stockpiles to the combustion plant. Run-off of contaminated water is limited by appropriate design of coal storage facilities. All water is carefully treated before reuse or disposal.

5.3.4 Coal Preparation

As-mined coal (*run-of-mine coal*) contains a mixture of different size fractions, sometimes together with unwanted impurities such as rock and dirt. Thus, another sequence of events is necessary to make the coal of a consistent quality and salable. Such events are called *coal cleaning*. Effective preparation of coal prior to combustion improves the homogeneity of coal supplied, reduces transport costs, improves the utilization efficiency, produces less ash for disposal at the power plant, and may reduce the emissions of sulfur oxides.

Coal cleaning (*coal preparation, coal beneficiation*) is the stage in coal production when the run-of-mine coal is processed into a range of clean, graded, and uniform coal products suitable for the commercial market. In some cases, the run-of-mine coal is of such quality that it meets the user specification without the need for beneficiation, in which case the coal would merely be crushed and screened to deliver the specified product.

A number of physical separation technologies are used in the washing and beneficiation of coals. After the raw run-of-mine coal is crushed, it is separated into various size fractions for optimum treatment. Larger material (10–150 mm lumps) is usually treated using *dense medium separation*—the coal is separated from other impurities by being floated across a tank containing a liquid of suitable specific gravity, usually a suspension of finely ground magnetite. The coal, being lighter, floats and is separated off, while heavier rocks and other impurities sink and are removed as waste. Any magnetite mixed with the coal is separated using water sprays, and is then recovered, using magnetic drums, and recycled.

The smaller size fractions are treated in a variety of ways—usually based on gravity differentials. In the *froth flotation* method, coal particles are removed in a froth produced by blowing air into a water bath containing chemical reagents. The bubbles attract the coal but not the waste and are skimmed off to recover the coal fines. After treatment, the various size fractions are screened and dewatered or dried, and then recombined before going through final sampling and quality-control procedures.

Blending also enables selective purchasing of different grades of coal. More expensive, higher quality supplies can be carefully mixed with lower quality coals to produce an average blend suited to the plant needs, at optimum cost.

5.4 USES

The use of coal in various parts of the world dates to the Bronze Age, 2000 to 1000 B.C.

The Chinese began to use coal for heating and smelting in the Warring States Period (475–221 B.C.). They are credited with organizing production and consumption to the extent that by the year 1000 A.D. this activity could be called an industry. China remained the world's largest producer and consumer of coal until the eighteenth century.

Outcrop coal was used in Britain during the Bronze Age (2000–3000 B.C.), where it has been detected as forming part of the composition of funeral pyres (Britannica, 2004). It was also commonly used in the early period of the Roman occupation. Evidence of trade in coal (dated to about 200 A.D.) has been found at the inland port of Heronbridge, near Chester, and in the Fenlands of East Anglia, where coal from the Midlands was transported for use in drying grain (Salway, 2001). Coal cinders have been found in the hearths of villas and military forts, particularly in Northumberland, dated to around 400 A.D. In the west of England contemporary writers described the wonder of a permanent brazier of coal on the altar of Minerva at *Aquae Sulis* (modern day Bath) although in fact easily-accessible surface coal from what is now the Somerset coalfield was in common use in quite lowly dwellings locally (Forbes, 1966).

However, there is no evidence that coal was of significant importance in Britain before 1000 A.D. *Mineral coal* came to be referred to as *sea coal* because it came to many places in eastern England, including London, by sea or because it was found on beaches (especially in northeast England) having fallen from exposed coal seams above or washed out of underwater coal seam outcrops. By the thirteenth century, underground mining from shafts or adits was developed (Britannica, 2004). It was, however, the development of the industrial revolution that led to the large-scale use of coal, as the steam engine took over from the water wheel.

The earliest use of coal in the Americas was by the Aztecs who used coal not only for heat but as ornaments as well. Coal deposits were discovered by colonists in eastern North America in the eighteenth century.

In the modern world, coal is primarily used as a solid fuel to produce electricity (approximately 40 percent of the world electricity production uses coal) and heat through combustion (Fig. 5.9).

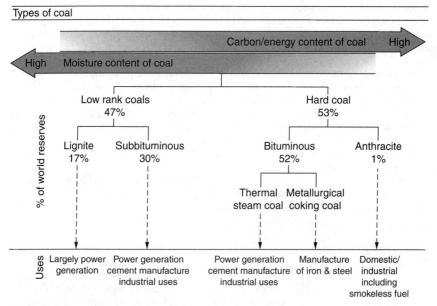

FIGURE 5.9 Uses of coal.

However, coal through the gasification process and the production of synthesis gas opens the way to a very wide range of products that include liquid fuels (Fig. 5.10).

When coal is used for electricity generation, it is usually pulverized and then burned in a furnace with a boiler. The furnace heat converts boiler water to steam, which is then used to

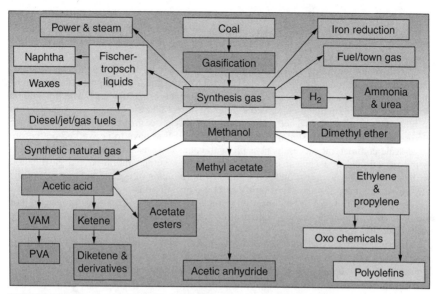

FIGURE 5.10 Illustration of the potential for coal through the gasification process. (*Source: Lynn Schloesser, L.: "Gasification Incentives," Workshop on Gasification Technologies, Ramkota, Bismarck, North Dakota, June 28–29, 2006.*)

spin turbines which turn generators and create electricity (Fig. 5.11). The thermodynamic efficiency of this process has been improved over time. Steam turbines have topped out with some of the most advanced reaching, about 35 percent thermodynamic efficiency for the entire process, which means 65 percent of the coal energy is rejected as waste heat into the surrounding environment. Old coal power plants are often less efficient and reject

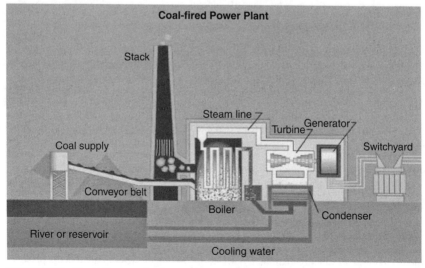

FIGURE 5.11 Electricity generation from coal.

higher levels of waste heat. The emergence of the supercritical turbine concept envisions running a boiler at extremely high temperature and pressure with projected efficiencies of 46 percent, with further theorized increases in temperature and pressure perhaps resulting in even higher efficiencies.

A more energy-efficient way of using coal for electricity production would be via solid-oxide fuel cells or molten-carbonate fuel cells (or any oxygen ion transport based fuel cells that do not discriminate between fuels, as long as they consume oxygen), which would be able to get 60 to 85 percent combined efficiency (direct electricity plus waste heat steam turbine). Currently these fuel cell technologies can only process gaseous fuels, and they are also sensitive to sulfur poisoning, issues which would first have to be worked out before large-scale commercial success is possible with coal. As far as gaseous fuels go, one idea is pulverized coal in a gas carrier, such as nitrogen. Another option is coal gasification with water, which may lower fuel cell voltage by introducing oxygen to the fuel side of the electrolyte, but may also greatly simplify carbon sequestration.

The potential for coal to be converted to fuels is dependent upon these properties, not the least of which is the carbon content (i.e., the chemical composition) and the energy value (calorific value).

Chemically, coal is a hydrogen-deficient hydrocarbon with an atomic hydrogen-to-carbon ratio near 0.8, as compared to petroleum hydrocarbons, which have an atomic hydrogen-to-carbon ratio approximately equal to 2, and methane that has an atomic carbon-to-hydrogen ratio equal to 4. For this reason, any process used to convert coal to alternative fuels must add hydrogen.

The chemical composition of the coal is defined in terms of its proximate and ultimate (elemental) analyses (Speight, 1994). The parameters of proximate analysis are moisture, volatile matter, ash, and fixed carbon. Elemental or ultimate analysis encompasses the quantitative determination of carbon, hydrogen, nitrogen, sulfur, and oxygen within the coal. Additionally, specific physical and mechanical properties of coal and particular carbonization properties are also determined.

The calorific value Q of coal is the heat liberated by its complete combustion with oxygen. Q is a complex function of the elemental composition of the coal. Q can be determined experimentally using calorimeters. Dulong suggests the following approximate formula for Q when the oxygen content is less than 10 percent:

$$Q = 337C + 1442(H - O/8) + 93S$$

C is the mass percent of carbon, H is the mass percent of hydrogen, O is the mass percent of oxygen, and S is the mass percent of sulfur in the coal. With these constants, Q is given in kilojoules per kilogram (1 kJ/kg = 2.326 Btu/lb).

The production of fuels from coal in relation to fuels from other energy technologies is dependent upon the cost of fuels from other sources and, most important, the degree of self sufficiency required by various level of government. The nature of coal is a major factor (assuming an ample supply of coal reserves) and the need for desulfurization of the products as well as the various steps leading from the mining of coal to its end use. Nevertheless, the production of fuels from coal is an old concept having been employed since it was first discovered that the *strange black rock* would burn when ignited and produce heat.

Coal can be liquefied by either direct or indirect processes (i.e., by using the gaseous products obtained by breaking down the chemical structure of coal) to produce liquid products. Four general methods are used for liquefaction: (a) pyrolysis and hydrocarbonization (coal is heated in the absence of air or in a stream of hydrogen), (b) solvent extraction (coal hydrocarbons are selectively dissolved and hydrogen is added to produce the desired liquids), (c) catalytic liquefaction (hydrogenation takes place in the presence of a catalyst), and (d) indirect liquefaction (carbon monoxide and hydrogen are combined in the presence of a catalyst).

The carbon monoxide and hydrogen are produced by the gasification of coal in which a mixture of gases is produced. In addition to carbon monoxide and hydrogen, methane and other hydrocarbons are also produced depending on the conditions involved. Gasification may be accomplished either in situ or in processing plants. In situ gasification is accomplished by controlled, incomplete burning of a coal bed underground while adding air and steam. The gases are withdrawn and may be burned to produce heat, generate electricity, or are used as synthesis gas in indirect liquefaction or the production of chemicals.

Producing diesel and other fuels from coal can be done through converting coal to syngas, a combination of carbon monoxide, hydrogen, carbon dioxide, and methane. The syngas is reacted through the Fischer-Tropsch synthesis to produce hydrocarbons that can be refined into liquid fuels. Research into the process of increasing the quantity of high-quality fuels from coal while reducing the costs could help ease the dependence on ever-increasing cost but depleting stock of petroleum.

Furthermore, by improving the catalysts used in directly converting coal into liquid hydrocarbons, without the generation of the intermediate syngas, less power could be required to produce a product suitable for upgrading in existing petroleum refineries. Such an approach could reduce energy requirements and improve yields of desired products.

While coal is an abundant natural resource, its combustion or gasification produces both toxic pollutants and greenhouse gases. By developing adsorbents to capture the pollutants (mercury, sulfur, arsenic, and other harmful gases), our researchers are striving not only to reduce the quantity of emitted gases but also to maximize the thermal efficiency of the cleanup.

5.5 GASEOUS FUELS

The gasification of coal or a derivative (i.e., char produced from coal) is the conversion of coal (by any one of a variety of processes) to produce gaseous products that are combustible. With the rapid increase in the use of coal from the fifteenth century onward (Nef, 1957; Taylor and Singer, 1957), it is not surprising that the concept of using coal to produce a flammable gas, especially the use of the water and hot coal (van Heek and Muhlen, 1991), became commonplace (Elton, 1958). In fact, the production of gas from coal has been a vastly expanding area of coal technology, leading to numerous research and development programs. As a result, the characteristics of rank, mineral matter, particle size, and reaction conditions are all recognized as having a bearing on the outcome of the process; not only in terms of gas yields but also on gas properties (Massey, 1974; van Heek and Muhlen, 1991). The products from the gasification of coal may be of low, medium, or high heat-content (high-Btu) as dictated by the process as well as by the ultimate use for the gas (Fryer and Speight, 1976; Mahajan and Walker, 1978; Anderson and Tillman, 1979; Cavagnaro, 1980; Bodle and Huebler, 1981; Argonne, 1990; Baker and Rodriguez, 1990; Probstein and Hicks, 1990; Lahaye and Ehrburger, 1991; Matsukata et al., 1992).

Coal gasification offers one of the most clean and versatile ways to convert the energy contained in coal into electricity, hydrogen, and other sources of power. Turning coal into synthetic gas isn't a new concept; in fact the basic technology dates back to World War II.

Coal gasification plants are cleaner than standard pulverized coal combustion facilities, producing fewer sulfur and nitrogen byproducts, which contribute to smog and acid rain. For this reason, gasification appeals as a way to utilize relatively inexpensive and expansive coal reserves, while reducing the environment impact.

The mounting interest in coal gasification technology reflects a convergence of two changes in the electricity generation marketplace: (a) the maturity of gasification technology, and (b) the extremely low emissions from *integrated gasification combined cycle* (IGCC) plants, especially air emissions, and the potential for lower-cost control of greenhouse gases than other coal-based systems. Fluctuations in the costs associated with natural-gas-based power, which is viewed as a major competitor to coal based power, can also play a role.

An IGCC plant includes three main processes (Fig. 5.12). The gasifier turns coal into fuel-gas, which has about half the energy of natural gas; standard cleaning processes remove any sulfur in the gas. The fuel-gas then burns in the combustion chamber of a gas turbine, which turns an alternator to generate electricity. In the third process, heat recovered from the gasifier and from the turbine exhaust converts water into steam which drives a steam turbine, which turns another alternator to generate yet more electricity.

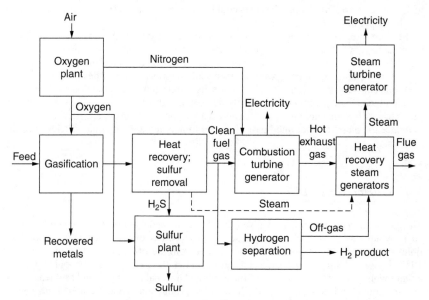

FIGURE 5.12 IGCC schematic.

The *combined cycle* links the gas turbine cycle with its high inlet temperature to the steam turbine cycle with its low outlet temperature; the wider the temperature range, the higher is the overall efficiency of converting fuel into electricity. An IGCC plant may achieve efficiency well above 40 percent, compared with around 36 percent from a conventional power station.

IGCC design is suitable for staged construction and is inherently modular: electrical utilities can therefore increase the size of a plant gradually, in line with demand. They can begin by installing a gas turbine burning oil or natural gas. As the load increases, the utility can add a boiler to recover heat from the turbine exhaust to raise steam, and it can add a steam turbine to generate more electricity. When the premium fuel becomes too expensive, the supplier can add a coal gasifier.

5.5.1 Gasifiers

Four types of gasifier are currently available for commercial use: countercurrent fixed bed, cocurrent fixed bed, fluid bed, and entrained flow. In all cases, the oxygen supplied is insufficient to produce complete combustion and the presence of water as steam favors the reactions that yield fuel-gas. However, to raise the efficiency of a power plant, commercial gasifiers must recover the heat released during gasification. Most designs do this by cooling the gasifier with water, to produce steam and fuel-gas. The fuel-gas itself may have to be cooled and reheating reduces the overall efficiency of the plant.

The *countercurrent fixed bed (up draft) gasifier* consists of a fixed bed of carbonaceous fuel (e.g., coal or biomass) through which the "gasification agent" (steam, oxygen, and/or air) flows in countercurrent configuration. The ash is either removed dry or as a slag. The slagging gasifiers require a higher ratio of steam and oxygen to carbon in order to reach temperatures higher than the ash fusion temperature. The nature of the gasifier means that the fuel must have high mechanical strength and must be noncaking so that it will form a permeable bed, although recent developments have reduced these restrictions to some extent. The throughput for this type of gasifier is relatively low. Thermal efficiency is high as the gas exit temperatures are relatively low. However, this means that tar and methane production is significant at typical operation temperatures, so product gas must be extensively cleaned before use or recycled to the reactor.

The major advantages of this type of gasifier are its simplicity, high charcoal burn-out, and internal heat exchange leading to low gas exit temperatures and high gasification efficiency. In this way, also fuels with high moisture content (up to 50 percent by weight) can be used.

Major drawbacks are the high amounts of tar and pyrolysis products, because the pyrolysis gas is not led through the oxidation zone. This is of minor importance if the gas is used for direct heat applications, in which the tars are simply burnt. In case the gas is used for engines, gas cleaning is required, resulting in problems of tar-containing condensates.

The *cocurrent fixed bed (down draft) gasifier* is similar to the countercurrent type, but the gasification agent gas flows in cocurrent configuration with the fuel (downward, hence the name *down draft gasifier*). Heat needs to be added to the upper part of the bed, either by combusting small amounts of the fuel or from external heat sources. The produced gas leaves the gasifier at a high temperature, and most of this heat is often transferred to the gasification agent added in the top of the bed, resulting in energy efficiency on level with the countercurrent type. Since all tars must pass through a hot bed of char in this configuration, tar levels are much lower than the countercurrent type.

Drawbacks of the downdraft gasifier are: (a) the high amounts of ash and dust particles in the gas, (b) the inability to operate on a number of unprocessed fuels, often pelletization or briquetting of the biomass is necessary (c) the outlet gas has a high temperature leading to a lower gasification efficiency, (d) the moisture content of the biomass must be less than 25 percent to maintain the high temperature and the mineral content (ash yield) should also be low and nonslagging, and (e) the feed must have uniform particle size.

In the *fluid bed gasifier*, the fuel is fluidized in oxygen (or air) and steam. The ash is removed dry or as heavy agglomerates that defluidize. The temperatures are relatively low in dry ash gasifiers, so the fuel must be highly reactive; low-grade coals are particularly suitable. Fluidized bed reactors feature extremely good mixing with good heat and mass transfer. Gasification is efficient and typically exceeds 90 percent of the feedstock, often falling in 95 to 99 percent range of carbon being converted. Ash is carried with gas and separated from the gas in cyclones.

The agglomerating gasifiers have slightly higher temperatures, and are suitable for higher rank coals. Fuel throughput is higher than for the fixed bed, but not as high as for the entrained flow gasifier. The conversion efficiency is rather low, so recycle or subsequent combustion of solids is necessary to increase conversion. Fluidized bed gasifiers are most useful for fuels that form highly corrosive ash that would damage the walls of slagging gasifiers. Biomass generally contains high levels of ash-forming constituents.

In the *entrained flow gasifier* a dry pulverized solid, an atomized liquid fuel, or a fuel slurry is gasified with oxygen (much less frequent: air) in cocurrent flow. The gasification reactions take place in a dense cloud of very fine particles. Most coals are suitable for this type of gasifier because of the high operating temperatures and because the coal particles are well separated from one another. The high temperatures and pressures also mean that a higher throughput can be achieved; however, thermal efficiency is somewhat lower as the gas must be cooled before it can be cleaned with existing technology. The high temperatures also mean that tar and methane are not present in the product gas; however the oxygen requirement is higher than for the other types of gasifiers. All entrained flow gasifiers remove the major part of the ash as a slag as the operating temperature is well above the ash fusion temperature. A smaller fraction of the ash is produced either as a very fine dry fly ash or as black colored fly ash slurry. Some fuels, in particular certain types of biomasses, can form slag that is corrosive for ceramic inner walls that serve to protect the gasifier's outer wall. However some entrained bed type of gasifiers do not possess a ceramic inner wall but have an inner water- or steam-cooled wall covered with partially solidified slag.

These types of gasifiers do not suffer from corrosive slag. Some fuels have ashes with very high ash fusion temperatures. In this case mostly limestone is mixed with the fuel prior to gasification. Addition of a little limestone will usually suffice for lowering the fusion temperatures. The fuel particles must be much smaller than for other types of gasifiers. This means the fuel must be pulverized, which requires somewhat more energy than for the other types of gasifiers. By far the most energy consumption related to entrained bed gasification is not the milling of the fuel but the production of oxygen used for the gasification.

In high temperature conditions (typically 1300–1400°C) high boiling oils and tars are almost completely destroyed. This type of gasifier was developed for coal and limited experience with biomass is available.

A more recent development is the *open core gasifier* design for gasification of small-sized biomass with high ash content. However, the producer gas is not tar free.

In the open core gasifier the air is sucked over the whole cross section from the top of the bed. This facilitates better oxygen distribution since the oxygen will be consumed over the whole cross section, so that the solid bed temperature will not reach the local extremes (hot spots) observed in the oxidation zone of conventional gasifiers due to poor heat transfer. Moreover, the air nozzles in conventional gasifiers generate caves and create obstacles that may obstruct solid flow especially for solids of low bulk (e.g., rice husk). On the other hand, the entry of air through the top of the bed creates a downward flow of the product gases thereby transporting the tar products to the combustion zone.

Finally, air-blown reactors (of any of the above type) produce reaction heat by partial oxidation inside the reactor and the product gas is diluted with nitrogen. The use of oxygen (or oxygen-enriched air) results in more concentrated gases. Indirect heating of the reactor is achieved by means of hot solids or through heat exchanger walls.

5.5.2 Gaseous Products

The products of coal gasification are varied insofar as the gas composition varies with the system employed and the predetermined follow-up to the production of products (Fig. 5.13). It is emphasized that the gas product must be first freed from any pollutants such as particulate matter and sulfur compounds before further use, particularly when the intended use is a water gas shift or methanation (Cusumano et al., 1978; Probstein and Hicks, 1990).

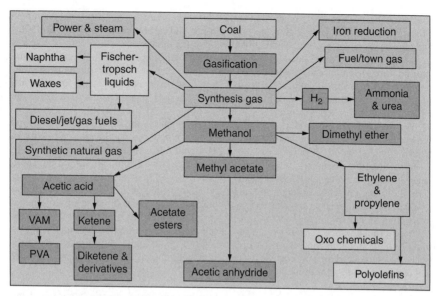

FIGURE 5.13 Potential products from coal gasification.

Low Heat-Content (Low-Btu) Gas. During the production of coal gas by oxidation with air, the oxygen is not separated from the air and, as a result, the gas product invariably has a low heat-content (150–300 Btu/ft^3; 5.6–11.2 MJ/m^3). Low heat-content gas is also the usual product of in situ gasification of coal (Sec. 5.5.5) which is used essentially as a method for obtaining energy from coal without the necessity of mining the coal, especially if the coal cannot be mined or if mining is uneconomic.

Several important chemical reactions, and a host of side reactions, are involved in the manufacture of low heat-content gas under the high temperature conditions employed. Low heat-content gas contains several components, four of which are always major components present at levels of at least several percent; a fifth component, methane, is marginally a major component.

The nitrogen content of low heat-content gas ranges from somewhat less than 33 percent v/v to slightly more than 50 percent v/v and cannot be removed by any reasonable means; the presence of nitrogen at these levels makes the product gas *low heat-content* by definition. The nitrogen also strongly limits the applicability of the gas to chemical synthesis. Two other noncombustible components [water (H_2O) and carbon dioxide (CO_2)] further lower the heating value of the gas; water can be removed by condensation and carbon dioxide by relatively straightforward chemical means.

The two major combustible components are hydrogen and carbon monoxide; the H_2/CO ratio varies from approximately 2:3 to about 3:2. Methane may also make an appreciable contribution to the heat content of the gas. Of the minor components hydrogen sulfide is the most significant and the amount produced is, in fact, proportional to the sulfur content of the feed coal. Any hydrogen sulfide present must be removed by one, or more, of several procedures (Speight, 1993).

Low heat-content gas is of interest to industry as a fuel gas or even, on occasion, as a raw material from which ammonia, methanol, and other compounds may be synthesized.

Medium Heat-Content (Medium-Btu) Gas. Medium heat-content gas has a heating value in the range 300 to 550 Btu/ft^3 (11.2–20.5 MJ/m^3) and the composition is much like that of low heat-content gas, except that there is virtually no nitrogen. The primary combustible gases in medium heat-content gas are hydrogen and carbon monoxide (Kasem, 1979). Medium heat-content gas is considerably more versatile than low heat-content gas; like low heat-content gas, medium heat-content gas may be used directly as a fuel to raise steam, or used through a combined power cycle to drive a gas turbine, with the hot exhaust gases employed to raise steam, but medium heat-content gas is especially amenable to synthesize methane (by methanation), higher hydrocarbons (by Fischer-Tropsch synthesis), methanol, and a variety of synthetic chemicals.

The reactions used to produce medium heat-content gas are the same as those employed for low heat-content gas synthesis, the major difference being the application of a nitrogen barrier (such as the use of pure oxygen) to keep diluent nitrogen out of the system.

In medium heat-content gas, the H_2/CO ratio varies from 2:3 to 3:1 and the increased heating value correlates with higher methane and hydrogen contents as well as with lower carbon dioxide contents. Furthermore, the very nature of the gasification process used to produce the medium heat-content gas has a marked effect upon the ease of subsequent processing. For example, the carbon-dioxide-acceptor product is quite amenable to use for methane production because it has (a) the desired H_2/CO ratio just exceeding 3:1, (b) an initially high methane content, and (c) relatively low water and carbon dioxide contents. Other gases may require appreciable shift reaction and removal of large quantities of water and carbon dioxide prior to methanation.

High Heat-Content (High-Btu) Gas. High heat-content gas is essentially pure methane and often referred to as *synthetic natural gas* or *substitute natural gas* (SNG) (Kasem, 1979; cf Speight, 1990). However, to qualify as substitute natural gas, a product must contain at least 95 percent methane; the energy content of synthetic natural gas is 980 to 1080 Btu/ft^3 (36.5–40.2 MJ/m^3).

The commonly accepted approach to the synthesis of high heat-content gas is the catalytic reaction of hydrogen and carbon monoxide:

$$3H_2 + CO \rightarrow CH_4 + H_2O$$

To avoid catalyst poisoning, the feed gases for this reaction must be quite pure and, therefore, impurities in the product are rare. The large quantities of water produced are removed by condensation and recirculated as very pure water through the gasification system. The hydrogen is usually present in slight excess to ensure that the toxic carbon monoxide is reacted; this small quantity of hydrogen will lower the heat content to a small degree.

The carbon monoxide/hydrogen reaction is somewhat inefficient as a means of producing methane because the reaction liberates large quantities of heat. In addition, the methanation catalyst is troublesome and prone to poisoning by sulfur compounds and the decomposition of metals can destroy the catalyst. Thus, hydrogasification may be employed to minimize the need for methanation:

$$[C]_{coal} + 2H_2 \rightarrow CH_4$$

The product of hydrogasification is far from pure methane and additional methanation is required after hydrogen sulfide and other impurities are removed.

5.5.3 Physicochemical Aspects

Coal varies widely in chemical composition. The most important constituents are carbon (C), hydrogen (H), and oxygen (O), with some sulfur and nitrogen, bound together in complex arrangements. If coal is heated in an inert atmosphere, this intricate molecular structure breaks down. Volatile matter, including hydrogen, water, methane, and higher-boiling tar is produced. This leaves a char consisting of almost pure carbon intermingled with inert, incombustible ash. Among the important attributes of a particular coal are the proportions of volatiles, ash, and sulfur; how chemically reactive it is; and whether it tends to swell or stick together, or cake when burned. Different coals may exhibit very different physical and chemical behavior.

Using very general equations, when coal is heated in the presence of oxygen and/or water, carbon monoxide (CO) and carbon dioxide (CO_2) are also formed as seven reactions compete:

$$[C]_{coal} + O_2 \rightarrow CO_2$$

$$2[C]_{coal} + O_2 \rightarrow 2CO$$

$$[C]_{coal} + CO_2 \rightarrow 2CO$$

$$[C]_{coal} + 2H_2 \rightarrow CH_4$$

$$[C]_{coal} + H_2O \rightarrow CO + H_2$$

$$CO + H_2O \rightarrow CO_2 + H_2$$

$$CO + 3H_2 \rightarrow CH_4 + H_2O$$

The rates of these different reactions depend on the temperature and pressure, the characteristics and physical form of the coal, the amounts of oxygen and steam available, and the configuration of the reacting materials. All the reactions (except the coal-carbon-dioxide reaction) are exothermic and release energy as heat.

However, if the process is to produce fuel gas, the coal-oxygen reaction to produce carbon dioxide must not dominate because all the energy in the coal will be released as heat and there will be none left in the gas. The gases needed are carbon monoxide, hydrogen, and methane. The reaction conditions must, therefore, be carefully controlled by controlling the amount of oxygen in the combustion chamber so that other reactions can occur. Hence, coal gasifiers are designed to favor the reactions that produce energy-rich gases, such as the coal-oxygen reaction to form carbon monoxide reactions and the coal stream reaction to produce synthesis gas (a mixture of carbon monoxide and hydrogen).

Sulfur in the coal will also react with oxygen or hydrogen to release energy and to form either sulfur oxides or hydrogen sulfide. Nitrogen present in the air of the combustion chamber may also form nitrogen oxides.

Reactions. Coal gasification involves the thermal decomposition of coal and the reaction of the coal carbon and other pyrolysis products with oxygen, water, and fuel gases such as methane.

The presence of oxygen, hydrogen, water vapor, carbon oxides, and other compounds in the reaction atmosphere during pyrolysis may either support or inhibit numerous reactions with coal and with the products evolved. The distribution of weight and chemical composition of the products are also influenced by the prevailing conditions (i.e., temperature, heating rate, pressure, residence time, and any other relevant parameters) and, not less importantly, by the nature of the feedstock (Wang and Mark, 1992).

If air is used as the means of combustion, the product gas will have a heat content of 150 to 300 Btu/ft^3 (5.6–11.2 MJ/m^3) (depending on process design characteristics) and will contain undesirable constituents such as carbon dioxide, hydrogen sulfide, and nitrogen. The use of pure oxygen, although expensive, results in a product gas having a heat content of 300 to 400 Btu/ft^3 (11.2–14.9 MJ/m^3) with carbon dioxide and hydrogen sulfide as by-products [both of which can be removed from low or medium heat-content (low- or medium-Btu) gas by any of several available processes].

If a high heat-content (high-Btu) gas [900–1000 Btu/ft^3 (33.5–37.3 MJ/m^3)] is required, efforts must be made to increase the methane content of the gas. The reactions which generate methane are all exothermic and have negative values but the reaction rates are relatively slow and catalysts may, therefore, be necessary to accelerate the reactions to acceptable commercial rates. Indeed, the overall reactivity of coal and char may be subject to catalytic effects. It is also possible that the mineral constituents of coal and char may modify the reactivity by a direct catalytic effect (Cusumano et al., 1978; Davidson, 1983; Baker and Rodriguez, 1990; Mims, 1991; Martinez-Alonso and Tascon, 1991).

Process Concepts. While there has been some discussion of the influence of physical process parameters and the effect of coal type on coal conversion, a note is warranted here regarding the influence of these various parameters on the gasification of coal.

Most notable effects are those due to coal character, and often to the maceral content. In regard to the maceral content, differences have been noted between the different maceral groups with inertinite being the most reactive (Huang et al., 1991). In more general terms of the character of the coal, gasification technologies generally require some initial processing of the coal feedstock with the type and degree of pretreatment a function of the process and/or the type of coal. For example, the Lurgi process will accept lump coal (1 in, 25 mm, to 28 mesh), but it must be noncaking coal with the fines removed. The caking, agglomerating coals tend to form a plastic mass in the bottom of a gasifier and subsequently plug up the system thereby markedly reducing process efficiency. Thus, some attempt to reduce caking tendencies is necessary and can involve preliminary partial oxidation of the coal thereby destroying the caking properties.

Depending on the type of coal being processed and the analysis of the gas product desired, pressure also plays a role in product definition. In fact, some (or all) of the following processing steps will be required: (a) pretreatment of the coal (if caking is a problem); (b) primary gasification of the coal; (c) secondary gasification of the carbonaceous residue from the primary gasifier; (d) removal of carbon dioxide, hydrogen sulfide, and other acid gases; (e) shift conversion for adjustment of the carbon monoxide-to-hydrogen mole ratio to the desired ratio; and (f) catalytic methanation of the carbon monoxide/hydrogen mixture to form methane. If high heat-content (high-Btu) gas is desired, all of these processing steps are required since coal gasifiers do not yield methane in the concentrations required (Mills, 1969; Graff et al., 1976; Cusumano et al., 1978; Mills, 1982).

Pretreatment. Some coals display caking, or agglomerating, characteristics when heated and these coals are usually not amenable to treatment by gasification processes employing fluidized bed or moving-bed reactors; in fact, caked coal is difficult to handle in fixed-bed reactors. The pretreatment involves a mild oxidation treatment which destroys the caking characteristics of coals and usually consists low-temperature heating of the coal in the presence of air or oxygen.

Primary Gasification. Primary gasification involves thermal decomposition of the raw coal via various chemical processes and many schemes involve pressures ranging from atmospheric to 1000 psi (6.9 MPa). Air or oxygen may be admitted to support combustion to provide the necessary heat. The product is usually a low heat-content (low-Btu) gas ranging from a carbon monoxide/hydrogen mixture to mixtures containing varying amounts of carbon monoxide, carbon dioxide, hydrogen, water, methane, hydrogen sulfide, nitrogen, and typical products of thermal decomposition such as tar (themselves being complex mixtures; see Dutcher et al., 1983), hydrocarbon oils, and phenols.

A solid char product may also be produced, and may represent the bulk of the weight of the original coal. This type of coal being processed determines (to a large extent) the amount of char produced and the analysis of the gas product.

Secondary Gasification. Secondary gasification usually involves the gasification of char from the primary gasifier. This is usually done by reacting the hot char with water vapor to produce carbon monoxide and hydrogen:

$$[C]_{char} + H_2O \rightarrow CO + H_2$$

Shift Conversion. The gaseous product from a gasifier generally contains large amounts of carbon monoxide and hydrogen, plus lesser amounts of other gases. Carbon monoxide and hydrogen (if they are present in the mole ratio of 1:3) can be reacted in the presence of a catalyst to produce methane (Cusumano et al., 1978). However, some adjustment to the ideal (1:3) is usually required and, to accomplish this, all or part of the stream is treated according to the waste gas shift (shift conversion) reaction. This involves reacting carbon monoxide with steam to produce a carbon dioxide and hydrogen whereby the desired 1:3 mole ratio of carbon monoxide to hydrogen may be obtained:

$$CO + H_2O \rightarrow CO_2 + H_2$$

Methanation. Several exothermic reactions may occur simultaneously within a methanation unit. A variety of metals have been used as catalysts for the methanation reaction; the most common, and to some extent the most effective methanation catalysts, appear to be nickel and ruthenium, with nickel being the most widely used (Seglin, 1975; Cusumano et al., 1978; Tucci and Thompson, 1979; Watson, 1980). The synthesis gas must be desulfurized before the methanation step since sulfur compounds will rapidly deactivate (poison) the catalysts (Cusumano et al., 1978). A problem may arise when the concentration of carbon monoxide is excessive in the stream to be methanated since large amounts of heat must be removed from the system to prevent high temperatures and deactivation of the catalyst by sintering as well as the deposition of carbon (Cusumano et al., 1978). To eliminate this problem temperatures should be maintained below 400°C (752°F).

Hydrogasification. Not all high heat-content (high-Btu) gasification technologies depend entirely on catalytic methanation and, in fact, a number of gasification processes use hydrogasification, that is, the direct addition of hydrogen to coal under pressure to form methane (Anthony and Howard, 1976):

$$[C]_{coal} + 2H_2 \rightarrow CH_4$$

The hydrogen-rich gas for hydrogasification can be manufactured from steam by using the char that leaves the hydrogasifier. Appreciable quantities of methane are formed directly in the primary gasifier and the heat released by methane formation is at a sufficiently high temperature to be used in the steam-carbon reaction to produce hydrogen so that less oxygen is used to produce heat for the steam-carbon reaction. Hence, less heat is lost in the low-temperature methanation step, thereby leading to higher overall process efficiency.

5.5.4 Gasification Processes

Gasification processes are segregated according to the bed types, which differ in their ability to accept (and use) caking coals and are generally divided into four categories based on reactor (bed) configuration: (a) fixed bed, (b) fluidized bed, (c) entrained bed, and (d) molten salt.

In a fixed-bed process the coal is supported by a grate and combustion gases (steam, air, oxygen, etc.) pass through the supported coal whereupon the hot produced gases exit from the top of the reactor. Heat is supplied internally or from an outside source, but caking coals cannot be used in an unmodified fixed-bed reactor.

The fluidized bed system uses finely sized coal particles and the bed exhibits liquid-like characteristics when a gas flows upward through the bed. Gas flowing through the coal produces turbulent lifting and separation of particles and the result is an expanded bed having greater coal surface area to promote the chemical reaction, but such systems have a limited ability to handle caking coals.

An entrained-bed system uses finely sized coal particles blown into the gas steam prior to entry into the reactor and combustion occurs with the coal particles suspended in the gas phase; the entrained system is suitable for both caking and noncaking coals.

The molten salt system employs a bath of molten salt to convert coal (Cover et al., 1973; Howard-Smith and Werner, 1976; Koh et al., 1978).

Fixed-Bed Processes

The Lurgi Process. The Lurgi process was developed in Germany before World War II and is a process that is adequately suited for large-scale commercial production of synthetic natural (Verma, 1978).

The older Lurgi process uses a dry ash gasifier (Fig. 5.14) which differs significantly from the more recently developed slagging process (Baughman, 1978; Massey, 1979). The dry ash Lurgi gasifier is a pressurized vertical kiln which accepts crushed [¼ × 1¾ in (6 × 44 mm)] noncaking coal and reacts the moving bed of coal with steam and either air

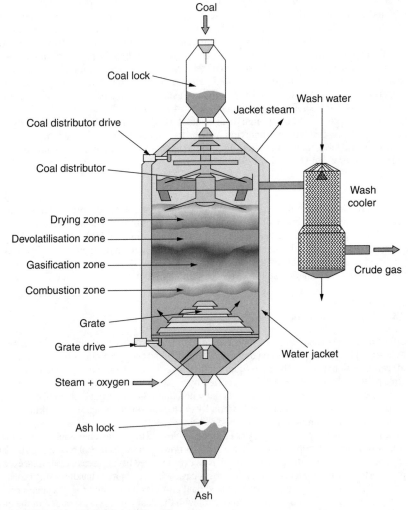

FIGURE 5.14 The Lurgi dry-ash gasifier.

or oxygen. The coal is gasified at 350 to 450 psi (2.4–3.1 MPa) and devolatilization takes place in the temperature range 615 to 760°C (1139–1400°F); residence time in the reactor is approximately 1 hour. Hydrogen is supplied by injected steam and the necessary heat is supplied by the combustion of a portion of the product char. The revolving grate, located at the bottom of the gasifier supports the bed of coal, removes the ash, and allows steam and oxygen (or air) to be introduced.

The Lurgi product gas has high methane content relative to the products from non-pressurized gasifiers. With oxygen injection, the gas has a heat content of approximately 450 Btu/ft^3 (16.8 MJ/m^3). The crude gas which leaves the gasifier contains tar, oil, phenols, ammonia, coal fines, and ash particles. The steam is first quenched to remove the tar and oil and, prior to methanation, part of the gas passes through a shift converter and is then washed to remove naphtha and unsaturated hydrocarbons; a subsequent step removes the acid gases. The gas is then methanated to produce a high heat-content pipeline quality product.

The Wellman Galusha Process. The Wellman Galusha process has been in commercial use for more than 50 years (Howard-Smith and Werner, 1976). These are two types of gasifiers, the standard type and the agitated type and the rated capacity of an agitated unit may be 25 percent (or more) higher than that of a standard gasifier of the same size. In addition, an agitated gasifier is capable of treating volatile caking bituminous coals.

The gasifier is water-jacketed and, therefore, the inner wall of the vessel does not require a refractory lining. Agitated units include a varying speed revolving horizontal arm which also spirals vertically below the surface of the coal bed to minimize channeling and to provide a uniform bed for gasification. A rotating grate is located at the bottom of the gasifier to remove the ash from the bed uniformly. Steam and oxygen are injected at the bottom of the bed through tuyeres. Crushed coal is fed to the gasifier through a lock hopper and vertical feed pipes. The fuel valves are operated so as to maintain a relatively constant flow of coal to the gasifier to assist in maintaining the stability of the bed and, therefore, the quality of the product gas.

Entrained-Bed Processes. The Koppers-Totzek Process (Baughman, 1978; Michaels and Leonard, 1978; van der Burgt, 1979) is, perhaps, the best known of the entrained-solids processes and operates at atmospheric pressure. The reactor is a relatively small, cylindrical, refractory-lined vessel into which coal, oxygen, and steam are charged. The reactor typically operates at an exit temperature of about 1480°C (2696°F) and the pressure is maintained slightly above atmospheric pressure.

Gases and vaporized hydrocarbons produced by the coal at medium temperatures immediately pass through a zone of very high temperature in which they decompose so rapidly that coal particles in the plastic stage do not agglomerate, and thus any type of coal can be gasified irrespective of caking tendencies, ash content, or ash fusion temperature. The gas product contains no ammonia, tars, phenols, or condensables and can be upgraded to synthesis gas by reacting all or part of the carbon monoxide content with steam to produce additional hydrogen plus carbon dioxide.

Molten Salt Processes. Molten salt processes feature the use of a molten bath [>1550°C (>2822°F)] into which coal, steam, and oxygen are injected (Karnavos et al., 1973; La Rosa and McGarvey, 1975). The coal devolatilizes with some thermal cracking of the volatile constituents. The product gas, which leaves the gasifier, is cooled, compressed, and fed to a shift converter where a portion of the carbon monoxide is reacted with steam to attain a CO/H$_2$ ratio of 1:3. The carbon dioxide so produced is removed and the gas is again cooled and enters a methanator where carbon monoxide and hydrogen react to form methane.

5.5.5 Underground Gasification

The aim of underground (or in situ) gasification of coal is to convert the coal into combustible gases by combustion of a coal seam in the presence of air, oxygen, or oxygen and steam. Thus, seams that were considered to be inaccessible, unworkable, or uneconomical to mine could be put to use. In addition, strip mining and the accompanying environment impacts, the problems of spoil banks, acid mine drainage, and the problems associated with use of high-ash coal are minimized or even eliminated.

The principles of underground gasification are very similar to those involved in the above-ground gasification of coal. The concept involves the drilling and subsequent linking of two boreholes so that gas will pass between the two (Fig. 5.15) (King and Magee, 1979). Combustion is then initiated at the bottom of one bore-hole (injection well) and is maintained by the continuous injection of air. In the initial reaction zone (combustion zone), carbon dioxide is generated by the reaction of oxygen (air) with the coal:

$$[C]_{coal} + O_2 \rightarrow CO_2$$

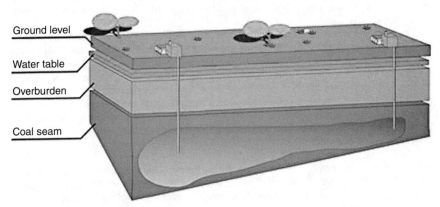

FIGURE 5.15 Underground coal gasification.

The carbon dioxide reacts with coal (partially devolatilized) further along the seam (reduction zone) to produce carbon monoxide:

$$[C]_{coal} + CO_2 \rightarrow 2CO$$

In addition, at the high temperatures that can frequently occur, moisture injected with oxygen or even moisture inherent in the seam may also react with the coal to produce carbon monoxide and hydrogen:

$$[C]_{coal} + H_2O \rightarrow CO + H_2$$

The gas product varies in character and composition but usually falls into the low-heat (low-Btu) category ranging from 125 to 175 Btu/ft^3 (4.7–6.5 MJ/m^3) (King and Magee, 1979).

5.5.6 The Future

Fuel in the form of gas has major advantages. Unlike electricity, a gaseous fuel can be stored until it is required and, unlike liquid fuel, gas is clean and leaves no residue in vessels used for storage or transport. Unlike solid fuel, a gas can be distributed continuously and delivered at precise rates to exact locations, at any scale from pilot light to power station. Operators can control the flow from moment to moment, and measure the amount used accurately for monitoring and billing. They can vary the composition and characteristics of a gaseous fuel readily, for instance, by blending natural gas with gas from coal.

Furthermore, a gaseous fuel does not have to be supported on a grate or in a combustion chamber, nor does it leave a solid residue that operators must remove and dispose.

Burning premium gaseous fuels, predominantly natural gas, in combined cycle gas turbines (CCGTs) has had a substantial impact in recent years on the efficiency, cleanliness and cost of power generation. Future resource projections and strategic considerations, however, have prompted renewed interest in advanced coal technologies in addition to combustion of renewable energy sources such as biomass-derived fuels.

An important component in the future exploitation of coal is its efficient gasification to a mixture of hydrogen and carbon monoxide, together with an inert ballast of nitrogen, carbon dioxide, and water. Unlike natural gas (or methane), which has attracted a wide range of combustion applications and for which an extensive combustion database has been developed, coal-derived gaseous fuel mixtures have not been widely investigated.

Stable, lean-burning combustion systems present a considerable design challenge, even for premium fuels. The low levels of oxides of nitrogen (NO_x) emitted by current natural gas-fired turbines, typically less than 10 ppm in many applications, have been achieved with highly refined strategies for fuel-air mixture preparation based on a combination of empiricism and numeric simulation. The starting point for similar combustor developments in relation to coal-derived gaseous fuel mixtures is much more poorly defined. Differences in the calorific value of the fuel, reflecting the level of inflammability and flame stability, introduce changes to combustion characteristics that must be accommodated in any design.

The successful exploitation of coal-derived gasified fuels in power generation using combined cycle gas turbines will require that the emissions performance and operability of such plant be broadly comparable with those presently demonstrated with natural gas.

5.6 LIQUID FUELS

One of the early processes for the production of liquid fuels from coal involved the Bergius process. In the process, lignite or subbituminous coal is finely ground and mixed with heavy oil recycled from the process. Catalyst is typically added to the mixture and the mixture is pumped into a reactor. The reaction occurs at between 400 to 500°C and 20 to 70 MPa hydrogen pressure. The reaction produces heavy oil, middle oil, gasoline, and gas:

$$n[C]_{coal} + (n+1)H_2 \rightarrow C_nH_{2n+2}$$

A number of catalysts have been developed over the years, including catalysts containing tungsten, molybdenum, tin, or nickel.

The different fractions can be sent to a refinery for further processing to yield of synthetic fuel or a fuel-blending stock of the desired quality. It has been reported that as much as 97 percent of the coal carbon can be converted to synthetic fuel but this very much depends on the coal type, the reactor configuration, and the process parameters.

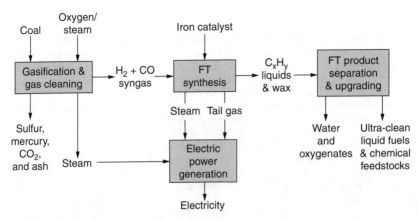

FIGURE 5.16 Production of liquid fuels from coal.

More recently, other processes have been developed for the conversion of coal to liquid fuels. The Fischer-Tropsch process of indirect synthesis of liquid hydrocarbons (Fig. 5.16) was used in Nazi Germany for many years and is today used by Sasol in South Africa. Coal would be gasified to make syngas (a balanced purified mixture of CO and H_2 gas) and the syngas condensed using Fischer-Tropsch catalysts to make light hydrocarbons which are further processed into gasoline and diesel. Syngas can also be converted to methanol, which can be used as a fuel, fuel additive, or further processed into gasoline via the Mobil M-gas process.

In fact, the production of liquid fuels from coal is not new and has received considerable attention (Berthelot, 1869; Batchelder, 1962; Stranges, 1983; Stranges, 1987) since the concept does represent alternate pathways to liquid fuels (Donath, 1963; Anderson and Tillman, 1979; Whitehurst et al., 1980; Gorin, 1981; Argonne, 1990). In fact, the concept is often cited as a viable option for alleviating projected shortages of liquid fuels as well as offering some measure of energy independence for those countries with vast resources of coal who are also net importers of crude oil.

In spite of the interest in coal liquefaction processes that emerged during the 1970s and the 1980s, petroleum prices always remained sufficiently low to ensure that the initiation of a synthetic fuels industry based on nonpetroleum sources would not become a commercial reality.

5.6.1 Physicochemical Aspects

The thermal decomposition of coal to a mix of solid, liquid, and gaseous products is usually achieved by the use of temperatures up to 1500°C (2732°F) (Wilson and Wells, 1950; McNeil, 1966; Gibson and Gregory, 1971). But, coal carbonization is not a process which has been designed for the production of liquids as the major products.

The chemistry of coal liquefaction is also extremely complex, not so much from the model compound perspective but more from the interactions that can occur between the constituents of the coal liquids. Even though many schemes for the chemical sequences, which ultimately result in the production of liquids from coal, have been formulated, the exact chemistry involved is still largely speculative, largely because the interactions of the constituents with each other are generally ignored. Indeed, the so-called structure of coal itself is still only speculative.

Hydrogen can represent a major cost item of the liquefaction process and, accordingly, several process options have been designed to limit (or control) the hydrogen consumption or even to increase the hydrogen-to-carbon atomic ratio without the need for added gas phase hydrogen (Speight, 1994). Thus, at best, the chemistry of coal liquefaction is only speculative. Furthermore, various structures have been postulated for the structure of coal (albeit with varying degrees of uncertainty) but the representation of coal as any one of these structures is extremely difficult and, hence, projecting a thermal decomposition route and the accompanying chemistry is even more precarious.

The majority of the coal liquefaction processes involve the addition of a coal-derived solvent prior to heating the coal to the desired process temperature. This is, essentially, a means of facilitating the transfer of the coal to a high-pressure region (usually the reactor) and also to diminish the sticking that might occur by virtue of the plastic properties of the coal.

5.6.2 Liquefaction Processes

The process options for coal liquefaction can generally be divided into four categories: (a) pyrolysis, (b) solvent extraction, (c) catalytic liquefaction, and (d) indirect liquefaction.

Pyrolysis Processes. Pyrolysis of coal dates back to the eighteenth century, using temperatures below 700°C in fixed- or moving-bed reactors. The primary product was a low-volatile smokeless domestic fuel, although the value of the liquid products was also soon recognized. During the 1920s and 1930s there was a great deal of interest in low-temperature processes, but interest died in the mid-1940s when gas and oil became readily available at low prices. With the oil embargo and increased oil prices of the early 1970s, interest renewed in coal pyrolysis, but in the 1980s interest has again declined along with petroleum prices (Khan and Kurata, 1985).

The first category of coal liquefaction processes, pyrolysis processes, involves heating coal to temperatures in excess of 400°C (752°F), which results in the conversion of the coal to gases, liquids, and char. The char is hydrogen deficient thereby enabling intermolecular or intramolecular hydrogen transfer processes to be operative, resulting in relatively hydrogen-rich gases and liquids. Unfortunately, the char produced often amounts to more than 45 percent by weight of the feed coal and, therefore, such processes have often been considered to be uneconomic or inefficient use of the carbon in the coal.

In the presence of hydrogen (*hydrocarbonization*) the composition and relative amounts of the products formed may vary from the process without hydrogen but the yields are still very much dependent upon the process parameters such as heating rate, pressure, coal type, coal (and product) residence time, coal particle size, and reactor configuration. The operating pressures for pyrolysis processes are usually less than 100 psi [690 kPa; more often between 5 and 25 psi (34–172 kPa)] but the hydrocarbonization processes require hydrogen pressures of the order of 300 to 1000 psi (2.1–6.9 MPa). In both categories of process, the operating temperature can be as high as 600°C (1112°F).

There are three types of pyrolysis reactors that are of interest: (a) a mechanically agitated reactor, (b) an entrained-flow reactor, and (c) a fluidized bed reactor.

The agitated reactor may be quite complex but the entrained-flow reactor has the advantage of either down- or up-flow operation and can provide short residence times. In addition, the coal can be heated rapidly, leading to higher yields of liquid (and gaseous) products that may well exceed the volatile matter content of the coal as determined by the appropriate test (Kimber and Gray, 1967). The short residence time also allows a high throughput of coal and the potential for small reactors. Fluidized reactors are reported to have been successful for processing noncaking coals but are not usually recommended for caking coals.

Solvent Extraction Processes. Solvent extraction processes are those processes in which coal is mixed with a solvent (*donor* solvent) that is capable of providing atomic or molecular hydrogen to the system at temperatures up to 500°C (932°F) and pressures up to 5000 psi (34.5 MPa). High-temperature solvent extraction processes of coal have been developed in three different process configurations: (a) extraction in the absence of hydrogen but using a recycle solvent that has been hydrogenated in a separate process stage; (b) extraction in the presence of hydrogen with a recycle solvent that has not been previously hydrogenated; and (c) extraction in the presence of hydrogen using a hydrogenated recycle solvent. In each of these concepts, the distillates of process-derived liquids have been used successfully as the recycle solvent which is recovered continuously in the process.

The overall result is an increase (relative to pyrolysis processes) in the amount of coal that is converted to lower molecular weight (i.e., soluble) products. More severe conditions are more effective for sulfur and nitrogen removal to produce a lower-boiling liquid product that is more amenable to downstream processing. A more novel aspect of the solvent extraction process type is the use of tar sand bitumen and/or heavy oil as process solvents (Moschopedis et al., 1980, 1982; Curtis et al., 1987; Schulman et al., 1988; Curtis and Hwang, 1992; Rosal et al., 1992).

Catalytic Liquefaction Processes. The final category of direct liquefaction process employs the concept of catalytic liquefaction in which a suitable catalyst is used to add hydrogen to the coal. These processes usually require a liquid medium with the catalyst dispersed throughout or may even employ a fixed-bed reactor. On the other hand, the catalyst may also be dispersed within the coal whereupon the combined coal-catalyst system can be injected into the reactor.

Many processes of this type have the advantage of eliminating the need for a hydrogen donor solvent (and the subsequent hydrogenation of the spent solvent) but there is still the need for an adequate supply of hydrogen. The nature of the process also virtually guarantees that the catalyst will be deactivated by the mineral matter in the coal as well as by coke lay-down during the process. Furthermore, in order to achieve the direct hydrogenation of the coal, the catalyst and the coal must be in intimate contact, but if this is not the case, process inefficiency is the general rule.

Indirect Liquefaction Processes. The other category of coal liquefaction processes invokes the concept of the indirect liquefaction of coal.

In these processes, the coal is not converted directly into liquid products but involves a two-stage conversion operation in which coal is first converted (by reaction with steam and oxygen) to produce a gaseous mixture that is composed primarily of carbon monoxide and hydrogen (syngas; synthesis gas). The gas stream is subsequently purified (to remove sulfur, nitrogen, and any particulate matter) after which it is catalytically converted to a mixture of liquid hydrocarbon products.

The synthesis of hydrocarbons from carbon monoxide and hydrogen (synthesis gas) (the Fischer-Tropsch synthesis) is a procedure for the indirect liquefaction of coal (Storch et al., 1951; Batchelder, 1962; Dry, 1976; Anderson, 1984; Jones et al., 1992). This process is the only coal liquefaction scheme currently in use on a relatively large commercial scale; South Africa is currently using the Fischer-Tropsch process on a commercial scale in their SASOL complex.

Thus, coal is converted to gaseous products at temperatures in excess of 800°C (1472°F), and at moderate pressures, to produce synthesis gas:

$$[C]_{coal} + H_2O \rightarrow CO + H_2$$

FUELS FROM COAL 161

The gasification may be attained by means of any one of several processes or even by gasification of coal in place (underground, or in situ, gasification of coal, Sec. 5.5.5).

In practice, the Fischer-Tropsch reaction is carried out at temperatures of 200 to 350°C (392–662°F) and at pressures of 75 to 4000 psi (0.5–27.5 MPa). The H_2/CO ratio is usually 2.2:1 or 2.5:1. Since up to three volumes of hydrogen may be required to achieve the next stage of the liquids production, the synthesis gas must then be converted by means of the water-gas shift reaction to the desired level of hydrogen:

$$CO + H_2O \rightarrow CO_2 + H_2$$

After this, the gaseous mix is purified and converted to a wide variety of hydrocarbons:

$$nCO + (2n + 1)H_2 \rightarrow C_nH_{2n+2} + nH_2O$$

These reactions result primarily in low- and medium-boiling aliphatic compounds suitable for gasoline and diesel fuel.

Reactors. Several types of reactor are available for use in liquefaction processes and any particular type of reactor can exhibit a marked influence on process performance.

The simplest type of reactor is the noncatalytic reactor which consists, essentially, of a vessel (or even an open tube) through which the reactants pass. The reactants are usually in the fluid state but may often contain solids such as would be the case for coal slurry. This particular type of reactor is usually employed for coal liquefaction in the presence of a solvent.

The second type of noncatalytic reactor is the continuous-flow, stirred-tank reactor, which has the notable feature of encouraging complete mixing of all of the ingredients, and if there is added catalyst (suspended in the fluid phase) the reactor may be referred to as a slurry reactor.

The fixed-bed catalytic reactor contains a bed of catalyst particles through which the reacting fluid flows; the catalysis of the desired reactions occurs as the fluid flows through the reactor. The liquid may pass through the reactor in a downward flow or in an upward flow but the problems that tend to accompany the latter operation (especially with regard to the heavier, less conventional feedstocks) must be recognized. In the downward-flowing mode, the reactor may often be referred to as a trickle-bed reactor.

Another type of reactor is the fluidized bed reactor, in which the powdered catalyst particles are suspended in a stream of up-flowing liquid or gas. A form of this type of reactor is the ebullating-bed reactor. The features of these two types of reactor are the efficient mixing of the solid particles (the catalyst) and the fluid (the reactant) that occurs throughout the whole reactor.

The final type of reactor to be described is the entrained-flow reactor in which the solid particles travel with the reacting fluid through the reactor. Such a reactor has also been described as a dilute or lean-phase fluidized bed with pneumatic transport of solids.

5.6.3 Products

Liquid products from coal are generally different from those produced by petroleum refining, particularly as they can contain substantial amounts of phenols. Therefore, there will always be some question about the place of coal liquids in refining operations. For this reason, there have been some investigations of the characterization and *next-step* processing of coal liquids.

As a first step in the characterization of coal liquids, it is generally recognized that some degree of fractionation is necessary (Whitehurst et al., 1980) followed by one, or more, forms of chromatography to identify the constituents (Kershaw, 1989; Philp and de las Heras, 1992). The fractionation of coal liquids is based largely on schemes developed for the characterization of petroleum (Dooley et al., 1978; Speight, 2007), but because of the difference between coal liquids and petroleum, some modification of the basic procedure is usually required to make the procedure applicable to coal liquids (Ruberto et al., 1976; Bartle, 1989).

The composition of coal liquids produced from coal depends very much on the character of the coal and on the process conditions and, particularly, on the degree of *hydrogen addition* to the coal (Aczel et al., 1978; Schiller, 1978; Schwager et al., 1978; Wooton et al., 1978; Whitehurst et al., 1980; Kershaw, 1989).

Current concepts for refining the products of coal liquefaction processes rely for the most part on the already existing petroleum refineries, although it must be recognized that the acidity (i.e., phenol content) of the coal liquids and their potential incompatibility with conventional petroleum (including heavy oil) may pose new issues within the refinery system (Speight, 1994, 2007).

5.6.4 The Future

With the continuing violence in the Middle East and (at the time of writing) oil prices in excess of $90 per barrel, the issue of energy security is again of prime importance. Part of this energy security can come by the production of liquid fuels from coal. As always, the real question relates to economics—that is tied to the price of oil.

The Sasol technology, a third-generation Fischer-Tropsch process, was developed in Germany and used in World War II, and later in South Africa. In the process, steam and oxygen are passed over coke at high temperatures and pressures. Synthesis gas (a mixture of carbon monoxide and hydrogen) is produced and then reassembled into liquid fuels through the agency of the Fischer-Tropsch reaction (Chap. 7).

Although considered too expensive to compete with the production of liquid fuels from petroleum, the liquid fuels produced from coal (by gasification and the conversion of the synthesis gas) have been environmentally friendly insofar as they have no sulfur and, therefore, are more in keeping with the recent laws regarding ultra-low sulfur fuels.

The coal-to-liquid technology would complement the expanding tar-sands technology (Chap. 4) allowing the long-predicted decline in petroleum production to be delayed for decades and the geopolitics of energy would be rewritten.

Overall, the coal-to-liquids technology is one element of an integrated program that is necessary to deal with energy fuel security and assurance of supply.

In fact, the technologies required to produce large-scale supplies of clean liquid fuels from coal are not on the drawing boards or in laboratories. They are in use around the world today, from countries such as South Africa, which has long relied on coal liquefaction to provide a substantial percentage of its transportation fuels, to China, India, Indonesia, and the Philippines.

The production of liquid fuels from coal begins with coal as a raw material or feedstock. Both indirect and direct liquefaction processes have been proven.

In indirect coal liquefaction, coal is subjected to intense heat and pressure to create a synthesis gas comprised of hydrogen and carbon monoxide. The synthetic gas is treated to remove impurities and unwanted compounds such as mercury and sulfur. This clean gas enters a second stage (Fischer-Tropsch process) which converts the synthesis gas into clean liquid fuels and other chemical products. Diesel fuel produced by Fischer-Tropsch synthesis is virtually sulfur-free with low aromatics and a high cetane value and is cleaner than conventional diesel. It burns more completely and emissions are significantly lower than low-sulfur diesel. Most of the CO_2 is already concentrated and ready for capture and possible sequestration or for use in enhanced oil or gas recovery.

In direct coal liquefaction, coal is pulverized and mixed with oil and hydrogen in a pressurized environment. This process converts the coal into a synthetic crude oil that can then be refined into a variety of fuel products. The direct coal liquefaction technology has been demonstrated in the United States and is now being commercially deployed in China and other countries.

The obvious drawback relates to environment issues. The process of converting coal into liquid and using it for transportation releases nearly twice as much carbon dioxide as burning diesel made from crude oil does. In a world conscious of climate change, that excess carbon is a major issue. One way round this problem might be to take the carbon dioxide and bury it underground. Another would be to replace fossil-fuel feedstock with biomass (Chap. 8).

Expense is another issue. The Fischer-Tropsch process has always been relatively expensive. However, the need may have to justify the expense! The first major use of Fischer Tropsch technology was during World War II, when Germany produced about 90 percent of the diesel and aviation fuel. South Africa began liquefying coal in response to apartheid-era sanctions, and in part as a result of its investment back then, continues to derive about 30 percent of its fuel from liquefied coal.

5.7 SOLID FUELS

5.7.1 Coke

In the current context, the term *solid fuel* refers to various types of solid material that is used as fuel to produce energy and provide heating, usually released through combustion. Coal itself is not included in this definition.

The most common coal-based solid fuel, coke, is a solid carbonaceous residue derived from low-ash, low-sulfur bituminous coal from which the volatile constituents are driven off by baking in an oven without oxygen at temperatures as high as 1000°C (1832°F) so that the fixed carbon and residual ash are fused together.

Coke is used as a fuel and as a reducing agent in smelting iron ore in a blast furnace. Coke from coal is grey, hard, and porous and has a heating value of 24.8 million Btu/ton (29.6 MJ/kg). By-products of this conversion of coal to coke include coal tar, ammonia, light oils, and coal-gas. On the other hand, petroleum coke is the solid residue obtained in oil refining but at lower temperature, which resembles coke but contains too many impurities to be useful in metallurgic applications.

Next to combustion, carbonization for the production of coke represents one of the most popular, and oldest, uses of coal (Armstrong, 1929; Forbes, 1950).

The thermal decomposition of coal on a commercial scale is often more commonly referred to as carbonization and is more usually achieved by the use of temperatures up to 1500°C (2732°F). The degradation of the coal is quite severe at these temperatures and produces (in addition to the desired coke) substantial amounts of gaseous products.

The original process of heating coal (in rounded heaps; the hearth process) remained the principal method of coke production for many centuries, although an improved oven in the form of a beehive shape was developed in the Newcastle area of England in about 1759. The method of coke production was initially the same as for the production of charcoal, that is, stockpiling coal in round heaps (known as milers), igniting the piles, and then covering the sides with a clay-type soil. The smoke generated by the partial combustion of coal tars and gases was soon a major problem near residential areas, and the coking process itself could not be controlled because of climatic elements: rain, wind, and ice. It was this new modification of the charring-coking kiln which laid the foundation for the modern coke blast furnace: the beehive process.

Carbonization is essentially a process for the production of a carbonaceous residue by thermal decomposition (with simultaneous removal of distillate) of organic substances.

$$[C]_{\text{organic carbon}} \rightarrow [C]_{\text{coke/char/carbon}} + \text{liquids} + \text{gases}$$

The process is a complex sequence of events which can be described in terms of several important physicochemical changes, such as the tendency of the coal to soften and flow when heated (plastic properties) or the relationship to carbon-type in the coal. In fact, some coals become quite fluid at temperatures on the order of 400 to 500°C (752–932°F) and there is a considerable variation in the degree of maximum plasticity, the temperature of maximum plasticity, as well as the plasticity temperature range for various types of coal (Kirov and Stephens, 1967; Mochida et al., 1982; Royce et al., 1991). The yields of tar and low molecular weight liquids are to some extent variable but are greatly dependent on the process parameters, especially temperature, as well as the type of coal (Cannon et al., 1944; Davis, 1945; Poutsma, 1987; Ladner, 1988; Wanzl, 1988).

5.7.2 Charcoal

Charcoal is the blackish residue consisting of impure carbon obtained by removing water and other volatile constituents from animal and vegetation substances such as wood (Chap. 10) and, on occasion, from coal.

Charcoal is usually produced by heating wood, sugar, bone char, or other substances in the absence of oxygen. The soft, brittle, lightweight, black, porous material resembles coal and is 85 to 98 percent carbon with the remainder consisting of volatile chemicals and ash.

The first part of the word is of obscure origin, but the first use of the term "coal" in English was as a reference to charcoal. In this compound term, the prefix "chare-" meant "turn," with the literal meaning being "to turn to coal." The independent use of "char," meaning to scorch, to reduce to carbon, is comparatively recent and must be a back-formation from the earlier charcoal. It may be a use of the word *charren* or *churn*, meaning to turn, that is, wood changed or turned to coal, or it may be from the French *charbon*. A person who manufactured charcoal was formerly known as a collier (also as a wood collier). The word "collier" was also used for those who mined or dealt in coal, and for the ships that transported it.

5.8 ENVIRONMENTAL ASPECTS OF COAL UTILIZATION

During the combustion or conversion of coal, many compounds can be produced, some of which are carcinogenic. Coal also contains nitrogen, oxygen, and sulfur as well as volatile constituents and the minerals that remain as ash when the coal is burned. Thus some of the products of coal combustion can have detrimental effects on the environment. For example, coal combustion produces carbon dioxide (CO_2) and it has been postulated that the amount of carbon dioxide in the earth's atmosphere could increase to such an extent that changes in the earth's climate will occur. Also, sulfur and nitrogen in the coal form oxides [sulfur dioxide (SO_2), sulfur trioxide (SO_3), nitric oxide (NO), and nitrogen dioxide (NO_2)] during combustion that can contribute to the formation of acid rain.

In addition, coal combustion produces particulate matter (*fly ash*) that can be transported by winds for many hundreds of kilometers and solids (*bottom ash* and *slag*) that must be disposed. Trace elements originally present in the coal may escape as volatile material (e.g., chlorine and mercury) or be concentrated in the ash (e.g., arsenic and barium). Using such devices as electrostatic precipitators, baghouses, and scrubbers can trap some of these pollutants. Alternate means for combustion (e.g., fluidized bed combustion, magnetohydrodynamics, and low nitrogen dioxide burners) are also available to provide more efficient and environmentally attractive methods for converting coal to energy.

When carbon dioxide is released into the atmosphere it does not prevent the shorter-wavelength rays from the sun from entering the atmosphere but does prevent much of the long-wave radiation reradiated from the earth's surface from escaping into the space. The carbon dioxide absorbs this upward-propagating infrared radiation and reemits a portion of it downward, causing the lower atmosphere to remain warmer than it would otherwise be (the greenhouse effect). Other gases, such as methane and ozone also cause this effect and, consequently, are known as *greenhouse gases*.

5.9 CLEAN COAL TECHNOLOGIES

Clean coal technologies (CCTs) are a new generation of advanced coal utilization processes that are designed to enhance both the efficiency and the environment acceptability of coal extraction, preparation, and use. These technologies reduce emissions waste and increase the amount of energy gained from coal. The goal of the program was to foster development of the most promising CCTs such as improved methods of cleaning coal, fluidized bed combustion, integrated gasification combined cycle, furnace sorbent injection, and advanced flue-gas desulfurization.

Acid rain is the result of a series of complex reactions involving chemicals and compounds from many industrial, transportation, and natural sources. Sulfur dioxide emissions from new coal-fired facilities have been controlled since the 1970s by the various government regulations. The goal of the various regulations is to ensure the continued decrease in emissions of nitrogen and sulfur oxides from coal combustion into the atmosphere. As a result, emissions of sulfur dioxide have dropped even though coal use has increased.

5.10 REFERENCES

Aczel, T., R. B.Williams, R. A. Brown, and R. J. Pancirov: In *Analytical methods for Coal and Coal Products*, vol. 1, C. Karr Jr. (ed.), Academic Press Inc., New York, 1978, chap. 17.

Anderson, L. L. and D. A. Tillman: *Synthetic Fuels from Coal: Overview and Assessment,* John Wiley and Sons Inc., New York, 1979.

Anderson, R. B.: In *Catalysis on the Energy Scene,* S. Kaliaguine and A. Mahay (ed.). Elsevier, Amsterdam, Netherlands, Netherlands, 1984, p. 457.

Anthony, D. B. and J. B. Howard: *AIChE Journal,* 22, 1976, p. 625.

Argonne: *Environmental Consequences of, and Control Processes for, Energy Technologies,* Argonne National Laboratory. Pollution Technology Review No. 181, Noyes Data Corp., Park Ridge, N.J., 1990, chap. 6.

Armstrong, J: *Carbonization Technology and Engineering,* Lippincott Publishers, London, 1929.

Baker, R. T. K. and N. M. Rodriguez: In *Fuel Science and Technology Handbook,* J.G. Speight (ed.), Marcel Dekker Inc, New York, 1990, chap. 22.

Bartle, K. D: In *Spectroscopic Analysis of Coal Liquids,* J. Kershaw (ed.) Elsevier, Amsterdam, 1989, chap. 2.

Batchelder, H. R.: In *Advances in Petroleum Chemistry and Refining,* vol. 5, J. J. McKetta Jr. (ed.), Interscience Publishers Inc., New York. 1962. chap. 1.

Baughman, G. L.: *Synthetic Fuels Data Handbook,* Cameron Engineers, Denver, Colo., 1978.

Berthelot, M.: *Bull. Soc. Chim. Fra.,* 11, 1869, p. 278.

Bodle, W. W. and J. Huebler: In *Coal Handbook,* R. A. Meyers (ed.), Marcel Dekker Inc., New York. 1981, chap. 10.

BP: "Statistical Review of World Energy 2007," British Petroleum Company Ltd., London, England, June, 2007.

Britannica: "Coal Mining: Ancient Use of Outcropping Coal," Encyclopedia Britannica Inc., Chicago, Ill., 2004.

Cannon, C. G., M. Griffith, and W. Hirst: Proceedings, Conference on the Ultrafine Structure of Coals and Cokes. British Coal Utilization Research Association, Leatherhead, Surrey, England, 1944, p. 131.

Cavagnaro, D. M.: *Coal Gasification Technology,* National Technical Information Service, Springfield, Va., 1980.

Cover, A. E., W. C. Schreiner, and G. T. Skapendas: *Chem. Eng. Progr.,* 69(3), 1973, p. 31.

Curtis, C. W. and J. S. Hwang: *Fuel Process. Techno.,* 30, 1992, p. 47.

Curtis, C. W., J. A. Guin, M. C. Pass, and K. J. Tsai: *Fuel Sc. Techno. Int.,* 5, 1987, p. 245.

Cusumano, J. A., R. A. Dalla Betta, and R. B. Levy: *Catalysis in Coal Conversion,* Academic Press Inc., New York, 1978.

Davidson, R. M.: "Mineral Effects in Coal Conversion," Report No. ICTIS/TR22, International Energy Agency, London, England, 1983.

Davis, J. D.: In *Chemistry of Coal Utilization,* H. H. Lowry (ed.), John Wiley & Sons Inc., New York, 1945, chap. 22.

Donath, E. E.: In *Chemistry of Coal Utilization,* supp. vol., H. H. Lowry (ed.), John Wiley & Sons Inc., New York, 1963, chap. 22.

Dooley, J. E., C. J. Thompson, and S. E. Scheppele: In *Analytical Methods for Coal and Coal Products,* vol. 1, C. Karr Jr. (ed.), Academic Press Inc., New York, 1978, chap. 16.

Dry, M. E.: "Advances in Fischer–Tropsch chemistry," *Ind. Eng. Chem. Prod. Res. Dev.* 15(4), 1976, p. 282.

Dutcher, J. S., R. E. Royer, C. E. Mitchell, and A. R. Dahl: In *Advanced Techniques in Synthetic Fuels Analysis,* C. W. Wright, W. C. Weimer, and W. D. Felic (eds.), Technical Information Center, United States Department of Energy, Washington, D.C., 1983, p. 12.

EIA: *Net Generation by Energy Source by Type of Producer,* Energy Information Administration, United States Department of Energy, Washington, D.C., 2007, http://www.eia.doe.gov/cneaf/electricity/epm/table1_1.html.

Elton, A.: In *A History of Technology,* vol. 4, C. Singer, E. J. Holmyard, A. R. Hall, and T. I. Williams (eds.), Clarendon Press, Oxford, England, 1958, chap. 9.

Forbes, R. J.: *Metallurgy in Antiquity,* Brill Academic Publishers, Leiden, Netherlands, 1950.

Forbes, R. J.: *Studies in Ancient Technology,* Brill Academic Publishers, Leiden, Netherlands, 1966.

Fryer, J. F. and Speight, J. G.: "Coal Gasification: Selected Abstract and Titles," Information Series No. 74, Alberta Research Council, Edmonton, Canada, 1976.

Gibson, J. and D. H. Gregory: *Carbonization of Coal,* Mills and Boon, London, England, 1971.

Gorin, E: In *Chemistry of Coal Utilization,* Second Supplementary Volume. M. A. Elliott (ed.), John Wiley & Sons Inc., New York, 1981, chap. 27.

Graff, R. A. A., Dobner, S., and A. M. Squires: 55, 1976, p. 109.

Howard-Smith, I. and G. J. Werner: *Coal Conversion Technology,* Noyes Data Corp., Park Ridge, N.J., 1976, p.71.

Huang, Y. H., H. Yamashita, and A. Tomita: *Fuel Process. Techno.,* 29, 1991, p. 75.

Jones, C. J., B. Jager, and M. D. Dry: *Oil and Gas Journal,* 90(3), 1992, p. 53.

Karnavos, J. A., P. J. LaRosa, and E. A. Pelczarski: *Chem. Eng. Progr.,* 69(3), 1973, p. 54.

Kasem, A.: *Three Clean Fuels from Coal: Technology and Economics,* Marcel Dekker Inc., New York, 1979.

Kershaw, J.: In *Spectroscopic Analysis of Coal Liquids,* J. Kershaw (ed.), Elsevier, Amsterdam, 1989, chap. 6.

Khan, M. R. and T. M. Kurata: "The Feasibility of Mild Gasification of Coal: Research Needs," Technical Note DOE/METC-85/4019, U.S. Department of Energy, Washington, D.C., 1985.

Kimber, G. M. and M. D. Gray: *Combustion and Flame,* 11, 1967, p. 360.

King, R. B. and R. A. Magee: In *Analytical Methods for Coal and Coal Products,* vol. 3, C. Karr Jr. (ed.), Academic Press Inc., New York, 1979, chap. 41.

Kirov, N. Y. and J. N. Stephens: *Physical Aspects of Coal Carbonization,* National Coal Board Advisory Committee, Australia, 1967.

Koh, A. L., R. B. Harty, and J. G. Johnson: *Chem. Eng. Progr.,* 74(8), 1978, p.73.

Ladner, W. R.: *Fuel Process. Techno.,* 20, 1988, p. 207.

Lahaye, J. and P. Ehrburger (eds.): *Fundamental Issues in Control of Carbon Gasification Reactivity,* Kluwer Academic Publishers, Dordrecht, Netherlands. 1991.

La Rosa, P. and R. J. McGarvey: "Clean Fuels from Coal," Proceedings, Symposium II, Institute of Gas Technology, Chicago, Ill., 1975.

Lynn Schloesser, L.: "Gasification Incentives," Workshop on Gasification Technologies, Ramkota, Bismarck, North Dakota, June 28–29, 2006.

Mahajan, O. P. and P. L. Walker Jr.: In *Analytical Methods for Coal and Coal Products,* vol. 2, C. Karr Jr. (ed.), Academic Press Inc., New York, 1978, chap. 32.

Martinez-Alonso, A. and J. M. D. Tascon: In *Fundamental Issues in Control of Carbon Gasification Reactivity,* J. Lahaye and P. Ehrburger (eds.), Kluwer Academic Publishers, Dordrecht, Netherlands, 1991.

Massey, L.G. (ed.): "Coal Gasification," Advances in Chemistry Series No. 131. American Chemical Society, Washington, D.C., 1974.

Massey, L.G.: In *Coal Conversion Technology,* C. Y. Wen and E. S. Lee (ed.), Addison-Wesley Publishers Inc., Reading, Mass. 1979, p. 313.

Matsukata, M., E. Kikuchi, and Y. Morita: *Fuel,* 71, 1992, p. 819.

McNeil, D.: *Coal Carbonization Products,* Pergamon Press, London, England, 1966.

Michaels, H. J. and H. F. Leonard: *Chem. Eng. Progr.,* 74(8), 1978, p. 85.

Mills, G. A.: *Ind. Eng. Chem.,* 61(7), 1969, p. 6.

Mills, G. A.: *Chemtech.,* 12, 1982, p. 294.

Mims, C. A.: In *Fundamental Issues in Control of Carbon Gasification Reactivity,* J. Lahaye and P. Ehrburger (ed.), Kluwer Academic Publishers, Dordrecht, Netherlands, 1991, p. 383.

Mochida, I., Y. Korai, H. Fujitsu, K., Takeshita, K. Komatsubara, and K. Koba: *Fuel,* 61, 1982, p. 1083.

Moschopedis, S. E., R. W. Hawkins, J. F. Fryer, and J. G. Speight: *Fuel,* 59. 1980, p. 647.

Moschopedis, S. E., R. W. Hawkins, and J. G. Speight: *Fuel Process. Techno.,* 5, 1982, p. 213.

Nef, J. U.: In *A History of Technology,* vol. 3, C. Singer, E. J. Holmyard, A. R. Hall, and T. I. Williams (eds.). Clarendon Press, Oxford, England, 1957, chap. 3.

Philp, R. P. and F. X. de las Heras: In Chromatography, 5th ed., Part B: Applications, E. Heftmann (ed.), Elsevier, Amsterdam, Netherlands, 1992, chap. 21.

Poutsma, M. L.: "A Review of the Thermolysis of Model Compounds Relevant to Coal Processing," Report No. ORNL/TM10637, DE88 003690, Oak Ridge National Laboratory, Oak Ridge, Tenn., 1987.

Probstein, R. F. and R. E. Hicks: *Synthetic Fuels,* pH Press, Cambridge, Mass., 1990, chap. 4.

Rosal, R., L. F. Cabo, F. V. Dietz, and H. Sastre: *Fuel Process. Techno.,* 31, 1992, p. 209.

Royce, A. J., P. J. Readyhough, and P. L. Silveston: In *Coal Science II,* Symposium Series No. 461, H. H. Schobert, K. D. Bartle, and L. J. Lynch (ed.), American Chemical Society, Washington, D.C. 1991, chap. 24.

Ruberto, R. G., D. M. Jewell, R. K. Jensen, and D. C. Cronauer: In *Shale Oil, Tar Sands, and Related Fuel Sources,* T. F. Yen (ed.), Advances in Chemistry Series No. 151, American Chemical Society, Washington, D.C., 1976, chap. 3.

Salway, P: *A History of Roman Britain,* Oxford University Press, Oxford, England, 2001.

Schiller, J. E.: In *Analytical Chemistry of Liquid Fuel Sources: Tar Sands, Oil Shale, Coal, and Petroleum,* P. C. Uden, S. Siggia, and H. B. Jensen (eds.), Advances in Chemistry Series No. 170, American Chemical Society, Washington, D.C., 1978 chap. 4.

Schulman, B. L., F. E. Biasca, R. L. Dickenson, and D. R. Simbeck: Report No. DOE/FE/60457-H3. Contract No. DE-AC01-84FE60457, United States Department of Energy, Washington, D.C., 1988.

Schwager, I., P. A. Farmanian, and T. F. Yen: In *Analytical Chemistry of Liquid Fuel Sources: Tar Sands, Oil Shale, Coal, and Petroleum,* P. C. Uden, S. Siggia, and H. B. Jensen (ed.), Advances in Chemistry Series No. 170, American Chemical Society, Washington, D.C., 1978, chap. 5.

Seglin, L. (ed.): *Methanation of Synthesis Gas,* Advances in Chemistry Series No. 146. American Chemical Society, Washington, D.C., 1975.

Speight, J. G.: In *Fuel Science and Technology Handbook,* J. G. Speight (ed.), Marcel Dekker Inc., New York, 1990, chap. 33.

Speight, J. G.: *Gas Processing: Environmental Aspects and Methods,* Butterworth Heinemann, Oxford, England, 1993.

Speight, J. G.: *The Chemistry and Technology of Coal,* 2d ed.. Marcel Dekker Inc., New York, 1994.

Speight, J. G. *The Chemistry and Technology of Petroleum,* 4th ed., CRC-Taylor and Francis Group, Boca Raton, Fla., 2007.

Storch, H. H., Golumbic, N., and Anderson, R. B.: *The Fischer Tropsch and Related Syntheses,* John Wiley & Sons Inc., New York, 1951.

Stranges, A. N.: *J. Chem. Educ.,* 60, 1983, p. 617.

Stranges, A. N.: *Fuel Process. Techno.,* 16, 1987, p. 205.

Taylor, F. S. and C. Singer: In *A History of Technology,* vol 2, C. Singer, E. J. Holmyard, A. R. Hall, and T. I. Williams (eds.), Clarendon Press, Oxford, England, 1957, chap. 10.

Tucci, E. R. and W. J. Thompson: *Hydrocarbon Processing,* 58(2), 1979, p. 123.

Van der Burgt, M. J.: *Hydrocarbon Processing,* 58(1), 1979, p. 161.

Van Heek, K. H. and H. J. Muhlen: In *Fundamental Issues in Control of Carbon Gasification Reactivity,* J. Lahaye and P. Ehrburger (eds.), Kluwer Academic Publishers Inc., Netherlands. 1991, p. 1.

Verma, A.: *Chemtech.,* 8, 1978, p. 372 and 626.

Wang, W. and T. K. Mark: *Fuel,* 71, 1992, p. 871.

Wanzl, W.: *Fuel Process. Techno.,* 20, 1988, p. 317.

Watson, G. H.: "Methanation Catalysts," Report ICTIS/TR09, International Energy Agency, London, England, 1980.

Whitehurst, D. D., T. O. Mitchell, and M. Farcasiu: *Coal Liquefaction: The Chemistry and Technology of Thermal Processes,* Academic Press Inc., New York, 1980.

Wilson, P. J. Jr. and J. H. Wells: *Coal, Coke, and Coal Chemicals,* McGraw-Hill Inc., New York, 1950.

Wooton, D. L., W. M. Coleman, T. E. Glass, H. C. Dorn, and L. T. Taylor: In *Analytical Chemistry of Liquid Fuel Sources: Tar Sands, Oil Shale, Coal, and Petroleum,* P. C. Uden, S. Siggia, and H. B. Jensen (eds.), Advances in Chemistry Series No. 170. American Chemical Society, Washington, D.C. 1978, chap. 3.

CHAPTER 6
FUELS FROM OIL SHALE

Oil shale is an inorganic, nonporous sedimentary marlstone rock containing various amounts of solid organic material (known as *kerogen*) that yields hydrocarbons, along with non-hydrocarbons, and a variety of solid products, when subjected to pyrolysis (a treatment that consists of heating the rock at high temperature).

Thus, by definition, *kerogen* is naturally occurring insoluble organic matter found in shale deposits. However, *shale oil* is the synthetic fuel produced by the thermal decomposition of *kerogen* at high temperature [>500°C (>932°F)]. Shale oil is referred to as *synthetic crude oil* after hydrotreating.

The oil shale deposits in the western United States contain approximately 15 percent organic material by weight. By heating oil shale to high temperatures, kerogen can be released and converted to a liquid that, once upgraded, can be refined into a variety of liquid fuels, gases, and high-value chemical and mineral by-products. The United States has vast known oil shale resources that could translate into as much as 2.2 trillion barrels of known kerogen *oil-in-place*. Oil shale deposits concentrated in the Green River Formation in the states of Colorado, Wyoming, and Utah account for nearly three-quarters of this potential.

Because of the abundance and geographic concentration of the known resource, oil shale has been recognized as a potentially valuable U.S. energy resource since as early as 1859, the same year Colonel Drake completed his first oil well in Titusville, Pennsylvania (Chap. 1). Early products derived from shale oil included kerosene and lamp oil, paraffin, fuel oil, lubricating oil and grease, naphtha, illuminating gas, and ammonium sulfate fertilizer.

Since the beginning of the twentieth century, when the U.S. Navy converted its ships from coal to fuel oil, and the nation's economy was transformed by gasoline-fueled automobiles and diesel-fueled trucks and trains, concerns have been raised about assuring adequate supplies of liquid fuels at affordable prices to meet the growing needs of the nation and its consumers.

America's abundant resources of oil shale were initially eyed as a major source for these fuels. Numerous commercial entities sought to develop oil shale resources. The Mineral Leasing Act of 1920 made petroleum and oil shale resources on federal lands available for development under the terms of federal mineral leases. Soon, however, discoveries of more economically producible and refinable liquid crude oil in commercial quantities caused interest in oil shale to decline.

Interest resumed after World War II, when military fuel demand and domestic fuel rationing and rising fuel prices made the economic and strategic importance of the oil shale resource more apparent. After the war, the booming post–World War II economy drove demand for fuels ever higher. Public and private research and development efforts were commenced, including the 1946 U.S. Bureau of Mines Anvil Point, Colorado oil shale demonstration project. Significant investments were made to define and develop the resource and to develop commercially viable technologies and processes to mine, produce, retort, and upgrade oil shale into viable refinery feedstocks and by-products. Once again, however, major crude oil discoveries in the lower-48 United States, off-shore, and in Alaska, as well as other parts of the world reduced the foreseeable need for shale oil and interest and associated activities again diminished. Lower-48 U.S. crude oil reserves peaked in 1959 and lower-48 production peaked in 1970.

By 1970, oil discoveries were slowing, demand was rising, and crude oil imports, largely from Middle Eastern states, were rising to meet demand. Global oil prices, while still relatively low, were also rising reflecting the changing market conditions. Ongoing oil shale research and testing projects were reenergized and new projects were envisioned by numerous energy companies seeking alternative fuel feedstocks (Table 6.1). These efforts were significantly amplified by the impacts of the 1973 Arab Oil Embargo which demonstrated the nation's vulnerability to oil import supply disruptions, and were underscored by a new supply disruption associated with the 1979 Iranian Revolution.

TABLE 6.1 Timeline of Oil Shale Projects in the United States

Year	Event
1909	U.S. Government creates U.S. Naval Oil Shale Reserve
1910	Oil shale lands "claim-staked"
1916	U.S. Geological Survey (USGS) estimates 40 billion barrels of shale oil in Green River Formation in Colo., Wyo., and Utah.
1917	First oil shale retort kiln in DeBeque, Co.
1918	First oil shale boom begins with over 30,000 mining clams; lasts until 1925
1920	Mineral Leasing Act requires shale lands be leased through the Secretary of Interior
1929	Test retort at Rulison stops at 3600 bbl after oil discoveries in Calif., Tex., and Okla.
1944	U.S. Synthetic Liquid Fuels Act provides $18 million for experiments at Anvil Points
1950s	Gulf Oil and Shell Oil both purchase oil shale lands in Green River Formation
1956	Anvil Points operations cease after testing three experimental retort processes
1961	Unocal shuts down Parachute Creek "Union A" retort after 18 months and 800 bbl/day due to cost
1964	Colorado School of Mines leases Anvil Points facility to conduct research on U.S. Bureau of Mines Gas Combustion Retorts
1967	CER and U.S. AEC abandon plans for "Project Bronco" atomic subsurface retort
1972	Tosco, Sohio, and Cleveland Cliffs halt Colony oil shale project begun in 1964 after 270,000 bbl of production
1972	Occidental Petroleum conducts first of six in situ oil shale experiments at Logan Wash
1972	Paraho is formed as a consortium of 17 companies, obtains a lease of Anvil Points facility and builds and operates 24 ton/day pilot plant and 240 ton/day semi-works plant
1970s	Shell researches Piceance Creek in situ steam injection process for oil shale and nahcolite
1974	Four oil shale leases issued by BLM under Interior's Prototype Leasing Program
1974	Unocal develops new "Union B" retort process; Shell and Ashland join Colony Project
1976	Navy contracts with Paraho to produce 100,000 bbl of shale oil for testing as a military fuel
1976	Unocal begins planning commercial scale plant at Parachute Creek to be built when investment is economic; imported oil prices reach $41 per barrel
1977	Superior Oil abandons plan for Meeker oil shale plant planned since 1972
1979	Shell, Ashland, Cleveland Cliffs, and Sohio sell interests in Colony to ARCO and Tosco; Shell sells leases to Occidental and Tenneco
1979	Congress passes Energy Security Act, establishing U.S. Synthetic Fuels Corporation; authorizes up to $88 billion for synthetic fuels projects, including oil shale
1980	Exxon buys Arco's Colony interest and in 1981 begins Colony II construction, designed for 47,000 bbl/day using Tosco II retort process
1980	Congress approves $14 billion for synthetic fuels development
1980	Unocal plans Long Ridge 50,000 bbl/day plant applying "Union B" retort; begins construction in 1981
1980	Amoco Rio Blanco produces 1900 bbl of in situ oil at C-a tract
1981	Exxon begins to build Battlement Mesa company town for oil shale workers
1981	Second Rio Blanco in situ retort demonstration produces 24,400 bbl of shale oil

TABLE 6.1 Timeline of Oil Shale Projects in the United States (*Continued*)

1982	Oil demand falls and crude oil prices collapse
1982	Exxon Black Sunday: announces closure of Colony II due to cost and lower demand
1982	Shell continues in situ experiments at Red Pinnacle and laboratories through 1983
1985	Congress abolishes Synthetic Liquid Fuels Program after 40 years and $8 billion
1987	Shell purchases Ertl-Mahogany and Pacific tracts in Colorado
1987	Paraho reorganizes as New Paraho and begins production of SOMAT asphalt additive used in test strips in five states
1990	Exxon sells Battlement Mesa for retirement community
1991	Occidental closes C-b tract project before first retort begins operation
1991	Unocal closes Long Ridge after 5 million barrels and 10 years for operational issues and losses
1991	LLNL plans $20 million experiment plant at Parachute; Congress halts test funds in 1993
1991	New paraho reports successful tests of SOMAT shale oil asphalt additive
1997	DOE cedes oil shale lands to DOI/BLM
1997	Shell tests in situ heating on Mahogany property; defers further work on economic basis
2000	BLM seeks public comment on management of oil shale lands
2000	Shell returns to Mahogany with expanded in situ heating technology research plan (ongoing)
2004	DOE Office of Naval Petroleum and Oil Shale Reserves initiates study of the strategic significance of America's oil shale resources

By 1982, however, technology advances and new discoveries of offshore oil resources in the North Sea and elsewhere provided new and diverse sources for oil imports into the United States, and dampened global energy prices. Global political shifts promised to open previously restricted provinces to oil and gas exploration, and led economists and other experts to predict a long future of relatively low and stable oil prices. Despite significant investments by energy companies, numerous variations and advances in mining, restoration, retorting, and in situ processes, the costs of oil shale production relative to foreseeable oil prices made continuation of most commercial efforts impractical. During this time, numerous projects that were initiated and then terminated, primarily due to economic infeasibility relative to expected world oil prices or project design issues.

Several projects failed for technical and design reasons. Federal research and development, leasing, and other activities were significantly curtailed, and most commercial projects were abandoned. The collapse of world oil prices in 1984 seemed to seal the fate of oil shale as a serious player in the nation's energy strategy.

Oil shale is a complex and intimate mixture of organic and inorganic materials that vary widely in composition and properties. In general terms, oil shale is a fine-grained sedimentary rock that is rich in organic matter and yields oil when heated. Some oil shale is genuine shale but others have been misclassified and are actually siltstones, impure limestone, or even impure coal. Oil shale does not contain oil and only produces oil when it is heated to about 500°C (932°F), when some of the organic material is transformed into a distillate similar to crude oil.

There is no scientific definition of *oil shale* and the current definition is based on economics. Generally, oil shale is a mixture of carbonaceous molecules dispersed in an inorganic (mineral) matrix. It is called shale because it is found in a layered structure typical of sedimentary rocks, but the mineral composition can vary from true aluminosilicate shale to carbonate minerals. Thus, *oil shale* is a compact, laminated rock of sedimentary origin that yields over 33 percent of ash and contains insoluble organic matter that yields oil when distilled.

The amount of kerogen in the shale varies with depth, with the richer portions appearing much darker. For example, in Colorado (U.S.), the richest layers are termed the *Mahogany Zone* after the rich brown color.

Oil production potential from oil shale is measured by a laboratory pyrolysis method called Fischer assay (Speight, 1994) and is reported in barrels (42 gal) per ton. Rich zones can yield more than 40 gal/ton, while most shale falls in the range of 10 to 25 gal/ton. Shale oil yields of more than 25 gal/ton are generally viewed as the most economically attractive, and hence, the most favorable for initial development.

Despite the huge resources, oil shale is an under-utilized energy resource. In fact, one of the issues that arise when dealing with fuels from oil shale is the start-stop-start episodic nature of the various projects. The projects have varied in time and economic investment and viability. The reasons comprise competition from cheaper energy sources, heavy front-end investments and, of late, an unfavorable environment record. Oil shale has, though, a definite potential for meeting energy demand in an environmentally acceptable manner (Bartis et al., 2005; Andrews, 2006).

Just like the term *oil sand* (tar sand in the United States), the term *oil shale* is a misnomer since the mineral does not contain oil nor is it always *shale*. The organic material is chiefly *kerogen* and the *shale* is usually a relatively hard rock, called marl. Properly processed, kerogen can be converted into a substance somewhat similar to petroleum which is often better than the lowest grade of oil produced from conventional oil reservoirs but of lower quality than conventional light oil.

Shale oil, sometimes termed *retort oil*, is the liquid oil condensed from the effluent in oil shale retorting and typically contains appreciable amounts of water and solids, as well as having an irrepressible tendency to form sediments. However, shale oils are sufficiently different from crude oil that processing shale oil presents some unusual problems.

Retorting is the process of heating oil shale in order to recover the organic material, predominantly as a liquid. To achieve economically attractive recovery of product, temperatures of 400 to 600°C (752–1112°F) are required. A retort is simply a vessel in which the oil shale is heated from which the product gases and vapors can escape to a collector.

6.1 ORIGIN

In the creation of petroleum, source rocks are buried by natural geologic processes and, over geologic time, converts the organic materials to liquids and gases that can migrate through cracks and pores in the rocks until it reaches the surface or is trapped by a tight overhead formation. The result is an oil and/or gas reservoir.

Oil shale was deposited in a wide variety of environments including freshwater to saline ponds and lakes, epicontinental marine basins, and related subtidal shelves. They were also deposited in shallow ponds or lakes associated with coal-forming peat in limnic and coastal swamp depositional environments. It is not surprising, therefore, that oil shale exhibit a wide range in organic and mineral composition. Most oil shale contain organic matter derived from varied types of marine and lacustrine algae, with some debris of land plants, depending upon the depositional environment and sediment sources.

Oil shale does not undergo that natural maturation process but produces the material that has come to be known as kerogen (Scouten, 1990). In fact, there are indications that kerogen, being different to petroleum, may be a by-product of the maturation process. The kerogen residue that remains in oil shale is formed during maturation and is then rejected from the organic matrix because of its insolubility and relative unreactivity under the maturation conditions (Speight, 2007; Chap. 4). Furthermore, the fact that kerogen, under the conditions imposed upon it in the laboratory by high-temperature pyrolysis, forms hydrocarbon products does not guarantee that the kerogen of oil shale is a precursor to petroleum.

Oil shale ranging from Cambrian to Tertiary in age occurs in many parts of the world (Table 6.2). Deposits range from small occurrences of little or no economic value to those

TABLE 6.2 Estimate of Oil Shale Reserves (million metric tons)

Region	Shale reserves	Kerogen reserves	Kerogen in place
Africa	12,373	500	5,900
Asia	20,570	1,100	–
Australia	32,400	1,700	37,000
Europe	54,180	600	12,000
Middle East	35,360	4,600	24,000
North America	3,340,000	80,000	140,000
South America	–	400	10,000

Source: World Energy Council. 2004. WEC Survey of Energy Resources. London, England.
To convert metric ton to barrels, multiply by 7 indicating approximately 620 billion barrels of known recoverable kerogen, which has been estimated to be capable of producing 2,600 billion barrels of shale oil. This compares with 1,200 billion barrels of known worldwide petroleum reserves (*Source: BP Statistical Review of World Energy, 2006*).

of enormous size that occupy thousands of square miles and contain many billions of barrels of potentially extractable shale oil. Total world resources of oil shale are conservatively estimated at 2.6 trillion barrels. However, petroleum-based crude oil is cheaper to produce today than shale oil because of the additional costs of mining and extracting the energy from oil shale. Because of these higher costs, only a few deposits of oil shale are currently being exploited in China, Brazil, and Estonia. However, with the continuing decline of petroleum supplies, accompanied by increasing costs of petroleum-based products, oil shale presents opportunities for supplying some of the fossil energy needs of the world in the years ahead.

6.2 HISTORY

The use of oil shale can be traced back to ancient times. By the seventeenth century, oil shale was being exploited in several countries. As early as 1637, alum shale was roasted over wood fires to extract potassium aluminum sulfate, a salt used in tanning leather and for fixing colors in fabrics. The modern use of oil shale to produce oil dates to Scotland in the 1850s. In 1847 Dr James Young prepared lighting oil, lubricating oil, and wax from coal. Then he moved his operations to Edinburgh where oil shale deposits were found. In 1850 he patented the process of cracking the oil into its constituent parts. Oil from oil shale was produced in that region from 1857 until 1962 when production was cancelled due to the much lower cost of petroleum. Late in the 1800s, oil shale was being retorted on a small-scale for hydrocarbon production and this was terminated in 1966 because of the availability of cheaper supplies of petroleum crude oil.

Estonia first used oil shale as a low-grade fuel in 1838 after attempts to distill oil from the material failed. However it was not exploited until fuel shortages during World War I. Mining began in 1918 and has continued since, with the size of operation increasing with demand. After World War II, Estonian-produced oil shale gas was used in Leningrad and the cities in North Estonia as a substitute for natural gas. Two large oil shale-fired power stations were opened, a 1400 MW plant in 1965 and a 1600 MW plant in 1973. Oil shale production peaked in 1980 at 31.35 million tonnes. However, in 1981 the fourth reactor of the Sosnovy Bor nuclear power station opened in the nearby in Leningrad Oblast of Russia, reducing demand for Estonian shale. Production gradually decreased until 1995, since when production has increased again albeit only slightly. In 1999 the country used

11 million tonnes of shale in energy production, and plans to cut oil shale's share of primary energy production from 62 to 47–50 percent in 2010.

Australia mined 4 million tonnes of oil shale between 1862 and 1952, when government support of mining ceased. More recently, from the 1970s on, oil companies have been exploring possible reserves. Since 1995 Southern Pacific Petroleum N.L. and Central Pacific Minerals N.L. (SPP/CPM) (at one time joined by the Canadian company Suncor) has been studying the Stuart deposit near Gladstone, Queensland, which has a potential to produce 2.6 billion barrels of oil. From June 2001 through to March 2003, 703,000 barrels of oil, 62,860 barrels of light fuel oil, and 88,040 barrels of ultra-low sulfur naphtha were produced from the Gladstone area. Once heavily processed, the oil produced will be suitable for production of low-emission petrol. SPP was placed in receivership in 2003, and by July 2004, Queensland Energy Resources announced an end to the Stuart Shale Oil project in Australia.

Brazil has produced oil from oil shale since 1935. Small demonstration oil-production plants were built in the 1970s and 1980s, with small-scale production continuing today. China has been mining oil shale to a limited degree since the 1920s near Fushun, but the low price of crude oil has kept production levels down. Russia has been mining its reserves on a small-scale basis since the 1930s.

Oil shale retorting was also carried out in Fushun, Manchuria in 1929 and, while under the control of the Japanese during World War II, production reached a rate of 152,000 gal/day (575,000 L/day) of crude shale oil. It has been estimated that production of the Chinese oil shale industry in the mid-1970s was expanded from 1.7 to 2.7 million gallons per day (6.5–9.5 million liters per day).

In the United States many pilot-retorting processes have been tested for short periods. Among the largest were a semicommercial-size retort operated by Union Oil in the late 1950s, which processed 1100 tons/day of high-grade shale. A pilot plant operated by TOSCO (The Oil Shale Corporation) processed 900 tons/day of high-grade shale in the early 1970s. For a shale grade of 37 gal/ton (140 L/day), these feed rates correspond, respectively, to production of 43,500 gal/day (165,000 L/day) and 357,000 gal/day (135,000 L/day) of crude shale oil.

Oil distilled from shale was first burnt for horticultural purposes in the nineteenth century, but it was not until the 1900s that larger investigations were made and the Office of Naval Petroleum and Oil Shale Reserves was established in 1912. The reserves were seen as a possible emergency source of fuel for the military, particularly the U.S. Navy, which had, at the beginning of the twentieth century, converted its ships from coal to fuel oil, and the nation's economy was transformed by gasoline-fueled automobiles and diesel-fueled trucks and trains. Concerns have been raised about assuring adequate supplies of liquid fuels at affordable prices to meet the growing needs of the nation and its consumers.

6.3 OCCURRENCE AND DEVELOPMENT

Oil shale is sedimentary marlstone rock that is embedded with rich concentrations of organic material known as kerogen. The western oil shale of the United States contains approximately 15 percent organic material, by weight. By heating oil shale to high temperatures, kerogen can be released and converted to a liquid that, once upgraded, can be refined into a variety of liquid fuels, gases, and high-value chemical and mineral byproducts.

Oil shale represents a large and mostly untapped source of hydrocarbon fuels. Like oil sands, it is an unconventional or alternate fuel source and it does not contain oil. Oil is produced by thermal decomposition of the kerogen, which is intimately bound within the shale matrix and is no readily extractable.

Many estimates have been published for oil shale reserves (in fact resources), but the rank of countries vary with time and authors, except that United States is always on number one over 60 percent. Brazil is the most frequent number two. The United States has vast known oil shale resources that could translate into as much as 2.2 trillion barrels of known kerogen *oil-in-place*. In fact, the largest known oil shale deposits in the world are in the Green River Formation, which covers portions of Colorado, Utah, and Wyoming. Estimates of the oil resource in place within the Green River Formation range from 1.5 to 1.8 trillion barrels. However, not all resources in place are recoverable. For potentially recoverable oil shale resources, there is an approximate upper boundary of 1.1 trillion barrels of oil and a lower boundary of about 500 billion barrels. For policy planning purposes, it is enough to know that any amount in this range is very high. For example, the midpoint in the estimate (800 billion barrels) is more than triple the proven oil reserves of Saudi Arabia. With present demand for petroleum products in the United States at approximately 20 million barrels per day, oil shale by only meeting a quarter of that demand would last for more than 400 years.

Oil shale represents a large and mostly untapped hydrocarbon resource. Like tar sand (*oil sand* in Canada), oil shale is considered unconventional because oil cannot be produced directly from the resource by sinking a well and pumping. Oil has to be produced thermally from the shale. The organic material contained in the shale is called *kerogen*, a solid material intimately bound within the mineral matrix.

Oil shale occurs in nearly 100 major deposits in 27 countries worldwide (Duncan and Swanson, 1965; Culbertson and Pitman, 1973). It is generally shallower (<3000 ft) than the deeper and warmer geologic zones required to form oil. Worldwide, the oil shale resource base is believed to contain about 2.6 trillion barrels, of which the vast majority, or about 2 trillion barrels, (including eastern and western shale), is located within the United States.

6.3.1 Australia

Australia has reported a proved amount in place of 32.4 billion metric tons of oil shale, with proved recoverable reserves of oil put at 1725 million metric tons. Additional reserves of shale oil are huge: in excess of 35 billion metric tons.

Production from oil shale deposits in southeastern Australia began in the 1860s, coming to an end in the early 1950s when government funding ceased. Between 1865 and 1952 some 4 million metric tons of oil shale were processed. During the 1970s and early 1980s a modern exploration program was undertaken by two Australian companies, Southern Pacific Petroleum (SPP) and Central Pacific Minerals (CPM). The aim was to find high-quality oil shale deposits amenable to open-pit mining operations in areas near infrastructure and deepwater ports. The program was successful in finding a number of silica-based oil shale deposits of commercial significance along the coast of Queensland.

In 1995, SPP and CPM signed a joint venture agreement with the Canadian company Suncor Energy Inc., to commence development of one of the oil shale deposits, the Stuart deposit. Located near Gladstone, it has a total in situ shale oil resource of 2.6 billion barrels and the capacity to produce more than 200,000 bbl/day. Suncor had had the role of operator of the Stuart project. In April 2001, SPP and CPM purchased Suncor's interest.

The Stuart project is conceived as processing (stage 3) 125,000 ton/day of oil shale to give 65,000 bbl/day of shale oil products, bringing total Stuart production to about 85,000 bbl/day by 2009.

The raw shale oil produced will constitute relatively light crude with 42° API gravity (American Petroleum Institute), 0.4 percent by weight of sulfur and 1.0 percent by weight nitrogen. To meet the needs of the market, the raw oil requires further processing, resulting in raw low-sulfur naphtha and medium shale oil. It is planned that the medium shale oil will be sent directly to tankage for marketing as 27° API gravity, 0.4 percent by weight of sulfur fuel oil cutter stock, while the raw naphtha will be hydrotreated to remove nitrogen and sulfur to below 1 ppm. The hydrotreated naphtha may also provide a feedstock for the manufacture of clean gasoline with low emissions characteristics.

6.3.2 Brazil

The oil shale resource base is one of the largest in the world and was first exploited in the late nineteenth century in the State of Bahia.

Brazil started production a long time ago in 1881 and has the second rank after the United States for resources (well distributed) and after Estonia for production. In 1935, shale oil was produced at a small plant in Sìo Mateus do Sul in the State of Paran and in 1950, following government support, a plant capable of producing 10,000 bbl/day shale oil was proposed for Trememb, Sìo Paulo.

Following the formation of Petrobras in 1953, the company developed the Petrosix process for shale transformation. Concentrating its operations on the reservoir of Sìo Mateus do Sul, the company brought a pilot plant (8 in internal diameter retort) into operation in 1982. Its purpose is for oil shale characterization, retorting tests, and developing data for economic evaluation of new commercial plants. A 6-ft retort (internal diameter) demonstration plant followed in 1984 and is used for the optimization of the Petrosix technology.

A 2200 ton/day, 18 ft (internal diameter) semi-works retort (the Irati Profile Plant), originally brought on line in 1972, began operating on a limited commercial scale in 1981 and a further commercial plant, the 36 ft (internal diameter) Industrial Module retort was brought into service in December 1991. Together the two commercial plants process some 7800 t of bituminous shale daily. The retort process (Petrosix), where the shale undergoes pyrolysis, yields a nominal daily output of 3870 bbl shale oil, 120 t of fuel gas, 45 t of liquefied shale gas and 75 t of sulfur. Output of shale oil in 1999 was 195.2 thousand metric tons.

Brazil developed the world's largest surface oil shale pyrolysis reactor being the Petrosix 11-m vertical shaft gas combustion retort (GCR). However, it seems that the Brazilian success in oil and in biomass to liquids has put oil shale in the shade, as no new plan is found on the web. The production of shale oil dropped from 3900 bbl/day in 1999 to 3100 bbl/day in 2002. Late figures are not known with any degree of certainty.

6.3.3 Canada

Oil shale occurs throughout the country, with the best known and most explored deposits being those in the provinces of Nova Scotia and New Brunswick. Of the areas in Nova Scotia known to contain oil shale, development has been attempted at two—Stellarton and Antigonish. Mining took place at Stellarton from 1852 to 1859 and 1929 to 1930 and at Antigonish around 1865. The Stellarton Basin is estimated to hold some 825 million metric tons of oil shale, with an in situ oil content of 168 million barrels. The Antigonish Basin has the second largest oil shale resource in Nova Scotia, with an estimated 738 million metric tons of shale and 76 million barrels of oil in situ.

Investigations into retorting and direct combustion of Albert Mines shale (New Brunswick) have been conducted, including some experimental processing in 1988 at the Petrobras plant in Brazil. Interest has been shown in the New Brunswick deposits for the potential they might offer to reduce sulfur emissions by cocombustion of carbonate-rich shale residue with high-sulfur coal in power stations.

6.3.4 China

Fushun, a city in the northeastern province of Liaoning, is where oil shale from the Eocene Jijuntun Formation is mined.

The average thickness of the Jijuntun Formation is estimated to be 115 m (within a range of 48–190 m). The oil shale in the formation can be divided into two parts of differing composition: the lower 15 m of light-brown oil shale of low-grade and the upper 100 m of brown to dark-brown, finely laminated oil shale. The oil content of the low-grade oil shale is less than 4.7 percent by weight and the richer upper grade is greater than 4.7 percent. However, depending on the exact location of the deposit, the maximum oil content can be as high as 16 percent. It has been reported that the average oil content is 7 to 8 percent which would produce in the region of 78 to 89 L of oil per metric ton of oil shale (assuming a 0.9 specific gravity).

In 1961, China was producing one-third of its total oil from shale oil. In 1983, the Chinese reported that the oil shale resources in the area of the West Open-Pit mine were 260 million metric tons, of which 235 million metric tons were considered mineable. It has also been reported that the entire Fushun area has a resource of approximately 3.6 billion metric tons.

The commercial extraction of oil shale and the operation of heating retorts for processing the oil shale were developed in Fushun between 1920 and 1930. After World War II, Refinery No. 1 had 200 retorts, each with a daily throughput of 100 to 200 t of oil shale. It continued to operate and was joined by the Refinery No. 2 starting up in 1954. In Refinery No. 3 shale oil was hydrotreated for producing light liquid fuels. Shale oil was also open-pit mined in Maoming, Guangdong Province and 64 retorts were put into operation there in the 1960s.

At the beginning of the 1960s, 266 retorts were operating in Fushun Refinery Nos. 1 and 2. However, by the early 1990s the availability of much cheaper crude oil had led to the Maoming operation and Fushun Refinery Nos. 1 and 2 being shut down.

A new facility—the Fushun Oil Shale Retorting Plant—came into operation in 1992 under the management of the Fushun Bureau of Mines. Its 60 retorts annually produce 60,000 t of shale oil to be sold as fuel oil, with carbon black as a by-product.

6.3.5 Estonia

Oil shale was first scientifically researched in the eighteenth century. In 1838, work was undertaken to establish an opencast pit near the town of Rakvere and an attempt was made to obtain oil by distillation. Although it was concluded that the rock could be used as solid fuel and, after processing, as liquid or gaseous fuel, the *kukersite* (derived from the name of the locality) was not exploited until the fuel shortages created by World War I began to impact.

The Baltic Oil Shale Basin is situated near the northwestern boundary of the East European Platform. The Estonia and Tapa deposits are both situated in the west of the Basin, the former being the largest and highest-quality deposit within the Basin.

Since 1916, oil shale has had an enormous influence on the energy economy, particularly during the period of Soviet rule and then under the reestablished Estonian Republic. At a very early stage, an oil shale development program declared that kukersite could be used directly as a fuel in the domestic, industrial, or transport sectors. Moreover, it is easily mined and could be even more effective as a combustible fuel in power plants or for oil distillation. Additionally kukersite ash could be used in the cement and brick-making industries.

Permanent mining began in 1918 and has continued until the present day, with capacity (both underground and opencast mining) increasing as demand rose. By 1955, oil shale output had reached 7 million metric tons and was mainly used as a power station/chemical plant fuel and in the production of cement. The opening of the 1400 MW Baltic Thermal

Power Station in 1965 followed; in 1973, the 1600 MW Estonian Thermal Power Station again boosted production, and by 1980 (the year of maximum output) the figure had risen to 31.35 million metric tons.

In 1981, the opening of a nuclear power station in the Leningrad district of Russia signaled the beginning of the decline in Estonian oil shale production. No longer were vast quantities required for power generation and the export of electricity. The decline lasted until 1995, with some small annual increases thereafter.

The Estonian government has taken the first steps toward privatization of the oil-shale industry and is beginning to tackle the air and water pollution problems that nearly a century of oil shale processing has brought. In 1999, the oil shale produced was 10.7 million metric tons. Imports amounted to 1.4 million metric tons, 0.01 million metric tons were exported, 11.1 million metric tons used for electricity and heat generation, and 1.3 million metric tons were distilled to produce 151,000 t of shale oil.

Estonian oil shale resources are currently put at 5 billion metric tons including 1.5 billion metric tons of active (mineable) reserves. It is possible that the power production part of the industry will disappear by 2020 and that the resources could last for 30 to 50 years but scenarios abound on the replacement of oil shale by alternative resources.

Until recently only 16 percent of Estonian shale was used for petroleum and chemical manufacturing. However, because of the environmental problems the goal is to decrease oil shale production.

6.3.6 Germany

A minimal quantity (0.5 million metric tons per annum) of oil shale is produced for use at the Rohrback cement works at Dotternhausen in southern Germany, where it is consumed directly as a fuel for power generation, the residue being used in the manufacture of cement.

6.3.7 Israel

Sizeable deposits of oil shale have been discovered in various parts of Israel, with the principal resources located in the north of the Negev desert. Israeli reported in 1998 that the proved amount of oil shale in place exceeded 15 billion metric tons, containing proved recoverable reserves of 600 million metric tons of shale oil. The largest deposit (Rotem Yamin) has shale beds with a thickness of 35 to 80 m, yielding 60 to 71 L of oil per metric ton. Israeli oil shale is generally relatively low in heating value and oil yield, and high in sulfur content, compared with other major deposits. A pilot power plant fuelled by oil shale has been technically proven in the Negev region. Annual production of oil shale has averaged 450,000 t in recent years.

6.3.8 Jordan

There are extremely large proven and exploitable reserves of oil shale in the central and northwestern regions of the country. The proved amount of oil shale in place is reported to be 40 billion metric tons; proved recoverable reserves of shale oil are put at 4 billion metric tons, with estimated additional reserves of 20 billion metric tons.

Jordanian shale is generally of quite good quality, with relatively low ash and moisture content. Gross calorific value (7.5 MJ/kg) and oil yield (8–12 percent) are on a par with those of western Colorado (U.S.) shale; however, Jordanian shale has exceptionally high sulfur content (up to 9 percent by weight of the organic content). The reserves are exploitable by opencast mining and are easily accessible.

Suncor of Canada has conducted limited exploration in the Lajjun area, southwest of Amman. During 1999, the company was engaged in discussions on the possible development of an oil shale extraction facility.

The eventual exploitation of the only substantial fossil fuel resource to produce liquid fuels and/or electricity, together with chemicals and building materials, would be favored by three factors: (a) the high organic-matter content of Jordanian oil shale, (b) the suitability of the deposits for surface mining, and (c) their location near potential consumers (i.e., phosphate mines, potash, and cement works).

6.3.9 Morocco

Morocco has very substantial oil shale reserves but to date they have not been exploited. During the early 1980s, Shell and the Moroccan state entity ONAREP conducted research into the exploitation of the oil-shale reserves at Tarfaya, and an experimental shale-processing plant was constructed at another major deposit (Timahdit). At the beginning of 1986, however, it was decided to postpone shale exploitation at both sites and to undertake a limited program of laboratory and pilot-plant research.

Morocco quotes the proved amount of oil shale in place as 12.3 billion metric tons, with proved recoverable reserves of shale oil amounting to 3.42 billion barrels (equivalent to approximately 500 million metric tons).

6.3.10 Russian Federation

There are oil shale deposits in Leningrad Oblast, across the border from those in Estonia. Annual output is estimated to be about 2 million metric tons, most of which is exported to the Baltic power station in Narva, Estonia. In 1999, Estonia imported 1.4 million metric tons of Russian shale but is aiming to reduce the amount involved, or eliminate the trade entirely. There is another oil-shale deposit near Syzran on the river Volga.

The exploitation of Volga Basin shale, which has a higher content of sulfur and ash, began in the 1930s. Although the use of such shale as a power station fuel has been abandoned owing to environment pollution, a small processing plant may still be operating at Syzran, with a throughput of less than 50,000 tonnes of shale per annum.

6.3.11 Thailand

Some exploratory drilling by the government was made as early as 1935 near Mae Sot in Tak Province on the Thai-Burmese border. The oil-shale beds are relatively thin and the structure of the deposit is complicated by folding and faulting.

Some 18.7 billion metric tons of oil shale have been identified in Tak Province but to date it has not been economic to exploit the deposits. Proved recoverable reserves of shale oil are put at 810 million metric tons.

6.3.12 United States

It is estimated that nearly 62 percent of the world's potentially recoverable oil shale resources are concentrated in the United States. The largest of the deposits is found in the 42,700 km^2 Eocene Green River Formation in northwestern Colorado, northeastern Utah, and southwestern Wyoming. The richest and most easily recoverable deposits are located in the Piceance Creek Basin in western Colorado and the Uinta Basin in eastern Utah. The shale oil can be extracted by surface and in situ methods of retorting: depending upon the methods of mining and processing used, as much as one-third or more of this resource might be recoverable. There is also the Devonian-Mississippian black shale in the eastern United States.

Data reported for the present survey indicate the vastness of U S. oil shale resources: the proved amount of shale in place is put at 3340 billion metric tons, with a shale oil content of 242 billion metric tons, of which about 89 percent is located in the Green River deposits and 11 percent in the Devonian black shale. Recoverable reserves of shale oil are estimated to be within the range of 60 to 80 billion metric tons, with additional resources put at 62 billion metric tons.

Oil distilled from shale was burnt and used horticulturally in the second half of the nineteenth century in Utah and Colorado but very little development occurred at that time. It was not until the early 1900s that the deposits were first studied in detail by USGS (U.S. Geological Survey) and the government established the Naval Petroleum and Oil Shale Reserves, which for much of the twentieth century served as a contingency source of fuel for the military. These properties were originally envisioned as a way to provide a reserve supply of oil to fuel U.S. naval vessels.

Oil shale development had always been on a small-scale but the project that was to represent the greatest development of the shale deposits was begun immediately after World War II in 1946—the U.S. Bureau of Mines established the Anvils Point oil shale demonstration project in Colorado. However, processing plants had been small and the cost of production high. It was not until the United States had become a net oil importer, together with the oil crises of 1973 and 1979, that interest in oil shale was reawakened.

In the latter part of the twentieth century, military fuel needs changed and the strategic value of the shale reserves began to diminish. In the 1970s, ways to maximize domestic oil supplies were devised and the oil shale fields were opened up for commercial production. Oil companies led the investigations: leases were obtained and consolidated but one-by-one these organizations gave up their oil shale interests. Unocal was the last to do so in 1991.

Recoverable resources of shale oil from the marine black shale in the eastern United States were estimated in 1980 to exceed 400 billion barrels. These deposits differ significantly in chemical and mineralogic composition from Green River oil shale. Owing to its lower atomic hydrogen/carbon ratio, the organic matter in eastern oil shale yields only about one-third as much oil as Green River oil shale, as determined by conventional Fischer assay analyses. However, when retorted in a hydrogen atmosphere, the oil yield of eastern oil shale increases by as much as 2.0 to 2.5 times the Fischer assay yield.

Green River oil shale contains abundant carbonate minerals including dolomite, nahcolite, and dawsonite. The latter two minerals have potential by-product value for their soda ash and alumina content, respectively. The eastern oil shale are low in carbonate content but contain notable quantities of metals, including uranium, vanadium, molybdenum, and others which could add significant by-product value to these deposits.

All field operations have ceased and at the present time shale oil is not being produced in the United States. Large-scale commercial production of oil shale is not anticipated before the second or third decade of the twenty-first century.

The most economically attractive deposits, containing an estimated 1.5 trillion barrels (richness of >10 gal/ton) are found in the Green River Formation of Colorado (Piceance Creek Basin), Utah (Uinta Basin), and Wyoming (Green River and Washakie Basins). Eastern oil shale underlies 850,000 acres of land in Kentucky, Ohio, and Indiana. In the Kentucky Knobs region in the Sunbury shale and the New Albany/Ohio shale, 16 billion barrels, at a minimum grade of 25 gal/ton, are located. Due to differences in kerogen type (compared to western shale) eastern oil shale requires different processing. Potential oil yields from eastern shale could someday approach yields from western shale, with processing technology advances (Johnson et al., 2004).

However, in spite of all of the numbers and projections, it is difficult to gather production data (given either in shale oil or oil shale in weight or in volume) and few graphs are issued. There are large discrepancies between percentages in reserve and in production because of the assumptions of *estimates* of the total resource and recoverable reserves. Thus, use of the data requires serious review.

6.4 LIQUID FUELS

In the United States, there are two principal oil shale types, the shale from the Green River Formation in Colorado, Utah, and Wyoming, and the Devonian-Mississippian black shale of the East and Midwest (Table 6.3) (Baughman, 1978). The Green River shale is considerably richer, occur in thicker seams, and has received the most attention for synthetic fuel manufacture and is, unless otherwise stated, the shale referenced in the following text.

TABLE 6.3 Composition (Percent By Weight) of the Organic Matter in Mahogany Zone and New Albany shale

Component	Green River Mahogany zone	New Albany
Carbon	80.5	82.0
Hydrogen	10.3	7.4
Nitrogen	2.4	2.3
Sulfur	1.0	2.0
Oxygen	5.8	6.3
Total	100.0	100.0
H/C atom ratio	1.54	1.08

Source: Baughman, G. L.: *Synthetic Fuels Data Handbook*, 2d ed., Cameron Engineers, Inc., Denver, Colo., 1978.

The common property of these two shale deposits is the presence of the kerogen. The chemical composition of the kerogen has been the subject of many studies (Scouten, 1990) but whether or not the findings are indicative of the true nature of the kerogen is extremely speculative. It is, however, a reasonable premise that kerogen from different shale samples varies in character, similar to petroleum from different reservoirs varying in quality and composition.

6.4.1 Thermal Decomposition of Oil Shale

The active devolatilization of oil shale begins at about 350 to 400°C, with the peak rate of oil evolution at about 425°C, and with devolatilization essentially complete in the range of 470 to 500°C (Hubbard and Robinson, 1950; Shih and Sohn, 1980). At temperatures near 500°C, the mineral matter, consisting mainly of calcium/magnesium and calcium carbonates, begins to decompose yielding carbon dioxide as the principal product. The properties of crude shale oil are dependent on the retorting temperature, but more importantly on the temperature-time history because of the secondary reactions accompanying the evolution of the liquid and gaseous products. The produced shale oil is dark brown, odoriferous, and tending to waxy oil.

Kinetic studies (Scouten, 1990) indicate that below 500°C the kerogen (organic matter) decomposes into bitumen with subsequent decomposition into oil, gas, and carbon residue. The actual kinetic picture is influenced by the longer time required to heat the organic material which is dispersed throughout the mineral matrix, and to the increased resistance to the outward diffusion of the products by the matrix which does not decompose. From the practical standpoint of oil shale retorting, the rate of oil production is the important aspect of kerogen decomposition.

The processes for producing oil from oil shale involve heating (retorting) the shale to convert the organic kerogen to a raw shale oil (Janka and Dennison, 1979; Rattien and Eaton, 1976; Burnham and McConaghy, 2006). Conversion of kerogen to oil without the agency of heat has not yet been proven commercially, although there are schemes for accomplishing such a task but, in spite of claims to the contrary, these have not moved into the viable commercial or even demonstration stage.

Thus, there are two basic oil shale retorting approaches: (a) mining followed by retorting at the surface and (b) in situ retorting, that is, heating the shale in place underground. Each method, in turn, can be further categorized according to the method of heating (Table 6.4) (Burnham and McConaghy 2006).

TABLE 6.4 Categorization of the Various Oil Shale Retorting Methods

Heading method	Above ground	Below ground
Conduction through a wall (various fuels)	Pumpherston, Fischer assay, ATP, Oil-Tech	Shell ICP (primary method)
Externally generated hot gas	Union B, Paraho Indirect, Superior Indirect	
Internal combustion	Union A, Paraho Direct, Superior Direct, Kiviter, Petrosix	Oxy MIS, LLNL RISE, Geokinetics Horizontal, Rio Blanco
Hot recycled solids (inert or burned shale)	Galoter, Lurgi, Chevron STB, LLNL HRS, Shell Spher	
Reactive fluids	IGT Hytort (high-pressure H_2), Donor solvent processes	Shell ICP (some embodiments)
Volumetric heating		ITTRI and LLNL radio-frequency

6.4.2 Mining and Retorting

With the exception of in situ processes, oil shale must be mined before it can be converted to shale oil. Depending on the depth and other characteristics of the target oil shale deposits, either surface or underground mining methods may be used.

Open-pit mining has been the preferred method whenever the depth of the target resource is favorable to access through overburden removal. In general, open-pit mining is viable for resources where the over burden is less than 150 ft in thickness and where the ratio of overburden-to-deposit thickness ratio is less than 1/1. Removing the ore may require blasting if the resource rock is consolidated. In other cases, exposed shale seams can be bulldozed. The physical properties of the ore, the volume of operations, and project economics determine the choice of method and operation.

When the depth of the overburden is too great, underground mining processes are required. Underground mining necessitates a vertical, horizontal, or directional access to the kerogen-bearing formation. Consequently, a strong *roof formation* must exist to prevent collapse or cave-ins, ventilation must be provided, and emergency egress must also be planned.

Oil shale can be mined using one of two methods: (a) underground mining using the room and pillar method or (b) surface mining. Room and pillar mining has been the preferred underground mining option in the Green River formations. Technology currently allows for cuts up to 90 ft in height to be made in the Green River formation, where ore-bearing zones can be hundreds of meters thick. Mechanical *continuous miners* have also been selectively tested in this environment.

After mining, the oil shale is transported to a facility for retorting after which the oil must be upgraded by further processing before it can be sent to a refinery, and the spent shale must be disposed of, often by putting it back into the mine. Eventually, the mined land is reclaimed. Both mining and processing of oil shale involve a variety of environmental impacts, such as global warming and greenhouse gas emissions, disturbance of mined land, disposal of spent shale, use of water resources, and impacts on air and water quality. The development of a commercial oil shale industry in the United States would also have significant social and economic impacts on local communities. Other impediments to development of the oil shale industry in the United States include the relatively high cost of producing oil from oil shale (currently greater than $60 per barrel) and the lack of regulations to lease oil shale.

Surface retorting involves transporting mined oil shale to the retort facility, retorting and recovering the raw kerogen oil, upgrading the raw oil to marketable products and disposing of the *spent* shale (Fig. 6.1). Retorting processes require mining more than a ton of shale to produce 1 bbl oil. The mined shale is crushed to provide a desirable particle size, injected into a heated reactor (*retort*), where the temperature is increased to about 450°C (842°F). At this temperature, the kerogen decomposes to a mixture of liquid and gas. One way the various retorting processes differ is in how the heat is provided to the shale by hot gas, a solid heat carrier, or conduction through a heated wall.

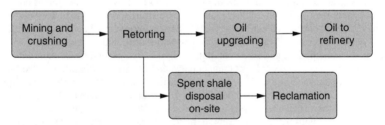

FIGURE 6.1 Process steps in mining and surface retorting.
(*Source: Bartis, J. T., T. LaTourrette, and L. Dixon: "Oil Shale Development in the United States: Prospects and Policy Issues," Prepared for the National Energy Technology of the United States Department of Energy, Rand Corporation, Santa Monica, Calif., 2005.*)

Advances in mining technology continue in other mineral exploitation industries, including the coal industry. Open-pit mining is a well established technology in coal, tar sand, and hard rock mining. Furthermore, room and pillar and underground mining have previously been proven at commercial scale for oil shale in the western United States. Costs for room and pillar mining will be higher than for surface mining, but these costs may be partially offset by having access to richer ore.

Current mining advances continue to reduce mining costs, lowering the cost of shale delivered to conventional retort facilities. Restoration approaches for depleted open-pit mines are demonstrated, both in oil shale operations and other mining industries.

The fundamental issue with all oil shale technologies is the need to provide large amounts of heat energy to decompose the kerogen to liquid and gas products. More than 1 t of shale must be heated to temperatures in the range 425 to 525°C (797–977°F) for each barrel of oil generated, and the heat supplied must be of relatively high quality to reach retorting temperature. Once the reaction is complete, recovering sensible heat from the hot rock is very desirable for optimum process economics.

This leads to three areas where new technology could improve the economics of oil recovery.

1. Recovering heat from the spent shale.
2. Disposal of spent shale, especially if the shale is discharged at temperatures where the char can catch fire in the air.
3. Concurrent generation of large volumes of carbon dioxide when the minerals contain limestone, as they do in Colorado and Utah.

Heat recovery from hot solids is generally not very efficient. The major exception to this generalization is in the field of fluidized bed technologies, where many of the lessons of fluids behavior can be applied. To apply fluidized bed technologies to oil shale would require grinding the shale to sizes less than about 1 mm, an energy intensive task that would result in an expensive disposal problem. However, such fine particles might be used in a lower temperature process for sequestering CO_2, with the costs of grinding now spread over to the solution of this problem.

Disposal of spent shale is also a problem that must be solved in economic fashion for the large-scale development of oil shale to proceed. Retorted shale contains carbon as a kind of char, representing more than half of the original carbon values in the shale. The char is potentially pyrophoric and can burn if dumped into the open air while hot. The heating process results in a solid that occupies more volume than the fresh shale because of the problems of packing random particles. A shale oil industry producing 100,000 bbl/day, about the minimum for a world-scale operation, would process more than 100,000 tons of shale (density about 3 g/cc) and result in more than 35 m^3 of spent shale; this is equivalent to a block more than 100 ft on a side (assuming some effort at packing to conserve volume). Unocal's 25,000 bbl/day project of the 1980s filled an entire canyon with spent shale over several years of operation. Some fraction of the spent shale could be returned to the mined-out areas for remediation, and some can potentially be used as feed for cement kilns.

Unocal's process relied on direct contact between hot gases passing downward through a rising bed of crushed shale. This required that the retorting shale be pumped upward against gravity. Retorted shale reaching the top of the retort spilled over the sides and was cooled as it left the vessel. Oil formed in the process trickled down through the bed of shale, exchanged its heat with fresh shale rising in the roughly conical retort, and was drawn from the bottom. Unocal produced 4.5 million barrels from 1980 until 1991 (AAPG, 2005) from oil shale averaging 34 gal/ton. The major problem that had to be overcome was formation of fine solids by decrepitation of the shale during retorting; the fines created problems in controlling solids flow in the retort and cooling shafts.

The TOSCO process used a rotating kiln that was reminiscent of a cement kiln in which heat was transferred to the shale by ceramic balls heated in an exterior burner. Retorted shale was separated from the balls using a coarse screen and the balls were recovered for recycle. Emerging vapors were cooled to condense product oil. The system was tested at the large pilot scale, but construction of a commercial retort was halted in 1982. One problem with the system was slow destruction of the ceramic balls by contact with the abrasive shale particles.

6.4.3 In Situ Technologies

In situ processes introduce heat to the kerogen while it is still embedded in its natural geologic formation. There are two general in situ approaches; true in situ, in which there is minimal or no disturbance of the ore bed, and modified in situ, in which the bed is given a rubblelike texture, either through direct blasting with surface up-lift or after partial mining to create void space. Recent technology advances are expected to improve the viability of oil shale technology, leading to commercialization.

Of importance in the in situ processing option is the characteristic that oil shale is relatively hard impermeable rocks through which fluids will not flow. Also of interest is the specific gravity of the oil shale since much of the inorganic material must be disposed. The specific gravity of the Green River kerogen is approximately 1.05 and the mineral fraction has an approximate value of 2.7 (Baughman, 1978).

In situ processes can be technically feasible where permeability of the rock exists or can be created through fracturing. The target deposit is fractured, air is injected, the deposit is ignited to heat the formation, and resulting shale oil is moved through the natural or man-made fractures to production wells that transport it to the surface (Fig. 6.2). However, difficulties in controlling the flame front and the flow of pyrolized oil can limit the ultimate oil recovery, leaving portions of the deposit unheated and portions of the pyrolized oil unrecovered.

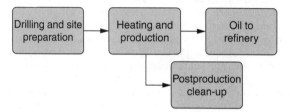

FIGURE 6.2 Process steps in thermal in situ conversion.
(*Source: Bartis, J. T., T. LaTourrette, and L. Dixon: "Oil Shale Development in the United States: Prospects and Policy Issues," Prepared for the National Energy Technology of the United States Department of Energy, Rand Corporation, Santa Monica, Calif., 2005.*)

Thus, in situ processes avoid the need to mine the shale but require that heat be supplied underground and that product be recovered from a relatively nonporous bed. As such, the in situ processes tend to operate slowly, behavior that the Shell in situ process (Fig. 6.3) exploits by heating the resource to around 650°F (343°C) over a period of 3 to 4 years.

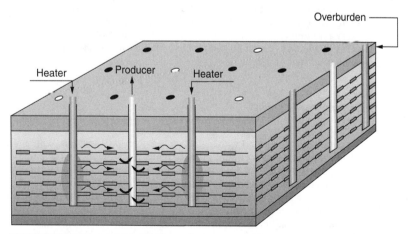

FIGURE 6.3 Schematic of the Shell in situ conversion process.
(*Source: Shell Exploration and Development Company, Houston, Texas.*)

The process involves use of ground-freezing technology to establish an underground barrier called "freeze wall" around the perimeter of the extraction zone. The freeze wall is created by pumping refrigerated fluid through a series of wells drilled around the extraction zone. The freeze wall prevents groundwater from entering the extraction zone, and keeps hydrocarbons and other products generated by the in situ retorting from leaving the project perimeter.

High yields of liquid products, with minimal secondary reactions, are anticipated (Mut, 2005; Karanikas et al., 2005).

In situ processes avoid the spent shale disposal problems because the spent shale remains where it is created but, on the other hand, the spent shale will contain uncollected liquids that can leach into ground water, and vapors produced during retorting can potentially escape to the aquifer (Karanikas et al., 2005).

Modified in situ processes attempt to improve performance by exposing more of the target deposit to the heat source and by improving the flow of gases and liquid fluids through the rock formation, and increasing the volumes and quality of the oil produced. Modified in situ involves mining beneath the target oil shale deposit prior to heating. It also requires drilling and fracturing the target deposit above the mined area to create void space of 20 to 25 percent. This void space is needed to allow heated air, produced gases, and pyrolized shale oil to flow toward production wells. The shale is heated by igniting the top of the target deposit. Condensed shale oil that is pyrolyzed ahead of the flame is recovered from beneath the heated zone and pumped to the surface.

The Occidental vertical modified in situ process was developed specifically for the deep, thick shale beds of the Green River Formation. About 20 percent of the shale in the retort area is mined; the balance is then carefully blasted using the mined out volume to permit expansion and uniform distribution of void space throughout the retort (Petzrick, 1995).

In this process, some of the shale was removed from the ground and explosively shattered the remainder to form a packed bed reactor within the mountain. Drifts (horizontal tunnels into the mountain) provided access to the top and bottom of the retort. The top of the bed was heated with burners to initiate combustion and a slight vacuum pulled on from the bottom of the bed to draw air into the burning zone and withdraw gaseous products. Heat from the combustion retorted the shale below, and the fire spread to the char left behind. Key to success was formation of shattered shale of relatively uniform particle size in the retort, at reasonable cost for explosives.

If the oils shale contains a high proportion of dolomite (a mixture of calcium carbonate and magnesium carbonate; for example, Colorado oil shale) the limestone decomposes at the customary retorting temperatures to release large volumes of carbon dioxide. This consumes energy and leads to the additional problem of sequestering the carbon dioxide to meet global climate change concerns.

6.5 REFINING SHALE OIL

Shale-retorting processes produce oil with almost no heavy residual fraction. With upgrading, shale oil is a light boiling premium product more valuable than most crude oils. However, the properties of shale oil vary as a function of the production (retorting) process. Fine mineral matter carried over from the retorting process and the high viscosity and instability of shale oil produced by present retorting processes have necessitated upgrading of the shale oil before transport to a refinery.

After fines removal the shale oil is hydrotreated to reduce nitrogen, sulfur, and arsenic content and improve stability; the cetane index of the diesel and heater oil portion is also improved. The hydrotreating step is generally accomplished in fixed catalyst bed processes under high

TABLE 6.5 Major Compound Types in Shale Oil

Saturates	Paraffins
	Cycloparaffins
Olefins	
Aromatics	Benzenes
	Indans
	Tetralins
	Naphthalenes
	Biphenyls
	Phenanthrenes
	Chrysenes
Heteroatom systems	Benzothiophenes
	Dibenzothiophenes
	Phenols
	Carbazoles
	Pyridines
	Quinolines
	Nitriles
	Ketones
	Pyrroles

hydrogen pressures, and hydrotreating conditions are slightly more severe than for comparable boiling range petroleum stocks, because of the higher nitrogen content of shale oil.

Shale oil contains a large variety of hydrocarbon compounds (Table 6.5) but also has high nitrogen content compared to a nitrogen content of 0.2 to 0.3 weight percent for a typical petroleum. In addition, shale oil also has a high olefin and diolefin content. It is the presence of these olefins and diolefins, in conjunction with high nitrogen content, which gives shale oil the characteristic difficulty in refining (Table 6.6) and the tendency to form insoluble sediment. Crude shale oil also contains appreciable amounts of arsenic, iron, and nickel that interfere with refining.

Upgrading, or partial refining, to improve the properties of a crude shale oil maybe carried out using different options. Hydrotreating is the option of choice to produce a stable product that is comparable to benchmark crude oils (Table 6.7). In terms of refining and catalyst activity, the nitrogen content of shale oil is a disadvantage. But, in terms of the use of shale oil residua as a modifier for asphalt, where nitrogen species can enhance binding

TABLE 6.6 Challenges for Oil Shale Processing

Particulates	Plugging on processing
	Product quality
Arsenic content	Toxicity
	Catalyst poison
High pour point	Oil not pipeline quality
Nitrogen content	Catalyst poison
	Contributes to instability
	Toxicity
Diolefins	Contributes to instability
	Plugging on processing

TABLE 6.7 Properties of Oil-Shale Distillates Compared with Benchmark Crude Oils

	API	%Sulfur
OTA reported oil-shale distillates properties[*]	19.4–28.4	0.59–0.92
Shell ICP oil-shale distillate[†]	34	0.8
Oil tech oil-shale distillate[‡]	30	no report
West Texas intermediate crude oil[§]	40	0.30
NYMEX deliverable grade sweet crude oil specification[¶]	37–42	<0.42
Alaska north slope crude oil[§]	29–29.5	1.10

[*] OTA, *An Assessment of Oil Shale Technologies*, Table 19, 1980.
[†] Energy Washington Week, "Shell Successfully Tests Pilot of New In Situ Oil Shale Technology," Oct. 12, 2005.
[‡] Jack Savage, Testimony Before the Subcommittee on Energy and Mineral Resources, June 23, 2005.
[§] *Platt's Oil Guide to Specifications*, 1999.
[¶] NYMEX, *Exchange Rulebook*, Light "Sweet" Crude Oil Futures Contract.
Source: Andrews, A.: "Oil Shale: History, Incentives, and Policy," Specialist, Industrial Engineering and Infrastructure Policy Resources, Science, and Industry Division, Congressional Research Service, the Library of Congress, Washington, D.C., 2006.

with the inorganic aggregate, the nitrogen content is beneficial. If not removed, the arsenic and iron in shale oil would poison and foul the supported catalysts used in hydrotreating.

Blending shale oil products with corresponding crude oil products, using shale oil fractions obtained from a very mildly hydrogen treated shale oil, yields kerosene and diesel fuel of satisfactory properties. Hydroprocessing shale oil products, either alone or in a blend with the corresponding crude oil fractions, is therefore necessary. The severity of the hydroprocessing has to be adjusted according to the particular property of the feed and the required level of the stability of the product.

Gasoline from shale oil usually contains a high percentage of aromatic and naphthenic compounds that are not affected by the various treatment processes. The olefin content, although reduced in most cases by refining processes, will still remain significant. It is assumed that diolefins and the higher unsaturated constituents will be removed from the gasoline product by appropriate treatment processes. The same should be true, although to a lesser extent, for nitrogen- and sulfur-containing constituents.

The sulfur content of raw shale oil gasoline may be rather high due to the high sulfur content of the shale oil itself and the frequently even distribution of the sulfur compounds in the various shale oil fractions. Not only the concentration but also the type of the sulfur compounds is of an importance when studying their effect on the gum-formation tendency of the gasoline containing them.

Sulfides (R-S-R), disulfides (R-S-S-R), and mercaptans (R-SH) are, among the other sulfur compounds, the major contributors to the gum formation in gasoline. Sweetening processes for converting mercaptans to disulfides should therefore not be used for shale oil gasoline; sulfur extraction processes are preferred.

Catalytic hydrodesulfurization processes are not a good solution for the removal of sulfur constituents from gasoline when high proportions of unsaturated constituents are present. A significant amount of the hydrogen would be used for hydrogenation of the unsaturated components. However, when hydrogenation of the unsaturated hydrocarbons is desirable, catalytic hydrogenation processes would be effective.

Gasoline derived from shale oil contains varying amounts of oxygen compounds. The presence of oxygen in a product, in which free radicals form easily, is a cause for concern. Free hydroxy radicals are generated and the polymerization chain reaction is quickly brought to its propagation stage. Unless effective means are provided for the termination of the polymerization process, the propagation stage may well lead to an uncontrollable generation of oxygen bearing free radicals leading to gum and other polymeric products.

Diesel fuel derived from oil shale is also subject to the degree of unsaturation, the effect of diolefins, the effect of aromatics, and to the effect of nitrogen and sulfur compounds.

Jet fuel produced from shale oil would have to be subjected to suitable refining treatments and special processes. The resulting product must be identical in its properties to corresponding products obtained from conventional crude oil. This can be achieved by subjecting the shale oil product to severe catalytic hydrogenation process with a subsequent addition of additives to ensure resistance to oxidation.

If antioxidants are used for a temporary reduction of shale oil instability, they should be injected into the shale oil (or its products) as soon as possible after production of the shale oil. The antioxidant types and their concentrations should be determined for each particular case separately.

The antioxidants combine with the free radicals or supply available hydrogen atoms to mitigate the progress of the propagation and branching processes. When added to the freshly produced unstable product, the antioxidants may be able to fulfill this purpose. However, when added after some delay, that is, after the propagation and the branching processes have advanced beyond controllable limits, the antioxidants would not be able to prevent formation of degradation products.

Exposure to oxygen is a major factor contributing to degradation product formation in shale oils. Peroxy radicals, that are readily formed when untreated shale oils or their products, are exposed to oxygen lead to rapid gum formation rate. Once oxygen is eliminated from such a system, the polymerization chain reaction tends to arrive to its termination stage. The termination stage of this polymerization chain reaction can take place by one of several ways, as for example exhaustion of the reactive monomers or a combination of two free radicals. Chain reaction termination can be so affected by radical combination or disproportionation.

In all cases free radicals have to be eliminated from the system. The chain termination can be also induced by certain constituents present naturally or added artificially in the form of antioxidants.

Thus, shale oil is different to conventional crude oils, and several refining technologies have been developed to deal with this. The primary problems identified in the past were arsenic, nitrogen, and the waxy nature of the crude. Nitrogen and wax problems were solved using hydroprocessing approaches, essentially classical hydrocracking and the production of high quality lube stocks, which require that waxy materials be removed or isomerized. However, the arsenic problem remains.

In general, oil-shale distillates have a much higher concentration of high boiling-point compounds that would favor production of middle-distillates (such as diesel and jet fuels) rather than naphtha. Oil-shale distillates also had a higher content of olefins, oxygen, and nitrogen than crude oil, as well as higher pour points and viscosities. Above-ground-retorting processes tended to yield a lower API gravity oil than the in situ processes (a 25° API gravity was the highest produced). Additional processing equivalent to hydrocracking would be required to convert oil-shale distillates to a lighter range hydrocarbon (gasoline). Removal of sulfur and nitrogen would, however, require hydrotreating.

By comparison, a typical 35° API gravity crude oil may be composed of up to 50 percent of gasoline and middle-distillate range hydrocarbons. West Texas Intermediate crude (benchmark crude for trade in the commodity futures market) has 0.3 percent by weight sulfur, and Alaska North Slope crude has 1.1 percent by weight sulfur. The New York Mercantile Exchange (NYMEX) specifications for light "sweet" crude limits sulfur content to 0.42 percent or less (ASTM D4294) and an API gravity between 37° and 42° (ASTM D287).

A conventional refinery distills crude oil into various fractions, according to boiling point range, before further processing. In order of their increasing boiling range and density, the distilled fractions are fuel gases, light and heavy straight-run naphtha (90–380°F), kerosene (380–520°F), gas-oil (520–1050°F), and residuum (above 1050°F) (Speight, 2007). Crude oil may contain 10 to 40 percent gasoline, and early refineries directly distilled a straight-run gasoline (light naphtha) of low-octane rating. A hypothetical refinery may "crack" a barrel of crude oil into two-thirds gasoline and one-third distillate fuel (kerosene, jet, and diesel), depending on the refinery's configuration, the slate of crude oils refined, and the seasonal product demands of the market.

Just as natural clay catalysts help transform kerogen to petroleum through catagenesis, metallic catalysts help transform complex hydrocarbons to lighter molecular chains in modern refining processes. The *catalytic-cracking* process developed during the World War II era enabled refineries to produce high-octane gasoline needed for the war effort. *Hydrocracking*, which entered commercial operation in 1958, improved on catalytic-cracking by adding hydrogen to convert residuum into high-quality motor gasoline and naphtha-based jet fuel. Many refineries rely heavily on *hydroprocessing* to convert low-value gas oils residuum to high-value transportation fuel demanded by the market. Middle-distillate range fuels (diesel and jet) can be blended from a variety of refinery processing streams. To blend jet fuel, refineries use desulfurized straight-run kerosene, kerosene boiling range hydrocarbons from a hydrocracking unit, and light coker gas-oil (cracked residuum). Diesel fuel can be blended from naphtha, kerosene, and light cracked-oils from coker and fluid catalytic cracking units. From the standard 42-gal barrel of crude oil, United States refineries may actually produce more than 44 gal of refined products through the catalytic reaction with hydrogen.

Oil derived from shale has been referred to as a synthetic crude oil and thus closely associated with synthetic fuel production. However, the process of retorting shale oil bears more similarities to conventional refining than to synthetic fuel processes. For the purpose of this report, the term *oil-shale distillate* is used to refer to middle-distillate range hydrocarbons produced by retorting oil shale. Two basic retorting processes were developed early on—aboveground retorting and underground, or in situ, retorting (qv). The retort is typically a large cylindrical vessel, and early retorts were based on rotary kiln ovens used in cement manufacturing. In situ technology involves mining an underground chamber that functions as a retort. A number of design concepts were tested from the 1960s through the 1980s.

Retorting essentially involves destructive distillation (*pyrolysis*) of oil shale in the absence of oxygen. Pyrolysis (temperatures above 900°F) thermally breaks down (*cracks*) the kerogen to release the hydrocarbons and then cracks the hydrocarbons into lower-weight hydrocarbon molecules. Conventional refining uses a similar thermal cracking process, termed *coking*, to break down high-molecular weight residuum.

As the demand for light hydrocarbon fractions constantly increases, there is much interest in developing economic methods for recovering liquid hydrocarbons from oil shale on a commercial scale. However, the recovered hydrocarbons from oil shale are not yet economically competitive against the petroleum crude produced. Furthermore, the value of hydrocarbons recovered from oil shale is diminished because of the presence of undesirable contaminants. The major contaminants are sulfurous, nitrogenous, and metallic (and organometallic) compounds, which cause detrimental effects to various catalysts used in the subsequent refining processes. These contaminants are also undesirable because of their disagreeable odor, corrosive characteristics, and combustion products that further cause environmental problems.

Accordingly, there is great interest in developing more efficient methods for converting the heavier hydrocarbon fractions obtained in a form of shale oil into lighter-molecular-weight hydrocarbons. The conventional processes include catalytic cracking, thermal cracking, and coking. It is known that heavier hydrocarbon fractions and refractory materials can be converted to lighter materials by hydrocracking. These processes are most commonly used on liquefied coals or heavy residual or distillate oils for the production of substantial yields of low-boiling saturated products, and to some extent on intermediates that are used as domestic fuels, and still heavier cuts that are used as lubricants. These destructive hydrogenation or hydrocracking processes may be operated on a strictly thermal basis or in the presence of a catalyst. Thermodynamically speaking, larger hydrocarbon molecules are broken into lighter species when subjected to heat.

The hydrogen/carbon atomic ratio of such molecules is lower than that of saturated hydrocarbons, and abundantly supplied hydrogen improves this ratio by saturating reactions, thus producing liquid species. These two steps may occur simultaneously. However, the application of hydrocracking process has been hampered by the presence of certain contaminants in such hydrocarbons. The presence of sulfur- and nitrogen-containing compounds along with organometallic compounds in crude shale oils and various refined petroleum products has long been considered undesirable. Desulfurization and denitrification processes have been developed for this purpose.

The thermal cracking process is directed toward the recovery of gaseous olefins as the primarily desired cracked product, in preference to gasoline range liquids. By this process, it is claimed that at least 15 to 20 percent of the feed shale oil is converted to ethylene, which is the most common gaseous product. Most of the feed shale oil is converted to other gaseous and liquid products. Other important gaseous products are propylene, 1,3-butadiene, ethane, and butanes. Hydrogen is also recovered as a valuable non-hydrocarbon gaseous product. Liquid products can comprise 40 to 50 weight percent or more of the total product. Recovered liquid products include benzene, toluene, xylene, gasoline-boiling-range liquids, and light and heavy oils. Coke is a solid product of the process and is produced by polymerization of unsaturated materials. Coke is typically formed in an oxygen-deficient environment via dehydrogenation and aromatization. Most of the formed coke is removed from the process as a deposit on the entrained inert heat carrier solids.

The thermal cracking reactor does not require a gaseous hydrogen feed. In the reactor, entrained solids flow concurrently through the thermal riser at an average riser temperature of 700 to 1400°C. The preferred high length-to-diameter (L-to-D) ratio is in the range of a high 4:1 to 40:1, or 5:1 to 20:1 preferably.

The moving bed hydroprocessing reactor is used to produce crude oil from oil shale or tar sands containing large amounts of highly abrasive particulate matter, such as rock dust and ash. The hydroprocessing takes place in a dual-function moving bed reactor, which simultaneously removes particulate matter by the filter action of the catalyst bed. The effluent from the moving bed reactor is then separated and further hydroprocessed in fixed bed reactors with fresh hydrogen added to the heavier hydrocarbon fraction to promote desulfurization.

A preferred way of treating the shale oil involves using a moving bed reactor followed by a fractionation step to divide the wide-boiling-range crude oil produced from the shale oil into two separate fractions. The lighter fraction is hydrotreated for the removal of residual metals, sulfur, and nitrogen, whereas the heavier fraction is cracked in a second fixed bed reactor normally operated under high-severity conditions.

The fluidized bed hydroretort process eliminates the retorting stage of conventional shale upgrading, by directly subjecting crushed oil shale to a hydroretorting treatment in an upflow, fluidized bed reactor such as that used for the hydrocracking of heavy petroleum residues. This process is a *single stage retorting and upgrading* process. Therefore, the process involves: (a) crushing oil shale, (b) mixing the crushed oil shale with a hydrocarbon liquid to provide a pumpable slurry, (c) introducing the slurry along with a hydrogen-containing gas into an upflow, fluidized bed reactor at a superficial fluid velocity sufficient to move the mixture upwardly through the reactor, (d) hydroretorting the oil shale, (e) removing the reaction mixture from the reactor, and (f) separating the reactor effluent into several components.

The mineral carbonate decomposition is minimized, as the process operating temperature is lower than that used in retorting. Therefore, the gaseous product of this process has a greater heating value than that of other conventional methods. In addition, owing to the exothermic nature of the hydroretorting reactions, less energy input is required per barrel of product obtained. Furthermore, there is practically no upper or lower limit on the grade of oil shale that can be treated.

Hydrocracking is a cracking process in which higher-molecular-weight hydrocarbons pyrolyze to lower-molecular-weight paraffins and olefins in the presence of hydrogen. The hydrogen saturates the olefins formed during the cracking process. Hydrocracking is used to process low-value stocks with high heavy metal content. It is also suitable for highly aromatic feeds that cannot be processed easily by conventional catalytic cracking. Shale oils are not highly aromatic, whereas coal liquids are very highly aromatic.

Middle-distillate (often called *mid-distillate*) hydrocracking is carried out with a noble metal catalyst. The average reactor temperature is 480°C, and the average pressure is around 130 to 140 atm. The most common form of hydrocracking is carried out as a two-stage operation. The first stage is to remove nitrogen compounds and heavy aromatics from the raw crude, whereas the second stage is to carry out selective hydrocracking reactions on the cleaner oil from the first stage. Both stages are processed catalytically. Once the hydrocracking stages are over, the products go to a distillation section that consists of a hydrogen sulfide stripper and a recycle splitter. Commercial hydrocracking processes include Gulf HDS, H-Oil, IFP Hydrocracking, Isocracking, LC-Fining, Microcat-RC (also known as M-Coke), Mild Hydrocracking, Mild Resid Hydrocracking (MRH), Residfining, Unicracking, and Veba Combi-Cracking (VCC).

Arsenic removed from the oil by hydrotreating remains on the catalyst, generating a material that is a carcinogen, an acute poison, and a chronic poison. The catalyst must be removed and replaced when its capacity to hold arsenic is reached. Unocal found that its disposal options were limited.

6.6 THE FUTURE

Oil shale still has a future and remains a viable option for the production of liquid fuels. Many of the companies involved in earlier oil shale projects still hold their oil shale technology and resource assets. The body of knowledge and understanding established by these past efforts provides the foundation for ongoing advances in shale oil production, mining, retorting, and processing technology and supports the growing worldwide interest and activity in oil shale development. In fact, in many cases, the technologies developed to produce and process kerogen oil from shale have not been abandoned, but rather *mothballed* for adaptation and application at a future date when market demand would increase and major capital investments for oil shale projects could be justified.

The fundamental problem with all oil shale technologies is the need to provide large amounts of heat energy to decompose the kerogen to liquid and gas products. More than 1 t of shale must be heated to temperatures in the range 454 to 537°C (850° to 1000°F) for each barrel of oil generated, and the heat supplied must be of relatively high quality to reach retorting temperature. Once the reaction is complete, recovering sensible heat from the hot rock is very desirable for optimum process economics. This leads to three areas where new technology could improve the economics of oil recovery: (a) recovering heat from the spent shale, (b) disposal of spent shale, especially if the shale is discharged at temperatures where the char can catch fire in the air, and (c) concurrent generation of large volumes of carbon dioxide.

The heat recovery from hot solids is generally not efficient, unless it is in the area of fluidized bed technology. However, to apply fluidized bed technology to oil shale would require grinding the shale to sizes less than about 1 mm, an energy intensive task that would result in an expensive disposal problem. However, such fine particles might be used in a lower temperature process for sequestering carbon dioxide. Disposal of spent shale is also a problem that must be solved in economic fashion for the large-scale development of oil shale to proceed.

Retorted shale contains carbon as char, representing more than half of the original carbon values in the shale. The char is potentially pyrophoric and can burn if dumped into the open air while hot. The heating process results in a solid that occupies more volume than the fresh shale because of the problems of packing random particles. A shale oil industry producing 100,000 bbl/day, about the minimum for a world-scale operation, would process more than 100,000 t of shale (density about 3 g/cc) and result in more than 35 m^3 of spent shale; this is equivalent to a block more than 100 ft on a side (assuming some effort at packing to conserve volume). Unocal's 25,000 bbl/day project of the 1980s filled an entire canyon with spent shale over several years of operation. Part of the spent shale could be returned to the mined-out areas for remediation, and some can potentially be used as feed for cement kilns.

In situ processes such as Shell's ICP avoid the spent shale disposal problems because the spent shale remains where it is created (Fletcher, 2005). In addition, ICP avoids carbon dioxide decomposition by operating at temperatures below about 350°C (662°F). On the other hand, the spent shale will contain uncollected liquids that can leach into groundwater, and vapors produced during retorting can potentially escape to the aquifer. Shell has gone to great efforts to design barrier methods for isolating its retorts to avoid these problems (Mut, 2005). Control of in situ operation is a challenge that Shell claims to have solved in its work (Mut, 2005; Karanikas et al., 2005).

Shale (such as the Colorado shale) that contains a high proportion of dolomitic limestone (a mixture of calcium and magnesium carbonates) thermally deposes under the conditions of retorting and releases large volumes of carbon dioxide. This consumes energy and leads to the additional problem of sequestering the carbon dioxide to meet global climate change concerns.

In addition, there are also issues with the produced shale oil that also need resolution.

Shale oil is different to conventional crude oils, and several technologies have been developed to deal with this. The primary problems identified were arsenic, nitrogen, and the waxy nature of the crude. Nitrogen and wax problems were solved by Unocal and other companies using hydroprocessing approaches, essentially classical hydrocracking. Since that time, Chevron and ExxonMobil have developed technologies aimed at making high-quality lube stocks, which require that waxy materials be removed or isomerized. These technologies are well adapted for shale oils. However, the arsenic problem remains (DOE, 2004b).

Unocal found that its shale oils contain several parts per million of arsenic. It developed a specialty hydrotreating catalyst and process, called for Shale Oil Arsenic Removal (SOAR). This process was demonstrated successfully in the 1980s and is now owned by UOP as part of the hydroprocessing package purchased from Unocal in the early 1990s. Unocal also patented other arsenic removal.

Arsenic removed from the oil by hydrotreating remains on the catalyst, generating a material that is a carcinogen, an acute poison, and a chronic poison. The catalyst must be removed and replaced when its capacity to hold arsenic is reached. Unocal found that its disposal options were limited. Today, regulations require precautions to be taken when a reactor is opened to remove a catalyst.

Thus several issues need to be resolved before an oil shale industry can be a viable option. These issues are not insurmountable but require the search for viable alternatives.

For example, an alternative not much explored involves chemical treatment of shale to avoid the high-temperature process. The analogy with coal liquefaction here is striking: liquids can be generated from coal in two distinct ways: (a) by pyrolysis, creating a char coproduct, or (b) by dissolving the coal in a solvent in the presence of hydrogen.

However, no similar "dissolution" approach to oil shale conversion is known, because the chemistry of kerogen is markedly different from the chemistry of coal.

As a first step in developing a direct route, some attempts were made in the 1970s to isolate kerogen from the oil shale by dissolving away the minerals. Acid treatment to dissolve the mineral carbonate followed by fluoride treatment to remove the aluminosilicate minerals might be considered. Such a scheme will only work if the kerogen is not chemically bonded to the inorganic matrix. However, if the kerogen is bonded to the inorganic matrix, the bonding arrangement must be defined for the scheme to be successful.

Opportunities for circumventing the arsenic problem include development of an in-reactor process for regenerating the catalyst, collecting arsenic in a safe form away from the catalyst, and development of a catalyst or process where the removed arsenic exits the reactor in the gas or liquid phase to be scrubbed and confined elsewhere.

Shale oil produced by both above-ground and in situ techniques in the 1970s and 1980s were rich in organic nitrogen. Nitrogen compounds are catalyst poisons in many common refinery processes such as fluid catalytic cracking, hydrocracking, isomerization, naphtha reforming, and alkylation. The standard method for handling nitrogen poisoning is hydrodenitrogenation (HDN).

HDN is a well-established high-pressure technology using nickel molybdenum catalysts. It can consume prodigious amounts of hydrogen, typically made by steam reforming of natural gas, with carbon dioxide as a by-product.

Thus, after a decline of production since 1980 and the current scenarios that face a petroleum-based economy, the perspectives for oil shale can be viewed with a moderately positive outlook. This perspective is prompted by the rising demand for liquid fuels, the rising demand for electricity, as well as the change of price relationships between oil shale and conventional hydrocarbons.

Experience in Estonia, Brazil, China, Israel, Australia, and Germany has already demonstrated that fuels and a variety of other products can be produced from oil shale at reasonable, if not competitive, cost. New technologies can raise efficiencies and reduce air and water pollution to sustainable levels and if innovative approaches are applied to waste remediation and carbon sequestration, oil shale technology take on a whole new perspective.

In terms of innovative technologies, both conventional and in situ retorting processes result in inefficiencies that reduce the volume and quality of the produced shale oil. Depending on the efficiency of the process, a portion of the kerogen that does not yield liquid is either deposited as coke on the host mineral matter, or is converted to hydrocarbon gases. For the purpose of producing shale oil, the optimal process is one that minimizes the regressive thermal and chemical reactions that form coke and hydrocarbon gases and maximizes the production of shale oil. Novel and advanced retorting and upgrading processes seek to modify the processing chemistry to improve recovery and/or create high-value by-products. Novel processes are being researched and tested in laboratory-scale environments. Some of these approaches include: lower heating temperatures; higher heating rates; shorter residence time durations; introducing scavengers such as hydrogen (or hydrogen transfer/donor agents); and introducing solvents (Baldwin, 2002).

Finally, the development of western oil shale resources will require water for plant operations, supporting infrastructure, and the associated economic growth in the region. While some oil shale technologies may require reduced process water requirements, stable and secure sources of significant volumes of water may still be required for large-scale oil shale development. The largest demands for water are expected to be for land reclamation and to support the population and economic growth associated with oil shale activity.

Nevertheless, if a technology can be developed to economically recover oil from oil shale, the potential is enormous. If the kerogen could be converted to oil, the quantities would be far beyond all known conventional oil reserves. Unfortunately, the prospects for oil shale development are uncertain (Bartis et al., 2005). The estimated cost of surface retorting remains high and many consider it unwise to move toward near-term commercial efforts.

However, advances in thermally conductive in situ conversion may cause shale-derived oil to be competitive with current high crude oil. If this becomes the case, oil shale development could soon occupy a very prominent position in the national energy agenda. Only when it is clear that at least one major private firm is willing to devote, without appreciable government subsidy, technical, management, and financial resources to oil shale development, will government decision makers address the policy issues related to oil shale development.

6.7 REFERENCES

AAPG: "Oil Shale," Energy minerals Division, American Association for Petroleum Geologists, Tulsa, Okla., 2005, http://emd.aapg.org/technical_areas/oil_shale.cfm.

Andrews, A.: "Oil Shale: History, Incentives, and Policy," Specialist, Industrial Engineering and Infrastructure Policy Resources, Science, and Industry Division, Congressional Research Service, the Library of Congress, Washington, D.C., 2006.

Baldwin, R. M.: *Oil Shale: A Brief Technical Overview,* Colorado School of Mines, Golden, Colo., July, 2002.

Bartis, J. T., T. LaTourrette, and L. Dixon: "Oil Shale Development in the United States: Prospects and Policy Issues," Prepared for the National Energy Technology of the United States Department of Energy, Rand Corporation, Santa Monica, Calif., 2005.

Baughman, G. L.: *Synthetic Fuels Data Handbook,* 2d ed., Cameron Engineers, Inc., Denver, Colo., 1978.

Burnham, A. K. and J. R. McConaghy: "Comparison of the Acceptability of Various Oil Shale Processes," Proceedings, AICHE 2006 Spring National Meeting, Orlando, Fla., Mar. 23–27, 2006.

Culbertson, W. C. and J. K. Pitman: "Oil Shale in United States Mineral Resources,"Paper No. 820, United States Geological Survey, Washington, D.C., 1973.

DOE: *Strategic Significance of America's Oil Shale Reserves, II. Oil Shale Resources, Technology, and Economics;* March, 2004, http://www.fe.doe.gov/programs/reserves/publications.

Duncan, D. C. and V. E. Swanson: "Organic-Rich Shale of the United States and World Land Areas," Circular No. 523. United States Geological Survey, Washington, D.C., 1965.

Fletcher, S.: "Efforts to Tap Oil Shale's Potential Yield Mixed Results," *Oil and Gas Journal,* Apr. 18, 2005.

Hubbard, A. B. and W. E. Robinson: "A Thermal Decomposition Study of Colorado Oil Shale," Report of Investigations No. 4744, United States Bureau of Mines, Washington, D.C., 1950.

Janka, J. C. and J. M. Dennison: "Devonian Oil Shale," Synthetic Fuels from Oil Shale Symposium, Institute of Gas Technology, Chicago, Ill., Dec. 3–6, 1979, pp. 21–116.

Johnson, H. R., P. M. Crawford, and J. W. Bunger: "Strategic Significance of America's Oil Shale Resource, Volume II, Oil Shale Resources, Technology and Economics," Office of Deputy Assistant Secretary for Petroleum Reserves, Office of Naval Petroleum and Oil Shale Reserves, United States Department of Energy, Washington, D.C., March, 2004.

Karanikas, J. M., E. P.de Rouffignac, H. J. Vinegar (Houston, Tex.), and S. Wellington: "In Situ Thermal Processing of An Oil Shale Formation While Inhibiting Coking," United States Patent 6877555, Apr., 12, 2005.

Mut, S.: "Oil Shale and Oil Sands Resources Hearing," Testimony before the United States Senate Energy and Natural Resources Committee, Tuesday, Apr. 12, 2005. http://energy.senate.gov/hearings/testimony.cfm?id=1445&wit_id=4139

Petzrick, P. A.: *Oil Shale and Tar Sand: Encyclopedia of Applied Physics,* vol. 12. VCH Publishers Inc., Berlin, Germany, 1995, pp. 77–99.

Rattien, S. and D. Eaton: In *Oil Shale: The Prospects and Problems of an Emerging Energy Industry,* vol. 1. J. M. Hollander and M. K. Simmons (eds.), Annual Review of Energy., 1976, pp. 183–212.

Scouten, C.: In *Fuel Science and Technology Handbook*, J. G. Speight (ed.), Marcel Dekker Inc., New York, 1990.

Shih, S. M. and H. Y. Sohn: *Ind. Eng. Chem. Process Des. Dev.,* 19, 1980, pp. 420–426.

Speight, J.G.: *The Chemistry and Technology of Coal,* 2d ed., 1994; Marcel Dekker, New York, 1994, p. 296.

Speight, J. G.: *The Chemistry and Technology of Petroleum,* 4th ed., CRC-Taylor and Francis Group, Boca Raton, Fla. 2007.

CHAPTER 7
FUELS FROM SYNTHESIS GAS

The hydrogen-to-carbon ratio of carbonaceous feedstocks such as heavy oil, tar sands, oil shale, coal, and biomass to liquid fuels is lower than the hydrogen to carbon ratio of conventional petroleum. In the process, the hydrogen/carbon atomic ratio must be adjusted to that of transportation fuels. One of the methods by which this can be achieved is through the application of the Fischer-Tropsch process to the conversion of synthesis gas to hydrocarbons.

Synthesis gas (*syngas*) is the name given to a gas mixture that contains varying amounts of carbon monoxide and hydrogen generated by the gasification of a carbon-containing fuel to a gaseous product with a heating value (Speight, 2007a and references cited therein). Examples include steam reforming of natural gas or liquid hydrocarbons to produce hydrogen, the gasification of coal, and in some types of waste-to-energy gasification facilities. The name comes from their use as intermediates in creating *synthetic natural gas* (SNG) and for producing ammonia or methanol. Synthesis gas is also used as an intermediate in producing synthetic petroleum for use as a fuel or lubricant via Fischer-Tropsch synthesis (Storch, 1945).

In the context of this book, synthesis gas, which can also be generated from biomass (Chap. 8), is not the same as biogas. Biogas is a clean and renewable form of energy generated from biomass that could very well substitute for conventional sources of energy. The gas is generally composed of methane (55–65 percent), carbon dioxide (35–45 percent), nitrogen (0–3 percent), hydrogen (0–1 percent), and hydrogen sulfide (0–1 percent) (Anunputtikul and Rodtong, 2004; Coelho et al., 2006).

Gasification to produce synthesis gas can proceed from just about any organic material, including biomass and plastic waste. The resulting syngas burns cleanly into water vapor and carbon dioxide. Alternatively, syngas may be converted efficiently to methane via the Sabatier reaction, or to a diesel-like synthetic fuel via the Fischer-Tropsch process. Inorganic components of the feedstock, such as metals and minerals, are trapped in an inert and environmentally safe form as char, which may have use as a fertilizer.

Regardless of the final fuel form, gasification itself and subsequent processing neither emits nor traps greenhouse gases such as carbon dioxide. Combustion of syngas or derived fuels does of course emit carbon dioxide.

Synthesis gas consists primarily of carbon monoxide, carbon dioxide, and hydrogen, and has less than half the energy density of natural gas. Synthesis gas is combustible and often used as a fuel source or as an intermediate for the production of other chemicals. Synthesis gas for use as a fuel is most often produced by gasification of coal or municipal waste mainly by the following paths:

$$C + O_2 \rightarrow CO_2$$

$$CO_2 + C \rightarrow 2CO$$

$$C + H_2O \rightarrow CO + H_2$$

When used as an intermediate in the large-scale, industrial synthesis of hydrogen and ammonia, it is also produced from natural gas (via the steam-reforming reaction) as follows:

$$CH_4 + H_2O \rightarrow CO + 3H_2$$

The synthesis gas produced in large waste-to-energy gasification facilities is used as fuel to generate electricity.

The manufacture of gas mixtures of carbon monoxide and hydrogen has been an important part of chemical technology for about a century. Originally, such mixtures were obtained by the reaction of steam with incandescent coke and were known as *water gas*. Used first as a fuel, water gas soon attracted attention as a source of hydrogen and carbon monoxide for the production of chemicals, at which time it gradually became known as synthesis gas. Eventually, steam reforming processes, in which steam is reacted with natural gas (methane) or petroleum naphtha over a nickel catalyst, found wide application for the production of synthesis gas.

A modified version of steam reforming known as autothermal reforming, which is a combination of partial oxidation near the reactor inlet with conventional steam reforming further along the reactor, improves the overall reactor efficiency and increases the flexibility of the process. Partial oxidation processes using oxygen instead of steam also found wide application for synthesis gas manufacture, with the special feature that they could utilize low-value feedstocks such as heavy petroleum residua. In recent years, catalytic partial oxidation employing very short reaction times (milliseconds) at high temperatures (850–1000°C) is providing still another approach to synthesis gas manufacture (Hickman and Schmidt, 1993). Nearly complete conversion of methane, with close to 100 percent selectivity to H_2 and CO, can be obtained with a rhenium monolith under well-controlled conditions. Experiments on the catalytic partial oxidation of *n*-hexane conducted with added steam give much higher yields of H_2 than can be obtained in experiments without steam, a result of much interest in obtaining hydrogen-rich streams for fuel cell applications.

The route of coal to synthetic automotive fuels, as practiced by Sasol, is technically proven and products with favorable environmental characteristics are produced. As is the case in essentially all coal conversion processes where air or oxygen is used for the utilization or partial conversion of the energy in the coal, the carbon dioxide burden is a drawback as compared to crude oil.

The uses of syngas include use as a chemical feedstock and in gas-to-liquid processes (Mangone, 2002), which use Fisher-Tropsch chemistry to make liquid fuels as feedstock for chemical synthesis, as well as being used in the production of fuel additives, including diethyl ether and methyl tertiary-butyl ether (MTBE), acetic acid, and its anhydride. Syngas could also make an important contribution to chemical synthesis through conversion to methanol (Olah et al., 2006). There is also the option in which stranded natural gas is converted to synthesis gas production followed by conversion to liquid fuels.

7.1 GASIFICATION OF COAL

Gasification is the conversion of a solid or liquid into a gas and excludes evaporation because the process involves chemical change. Thus, gasification is the process by which carbonaceous materials, such as coal, petroleum, or biomass, are converted into carbon monoxide and hydrogen by reacting the raw material at high temperatures with a controlled amount of oxygen (Chaps. 5 and 8). The resulting gas mixture is called *synthesis gas* or *syngas* and is itself a fuel.

Gasification is a very efficient method for extracting energy from many different types of organic materials, and also has applications as a clean waste disposal technique. In the process, coal or coal char is converted to gaseous products by reaction with steam, oxygen, air, hydrogen, carbon dioxide, or a mixture of these.

The gasification of coal or a derivative (i.e., char produced from coal) is, essentially, the conversion of coal (by any one of a variety of processes) to produce combustible gases. With the rapid increase in the use of coal from the fifteenth century onward (Nef, 1957; Taylor and Singer, 1957) it is not surprising the concept of using coal to produce a flammable gas, especially the use of the water and hot coal (van Heek and Muhlen, 1991), became commonplace (Elton, 1958). In fact, the production of gas from coal has been a vastly expanding area of coal technology, leading to numerous research and development programs. As a result, the characteristics of rank, mineral matter, particle size, and reaction conditions are all recognized as having a bearing on the outcome of the process; not only in terms of gas yields but also on gas properties (Massey, 1974; van Heek and Muhlen, 1991).

The products of coal gasification are varied insofar as the gas composition varies with the system employed (Fryer and Speight, 1976; Anderson and Tillman, 1979; Probstein and Hicks, 1990; Speight, 1994 and references cited therein). It is emphasized that the gas product must be first freed from any pollutants such as particulate matter and sulfur compounds before further use, particularly when the intended use is a water gas shift or methanation (Probstein and Hicks, 1990).

The advantage of gasification is that using the syngas is more efficient than direct combustion of the original fuel; more of the energy contained in the fuel is extracted. Syngas may be burned directly in internal combustion engines, used to produce methanol and hydrogen, or converted via the Fischer-Tropsch process into synthetic fuel. Gasification can also begin with materials that are not otherwise useful fuels, such as biomass or organic waste. In addition, the high-temperature combustion refines out corrosive ash elements such as chloride and potassium, allowing clean gas production from otherwise problematic fuels.

Gasification of coal has been, and continues to be, widely used on industrial scales to generate electricity. However, almost any type of organic carbonaceous material can be used as the raw material for gasification, such as wood, biomass, or even plastic waste. Thus, gasification may be an important technology for renewable energy. In particular biomass gasification is carbon neutral.

Another advantage of gasification-based energy systems is that when oxygen is used in the gasifier in place of air, the carbon dioxide produced by the process is in a concentrated gas stream, making it easier and less expensive to separate and capture. Once the carbon dioxide is captured, it can be sequestered and prevented from escaping to the atmosphere, where it could otherwise potentially contribute to the *greenhouse effect*.

7.1.1 Chemistry

Gasification relies on chemical processes at elevated temperatures >700°C, which distinguishes it from biologic processes such as anaerobic digestion that produce biogas.

In a gasifier, the carbonaceous material undergoes several different processes: (a) pyrolysis of carbonaceous fuels, (b) combustion, and (c) gasification of the remaining char.

Pyrolysis (*devolatilization*) is the thermal degradation of an organic substance in the absence of air to produce char, pyrolysis oil, and syngas, for example, the conversion of wood to charcoal. The pyrolysis process occurs as the carbonaceous feedstock heats up. Volatiles are released and char is produced, resulting in up to 70 percent weight loss for coal. The process is very dependent on the properties of the carbonaceous material and determines the structure and composition of the char, which will then undergo gasification reactions.

The *combustion* process occurs as the volatile products and some of the char reacts with oxygen to form carbon dioxide and carbon monoxide, which provides heat for the subsequent gasification reactions:

$$3C + 2O_2 \rightarrow CO_2 + 2CO$$

Gasification is the decomposition of hydrocarbons into a syngas by carefully controlling the amount of oxygen present, for example, the conversion of coal into town gas. The gasification process occurs as the char reacts with carbon dioxide and steam to produce carbon monoxide and hydrogen:

$$C + H_2O \rightarrow H_2 + CO$$

In addition, the gas phase water gas shift reaction reaches equilibrium very fast at the temperatures in a gasifier. This removes carbon dioxide from the reactor and provides water for the gasification reaction:

$$CO + H_2O \leftrightarrow CO_2 + H_2$$

In essence, a limited amount of oxygen or air is introduced into the reactor to allow some of the organic material to be burned to produce carbon monoxide and energy, which drives a second reaction that converts further organic material to hydrogen and additional carbon monoxide.

Coal gasification chemistry is reasonably simple and straightforward and, hence, is reasonably efficient. For many years, such processes were used to manufacture illuminating gas (coal gas) for gas lighting, before electric lighting became widely available. The simplest method, and the first used, was to heat coal in a retort in the absence of air, partially converting coal to gas with a residue of coke; the Scottish engineer William Murdock used this technique in pioneering the commercial gasification of coal in 1792. Murdock licensed his process to the Gas Light and Coke Company in 1813, and in 1816 the Baltimore Gas Company, the first coal gasification company in the United States, was established. The process of heating coal to produce coke and gas is still used in the metallurgical industry.

Currently, hydrogen is produced from coal by gasification and the subsequent processing of the resulting synthesis gas. In its simplest form, coal gasification works by first reacting coal with oxygen and steam under high pressures and temperatures to form a synthesis gas consisting primarily of carbon monoxide and hydrogen. This synthesis gas is cleaned of virtually all of its impurities and shifted to produce additional hydrogen. The clean gas is sent to a separation system to recover hydrogen.

The most complete conversion of coal or coke to gas that is feasible was achieved by reacting coal continuously in a vertical retort with air and steam. The gas obtained in this manner, called producer gas, has low heat content per unit volume of gas (100–150 Btu/ft^3). The development of a cyclic steam-air process in 1873 made possible the production of a gas of higher thermal content (300–350 Btu/ft^3), composed chiefly of carbon monoxide and hydrogen and known as water gas. By adding oil to the reactor, the thermal content of gas was increased to 500 to 550 Btu/ft^3; this became the standard for gas distributed to residences and industry. Since 1940, processes have been developed to produce continuously a gas equivalent to water gas; this involves the use of steam and essentially pure oxygen as a reactant. A more recently developed process reacts coal with pure oxygen and steam at an elevated pressure (450 psi) to produce a gas that may be converted to synthetic natural gas.

7.1.2 Processes

The most common modern coal gasification process uses lump coal in a vertical retort (Speight, 1994 and references cited therein). In the process, coal is fed at the top with air, and steam is introduced at the bottom. The air and steam rising up the retort heat the coal in its downward flow and react with the coal to convert it to gas. Ash is removed at the bottom of the retort. Using air and steam as reacting gases results in a producer gas; using oxygen and steam results in a water gas. Increasing operating pressure increases the productivity.

Two other processes currently in commercial use react finely powdered coal with steam and oxygen. For example, one of these, the Winkler process, uses a fluidized bed in which the powdered coal is agitated with the reactant gases.

In the *Winkler process* dried, crushed coal is fed to the fluidized bed gasifier through a variable-speed screw feeder whereupon the coal is contacted with steam and oxygen injected near the bottom of the vessel (Howard-Smith and Werner, 1976; Baughman, 1978). The upward flow of steam and oxygen maintains the bed in a fluidized state at a temperature of 815 to 980°C (1499–1796°F) with a pressure that is marginally higher than the atmospheric pressure. The high operating temperature reduces the amount of tars and other heavy hydrocarbons in the product (Nowacki, 1980).

Another example, the Koppers-Totzek process (Baughman, 1978; Michaels and Leonard, 1978; van der Burgt, 1979) is an entrained-solids process which operates at atmospheric pressure. The reactor is a relatively small, cylindrical, refractory-lined coal "burner" into which coal, oxygen, and steam are charged through at least two burner heads. The feed coal for the process is crushed (so that 70 percent will pass through a 200-mesh screen), mixed with oxygen and low-pressure steam, and injected into the gasifier through a burner head. The heads are spaced 180° or 90° apart (representing two-headed or four-headed opposed burner arrangements) and are designed such that steam envelopes the flame and protects the reactor walls from excessive heat.

The reactor typically operates at an exit temperature of 1480°C (2696°F) and the pressure is maintained just slightly above atmospheric pressure. Approximately 85 to 90 percent of the total carbon may be gasified in a single pass through the gasifier because carbon conversion is a function of the reactivity of the coal and approaches 100 percent for lignite.

The heat in the reactor causes the formation of slag from mineral ash and this is removed from the bottom of the gasifier through a water seal. Gases and vaporized hydrocarbons produced by the coal at medium temperatures immediately pass through a zone of very high temperature in which they decompose so rapidly that coal particles in the plastic stage do not agglomerate, and thus any type of coal can be gasified irrespective of caking tendencies, ash content, or ash fusion temperature.

In addition, the high operating temperature ensures that the gas product contains no ammonia, tars, phenols, or condensable hydrocarbons. The raw gas can be upgraded to synthesis gas by reacting all or part of the carbon monoxide content with steam to produce additional hydrogen plus carbon dioxide.

As petroleum supplies decrease, the desirability of producing gas from coal may increase, especially in those areas where natural gas is in short supply. It is also anticipated that costs of natural gas will increase, allowing coal gasification to compete as an economically viable process. Research in progress on a laboratory and pilot-plant scale should lead to the invention of new process technology by the end of the century, thus accelerating the industrial use of coal gasification.

Thus, the products of coal gasification consist of carbon monoxide, carbon dioxide, hydrogen, methane, and some other gases in proportions dependent upon the specific reactants and conditions (temperatures and pressures) employed within the reactors and the treatment steps which the gases undergo subsequent to leaving the gasifier. Similar chemistry can also be applied to the gasification of coke derived from petroleum and other sources.

The reaction of coal or coal char with air or oxygen to produce heat and carbon dioxide could be called gasification, but it is more properly classified as combustion. The principal purposes of such conversion are the production of synthetic natural gas as a substitute gaseous fuel and synthesis gases for production of chemicals and plastics.

In all cases of commercial interest, gasification with steam, which is endothermic, is an important chemical reaction. The necessary heat input is typically supplied to the gasifier by combusting a portion of the coal with oxygen added along with the steam. From the industrial viewpoint, the final product is synthesis gas, medium-Btu gas, or substitute natural gas. Each of the gas types has potential industrial applications.

In the chemical industry, synthesis gas from coal is a potential alternative source of hydrogen and carbon monoxide. This mixture is obtained primarily from the steam reforming of natural gas, natural gas liquids, or other petroleum liquids. Fuel users in the industrial sector have studied the feasibility of using medium-Btu gas instead of natural gas or oil for fuel applications. Finally, the natural gas industry is interested in substitute natural gas, which can be distributed in existing pipeline networks.

The conversion of the gaseous products of coal gasification processes to synthesis gas, a mixture of hydrogen (H_2) and carbon monoxide (CO), in a ratio appropriate to the application, needs additional steps, after purification. The product gases—carbon monoxide, carbon dioxide, hydrogen, methane, and nitrogen—can be used as fuels or as raw materials for chemical or fertilizer manufacture.

7.1.3 Gasifiers

The focal point of any gasification-based system is the gasifier. A gasifier converts hydrocarbon feedstock into gaseous components by applying heat under pressure in the presence of steam.

A gasifier differs from a combustor in that the amount of air or oxygen available inside the gasifier is carefully controlled so that only a relatively small portion of the fuel burns completely. The *partial oxidation* process provides the heat and rather than combustion, most of the carbon-containing feedstock is chemically broken apart by the heat and pressure applied in the gasifier resulting in the chemical reactions that produce synthesis gas. However, the composition of the synthesis gas will vary because of dependence upon the conditions in the gasifier and the type of feedstock.

Minerals in the fuel (i.e., the rocks, dirt, and other impurities which do not gasify) separate and leave the bottom of the gasifier either as an inert glass-like slag or other marketable solid products.

Sulfur impurities in the feedstock are converted to hydrogen sulfide (H_2S) and carbonyl sulfide (COS), from which sulfur can be extracted, typically as elemental sulfur. Nitrogen oxides (NO_x), other potential pollutants, are not formed in the oxygen-deficient (reducing) environment of the gasifier. Instead, ammonia (NH_3) is created by nitrogen-hydrogen reactions and can be washed out of the gas stream.

In Integrated Gasification Combined Cycle (IGCC) systems, the synthesis gas is cleaned of its hydrogen sulfide, ammonia, and particulate matter and is burned as fuel in a combustion turbine (much like natural gas is burned in a turbine). The combustion turbine drives an electric generator. And hot air from the combustion turbine can be channeled back to the gasifier or the air separation unit, while exhaust heat from the combustion turbine is recovered and used to boil water, creating steam for a steam turbine-generator.

The use of these two types of turbines—a combustion turbine and a steam turbine—in combination, known as a *combined cycle*, is one reason why gasification-based power systems can achieve unprecedented power generation efficiencies. Currently, commercially available gasification-based systems can operate at around 42 percent efficiencies; in the future, these systems may be able to achieve efficiencies approaching 60 percent. A conventional coal-based boiler plant, by contrast, employs only a steam turbine-generator and is typically limited to 33 to 40 percent efficiency.

Higher efficiency means that less fuel is required to generate the rated power, resulting in better economics (which can mean lower costs to the consumer) and the formation of fewer greenhouse gases—a 60 percent-efficient gasification power plant can cut the formation of carbon dioxide by 40 percent compared to a typical coal combustion plant.

7.2 GASIFICATION OF PETROLEUM FRACTIONS

One of the important aspects of petroleum refining is the supply of adequate amounts of hydrogen for the various hydrotreating processes, such as desulfurization and in hydroconversion processes, such as hydrocracking (Speight, 2007b).

As hydrogen use has become more widespread in refineries, hydrogen production has moved from the status of a high-tech specialty operation to an integral feature of most refineries. This has been made necessary by the increase in hydrotreating and hydrocracking, including the treatment of progressively heavier feedstocks. In fact, the use of hydrogen in thermal processes is perhaps the single most significant advance in refining technology during the twentieth century (Speight, 2007b and references cited therein).

In some refineries, the hydrogen needs can be satisfied by hydrogen recovery from catalytic reformer product gases, but other external sources are required. However, for the most part, many refineries now require on-site hydrogen production facilities to supply the gas for their own processes. Most of this non-reformer hydrogen is manufactured either by steam-methane reforming or by oxidation processes. However, other processes, as refineries and refinery feedstocks evolved during the last four decades, the demand for hydrogen has increased and reforming processes are no longer capable of providing the quantities of hydrogen that are adequate for feedstock hydrogenation.

In conjunction with hydrogen production, usually by partial oxidation processes, there is the concurrent production of carbon monoxide. Commonly, steam reforming of low molecular-weight hydrocarbons is the main method of hydrogen production which also produces synthesis gas.

The most common, and perhaps the best, feedstocks for steam reforming are low-boiling saturated hydrocarbons that have a low sulfur content, including natural gas, refinery gas, liquefied petroleum gas (LPG), and low-boiling naphtha.

Natural gas is the most common feedstock for hydrogen production since it meets all the requirements for reformer feedstock. Natural gas typically contains more than 90 percent methane and ethane with only a few percent of propane and higher-boiling hydrocarbons. Natural gas may (or most likely will) contain traces of carbon dioxide with some nitrogen and other impurities. Purification of natural gas, before reforming, is usually relatively straightforward (Speight, 2007a). Traces of sulfur must be removed to avoid poisoning the reformer catalyst; zinc oxide treatment in combination with hydrogenation is usually adequate.

Light *refinery gas*, containing a substantial amount of hydrogen, can be an attractive steam reformer feedstock since it is produced as a by-product. Processing of refinery gas will depend on its composition, particularly the levels of olefins and of propane and heavier hydrocarbons. Olefins, that can cause problems by forming coke in the reformer, are converted to saturated compounds in the hydrogenation unit. Higher-boiling hydrocarbons in refinery gas can also form coke, either on the primary reformer catalyst or in the preheater. If there is more than a few percent of C_3 and higher compounds, a promoted reformer catalyst should be considered, in order to avoid carbon deposits.

Refinery gas from different sources varies in suitability as hydrogen plant feed. Catalytic reformer off-gas, for example, is saturated, very low in sulfur, and often has high-hydrogen content. The process gases from a coking unit or from a fluid catalytic cracking unit are much less desirable because of the content of unsaturated constituents. In addition to olefins, these gases contain substantial amounts of sulfur that must be removed before the gas is used as feedstock. These gases are also generally unsuitable for direct hydrogen recovery, since the hydrogen content is usually too low. Hydrotreater off-gas lies in the middle of the range. It is saturated, so it is readily used as hydrogen plant feed. Content of hydrogen and heavier hydrocarbons depends to a large extent on the upstream pressure. Sulfur removal will generally be required.

As hydrogen use has become more widespread in refineries, hydrogen production has moved from the status of a high-tech specialty operation to an integral feature of most refineries. This has been made necessary by the increase in hydrotreating and hydrocracking, including the treatment of progressively heavier feedstocks (Speight, 2007b). The continued increase in hydrogen demand over the last several decades is a result of the conversion of petroleum to match changes in product slate and the supply of heavy, high-sulfur oil, and in order to make lower-boiling, cleaner, and more salable products. There are also many reasons other than product quality for using hydrogen in processes adding to the need to add hydrogen at relevant stages of the refining process, the most important being the availability of hydrogen.

Hydrogen has historically been produced during catalytic reforming processes as a by-product of the production of the aromatic compounds used in gasoline and in solvents. As reforming processes changed from fixed bed to cyclic to continuous regeneration, process pressures have dropped and hydrogen production per barrel of reformate has tended to increase. However, hydrogen production as a by-product is not always adequate to the needs of the refinery and other processes are necessary. Thus, hydrogen production by steam reforming or by partial oxidation of residua has also been used, particularly where heavy oil is available. Steam reforming is the dominant method for hydrogen production and is usually combined with pressure-swing adsorption (PSA) to purify the hydrogen to greater than 99 percent by volume.

The gasification of residua and coke to produce hydrogen and/or power may become an attractive option for refiners. The premise that the gasification section of a refinery will be the *garbage can* for deasphalter residues, high-sulfur coke, as well as other refinery wastes is worthy of consideration.

Of the processes that are available for the production of hydrogen, many can be considered dual processes insofar as they also produce carbon monoxide and, therefore, are considered as producers of synthesis gas. For example, most of the external hydrogen is manufactured by steam-methane reforming or by oxidation processes. Other processes such as ammonia dissociation, steam-methanol interaction, or electrolysis are also available for hydrogen production, but economic factors and feedstock availability assist in the choice between processing alternatives.

The processes described in this section are those gasification processes by which hydrogen is produced for use in other parts of the refinery.

7.2.1 Chemistry

In *steam reforming*, low-boiling hydrocarbons such as methane are reacted with steam to form hydrogen:

$$CH_4 + H_2O \rightarrow 3H_2 + CO \qquad \Delta H_{298\,K} = +97,400 \text{ Btu/lb}$$

where ΔH is the heat of the reaction. A more general form of the equation that shows the chemical balance for higher-boiling hydrocarbons is:

$$C_nH_m + nH_2O \rightarrow (n + m/2)H_2 + nCO$$

The reaction is typically carried out at approximately 815°C (1499°F) over a nickel catalyst packed into the tubes of a reforming furnace. The high temperature also causes the hydrocarbon feedstock to undergo a series of cracking reactions, plus the reaction of carbon with steam:

$$CH_4 \rightarrow 2H_2 + C$$

$$C + H_2O \rightarrow CO + H_2$$

Carbon is produced on the catalyst at the same time that hydrocarbon is reformed to hydrogen and carbon monoxide. With natural gas or similar feedstock, reforming predominates and the carbon can be removed by reaction with steam as fast as it is formed. When higher boiling feedstocks are used, the carbon is not removed fast enough and builds up, thereby requiring catalyst regeneration or replacement. Carbon buildup on the catalyst (when high-boiling feedstocks are employed) can be avoided by addition of alkali compounds, such as potash, to the catalyst, thereby encouraging or promoting the carbon-steam reaction.

However, even with an alkali-promoted catalyst, feedstock cracking limits the process to hydrocarbons with a boiling point less than 180°C (356°F). Natural gas, propane, butane, and light naphtha are most suitable. Pre-reforming, a process that uses an adiabatic catalyst bed, operating at a lower temperature, can be used as a pretreatment to allow heavier feedstocks to be used with lower potential for carbon deposition (coke formation) on the catalyst.

After reforming, the carbon monoxide in the gas is reacted with steam to form additional hydrogen (the *water-gas shift* reaction):

$$CO + H_2O \rightarrow CO_2 + H_2 \qquad \Delta H_{298\,K} = -16{,}500 \text{ Btu/lb}$$

This leaves a mixture consisting primarily of hydrogen and carbon monoxide that is removed by conversion to methane:

$$CO + 3H_2 \rightarrow CH_4 + H_2O$$

$$CO_2 + 4H_2 \rightarrow CH_4 + 2H_2O$$

The critical variables for steam-reforming processes are (a) temperature, (b) pressure, and (c) the steam/hydrocarbon ratio. Steam reforming is an equilibrium reaction, and conversion of the hydrocarbon feedstock is favored by high temperature, which in turn requires higher fuel use. Because of the volume increase in the reaction, conversion is also favored by low pressure, which conflicts with the need to supply the hydrogen at high pressure. In practice, materials of construction limit temperature and pressure.

On the other hand, and in contrast to reforming, shift conversion is favored by low temperature. The gas from the reformer is reacted over iron oxide catalyst at 315 to 370°C (599–698°F) with the lower limit being dictated activity of the catalyst at low temperature.

Hydrogen can also be produced by *partial oxidation* (POx) of hydrocarbons in which the hydrocarbon is oxidized in a limited or controlled supply of oxygen:

$$2CH_4 + O_2 \rightarrow 2CO + 4H_2 \qquad \Delta H_{298\,K} = -10{,}195 \text{ Btu/lb}$$

The shift reaction also occurs and a mixture of carbon monoxide and carbon dioxide is produced in addition to hydrogen. The catalyst tube materials do not limit the reaction temperatures in partial oxidation processes and higher temperatures may be used that enhance the conversion of methane to hydrogen. Indeed, much of the design and operation of hydrogen plants involves protecting the reforming catalyst and the catalyst tubes because of the extreme temperatures and the sensitivity of the catalyst. In fact, minor variations in feedstock composition or operating conditions can have significant effects on the life of the catalyst or the reformer itself. This is particularly true of changes in molecular weight of the feed gas, or poor distribution of heat to the catalyst tubes.

Since the high temperature takes the place of a catalyst, partial oxidation is not limited to the lower boiling feedstocks that are required for steam reforming. Partial oxidation processes were first considered for hydrogen production because of expected shortages of lower boiling feedstocks and the need to have a disposal method available for higher boiling, high-sulfur streams such as asphalt or petroleum coke.

Catalytic partial oxidation, also known as autothermal reforming, reacts oxygen with a light feedstock and by passing the resulting hot mixture over a reforming catalyst. The use of a catalyst allows the use of lower temperatures than in non-catalytic partial oxidation and which causes a reduction in oxygen demand.

The feedstock requirements for catalytic partial oxidation processes are similar to the feedstock requirements for steam reforming and light hydrocarbons from refinery gas to naphtha are preferred. The oxygen substitutes for much of the steam in preventing coking and a lower steam/carbon ratio is required. In addition, because a large excess of steam is not required, catalytic partial oxidation produces more carbon monoxide and less hydrogen than steam reforming. Thus, the process is more suited to situations where carbon monoxide is the more desirable product, for example, as synthesis gas for chemical feedstocks.

7.2.2 Processes

In spite of the use of low-quality hydrogen (that contain up to 40 percent by volume hydrocarbon gases), a high-purity hydrogen stream (95–99 percent by volume of hydrogen) is required for hydrodesulfurization, hydrogenation, hydrocracking, and petrochemical processes. Hydrogen, produced as a by-product of refinery processes (principally hydrogen recovery from catalytic reformer product gases) often is not enough to meet the total refinery requirements, necessitating the manufacturing of additional hydrogen or obtaining supply from external sources.

Heavy Residue Gasification and Combined Cycle Power Generation. Heavy residua are gasified and the produced gas is purified to clean fuel gas . As an example, solvent deasphalter residuum is gasified by partial oxidation method under pressure of about 570 psi (3930 kPa) and at temperature between 1300 and 1500°C (2372–2732°F). The high-temperature-generated gas flows into the specially designed waste heat boiler, in which the hot gas is cooled and high-pressure saturated steam is generated. The gas from the waste heat boiler is then heat exchanged with the fuel gas and flows to the carbon scrubber, where unreacted carbon particles are removed from the generated gas by water scrubbing.

The gas from the carbon scrubber is further cooled by the fuel gas and boiler feed water and led into the sulfur compound removal section, where hydrogen sulfide (H_2S) and carbonyl sulfide (COS) are removed from the gas to obtain clean fuel gas. This clean fuel gas is heated with the hot gas generated in the gasifier and finally supplied to the gas turbine at a temperature of 250 to 300°C (482–572°F).

The exhaust gas from the gas turbine having a temperature of about 550 to 600°C (1022–1112°F) flows into the heat recovery steam generator consisting of five heat exchange elements. The first element is a superheater in which the combined stream of the high-pressure saturated steam generated in the waste heat boiler and in the second element (high-pressure steam evaporator) is super heated. The third element is an economizer, the fourth element is a low pressure steam evaporator and the final or the fifth element is a deaerator heater. The off gas from heat recovery steam generator having a temperature of about 130°C is emitted into the air via stack.

In order to decrease the nitrogen oxide (NO_x) content in the flue gas, two methods can be applied. The first method is the injection of water into the gas turbine combustor. The second method is to selectively reduce the nitrogen oxide content by injecting ammonia gas in the presence of de-NO_x catalyst that is packed in a proper position of the heat recovery steam generator. The latter is more effective than the former to lower the nitrogen oxide emissions to the air.

Hybrid Gasification Process. In the hybrid gasification process, a slurry of coal and residual oil is injected into the gasifier where it is pyrolyzed in the upper part of the reactor to produce gas and chars. The chars produced are then partially oxidized to ash. The ash is removed continuously from the bottom of the reactor.

In this process, coal and vacuum residue are mixed together into slurry to produce clean fuel gas. The slurry fed into the pressurized gasifier is thermally cracked at a temperature of 850 to 950°C (1562–1742°F) and is converted into gas, tar, and char. The mixture oxygen and steam in the lower zone of the gasifier gasify the char. The gas leaving the gasifier is quenched to a temperature of 450°C (842°F) in the fluidized bed heat exchanger, and is then scrubbed to remove tar, dust, and steam at around 200°C (392°F).

The coal and residual oil slurry is gasified in the fluidized bed gasifier. The charged slurry is converted to gas and char by thermal cracking reactions in the upper zone of the fluidized bed. The produced char is further gasified with steam and oxygen that enter the gasifier just below the fluidizing gas distributor. Ash is discharged from the gasifier and indirectly cooled with steam and then discharged into the ash hopper. It is burned with an incinerator to produce process steam. Coke deposited on the silica sand is removed in the incinerator.

Hydrocarbon Gasification. The gasification of hydrocarbons to produce hydrogen is a continuous, noncatalytic process that involves partial oxidation of the hydrocarbon. Air or oxygen (with steam or carbon dioxide) is used as the oxidant at 1095 to 1480°C (2003–2696°F). Any carbon produced (2–3 percent by weight of the feedstock) during the process is removed as a slurry in a carbon separator and pelletized for use either as a fuel or as raw material for carbon-based products.

Hypro Process. The hypro process is a continuous catalytic method for hydrogen manufacture from natural gas or from refinery effluent gases. The process is designed to convert natural gas:

$$CH_4 \rightarrow C + 2H_2$$

Hydrogen is recovered by phase separation to yield hydrogen of about 93 percent purity; the principal contaminant is methane.

Pyrolysis Processes. There has been recent interest in the use of pyrolysis processes to produce hydrogen. Specifically the interest has focused on the pyrolysis of methane (natural gas) and hydrogen sulfide.

Natural gas is readily available and offers relatively rich stream of methane with lower amounts of ethane, propane, and butane also present. The thermocatalytic decomposition of natural gas hydrocarbons offers an alternate method for the production of hydrogen:

$$C_nH_m \rightarrow nC + m/2H_2$$

If a hydrocarbon fuel such as natural gas (methane) is to be used for hydrogen production by direct decomposition, then the process that is optimized to yield hydrogen production may not be suitable for production of high-quality carbon black by-product intended for the industrial rubber market. Moreover, it appears that the carbon produced from high-temperature [850–950°C (1562–1742°F)] direct thermal decomposition of methane is soot-like material with high tendency for the catalyst deactivation. Thus, if the object of methane decomposition is hydrogen production, the carbon by-product may not be marketable as high-quality carbon black for rubber and tire applications.

Hydrogen sulfide decomposition is a highly endothermic process and equilibrium yields are poor. At temperatures less than 1500°C (2732°F), the thermodynamic equilibrium is unfavorable toward hydrogen formation. However, in the presence of catalysts such as platinum-cobalt [at 1000°C (1832°F)], disulfides of molybdenum (Mo) or tungsten (W) at 800°C (1472°F), or other transition metal sulfides supported on alumina [at 500–800°C (932–1472°F)], decomposition of hydrogen sulfide proceeds rapidly. In the temperature range of about 800 to 1500°C (1472–2732°F), thermolysis of hydrogen sulfide can be treated simply:

$$H_2S \rightarrow H_2 + 1/xS_x \qquad \Delta H_{298\,K} = +34{,}300 \text{ Btu/lb}$$

where $x = 2$. Outside this temperature range, multiple equilibria may be present depending on temperature, pressure, and relative abundance of hydrogen and sulfur.

Above approximately 1000°C (1832°F), there is a limited advantage to using catalysts since the thermal reaction proceeds to equilibrium very rapidly. The hydrogen yield can be doubled by preferential removal of either hydrogen or sulfur from the reaction environment, thereby shifting the equilibrium. The reaction products must be quenched quickly after leaving the reactor to prevent reversible reactions.

Shell Gasification (Partial Oxidation) Process. The Shell Gasification Process is a flexible process for generating synthesis gas, principally hydrogen and carbon monoxide, for the ultimate production of high-purity, high-pressure hydrogen, ammonia, methanol, fuel gas, town gas or reducing gas by reaction of gaseous or liquid hydrocarbons with oxygen, air, or oxygen-enriched air.

The most important step in converting heavy residue to industrial gas is the partial oxidation of the oil using oxygen with the addition of steam. The gasification process takes place in an empty, refractory-lined reactor at temperatures of about 1400°C (2552°F) and pressures between 29 and 1140 psi (199–7860 kPa). The chemical reactions in the gasification reactor proceed without catalyst to produce gas containing carbon amounting to some 0.5 to 2 percent by weight, based on the feedstock. The carbon is removed from the gas with water, extracted in most cases with feed oil from the water and returned to the feed oil. The high reformed gas temperature is utilized in a waste heat boiler for generating steam. The steam is generated at 850 to 1565 psi (5860–10790 kPa). Some of this steam is used as process steam and for oxygen and oil preheating. The surplus steam is used for energy production and heating purposes.

Steam-Naphtha Reforming. Steam-naphtha reforming is a continuous process for the production of hydrogen from liquid hydrocarbons and is, in fact, similar to steam-methane reforming that is one of several possible processes for the production of hydrogen from low-boiling hydrocarbons other than ethane. A variety of naphtha-types in the gasoline boiling range may be employed, including feeds containing up to 35 percent aromatics. Thus, following pretreatment to remove sulfur compounds, the feedstock is mixed with steam and taken to the reforming furnace [675–815°C (1247–1499°F), 300 psi (2068 kPa)], where hydrogen is produced.

Texaco Gasification (Partial Oxidation) Process. The Texaco gasification process is a partial oxidation gasification process for generating synthetic gas, principally hydrogen and carbon monoxide. The characteristic of Texaco gasification process is to inject feedstock together with carbon dioxide, steam, or water into the gasifier. Therefore, solvent deasphalted residua, or petroleum coke rejected from any coking method can be used as feedstock for this gasification process. The produced gas from this gasification process can be used for the production of high-purity, high-pressurized hydrogen, ammonia, and methanol. The heat recovered from the high-temperature gas is used for the generation of steam in the waste heat boiler. Alternatively the less expensive quench type configuration is preferred when high-pressure steam is not needed or when a high degree of shift is needed in the downstream CO converter.

In the process, the feedstock, together with the feedstock carbon slurry recovered in the carbon recovery section, is pressurized to a given pressure, mixed with high-pressure steam and then blown into the gas generator through the burner together with oxygen.

The gasification reaction is a partial oxidation of hydrocarbons to carbon monoxide and hydrogen:

$$C_xH_{2y} + x/2O_2 \rightarrow xCO + yH_2$$

$$C_xH_{2y} + xH_2O \rightarrow xCO + (x+y)H_2$$

The gasification reaction is instantly completed, thus producing gas mainly consisting of H_2 and CO (H_2 + CO = >90 percent). The high-temperature gas leaving the reaction chamber of the gas generator enters the quenching chamber linked to the bottom of the gas generator and is quenched to 200 to 260°C (392–500°F) with water.

7.3 STEAM-METHANE REFORMING

Steam reforming of natural gas, sometimes referred to as steam-methane reforming is the most common method of producing commercial bulk hydrogen as well as synthesis gas.

Steam-methane reforming is a catalytic process that involves a reaction between natural gas or other light hydrocarbons and steam (van Beurden, 2004). Steam-methane reforming is the benchmark process that has been employed over a period of several decades for hydrogen production. The process involves reforming natural gas in a continuous catalytic process in which the major reaction is the formation of carbon monoxide and hydrogen from methane and steam.

The steam reforming of methane consists of three reversible reactions: the strongly endothermic reforming reactions [Eqs. (7.1) and (7.3)], and the moderately exothermic water-gas shift reaction [Eq. 7.2]:

$$CH_4 + H_2O \rightarrow CO + 3H_2 \quad \Delta H_{298\,K} = +97,400 \text{ Btu/lb} \quad (7.1)$$

In practice, it is necessary to terminate the reaction at the stage where the maximum yields of carbon monoxide and hydrogen are produced.

Due to its endothermic character, reforming is favored by high temperature and, because reforming is accompanied by a volume expansion, it is favored by low pressure. In contrast, the exothermic shift reaction is favored by low temperature, while unaffected by changes in pressure.

Increasing the amount of steam will enhance the methane conversion, but requires an additional amount of energy to produce the steam. In practice, steam-to-carbon ratios [i.e., $P(H_2O)/P(CH_4)$] of approximately 3 are applied. This value for the steam-to-carbon ratio will also suppress coke formation during the reaction (Rostrup-Nielsen, 1984; Rostrup-Nielsen and Bak-Hansen, 1993; Rostrup-Nielsen et al., 2002).

Higher molecular weight feedstocks can also be reformed to hydrogen:

$$C_3H_8 + 3H_2O \to 3CO + 7H_2$$

That is,

$$C_nH_m + nH_2O \to nCO + (0.5m + n)H_2 \tag{7.2}$$

In the actual process, the feedstock is first desulfurized by passage through activated carbon, which may be preceded by caustic and water washes. The desulfurized material is then mixed with steam and passed over a nickel-based catalyst [730–845°C (1346–1553°F) and 400 psi (2758 kPa)]. Effluent gases are cooled by the addition of steam or condensate to about 370°C (698°F), at which point carbon monoxide reacts with steam in the presence of iron oxide in a shift converter to produce carbon dioxide and hydrogen in which the carbon monoxide is then *shifted* with steam to form additional hydrogen and carbon dioxide in an exothermic (heat-releasing) reaction:

$$CO + H_2O = CO_2 + H_2 \qquad \Delta H_{298\,K} = -41.16 \text{ kJ/mol} \tag{7.3}$$

The carbon dioxide (usually by amine washing), leaving hydrogen, is separated for its commercial use; the hydrogen is usually a high-purity (>99 percent) material.

Since the presence of any carbon monoxide or carbon dioxide in the hydrogen stream can interfere with the chemistry of the catalytic application, a third stage is used to convert these gases to methane:

$$CO + 3H_2 \to CH_4 + H_2O$$

$$CO_2 + 4H_2 \to CH_4 + 2H_2O$$

For many refiners, sulfur-free natural gas (CH_4) is not always available to produce hydrogen by this process. In that case, higher-boiling hydrocarbons (such as propane, butane, or naphtha) may be used as the feedstock to generate hydrogen (qv).

The net chemical process for steam-methane reforming is then given by:

$$CH_4 + 2H_2O \to CO_2 + 4H_2 \qquad \Delta H_{298\,K} = +165.2 \text{ kJ/mol} \tag{7.4}$$

Indirect heating provides the required overall endothermic heat of reaction for the steam-methane reforming.

One way of overcoming the thermodynamic limitation of steam reforming is to remove carbon dioxide as it is produced, hence shifting the thermodynamic equilibrium toward the product side. The concept for sorption-enhanced methane-steam reforming is based on in situ removal of carbon dioxide by a sorbent such as calcium oxide (CaO).

$$CaO + CO_2 \to CaCO_3$$

Sorption enhancement enables lower reaction temperatures, which may reduce catalyst coking and sintering, while enabling use of less expensive reactor wall materials. In addition, heat release by the exothermic carbonation reaction supplies most of the heat required by the endothermic reforming reactions. However, energy is required to regenerate the sorbent to its oxide form by the energy-intensive calcination reaction, that is,

$$CaCO_3 \rightarrow CaO + CO_2$$

Use of a sorbent requires either that there be parallel reactors operated alternatively and out of phase in reforming and sorbent regeneration modes, or that sorbent be continuously transferred between the reformer/carbonator and regenerator/calciner.

In autothermal (or secondary) reformers, the oxidation of methane supplies the necessary energy and is carried out either simultaneously or in advance of the reforming reaction. The equilibrium of the methane-steam reaction and the water-gas shift reaction determines the conditions for optimum hydrogen yields. The optimum conditions for hydrogen production require: high temperature at the exit of the reforming reactor [800–900°C (1472–1652°F)], high excess of steam (molar steam-to-carbon ratio of 2.5 to 3), and relatively low pressures (below 450 psi). Most commercial plants employ supported nickel catalysts for the process.

At the operating temperatures, some of the methane may completely decompose and deposit a thick layer of inactive carbon on the catalyst surface (coke). Especially with nickel-based catalysts, steam reforming involves the risk of carbon formation, which may cause serious operational problems and catalyst deactivation. Generally, higher hydrocarbons are more prone to carbon formation than methane because the initial surface carbon intermediates are more readily formed. The concentration of these intermediates is an important factor, and is critical in influencing the delicate balance between carbon-forming and carbon-removing reactions. On nickel surfaces, carbon formation may take place mainly by three routes (Table 7.1) (Rostrup-Nielsen, 1984; Rostrup-Nielsen et al., 2002).

At lower temperatures (say 500°C and below), adsorbed hydrocarbons may accumulate on the surface and slowly may be transformed into a nonreactive polymer film ("gum"), blocking and deactivating the surface. This phenomenon can be retarded by hydrogen. Because of the endothermic nature of the steam-reforming reaction, high catalyst activity leads to a low temperature at the reaction site, resulting in a higher risk for carbon formation.

TABLE 7.1 Possible Routes to Deposition of Carbonaceous Products During Steam-Methane Reforming

Carbon type	Reactions involved	Phenomena	Critical parameters
Gum	$C_nH_m \rightarrow (CH_2)_n \rightarrow$ gum	Blocking of surface by polymerisation of adsorbed C_nH_m radicals: progressive deactivation	Low S/C ratio, absence of H_2, low temperature (below ~500°C), presence of aromatics
Whisker carbon, amorphous carbon	$CH_4 \rightarrow C + 2H_2$ $2CO \rightarrow C + CO_2$ $CO + H_2 \rightarrow C + H_2O$ $C_nH_m \rightarrow nC + {}^m/_2 H_2$	Break-up of catalyst pellet (whisker carbon: no deactivation of the surface)	Low S/C ratio, high temperature (above ~450°C), presence of olefins, aromatics
Pyrolytic coke	$C_nH_m \rightarrow$ olefins $\rightarrow$ coke	Encapsulation of catalyst pellet (deactivation), deposits on tube wall	High temperature (above ~600°C), high residence time, presence of olefins, sulfur poisoning

In many instances catalyst deactivation occurs during steam-methane reforming. There can be many reasons for catalyst deactivation (van Beurden, 2004). In addition, there is a distinction between poisoning and thermal deactivation. For example, if, on continued use, the activity decreases more rapidly than surface area, poisoning may be suspected whereas, if a decrease in surface area is concomitant with a decrease in activity, thermal deactivation is indicated.

Carbon formation depends on the kinetic balance between the surface reaction of the adsorbed hydrocarbons with oxygen species and the further dissociation of the hydrocarbon into adsorbed carbon atoms. In fact, for a given hydrocarbon feed, temperature, and pressure, carbon will be formed below a critical steam-to-carbon ratio (Twigg, 1989; Rostrup-Nielsen et al., 2002). This critical steam-to-carbon ratio increases with temperature and is dictated by thermodynamics. In practice however, carbon formation generally occurs before the thermodynamic limit is reached (e.g., by poisons, temperature and concentration gradients, etc.). By promotion of the catalyst it is possible to push the carbon formation limit to the thermodynamic limit. For instance, controlled passivation of the catalyst surface by sulfur, carbon deposition is inhibited (Udengaard et al., 1992). By using noble metal catalysts, it is possible to push the carbon limit even beyond the thermodynamic limit.

7.4 GASIFICATION OF OTHER FEEDSTOCKS

Gasification offers more scope for recovering products from waste than incineration. When waste is burnt in a modern incinerator the only practical product is energy, whereas the gases, oils, and solid char from pyrolysis and gasification can not only be used as a fuel but also purified and used as a feedstock for petrochemicals and other applications. Many processes also produce a stable granulate instead of an ash which can be more easily and safely utilized. In addition, some processes are targeted at producing specific recyclables such as metal alloys and carbon black. From waste gasification, in particular, it is feasible to produce hydrogen, which many see as an increasingly valuable resource.

Gasification can be used in conjunction with gas engines (and potentially gas turbines) to obtain higher conversion efficiency than conventional fossil fuel energy generation. By displacing fossil fuels, waste pyrolysis and gasification can help meet renewable energy targets, address concerns about global warming, and contribute to achieving Kyoto Protocol commitments. Conventional incineration, used in conjunction with steam-cycle boilers and turbine generators, achieves lower efficiency.

Many of the processes fit well into a modern integrated approach to waste management. They can be designed to handle the waste residues and are fully compatible with an active program of composting for the waste fraction that is subject to decay and putrefaction.

This, by analogy with coal, the high-temperature conversion of waste (Chap. 11) is a downdraft gasification process which gasifies the feed material within a controlled and limited oxygen supply. Combustion of the feed material is prevented by the limited oxygen supply. The temperature within the reactor reaches 2700°C, at which point molecular dissociation takes place. The pollutants that were contained within the feed waste material such as dioxins, furans, as well as pathogens are completely cracked into harmless compounds.

All metal components in the waste stream are converted into a castable iron alloy/pig iron for use in steel foundries. The mineral fraction is reduced to a nonleaching vitrified glass, used for road construction and/or further processed into a mineral wool for insulation. All of the organic material is fully converted to a fuel quality synthesis gas which can be used to produce electrical energy, heat, methanol, or used in the production of various other chemical compounds. The resultant syngas, with a hydrogen-to-carbon-monoxide ratio approximately equal to 1, is also capable of being used for the production of Fischer-Tropsch fuels. Under certain conditions, heat from the reactor could be used for district heating, industrial steam production, or water desalination plants.

A wide range of materials can be handled by gasification technologies and specific processes have been optimized to handle particular feedstock (e.g., tire pyrolysis and sewage sludge gasification), while others have been designed to process mixed wastes. For example, recovering energy from agricultural and forestry residues, household and commercial waste, and materials recycling (autoshredder residue, electrical and electronic scrap, tires, mixed plastic waste, and packaging residues) are feasible processes.

Biomass gasification could play a significant role in a renewable energy economy, because biomass production removes CO_2 from the atmosphere. While other biofuel technologies such as biogas and biodiesel are also carbon neutral, gasification runs on a wider variety of input materials, can be used to produce a wider variety of output fuels, and is an extremely efficient method of extracting energy from biomass. Biomass gasification is therefore one of the most technically and economically convincing energy possibilities for a carbon neutral economy.

However, a disadvantage cited against the use of using biomass as a feedstock in gasification reactors is that biomass generally contains high levels of corrosive ash and cause damage to the gasifier. The difference between the biomass minerals and coal minerals is not entirely clear as both can, obviously, lead to corrosion.

7.5 FISCHER-TROPSCH

Before synthesis gas can be used in the Fisher-Tropsch reaction there are several cleaning steps required lest the impure gas deactivate the catalyst.

The general gas cleaning protocol (after as cooling) involves: (a) a primary water scrubber, (b) a carbonyl sulfide, (c) a secondary water scrubber, (d) an olamine absorber, and (e) a combustor to convert sulfur compounds to sulfur dioxide (Speight, 2007a).

Other aspects of synthesis gas cleaning may involve other steps which include: (a) tar removal, (b) dust separation in a cyclone separator, (c) a bag filter, (d) carbonyl sulfide (COS) hydrolysis in the scrubber (not if amine washing is used) at 100 to 250°C, (e) tar condensation, (f) ammonia (NH_3) and hydrogen cyanide (HCN) scrubbing with sulfuric acid, and hydrogen sulfide scrubbing. The methods are readily available (Speight, 2007a).

Particulate removal systems from synthesis gas must remove ashes, char particles, and fluidized bed solids from the gas, because of potential erosion and emission problems. Techniques used are filtration (hot or cold) and scrubbing. Bag filters use plastic fibers for filtration up to 230°C; ceramic and metallic dust filters can be used at maximal running temperatures of about 800°C, or even more. Particles removed can be as small as 0.1 μm and cleaned gas contains less than 1 to 5 ppm. Alkali metal oxides condense at 550°C and stick to the surface of particles present and are removed together with them in particulate filter. Hydrogen sulfide present in the synthesis gas is removed by scrubbing with lime or caustic soda. Sulfur can be removed from gas by addition of ferrous sulfate and subsequent filtration and disposal of the dust. When the fluidized bed contains calcite and/or dolomite, sand reacts with sulfates and forms salts.

Although the focus of this section is the production of hydrocarbons from synthesis gas, it is worthy of note that all or part of the clean syngas can also be used as: (a) chemical *building blocks* to produce a broad range of chemicals (using processes well established in the chemical and petrochemical industry), (b) a fuel producer for highly efficient fuel cells (which run off the hydrogen made in a gasifier) or perhaps in the future, hydrogen turbines and fuel cell-turbine hybrid systems, and (c) a source of hydrogen that can be separated from the gas stream and used as a fuel or as a feedstock for refineries (which use the hydrogen to upgrade petroleum products).

However, the decreasing availability and increased price of petroleum has been renewed the worldwide interest in the production of liquid hydrocarbons from carbon monoxide and hydrogen using metal catalysts, also known as Fischer-Tropsch synthesis.

In the last decades, the interest in Fischer-Tropsch synthesis has changed as a result of environmental demands, technological developments, change in fossil energy reserves, and high oil prices.

The Fischer-Tropsch process is a catalyzed chemical reaction in which carbon monoxide and hydrogen are converted into liquid hydrocarbons of various forms. Typical catalysts used are based on iron and cobalt. The principal purpose of this process is to produce a synthetic petroleum substitute for use as synthetic lubrication oil or as synthetic fuel.

The development of pressurized Fischer-Tropsch synthesis goes 80 years back, and starts about 1925 in Germany when Prof. Franz Fischer, founding director of the Kaiser-Wilhelm Institute of Coal Research in Mälheim an der Ruhr, and his head of department, Dr. Hans Tropsch, applied for a patent describing a process to produce liquid hydrocarbons from carbon monoxide gas and hydrogen using metal catalysts.

The experiments took place in Franz Fischer's laboratory at the Kaiser Wilhelm Institute for Coal Research, and the concept resulted in an industry which produced almost 4 million barrels of synthetic fuel in 1945. At those times strategic reasons for liquid fuel production from coal exceeded economic aspects. A good example is the "oil-age" from 1955 to 1970 with plenty of cheap oil supply and as a result only a marginal interest in Fischer-Tropsch synthesis. High oil prices increase the focus at alternative fuels; likewise as carbon dioxide concentration concern arises, being related to global warming, the focus at new technologies rises. Today the driving forces are environmental concern, but also higher oil price, limited oil reserves, and increased focus at stranded gas.

The required gas mixture of carbon monoxide and hydrogen (synthesis gas) is created through a reaction of coke or coal with water steam and oxygen, at temperatures over 900°C. In the past, town gas and gas for lamps were a carbon monoxide-hydrogen mixture, made by gasifying coke in gas works. In the 1970s, it was replaced with imported natural gas (methane). Coal gasification and Fischer-Tropsch hydrocarbon synthesis together bring about a two-stage sequence of reactions which allows the production of liquid fuels like diesel and petrol out of the solid combustible coal.

The Fischer-Tropsch synthesis took its first serious place in industry in 1935 at Ruhrchemie in Oberhausen. By the beginning of the 1940s, some 600,000 t of liquid hydrocarbons were produced per year in German facilities, made from coal using Fischer-Tropsch synthesis. Licensed by Ruhrchemie, four facilities in Japan, as well as a plant in France and in Manchuria, were in service. After World War II, competition from crude oil made petrol production from coal unprofitable. The only new production facilities were in South Africa, for political reasons, built starting in 1950 in Sasolburg.

The Fischer-Tropsch synthesis is, in principle, a carbon chain building process, where methylene groups are attached to the carbon chain. The actual reactions that occur have been, and remain, a matter of controversy, as it has been the last century since 1930s.

Even though the overall Fischer-Tropsch process is described by the following chemical equation:

$$(2n+1)H_2 + nCO \rightarrow C_nH_{(2n+2)} + nH_2O$$

The initial reactants in the above reaction (i.e., CO and H_2) can be produced by other reactions such as the partial combustion of a hydrocarbon:

$$C_nH_{(2n+2)} + \tfrac{1}{2} nO_2 \rightarrow (n+1)H_2 + nCO$$

For example (when n = 1), methane (in the case of gas to liquids applications):

$$2CH_4 + O_2 \rightarrow 4H_2 + 2CO$$

Or by the gasification of any carbonaceous source, such as biomass:

$$C + H_2O \rightarrow H_2 + CO$$

The energy needed for this endothermic reaction is usually provided by (exothermic) combustion with air or oxygen:

$$2C + O_2 \rightarrow 2CO$$

The detailed behavior of these other reactions (Table 7.2) is not known with any degree of certainty and still remains somewhat speculative. The reactions are highly exothermic, and to avoid an increase in temperature, which results in lighter hydrocarbons, it is important to have sufficient cooling, to secure stable reaction conditions. The total heat of reaction amounts to approximately 25 percent of the heat of combustion of the synthesis gas, and lays thereby a theoretical limit on the maximal efficiency of the Fischer-Tropsch process.

TABLE 7.2 Fischer-Tropsch Reactions

Reaction	Reaction enthalpy: $\Delta H_{300 K}$ [kJ/mol]
$CO + 2H_2 \rightarrow -CH_2- + H_2O$	−165.0
$2CO + H_2 \rightarrow -CH_2- + CO_2$	−204.7
$CO + H_2O \rightarrow H_2 + CO_2$	−39.8
$3CO_2 + H_2 \rightarrow -CH_2- + 2CO_2$	−244.5
$CO_2 + 3H_2 \rightarrow -CH_2- + 2H_2O$	−125.2

The reaction is dependent on a catalyst, mostly an iron or cobalt catalyst where the reaction takes place (van Berge, 1995). There is either a low-temperature Fischer-Tropsch (LTFT) or high-temperature Fischer-Tropsch (HTFT) (with temperatures ranging between 200 and 240°C for LTFT and 300 and 350°C for HTFT). LTFT uses an iron catalyst, and HTFT either an iron or a cobalt catalyst. The different catalysts include also nickel-based and ruthenium-based catalysts, which also have enough activity for commercial use in the process. But the availability of ruthenium is limited and the nickel based catalyst has high activity but produces too much methane, and additionally the performance at high pressure is poor, due to production of volatile carbonyls. This leaves only cobalt and iron as practical catalysts, and this study will only consider these two. Iron is cheap, but cobalt has the advantage of higher activity and longer life, though it is on a metal basis 1000 times more expensive than iron catalyst.

For large-scale commercial Fischer-Tropsch reactors heat removal and temperature control are the most important design features to obtain optimum product selectivity and long catalyst lifetimes. Over the years, basically four Fischer-Tropsch reactor designs have been used commercially (Fig. 7.1).

These are the *multi-tubular fixed bed* reactor, the *slurry* reactor, or the *fluidized bed* reactor (with either fixed or circulating bed).

The *fixed bed reactor* consists of thousands of small tubes with the catalyst as surface-active agent in the tubes. Water surrounds the tubes and regulates the temperature by settling the pressure of evaporation. The slurry reactor is widely used and consists of fluid and solid elements, where the catalyst has no particular position, but flows around as small pieces of catalyst together with the reaction components. Overheating the catalyst causes the decrease of its activity and favors the deposition of carbon on the surface of particles.

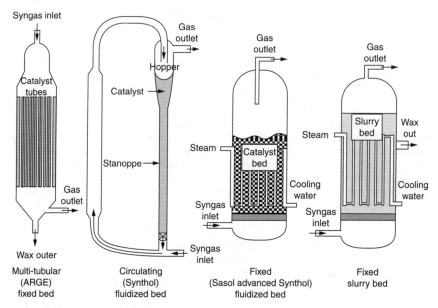

FIGURE 7.1 Types of Fischer-Tropsch reactors.

In the *slurry reactor*, the catalyst is suspended in the liquid and the gas is bubbled through the suspension. Generally this type of reactor gives (a) a more uniform temperature, (b) there is less catalyst present and consumed per ton of product, and (d) the differential pressure over bed is lower.

The *slurry reactor* and the *fixed bed reactor* are used in LTFT. The *fluidized bed reactors* are diverse, but characterized by the fluid behavior of the catalyst; the fluidized bed reactor is used in the HTFT.

Sasol in South Africa uses coal and natural gas as a feedstock, and produces a variety of synthetic petroleum products. The process was used in South Africa to meet its energy needs during its isolation under apartheid. This process has received renewed attention in the quest to produce low-sulfur diesel fuel in order to minimize the environmental impact from the use of diesel engines.

The Fischer-Tropsch technology as applied at Sasol can be divided into two operating regimes: (a) high-temperature Fischer Tropsch and (b) low-temperature Fischer-Tropsch.

The *high-temperature* Fischer Tropsch technology uses a fluidized catalyst at 300 to 330°C. Originally circulating fluidized bed units were used (Synthol reactors). Since 1989, a commercial scale classical fluidized bed unit has been implemented and improved upon.

The *low temperature* Fischer Tropsch technology has originally been used in tubular fixed bed reactors at 200 to 230°C. This produces a more paraffinic and waxy product spectrum than the *high-temperature* technology. A new type of reactor (the Sasol slurry phase distillate reactor) has been developed and is in commercial operation. This reactor uses a slurry phase system rather than a tubular fixed bed configuration and is currently the favored technology for the commercial production of synfuels.

The commercial Sasol Fischer-Tropsch reactors all use iron-based catalysts on the basis of the desired product spectrum and operating costs. Cobalt-based catalysts have also been known since the early days of this technology and have the advantage of higher conversion for low temperature cases. Cobalt is not suitable for high-temperature use due to excessive methane formation at such temperatures. For once, through maximum diesel production, cobalt has, despite its high cost, advantages and Sasol has also developed cobalt catalysts which perform very well in the slurry phase process.

The diesel produced by the slurry phase reactor has a highly paraffinic nature, giving a cetane number in excess of 70. The aromatic content of the diesel is typically below 3 percent and it is also sulfur-free and nitrogen-free. This makes it an exceptional diesel as such or it can be used to sweeten or to upgrade conventional diesels.

The Fischer-Tropsch process is an established technology and already applied on a large scale, although its popularity is hampered by high capital costs, high operation and maintenance costs, and the uncertain and volatile price of crude oil. In particular, the use of natural gas as a feedstock only becomes practical when using *stranded gas*, that is, sources of natural gas far from major cities which are impractical to exploit with conventional gas pipelines and liquefied natural gas technology; otherwise, the direct sale of natural gas to consumers would become much more profitable. It is suggested by geologists that supplies of natural gas will peak 5 to 15 years after oil does, although such predictions are difficult to make and often highly uncertain. Hence the increasing interest in coal as a source of synthesis gas.

Under most circumstances the production of synthesis gas by reforming natural gas will be more economical than from coal gasification, but site specific factors need to be considered. In fact, any technological advance in this field (such as better energy integration or the oxygen transfer ceramic membrane reformer concept) will speed up the rate at which the synfuels technology will become common practice.

There are large coal reserves which may increasingly be used as a fuel source during oil depletion. Since there are large coal reserves in the world, this technology could be used as an interim transportation fuel if conventional oil were to become more expensive. Furthermore, combination of biomass gasification and Fischer-Tropsch synthesis is a very promising route to produce transportation fuels from renewable or *green* resources.

Often a higher concentration of some sorts of hydrocarbons is wanted, which might be achieved by changed reaction conditions. Nevertheless, the product range is wide and infected with uncertainties, due to lack of knowledge of the details of the process and of the kinetics of the reaction. Since the different products have quite different characteristics such as boiling point, physical state at ambient temperature, and thereby different uses and ways of distribution, often only a few of the carbon chains is wanted. As an example the LTFT is used when longer carbon chains are wanted, because lower temperature increases the portion of longer chains.

The yield of diesel is therefore highly dependent on the chain growth probability, which again is dependent on pressure, temperature, feed gas composition, catalyst type, catalyst composition, and reactor design. The desire to increase the selectivity of some favorable products leads to a need of understanding the relation between reaction conditions and chain growth probability, which in turn request a mathematical expression for the growth probability in order to make a suitable model of the process.

There have been many attempts to model the product distribution of the Fischer-Tropsch process. As the knowledge of the process is limited, the modeling of the product distributions is not accurate. There are deviations between inspected product distributions and the different models when the conditions are changed. The general consensus seems to be that the product distribution follows an exponential function, with the probability of chain growth as an important factor.

In addition to application of the Fischer-Tropsch synthesis to the gasification products from petroleum residua, coal, biomass, wastes, and other carbonaceous feedstocks, application of the Fischer Tropsch process to the production of liquid fuels from natural gas is another established technology. Further expansion is planned for 2010, the bulk of this capacity being located in the Middle East (Qatar) (Chemical Market Reporter, 2004; IEA, 2004). The conversion efficiency is about 55 percent, with a theoretical maximum of about 78 percent. Due to the energy loss, this process makes only economic sense for cheap stranded gas. As the cost for liquefied natural gas transportation declines and demand increases, the importance or need for such options may also decline.

7.6 REFERENCES

Anderson, L. L. and D. A. Tillman: *Synthetic Fuels from Coal: Overview and Assessment,* John Wiley & Sons Inc., New York, 1979.

Anunputtikul, W. and S. Rodtong: "Laboratory Scale Experiments for Biogas Production from Cassava Tubers," Proceedings, Joint International Conference on Sustainable Energy and Environment, Hue Hen, Thailand, Dec. 1–3, 2004.

Baughman, G. L.: *Synthetic Fuels Data Handbook,* Cameron Engineers, Denver, Colo., 1978.

Chemical Market Reporter: "GTL could become major chemicals feedstock" Jan. 12, 2004.

Coelho, S. T., S. M. S. G. Velazquez, V. Pecora, and F. C. Abreu: "Energy Generation with Landfill Biogas," Proceedings, International RIO 6: World Climate & Energy Event, Rio de Janeiro, Brazil, Nov. 17–18, 2006.

Elton, A.: In *A History of Technology,* vol. 4, C. Singer, E. J. Holmyard, A. R. Hall, and T. I. Williams (eds.), Clarendon Press, Oxford, England, 1958, chap. 9.

Fryer, J. F. and J. G. Speight: "Coal Gasification: Selected Abstracts and Titles," Information Series No. 74, Alberta Research Council, Edmonton, Alberta, Canada, 1976.

Hickman, D. A. and L. D. Schmidt: *Science,* 259, 1993, p. 343.

Howard-Smith, I. and G. J. Werner: *Coal Conversion Technology,* Noyes Data Corp., Park Ridge, N.J., 1976.

IEA: "World Energy Outloook," IEA/OECD. Paris, France, 2004.

Mangone, C.: "Gas to Liquids—Conversions Produce Extremely Pure Base Oils," *Machinery Lubrication Magazine,* Independent Lubricant Manufacturers Association (ILMA), Nov., 2002.

Massey, L. G. (ed.): *Coal Gasification,* Advances in Chemistry Series No. 131, American Chemical Society, Washington, D.C., 1974.

Michaels, H. J. and H. F. Leonard: *Chem. Eng. Progr.,* 74(8), 1978, p. 85.

Nef, J. U.: In *A History of Technology,* vol. 3. C. Singer, E. J. Holmyard, A. R. Hall, and T. I. Williams (ed.), Clarendon Press, Oxford, England, 1957, chap. 3.

Nowacki, P.: *Lignite Technology,* Noyes Data Corporation, Park Ridge, N.J., 1980.

Olah, G. A., A. Goeppert, and G. K. S. Parkash: *Beyond Oil and Gas: The Methanol Economy,* Wiley-VCH, Weinheim, Germany, 2006, chap. 12.

Probstein, R. F. and R. E. Hicks: *Synthetic Fuels,* pH Press, Cambridge, Mass., 1990, chap. 4.

Rostrup-Nielsen, J. R.: "Catalytic Steam Reforming," in *Catalysis: Science and Technology,* J. R. Anderson and M. Boudart (eds.), Springer, New York, 1984.

Rostrup-Nielsen, J. R. and J. H. Bak Hansen: CO_2-Reforming of Methane over Transition Metals. *Journal of Catalysis,* 144, 1993, pp. 38–49.

Rostrup-Nielsen, J. R., J. Sehested, and J. K. Norskov: Hydrogen and Synthesis Gas by Steam- and CO_2-Reforming. *Advances in Catalysis,* 47, 2002, pp. 65–139.

Speight, J. G.: *The Chemistry and Technology of Coal,* 2d ed., Marcel Dekker Inc., New York, 1994.

Speight, J. G.: *Natural Gas: A Basic Handbook,* Gulf Publishing Company, Houston, Tex., 2007a.

Speight, J. G.: *The Chemistry and Technology of Petroleum,* 4th ed., CRC Press, Taylor and Francis Group, Boca Raton, Fla., 2007b.

Storch, H. H.: "Synthesis of Hydrocarbons from Water Gas," in *Chemistry of Coal Utilization,* vol. 2, H. H. Lowry (ed.), John Wiley & Sons Inc., New York, 1945, pp. 1797–1845.

Taylor, F. S. and C. Singer: In *A History of Technology,* vol. 2, C. Singer, E. J. Holmyard, A. R. Hall, and T. I. Williams (eds.), Clarendon Press, Oxford, England.. 1957, chap. 10.

Twigg, M. V.: *Catalyst Handbook,* Wolfe Publishing Ltd., London, England, 1989.

Udengaard, N. R., J. H. B. Hansen, and D.C. Hanson: *Oil & Gas Journal,* 90, 1992, p. 62.

van Berge, P. J.: "Cobalt as an alternative Fischer-Tropsch catalyst to iron for the production of middle distillates," 4th International Natural Gas Conversion Symposium, Kruger National Park, South Africa, Nov., 1995.

van Beurden, P. J.: "On The Catalytic Aspects of Steam-Methane Reforming: A Literature Survey," Report No. ECN-I-04-003, Energy Research Center of the Netherlands (ECN), Petten, Netherlands, 2004, http://www.ecn.nl/docs/library/report/2004/i04003.pdf.

van der Burgt, M. J.: *Hydrocarbon Processing,* 58(1), 1979, p. 161.

van Heek, K. H. and H. J. Muhlen: In *Fundamental Issues in Control of Carbon Gasification Reactivity,* J. Lahaye and P. Ehrburger (eds.), Kluwer Academic Publishers Inc., Amsterdam, Netherlands, 1991, p. 1.

CHAPTER 8
FUELS FROM BIOMASS

Fossil fuels are finite energy resources (Pimentel and Pimentel, 2006). Therefore, reducing national dependence of any country on imported crude oil is of critical importance for long-term security and continued economic growth. Supplementing petroleum consumption with renewable biomass resources is a first step toward this goal. The realignment of the chemical industry from one of petrochemical refining to a biorefinery concept is, given time, feasible has become a national goal of many oil-importing countries. However, clearly defined goals are necessary for increasing the use of biomass-derived feedstocks in industrial chemical production and it is important to keep the goal in perspective.

In this context, the increased use of biofuels should be viewed as one of a range of possible measures for achieving self sufficiency in energy, rather than a panacea (Crocker and Crofcheck, 2006; Worldwatch Institute, 2006; Freeman, 2007; Nersesian, 2007).

Biomass, a source of energy has been used since ancient times (Xiaohua and Zhenmin, 2004), is the collective name for *renewable materials* which includes: (a) energy crops grown specifically to be used as fuel, such as wood or various grasses, (b) agricultural residues and by-products, such as straw, sugarcane fiber, rice hulls animal waste, and (c) residues from forestry, construction, and other wood-processing industries (Brown, 2003; NREL 2003; Wright et al., 2006).

Biomass is a renewable resource, whose utilization has received great attention due to environmental considerations and the increasing demands of energy worldwide (Tsai et al., 2007). Biomass is clean for it has negligible content of sulfur, nitrogen, and ash-forming constituents, which give lower emissions of sulfur dioxide, nitrogen oxides, and soot than conventional fossil fuels. The main biomass resources include the following: forest and mill residues, agricultural crops and wastes, wood and wood wastes, animal wastes, livestock operation residues, aquatic plants, fast-growing trees and plants, and municipal and industrial wastes. The role of wood and forestry residues in terms of energy production is as old as fire itself and in many societies wood is still the major source of energy. In general, biomass can include anything that is not a fossil fuel and is bioorganic-based (Lucia et al., 2006).

There are many types of biomass resources that can be used and replaced without irreversibly depleting reserves and the use of biomass will continue to grow in importance as replacements for fossil materials are used as fuels and as feedstocks for a range of products (Narayan, 2007). Some biomass materials also have particular unique and beneficial properties which can be exploited in a range of products including pharmaceuticals and certain lubricants.

Following from this, a *biofuel* is any fuel that is derived from biomass, that is, recently living organisms or their metabolic by-products. *Biofuel* has also been defined as any fuel with an 80 percent minimum content by volume of materials derived from living organisms harvested within the 10 years preceding its manufacture.

However, for many staple food crops, a potentially large economic resource is effectively being thrown away. For example, the straw associated with the wheat crop in often ploughed back into the soil, even though only a small proportion is needed to maintain the level of organic matter. Thus, a huge renewable resource is not being usefully exploited since wheat straw contains a range of potentially useful chemicals. These include: (a) cellulose and related compounds which can be used for the production of paper and/or bioethanol, (b) silica compounds which can be used as filter materials such as those necessary for water purification, and (c) long-chain lipids which can be used in cosmetics or for other specialty chemicals.

However, for the purpose of this chapter, three distinct sources of biomass energy are: (a) agricultural crops, (b) wood, and (c) municipal and industrial wastes (Chaps. 9, 10, and 11, respectively). Landfill gas is also included in this work (Chap. 11).

This includes everything from *primary sources* of crops and residues harvested/collected directly from the land to *secondary sources* such as sawmill residuals, to *tertiary sources* of postconsumer residuals that often end up in landfills. A *fourth source*, although not usually categorized as such, includes the gases that result from anaerobic digestion of animal manures or organic materials in landfills (Chap. 9) (Wright et al., 2006).

Primary biomass is produced directly by photosynthesis and includes all terrestrial plants now used for food, feed, fiber, and fuel wood. All plants in natural and conservation areas (as well as algae and other aquatic plants growing in ponds, lakes, oceans, or artificial ponds and bioreactors) are also considered primary biomass. However, only a small portion of the primary biomass produced will ever be harvested as feedstock material for the production of bioenergy and by-products.

More generally, biomass feedstocks are recognized by the specific plant content of the feedstock or the manner in which the feedstocks is produced.

For example, *primary biomass feedstocks* are thus primary biomass that is harvested or collected from the field or forest where it is grown. Examples of primary biomass feedstocks currently being used for bioenergy include grains and oilseed crops used for transportation fuel production, plus some crop residues (such as orchard trimmings and nut hulls) and some residues from logging and forest operations that are currently used for heat and power production. In the future it is anticipated that a larger proportion of the residues inherently generated from food crop harvesting, as well as a larger proportion of the residues generated from ongoing logging and forest operations, will be used for bioenergy (Smith, 2006). Additionally, as the bioenergy industry develops, both woody and herbaceous perennial crops will be planted and harvested specifically for bioenergy and product end uses.

Secondary biomass feedstocks differ from primary biomass feedstocks in that the secondary feedstocks are a by-product of processing of the primary feedstocks. By *processing* it is meant that there is substantial physical or chemical breakdown of the primary biomass and production of by-products; *processors* may be factories or animals. Field processes such as harvesting, bundling, chipping, or pressing do not cause a biomass resource that was produced by photosynthesis (e.g., tree tops and limbs) to be classified as secondary biomass.

Specific examples of secondary biomass includes sawdust from sawmills, black liquor (which is a by-product of paper making), and cheese whey (which is a by-product of cheese-making processes). Manures from concentrated animal-feeding operations are collectable secondary biomass resources. Vegetable oils used for biodiesel that are derived directly from the processing of oilseeds for various uses are also a secondary biomass resource (Wright et al., 2006; Bourne, 2007).

Tertiary biomass feedstock includes postconsumer residues and wastes, such as fats, greases, oils, construction and demolition wood debris, other wood waste from the urban environments, as well as packaging wastes, municipal solid wastes, and landfill gases. The category, *other wood waste from the urban environment* includes trimmings from urban trees, which technically fits the definition of primary biomass. However, because this material is normally handled as a waste stream along with other postconsumer wastes from urban environments (and included in those statistics), it makes the most sense to consider it to be part of the tertiary biomass stream.

Biomass feedstocks and fuels exhibit a wide range of physical, chemical, and agricultural/process-engineering properties. Despite their wide range of possible sources, biomass feedstocks are remarkably uniform in many of their fuel properties, compared with competing feedstocks such as coal or petroleum.

For example, there are many types of coal and the gross heating value of these types varies from 8600 to 12,900 Btu/lb (20–30 GJ/t). However, nearly all kinds of biomass feedstocks destined for combustion fall in the range 6450 to 8200 Btu/lb (15–19 GJ/t) (Wright et al., 2006). For most agricultural residues, the heating values are even more uniform—approximately 6450 to 7300 Btu/lb (15–17 GJ/t); the values for most woody materials are 7750 to 8200 Btu/lb (8–19 GJ/t). Moisture content is probably the most important determinant of heating value. Air-dried biomass typically has about 15 to 20 percent moisture, whereas the moisture content for oven-dried biomass is around 0 percent. Moisture content is also an important characteristic of coals, varying in the range of 2 to 30 percent. However, the bulk density (and hence energy density) of most biomass feedstocks is generally low, even after densification, about 10 and 40 percent of the bulk density of most fossil fuels. Liquid biofuels have comparable bulk densities to fossil fuels.

8.1 BIOMASS FUELS

Solid biofuels such as wood (Chap. 9) or dried dung have been used since man learned to control fire.

On the other hand, liquid biofuels for industrial applications was used since the early days of the car industry. Nikolaus August Otto, the inventor of the combustion engine, conceived his invention to run on ethanol while Rudolf Diesel, the inventor of the diesel engine, conceived it to run on peanut oil. Henry Ford originally had designed the Ford Model T, a car produced between 1903 and 1926, to run completely on ethanol. Ford's desires to mass produce electric cars did not come to fruition. However, when crude oil began being cheaply extracted from deeper in the soil (thanks to oil reserves discovered in Pennsylvania and Texas), cars began using fuels from oil.

Nevertheless, before World War II, biofuels were seen as providing an alternative to imported oil in countries such as Germany, which sold a blend of gasoline with alcohol fermented from potatoes under the name *Reichskraftsprit*. In Britain, grain alcohol was blended with petrol by the Distillers Company Ltd. under the name *Discol* and marketed through Esso's affiliate Cleveland.

After the war, cheap Middle Eastern oil lessened interest in biofuels. Then, with the oil shocks of 1973 and 1979, there was an increase in interests from governments and academics in biofuels. However, interest decreased with the counter-shock of 1986 that made oil prices cheaper again. But since about 2000 with rising oil prices, concerns over the potential oil peak, *greenhouse gas* emissions (*global warming*), and instability in the Middle East are pushing renewed interest in biofuels. Government officials have made statements and given fiscal aid in favor of biofuels, which can play a major role in the future.

The supply of crude oil, the basic feedstock for refineries and for the petrochemical industry, is finite and its dominant position will become unsustainable as supply/demand issues erode its economic advantage over other alternative feedstocks. This situation will be mitigated to some extent by the exploitation of more technically challenging fossil resources and the introduction of new technologies for fuels and chemicals production from natural gas and coal.

However, the use of fossil resources at current rates will have serious and irreversible consequences for the global climate. Consequently, there is a renewed interest in the utilization of plant-based matter as a raw material feedstock for the chemicals industry. Plants accumulate carbon from the atmosphere via photosynthesis and the widespread utilization of these materials as basic inputs into the generation of power, fuels, and chemicals is a viable route to reduce greenhouse gas emissions.

Thus, the petroleum and petrochemical industries are coming under increasing pressure not only to compete effectively with global competitors utilizing more advantaged hydrocarbon feedstocks but also to ensure that its processes and products comply with increasingly stringent environmental legislation.

The production of fuels and chemicals from renewable plant-based feedstocks utilizing state-of-the-art conversion technologies presents an opportunity to maintain competitive advantage and contribute to the attainment of national environmental targets. Bioprocessing routes have a number of compelling advantages over conventional petrochemical production; however, it is only in the last decade that rapid progress in biotechnology has facilitated the commercialization of a number of plant-based chemical processes. It is widely recognized that further significant production of plant-based chemicals will only be economically viable in highly integrated and efficient production complexes producing a diverse range of chemical products. This biorefinery concept is analogous to conventional oil refineries and petrochemical complexes that have evolved over many years to maximize process synergies, energy integration, and feedstock utilization to drive down production costs.

8.1.1 Feedstock Types

Plants offer a unique and diverse feedstock for chemicals. Plant biomass can be gasified to produce synthesis gas, a basic chemical feedstock and also a source of hydrogen for a future hydrogen economy (Hocevar, 2007). In addition, the specific components of plants such as carbohydrates, vegetable oils, plant fiber, and complex organic molecules known as primary and secondary metabolites can be utilized to produce a range of valuable monomers, chemical intermediates, pharmaceuticals, and the following materials:

Carbohydrates (starch, cellulose, sugars) are readily obtained from wheat and potato, while cellulose is obtained from wood pulp. The structures of these polysaccharides can be readily manipulated to produce a range of biodegradable polymers with properties similar to those of conventional plastics such as polystyrene foams and polyethylene film. In addition, these polysaccharides can be hydrolyzed, catalytically or enzymatically, to produce sugars, a valuable fermentation feedstock for the production of ethanol, citric acid, lactic acid, and dibasic acids such as succinic acid.

Vegetable oils are obtained from seed oil plants such as palm, sunflower, and soya. The predominant source of vegetable oils in many countries is rapeseed oil. Vegetable oils are a major feedstock for the oleochemical industry (surfactants, dispersants, and personal care products) and are now successfully entering new markets such as diesel fuel, lubricants, polyurethane monomers, functional polymer additives and solvents.

Plant fibers (lignocellulosic fibers) are extracted from plants such as hemp and flax, and can replace cotton and polyester fibers in textile materials and glass fibers in insulation products.

Specialty products such as highly complex bioactive molecules, the synthesis of which is often beyond the ability and economics of laboratories, and a wide range of chemicals are currently extracted from plants for a wide range of markets, from crude herbal remedies to very high value pharmaceutical intermediates.

Fats and greases, the proper categorization may be debatable since those are by-products of the reduction of animal biomass into component parts. However, most fats and greases, and some oils, are not available for bioenergy use until after they become a postconsumer waste stream; it seems appropriate for them to be included in the tertiary biomass category. Vegetable oils derived from processing of plant components and used directly for bioenergy (e.g., soybean oil used in biodiesel) would be a secondary biomass resource, though amounts being used for bioenergy are most likely to be tracked together with fats, greases, and waste oils.

Many different types of biomass can be grown for the express purpose of energy production. Crops that have been used for energy include: sugar cane, corn, sugar beets, grains, elephant grass, kelp (seaweed), and many others. There are two main factors which determine whether a crop is suitable for energy use. Good energy crops have a very high yield of dry material per unit of land (dry metric tons/hectare). A high yield reduces land requirements and lowers the cost of producing energy from biomass. Similarly, the amount of energy which can be produced from a biomass crop must be less than the amount of energy required to grow the crop. In some circumstances like the heavily mechanized corn farms in the U.S. Midwest, the amount of ethanol which can be recovered from the corn is barely larger than the fuel required for tractors, fertilizers, and processing.

The simplest, cheapest, and most common method of obtaining energy from biomass is direct combustion. Any organic material with a water content that will not interfere with sustained combustion of the material can be burned to produce energy. The heat of combustion can be used to provide space or process heat, water heating, or, through the use of a steam turbine, electricity. In the developing world, many types of biomass such as dung and agricultural wastes are burned for cooking and heating.

Thus, almost all crops, whether grown for food, animal feed, fiber, or any other purpose, result in some form of organic residues after their primary use has been fulfilled. These organic residues, as well as animal wastes (excrement) can be used for energy production through direct combustion or biochemical conversion. Current worldwide production of crop residues is very large; but an increased scale of use for fuel may have significant environmental impacts, the most serious being those of lost soil fertility and soil erosion.

Most crop residues are returned to the soil, and the humus resulting from their decomposition helps maintain soil nutrients, soil porosity, water infiltration, and storage, as well as reducing soil erosion. Crop residues typically contain 40 percent of the nitrogen (N), 80 percent of the potassium (K), and 10 percent of the phosphorous (P) applied to the soil in the form of fertilizer. If these residues are subjected to direct combustion for energy, only a small percentage of the nutrients are left in the ash. Similarly, soil erosion will increase. Estimates for the United States indicate that 22 percent of crop residues could be removed without causing substantial soil erosion, providing energy equivalent to 5 percent of U.S. needs.

Biomass is a renewable energy source, unlike the fossil fuel resources (petroleum, coal, and natural gas) and, like the fossil fuels, biomass is a form of stored solar energy. The energy of the sun is captured through the process of photosynthesis in growing plants. One advantage of biofuel in comparison to most other fuel types is it is biodegradable, and thus relatively harmless to the environment if spilled.

Many different biomass feedstocks can be used to produce liquid fuels (Worldwatch Institute, 2006). They include crops specifically grown for bioenergy, and various agricultural residues, wood residues, and waste streams. Their costs and availability vary widely. Collection and transportation costs are often critical.

Sugarcane, sugar beet, corn, and sweet sorghum are agricultural crops presently grown commercially for both carbohydrate production and animal feeds. Sugarcane, corn, and sweet sorghum are efficient at trapping solar energy and use specific biochemical pathways to recycle and trap carbon dioxide that is lost through photorespiration. Sugar beets are efficient because they store their carbohydrate in the ground. Sugarcane was the basis for the World's first renewable biofuel program in Brazil (Bourne, 2007). Corn is the basis for the present renewable ethanol fuel industry in the United States.

The sugars produced by these crops are easily fermented by *Saccharomyces cerevisiae*. The sucrose produced by sugarcane, sugar beet, and sweet sorghum can be fermented directly after squeezing them from the crop. Corn traps its carbohydrate largely in the form of starch which must first be converted into glucose through saccharification with glucoamylase. The residues left over after removing fermentable sugars can also be utilized. In some cases they end up as animal feeds, but many agricultural residues can be converted into additional fermentable sugars through saccharification with cellulases and hemicellulases. The hemicellulose sugars are not fermentable by *S. cerevisiae*, and must be converted to ethanol by pentose fermenting yeasts or genetically engineered organisms.

Bioenergy crops include fast-growing trees such as hybrid poplar, black locust, willow, and silver maple in addition to annual crops such as corn, sweet sorghum, and perennial grasses such as switch grass.

Briefly, switch grass is a thin-stemmed, warm season, perennial grass that has shown high potential as a high yielding crop that can be readily grown in areas that are also suitable for crop production. In fact, there are many perennial crops (grass and tree species) that show high potential for production of cost-competitive cellulosic biomass. Switch grass can be viewed as a surrogate for many *perennial energy crops* when estimating biomass supply and availability.

Many other crops are possible and the optimal crop will vary with growing season and other environmental factors. Most fast-growing woody and annual crops are high in hemicellulose sugars such as xylose.

Corn stalks and *wheat straws* are the two agricultural residues produced in the largest quantities. However, many other residues such as potato and beet waste may be prevalent in some regions. In addition to quantity it is necessary to consider density and water content (which may restrict the feasibility of transportation) and seasonality which may restrict the ability of the conversion plant to operate on a year-round basis. Facilities designed to use seasonal crops will need adequate storage space and should also be flexible enough to accommodate alternative feedstocks such as wood residues or other wastes in order to operate year-around. Some agricultural residues need to be left in the field in order to increase tilth (the state of aggregation of soil and its condition for supporting plant growth and to reduce erosion) but some residues such as corncobs can be removed and converted without much difficultly.

Similar to herbaceous crops, straw usually has lower moisture content than woody biomass. Conversely, it has a lower calorific value, bulk density, ash melting point and higher content of ash, and problematic inorganic component such as chlorine, potassium, and sulfur, which cause corrosion and pollution. The last two drawbacks can be relatively easily overcome by leaving straw on the field for a while. In such a way rainfall "washes" it naturally from a large part of potassium and chlorine. Alternatively, fresh straw can be directly shipped to the gasification plant, where it is washed by dedicated facilities at moderate temperatures (50–60°C). Due to washing, the initially low moisture content of straw becomes higher in both cases and hence a mandatory drying is applied afterward. In both cases also the content of corrosive components is reduced, but not completely taken out. In order to decrease handling costs, straw and dedicated herbaceous energy crops are usually baled before being shipped to the gasification plant. The weight and the size of bales depend on the baling equipment and on the requirements of the gasification plant (Van Loo and Koppejan, 2003; NSCA, 2004).

Softwood residues are generally in high demand as feedstocks for paper production, but hardwood timber residues have less demand and fewer competing uses. In the past, as much as 50 percent of the tree was left on site at the time of harvest. Whole tree harvest systems for pulp chips recover a much larger fraction of the wood. Wood harvests for timber production often generates residues which may be left on the site or recovered for pulp production. Economics of wood recovery depend greatly on accessibility and local demand.

Underutilized wood species include Southern red oak, poplar, and various small diameter hardwood species. Unharvested dead and diseased trees can comprise a major resource in some regions. When such timber has accumulated in abundance, it comprises a fire hazard and must be removed. Such low grade wood generally has little value and is often removed by prescribed burns in order to reduce the risk of wildfires.

Waste streams can also be exploited for ethanol production. They are often inexpensive to obtain, and in many instances they have a negative value attributable to current disposal costs. Some principal waste streams currently under consideration include mixed paper from municipal solid waste, cellulosic fiber fines from recycled paper mills, bagasse from sugar manufacture, corn fiber, potato waste, and citrus waste, sulfite waste liquors, and hydrolysis streams from fiber board manufacture. Each waste stream has its own unique characteristics, and they generally vary from one source or time to another. Waste streams with lower lignin contents and smaller particle sizes are easier to deal with than those with higher lignin contents and larger particle sizes. Waste paper that has been treated by a chemical pulping process is much more readily converted than is native wood or herbaceous residue.

8.1.2 Feedstock Properties

The components of biomass include cellulose, hemicelluloses, lignin, lipids, proteins, simple sugars, starches, water, hydrocarbons, ash-forming constituents, and extractable compounds. These constituents influence the properties of biomass and, in turn, have a significant bearing on the thermal conversion of biomass. In addition, the high moisture, oxygen content, hydrogen content, and volatile matter content, and low energy density also influence biomass conversion.

The high oxygen and hydrogen contents account for the high proportion of volatile matter and consequent high yields of gases and liquids on pyrolysis. A relatively high water yield results from the high oxygen concentration in biomass, and which consumes considerable hydrogen. Consequently the advantages of the high hydrogen-to-carbon (H/C) ratio associated with biomass are not reflected in the products to the extent that might be expected. In fact, pyrolysis gases can be deficient in pure hydrogen and pyrolysis liquids are highly oxygenated, viscous tars.

An additional and significant source of water vapor in biomass gases is the high moisture content of the source materials. In countercurrent flow schemes such as the Lurgi moving bed gasifier, this water is evolved in the relatively low temperature drying and pyrolysis zones and does not partake in gas phase or carbon-steam gasification reactions. On the other hand, in fluidized bed systems the moisture is evolved in the high temperature well-mixed reaction zone and therefore does participate in the reactions. If the system is directly heated and air blown, the additional heat required to evaporate the water will result in more nitrogen being introduced, and more carbon dioxide being produced, reducing the calorific value of the product gas. As the gas from air-blown processes is, in any case, a low-calorific value product, this factor is probably of little consequence other than with very wet feedstock. In oxygen-blown systems, however, the additional pure oxygen is required and higher carbon dioxide content of the medium calorific value off-gas may be of sufficient impact to dictate some degree of drying as a pretreatment.

Apart from drying, additional beneficiation may be undertaken to yield a resource of higher energy density. These operations will normally be undertaken at the source, so transport and subsequent storage costs may be reduced as well. Beneficiation steps include size reduction and densification. Waste heat, if available, may be used for drying, while size reduction and compression to form pellets or briquettes is estimated to require less than 2 percent of the energy in the dry biomass. Nevertheless, these operations are time consuming, and can be either labor or capital intensive.

Some advantages of biomass over conventional fossil fuels are the low sulfur content and highly reactive char. In addition, biomass materials do not cake and can therefore be easily handled in both fluidized and moving bed reactors. Finally, catalyst poisons are not present in biomass in significant concentrations. This can be important for the initial thermal processing as well as for subsequent upgrading operations.

Generally, biomass feedstocks and fuels exhibit a wide range of physical, chemical, and agricultural/process engineering properties. However, despite their wide range of possible sources, biomass feedstocks are remarkably uniform in many of their fuel properties, compared with competing feedstocks such as coal or petroleum.

For example, there are many kinds of coals whose gross heating value ranges from 8600 to 12,900 Btu/lb. However, nearly all kinds of biomass feedstocks destined for combustion fall in the range 6450 to 8200 Btu/lb. For most agricultural residues, the heating values are even more uniform (6450 to 7300 Btu/lb); the values for most woody materials are 7750 to 8200 Btu/lb. Moisture content is probably the most important determinant of heating value. Air-dried biomass typically has about 15 to 20 percent moisture, whereas the moisture content for oven-dried biomass is around 0 percent. Moisture content is also an important characteristic of coals, varying in the range of 2 to 30 percent. However, the bulk density (and hence energy density) of most biomass feedstocks is generally low, even after densification, about 10 to 40 percent of the bulk density of most fossil fuels. Liquid biofuels have comparable bulk densities to fossil fuels.

Most biomass materials are easier to gasify than coal because they are more reactive with higher ignition stability. This characteristic also makes them easier to process thermochemically into higher-value fuels such as methanol or hydrogen. Ash content is typically lower than for most coals, and sulfur content is much lower than for many fossil fuels. Unlike coal ash, which may contain toxic metals and other trace contaminants, biomass ash may be used as a soil amendment to help replenish nutrients removed by harvest. A few biomass feedstocks stand out for their peculiar properties, such as high silicon or alkali metal contents; these may require special precautions for harvesting, processing, and combustion equipment. Note also that mineral content can vary as a function of soil type and the timing of feedstock harvest. In contrast to their fairly uniform physical properties, biomass fuels are rather heterogeneous with respect to their chemical elemental composition.

Among the liquid biomass fuels, biodiesel (vegetable oil ester) is noteworthy for its similarity to petroleum-derived diesel fuel, apart from its negligible sulfur and ash content. Bioethanol has only about 70 percent the heating value of petroleum distillates such as gasoline, but its sulfur and ash contents are also very low. Both of these liquid fuels have lower vapor pressure and flammability than their petroleum-based competitors and this is an advantage in some cases (e.g., use in confined spaces such as mines) but a disadvantage in others (e.g., engine starting at cold temperatures).

The most suitable biomass resources for thermal conversion are wood and the organic portion of municipal solid waste. Crop residues and grasses are of intermediate value, although thermal conversion might prove more effective than fermentation. If the versatility of a liquid fuel is desired, gasification may be combined with methanol or Fischer-Tropsch synthesis. Very wet resources such as aquatic biomass and animal wastes are not suited to thermal conversion, and are best reserved for anaerobic digestion (Chap. 9).

8.2 THE CHEMISTRY OF BIOMASS

Biomass is typically composed 75 to 90 percent by weight of sugar species, the other 10 to 25 weight percent being mainly lignin.

The energy in biomass is the chemical energy associated with the carbon and hydrogen atoms contained in oxidizable organic compounds which are the source of the carbon and hydrogen in carbon dioxide and water. The conversion by plants of carbon dioxide and water to a combustible organic form occurs by the process of photosynthesis in which solar energy and chlorophyll are the important players.

Chlorophyll, present in the cells of green plants, absorbs solar energy and makes it available for the photosynthesis, which may be represented by the simplified chemical reaction:

$$CO_2 + H_2O \rightarrow (CH_2O)_x + O_2$$

The oxidizable organic materials that are produced by photosynthesis and which determine the properties of the plant matter of relevance to biomass energy utilization are carbohydrates and lignin.

All of the carbohydrates present are *saccharides* (i.e., sugars) or polymers of sugars (i.e., *polysaccharides*) that fall into three types: (a) starch, (b) cellulose, and (c) hemicellulose.

The simple sugars include glucose, fructose, and the like while the polymeric sugars such as cellulose and starch (Fig. 8.1) can be readily broken down to their constituent monomers by hydrolysis, preparatory to conversion to ethanol or other chemicals.

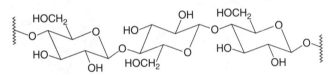

FIGURE 8.1 Generalized structure of cellulose.

Starch is a granular polysaccharide which accumulates in the storage tissues of plants such as seeds, tubers, roots, and stem pith. It is an important constituent of corn, potato, rice, and tapioca. Starch consists of 10 to 20 percent amylose, which is water soluble, and 80 to 90 percent amylopectin, which is insoluble in water. Both the constituents of starch are polymers of glucose, with amylose linked in chain structures, while amylopectin is a highly branched structure. Starch is not as chemically resistant as cellulose, and can be readily hydrolyzed by dilute acids and enzymes to fermentable sugars.

Hemicelluloses are polysaccharides that occur in association with cellulose. They are chemically different from cellulose, are amorphous, and have much lower molecular weight. While cellulose is built from the single sugar glucose, most hemicelluloses contain two to four different sugars as building blocks. Glucose is a component of some hemicelluloses, although xylose is a dominant sugar in hardwood hemicellulose, and mannose is important in softwood hemicellulose. Unlike the other sugars described so far, xylose contains only 5 carbon atoms and is a pentose.

The fraction of the cellulose containing xylose polymers is often referred to as *pentosan*. Hemicellulose is more soluble than cellulose, is dissolved by dilute alkaline solutions, and can be relatively readily hydrolyzed to fermentable sugars.

In contrast, lignin is a complex structure containing aromatic groups (Fig. 8.2) and is less readily degraded. Although lignocellulose is one of the cheapest and most abundant forms of biomass, it is difficult to convert this relatively unreactive material into sugars. Among other factors, the walls of lignocellulose are composed of lignin, which must be broken down in order to render the cellulose and hemicellulose accessible to acid hydrolysis. For this reason, many programs focused on ethanol production from biomass are based almost entirely on the fermentation of sugars derived from the starch in corn grain.

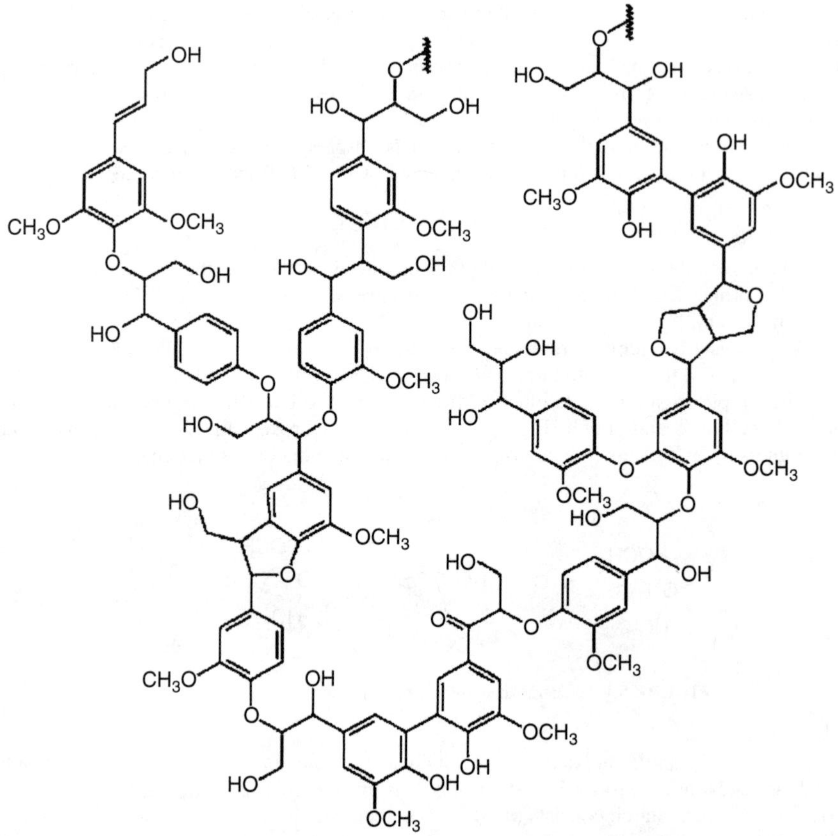

FIGURE 8.2 Hypothetical structure of lignin to illustrate the complexity of the molecule.

Lignin is the final major constituent of plant material important to biomass processing and it is a complex chemical compound that is most commonly derived from wood and is an integral part of the cell walls of plants, especially in tracheids, xylem fibers, and sclereids. The chemical structure of lignin is unknown and, at best, can only be represented by a hypothetical formula, the veracity of which is questionable.

Lignin is not a carbohydrate, but a polymer of single benzene rings linked with aliphatic chains; the phenolic compound p-hydroxyphenylpropane is an important monomer group in lignin. Like hemicellulose, lignin is amorphous and more soluble than cellulose. It may be removed from wood by steaming or by dissolving in hot aqueous or aqueous bisulfite solution. Lignin resists hydrolysis and is resistant to microbial degradation.

The term lignin was introduced in 1819 and is derived from the Latin word *lignum* (meaning *wood*). It is one of most abundant organic compounds on earth after cellulose and chitin. By way of clarification, chitin $(C_8H_{13}O_5N)_n$ is a long-chain polymeric polysaccharide of beta-glucose that forms a hard, semitransparent material found throughout the natural world. Chitin is the main component of the cell walls of fungi and is also a major component of the exoskeletons of arthropods, such as the crustaceans (e.g., crab, lobster, and shrimp) and the insects (e.g., ants, beetles, and butterflies), and of the beaks of cephalopods (e.g., squids and octopuses).

Lignin makes up about one-quarter to one-third of the dry mass of wood and is generally considered to be a large, cross-linked hydrophobis, aromatic macromolecule with molecular mass that is estimated to be in excess of 10,000. Degradation studies indicate that the molecule consists of various types of substructures which appear to repeat in random manner.

The biosynthesis of lignin begins with the synthesis of monolignols (e.g., coniferyl alcohol, sinapyl alcohol, and paracoumaryl alcohol) starting from an amino acid (phenylalanine). There are a number of other monolignols present in plants but different plants use different monolignols. The monolignols are synthesized as the respective glucosides which are water soluble and allows transportation through the cell membrane to the apoplast where the glucose moiety is removed after which the monolignols form lignin.

Lignin fills the spaces in the cell wall between cellulose, hemicellulose, and pectin components and is covalently linked to hemicellulose. Lignin also forms covalent bonds to polysaccharides and thereby crosslinks different plant polysaccharides. It confers mechanical strength to the cell wall (stabilizing the mature cell wall) and therefore to the entire plant.

8.3 PROCESSES

Biomass can be converted into commercial fuels, suitable substitute for fossil fuels (Narayan, 2007). These can be used for transportation, heating, electricity generation, or anything else fossil fuels are used for. The conversion is accomplished through the use of several distinct processes which include both biochemical conversion and thermal conversion to produce gaseous, liquid, and solid fuels which have high energy contents, are easily transportable, and are therefore suitable for use as commercial fuels.

8.3.1 Process Types

Biochemical conversion of biomass is completed through alcoholic fermentation to produce liquid fuels and anaerobic digestion or fermentation, resulting in biogas. Alcoholic fermentation of crops such as sugarcane and maize (corn) to produce ethanol for use in internal combustion engines has been practiced for years with the greatest production occurring in Brazil and the United States, where ethanol has been blended with gasoline for use in automobiles. With slight engine modifications, automobiles can operate on ethanol alone.

Anaerobic digestion of biomass has been practiced for almost a century, and is very popular in many developing countries such as China and India. The organic fraction of almost any form of biomass, including sewage sludge, animal wastes, and industrial effluents, can be broken down through anaerobic digestion into methane and carbon dioxide. This *biogas* is a reasonably clean burning fuel which can be captured and put to many different end uses such as cooking, heating, or electricity generation.

Thermal conversion offers a more effective means for the recovery or conversion of the energy content of wood and other lignocellulosic biomass. Wood and many other similar types of biomass which contain lignin and cellulose, (agricultural wastes, cotton gin waste, wood wastes, peanut hulls, etc.) can be converted through thermochemical processes into solid, liquid, or gaseous fuels. Pyrolysis, used to produce charcoal since the dawn of civilization, is still the most common thermochemical conversion of biomass to commercial fuel.

Fermentation. Traditional fermentation plants producing biogas are in routine use, ranging from farms to large municipal plants. As feedstock they use manure, agricultural residues, urban sewage, and waste from households, and the output gas is typically 64 percent methane. The biomass conversion process is accomplished by a large number of different agents, from the microbes decomposing and hydrolyzing plant material, over the acidophilic bacteria dissolving the biomass in aquatic solution, and to the strictly anaerobic methane bacteria responsible for the gas formation. Operating a biogas plant for a period of some months usually makes the bacterial composition stabilize in a way suitable for obtaining high conversion efficiency (typically above 60 percent, the theoretical limit being near 100 percent), and it is found important not to vary the feedstock compositions abruptly, if optimal operation is to be maintained. Operating temperatures for the bacterial processes are only slightly above ambient temperatures, for example, in the mesophilic region around 30°C.

A straightforward (but not necessarily economically optimal) route to hydrogen production would be to subject the methane generated to conventional steam reforming. The ensuing biomass-to-hydrogen conversion efficiency would in practice be about 45 percent. This scheme could be operated with present technology and thus forms a reference case for assessing proposed alternative hydrogen production routes.

One method is to select bacteria that produce hydrogen directly. Candidates would include *Clostridium* and *Rhodobacter* species. The best reactor-operating temperatures are often in the thermophilic interval or slightly above (50–80°C). Typical yields are 2 mol of hydrogen per mole of glucose, corresponding to 17 percent conversion efficiency. The theoretical maximum efficiency is around 35 percent, but there are also acetic or butanoic acids formed, which could be used to produce methane and thus additional energy, although not necessarily additional hydrogen.

Operation of this type of gas-producing plant would require pure feedstock biomass (here sugar), because of the specific bacteria needed for hydrogen production, and because contamination can cause decreased yields. Even the hydrogen produced has this negative effect and must therefore be removed continually.

Gasification. Gasification occurs through the thermal decomposition of biomass with the help of an oxidant such as pure oxygen or oxygen-enriched air to yield a combustible gas such as synthesis gas (syngas) rich in carbon monoxide and hydrogen (Albertazzi et al., 2007). The synthesis gas is posttreated, by steam-reforming or partial oxidation, to convert the hydrocarbons produced by gasification into hydrogen and carbon monoxide. The carbon monoxide is then put through the shift process to obtain a higher fraction of hydrogen, by carbon dioxide removal and methanation or by pressure swing adsorption (Mokhatab et al., 2007; Speight 2007).

In the gasification process, one or more reactants, such as oxygen, steam, or hydrogen, are introduced into the system. These chemical reactants combine with solid carbon at the higher gasification temperatures, so increasing the gas yield while consuming char. The amount of char by-product remaining on gasification is in fact essentially zero with biomass materials, while the small quantity of tars and oils evolved may be recycled to extinction. The introduced reactants also enter into gas phase reactions which, together with the shift in equilibrium and the change in relative reaction rates at the higher temperatures, results in a significantly better-quality gas than that obtained on pyrolysis. Important distinctions between pyrolysis and gasification are therefore the improved gas yield and the elimination of solid and liquid by-products.

One major advantage with gasification is the wide range of biomass resources available, ranging from agricultural crops, and dedicated energy crops to residues and organic wastes. The feedstock might have a highly various quality, but still the produced gas is quite standardized and produces a homogeneous product. This makes it possible to choose the feedstock that is the most available and economic at all times (Prins, 2005).

There are various types of gasifiers which, although already discussed (Chap. 5), deserve mention here.

The air-blown direct gasifiers operated at atmospheric pressure and used in power generation—fixed bed updraft and downdraft (Fig. 8.3) and fluidized bed bubbling

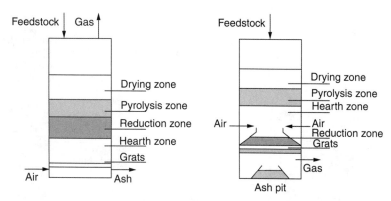

FIGURE 8.3 Updraft (left hand side) and downdraft (right hand side) fixed bed direct gasifiers. *(Source: Kavalov, B., and S. D. Peteves: "Status and Perspectives of Biomass-to-Liquid Fuels in the European Union," European Commission. Directorate General Joint Research Centre (DG JRC). Institute for Energy, Petten, Netherlands, 2005.)*

and circulating (Fig. 8.4)—are not suitable for biomass-to-liquids production. In addition, downdraft fixed bed gasifiers face severe constraints in scaling and are fuel inflexible, being able to process only fuels with well-defined properties. Updraft fixed bed gasifiers have fewer restrictions in scaling but the produced gas contains a lot of tars and methane.

Fluidized bed gasifiers generally do not encounter limitations in scaling and are more flexible concerning the particle size of fuels. Nevertheless, they still have limited fuel flexibility, due to a risk of slagging and fouling, agglomeration of bed material, and corrosion. The operating temperatures of air-blown fluidized bed gasifiers are therefore kept relatively low (800–1000°C), which implies incomplete decomposition of feedstocks, unless long residence times are used. Fluidized bed gasifiers (especially the bubbling bed ones) tend to contaminate the product gas with dust. The oxygen-blown atmospheric or pressurized

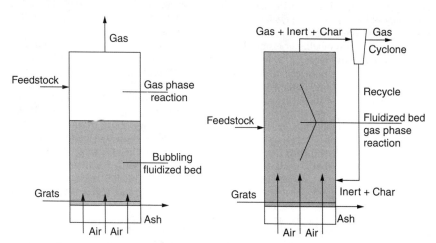

FIGURE 8.4 Bubbling (left hand side) and circulating (right hand side) fluidized bed direct gasifiers. *(Source: Kavalov, B., and S. D. Peteves: "Status and Perspectives of Biomass-to-Liquid Fuels in the European Union," European Commission. Directorate General Joint Research Centre (DG JRC). Institute for Energy, Petten, Netherlands, 2005.)*

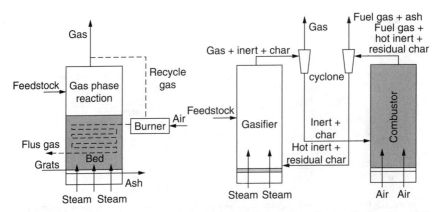

FIGURE 8.5 Gas (left hand side) and char (right hand side) indirect gasifiers. *(Source: Kavalov, B., and S. D. Peteves: "Status and Perspectives of Biomass-to-Liquid Fuels in the European Union," European Commission. Directorate General Joint Research Centre (DG JRC). Institute for Energy, Petten, Netherlands, 2005.)*

circulating fluidized bed gasifiers and the steam-blown gas or char indirect gasifiers (Fig. 8.5) are better solutions for biomass-to-liquids production. Both gasifying concepts reduce significantly the amount of nitrogen in the product gas. In the first case it is achieved via substituting air with oxygen. In the second case nitrogen ends up in the flue gas, but not in the product gas, because gasification and combustion are separated—the energy for the gasification is obtained by burning the chars from the first gasifier in a second reactor.

The gasification of biomass (e.g., wood scrap) is a well-known process, taking place in pyrolysis (oxygen supply far below what is required for complete combustion, the fraction called *equivalence ratio*) or fluidized-bed type of reactors. Conditions such as operating temperature determine whether hydrogen is consumed or produced in the process. Hydrogen evolution is largest for near-zero equivalence ratios, but the energy conversion efficiency is highest at an equivalence ratio around 0.25. The hydrogen fraction (in this case typically some 30 percent) must be separated for most fuel-cell applications, as well as for long-distance pipeline-transmission. In the pyrolysis-type application, gas production is low and most energy is in the oily substances that must be subsequently reformed in order to produce significant amounts of hydrogen. Typical operating temperatures are around 850°C. An overall energy conversion efficiency of around 50 percent is attainable, with considerable variations. Alternative concepts use membranes to separate the gases produced, and many reactor types uses catalysts to help the processes to proceed in the desired direction, notably at a lower temperature (down to some 500°C).

Each type of biomass has its own specific properties which determine performance as a feedstock in gasification plants. The most important properties relating to gasification are: (a) moisture content, (b) mineral content leading to ash production, (c) elemental composition, (d) bulk density and morphology, and (e) volatile matter content.

The moisture content of biomass is defined as the quantity of water in the material expressed as a percentage of the material's weight. For gasification processes like gasification, preference is given to relative dry biomass feedstocks because a higher quality gas is produced, that is, higher heating value, higher efficiency, and lower tar levels. Natural drying (i.e., on field) is cheap but requires long drying times. Artificial drying is more expensive but also more effective. In practice, artificial drying is often integrated with the gasification plant to ensure a feedstock of constant moisture content. Waste heat from the engine or exhaust can be used to dry the feedstock.

Ash content (which represents in inorganic content) of biomass is the inorganic oxides that remain after complete combustion of the feedstock. The amount of ash between different types of feedstocks differs widely (0.1 percent for wood up to 15 percent for some agricultural products) and influences the design of the reactor, particularly the ash removal system. The chemical composition of the ash is also important because it affects the melting behavior of the ash. Ash melting can cause slagging and channel formation in the reactor. Slags can ultimately block the entire reactor.

The *elemental composition* of the biomass feedstock is important with regard to the heating value of the gas and to the emission levels. The production of nitrogen and sulfur compounds is generally small in biomass gasification because of the low nitrogen and sulfur content in biomass.

The *bulk density* refers to the weight of material per unit of volume and differs widely between different types of biomass. Together with the heating value, it determines the energy density of the gasifier feedstock, that is, the potential energy available per unit volume of the feedstock. Biomass of low bulk density is expensive to handle, transport, and store. Apart from handling and storing behavior, the bulk density is important for the performance of the biomass as a fuel inside the reactor: a high void space tends to result in channeling, bridging, incomplete conversion, and a decrease in the capacity of the gasifier. The bulk density varies widely (100–1000 kg/m^3) between different biomass feedstocks not only because of the character of the biomass but also as a result of the way the biomass comes available (chips, loose, baled).

The amount of volatile products (*volatile matter content*) has a major impact on the tar production levels in gasifiers. Depending on the gasifier design, the volatile matter leaves the reactor at low temperatures (updraft gasifiers) or pass through a hot incandescent oxidation zone (downdraft gasifiers) where they are thermally cracked. For biomass materials the volatile matter content varies between 50 and 80 percent.

Feedstock preparation is required for almost all types of biomass materials because of large variety in physical, chemical, and morphological characteristics. The degree to which any specific pretreatment is desirable will depend upon the gasifier. For example, capacity and type of reactor (downdraft gasifiers are more strict to uniform fuel specifications than updraft gasifiers) are important aspects of biomass gasification.

Sizing of the feedstock may be necessary as different sizes are specified for different types of gasifiers. For small scale fixed bed gasifiers, cutting and/or sawing of wood blocks is the preferred form of fuel preparation. Chipped wood is preferred for larger scale applications. The size range of chips can be chosen by screening such that the fuel is acceptable for a specific gasifier type. For medium- and large-scale fixed bed gasifiers wood chips from forestry or wood processing industries are produced by crushers, hammer mills, shredders, and/or mobile chippers (particularly for thinnings from landscape conservation activities). Most of these sizing apparatus are provided with a screen. Dependent on the feedstock morphology and characteristics (hardness) the throughput or capacity varies considerably.

Drying of the biomass fuel is advisable if fresh wet materials (moisture content 50–60 percent on wet basis) are to be gasified. Lowering the moisture content of the feedstock is associated with a better performance of the gasifier. Utilizing the exhaust gases from an internal combustion engine is a very efficient way to do so. The sensible heat in engine exhaust is sufficient to dry biomass from 70 percent down to 10 percent. Rotary kilns are the most applied dryers. The energy costs of drying are high, but these can be outweighed by the lower downstream gas-cleaning requirements for dried feed.

Densification, such as *briquetting* or *pelletization*, are important techniques to densify biomass materials for increasing the particle size and bulk density. The reasons for biomass include: (a) densified biomass is less expensive to transport, (b) densified products are easier to store and handle, and (c) densification enables certain biomass feedstocks to be gasified in a specific gasifier.

The calorific value of the gas is the prime factor for power generation—the higher the value, the better. Hence, the availability in the gas of any compounds that increase calorific value is generally welcomed—product gas, which contains carbon monoxide (CO), hydrogen (H_2), various hydrocarbons [methane (CH_4), ethylene (C_2H_4), ethane (C_2H_6) tars, and chars]. The presence of inert components [water (H_2O), carbon dioxide (CO_2), and nitrogen (N_2)] is also acceptable, provided it is kept within certain limits.

Environmental concerns include disposal of associated tars and ashes, particularly for the fluidized bed reactors, where these substances must be separated from the flue gas stream (in contrast to the pyrolysis plants, where most tar and ash deposits at the bottom of the reactor). Concerns over biomass transportation are similar to those mentioned above for fermentation, and a positive fertilizer effect can also in many cases be derived from the gasification residues.

Biomass ash has also the potential to be used as a clarifying agent in water treatment, as a wastewater adsorbent, as a liquid waste adsorbent, as a hazardous waste solidification agent, as a lightweight fill for roadways, parking areas, and structures, as asphalt mineral filler, or as a mine spoil amendment.

Direct conversion processes, that is, combustion, of biomass encounters the same problems as those in case of coal or other solid fuels. The conversion of biomass into other useful forms such as synthetic gas or liquid fuels is considered as an alternative way to make use of biomass energy. Gasification is an effective and well-known technology which has been applied to coal but the technology developed for coal is not exactly suit to biomass utilization because of the different physical and chemical properties of coal and biomass.

Combustion. Combustion offers the most direct route for energy recovery, and is an effective means of utilizing the total energy content of whole wood and other biomass. Technology for the combustion of carbonaceous feedstocks is well developed and feasible.

Gasification results in only a partial oxidation of the carbon constituent. Much of the calorific value of the original fuel leaves the gasifier in the form of chemical energy in carbon monoxide, methane, and hydrogen.

In combustion, air containing sufficient oxygen to completely oxidize the hydrocarbon is used. The resulting gas (flue gas) is essentially all carbon dioxide, water vapor, and diluent nitrogen, and has no calorific value. The energy of the original fuel leaves the system as sensible heat that can be used for steam generation and, if required, electricity generation.

Thus, combustion offers a practical means for recovering the energy in biomass materials, using currently available technology.

However, due to the large land area over which biomass must be harvested, the seasonal nature of the supply, and the large volume of material to be transported and stored, the reliable provision of biomass to a large combustion plant can present considerable problems.

Pyrolysis. Pyrolysis is the thermal decomposition of materials in the absence of oxygen or when significantly less oxygen is present than required for complete combustion.

Pyrolysis dates back to at least ancient Egyptian times, when tar for caulking boats and certain embalming agents were made by pyrolysis. In the 1980s, researchers found that the pyrolysis liquid yield could be increased using fast pyrolysis where a biomass feedstock is heated at a rapid rate and the vapors produced are also condensed rapidly (Mohan et al., 2006).

Pyrolysis is the basic thermochemical process for converting biomass to a more useful fuel. Biomass is heated in the absence of oxygen, or partially combusted in a limited oxygen supply, to produce a hydrocarbon-rich gas mixture, an oil-like liquid, and a carbon-rich solid residue. The pyrolysis of biomass to produce hydrocarbons has been studied since the 1930s, and while relatively simple to perform, typically such processes are nonselective, producing a wide range of products. Thus, direct thermochemical conversion processes including pyrolysis to produce fuels from biomass are also possible resulting in the production of gaseous products (gaseous fuels), liquid products (liquid fuels), and charcoal (solid fuels) (Kavalov and Peteves, 2005; Demirbas, 2007).

The pyrolysis of biomass is a thermal treatment that results in the production of charcoal, liquid, and gaseous products. Thus, biomass is heated in the absence of air and breaks down into a complex mixture of liquids, gases, and a residual char. If wood is used as the feedstock, the residual char is what is commonly known as charcoal. With more modern technologies, pyrolysis can be carried out under a variety of conditions to capture all the components, and to maximize the output of the desired product be it char, liquid, or gas.

The liquid fraction of the pyrolysis products consists of two phases: an aqueous phase containing a wide variety of organooxygen compounds of low molecular weight and a nonaqueous phase containing insoluble organics of high molecular weight. This phase is called tar and is the product of greatest interest. The ratios of acetic acid, methanol, and acetone of the aqueous phase were higher than those of the nonaqueous phase. The point where the cost of producing energy from fossil fuels exceeds the cost of biomass fuels has been reached. With few exceptions, energy from fossil fuels will cost more money than the same amount of energy supplied through biomass conversion (Demirbas, 2007).

Wood contains 80 percent or more of volatile organic matter that may be recovered as a gas and tar on pyrolysis. Pyrolysis of wood was at one time used for obtaining creosote oils as well as acetic acid (wood vinegar) and some methanol (wood spirits). The gases evolved (hydrogen, methane, carbon monoxide, and carbon dioxide) had no value for illumination purposes and were used to supply the heat for the pyrolysis. Gasification of wood may be used to produce either a low- or a medium-heat content gas if a gaseous fuel is desired, or a synthesis gas that can be converted to liquid fuels using one of the indirect liquefaction processes. Gasification followed by a synthesis step is expected to yield a higher-quality liquid than can be obtained by pyrolysis.

The relative quantity of char, tars, and gases evolved on pyrolysis is strongly dependent on the feedstock then on the rate of heating and the final temperature attained. Slow-heating, low temperature pyrolysis favors char yields. After the initial decomposition, subsequent coking and cracking reactions can result in a char that contains oxygen and hydrogen in addition to the carbon.

Apart from char and water, some tars and gases are always produced. The watery distillate evolved from wood at 160 to 175°C is called *pyroligneous acid* and contains 5 to 10 percent acetic acid and 1.5 to 3 percent methanol. This used to be the source of methanol when it was produced by the destructive distillation (pyrolysis) of wood.

The higher-boiling tar fraction produced at higher temperatures is more important as a source of liquid fuels. It contains aromatic hydrocarbons and *creosote oil* evolved from the lignin, as well as aliphatic compounds. The creosote oil, which consists of high-molecular-weight phenols, used to be considered the most valuable constituent of the tar fraction, in part due to its potential for resin manufacture. As a whole, the tar fraction is a viscous, highly oxygenated oil that is unstable and corrosive.

The pyrolysis option for biomass is attractive because solid biomass and wastes can be readily converted into liquid products. These liquids, as crude bio oil or slurry of charcoal of water or oil, have advantages in transport, storage, combustion, retrofitting, and flexibility in production and marketing.

The mechanism of the pyrolysis of biomass has been studied by many investigators (Domburg et al., 1974; Adjaye et al., 1992; Demirbas, 2000) and is proposed to involve heat variations, associated with the thermal degradation reactions, which affect the pyrolysis route. Several endothermic and/or exothermic peaks for biomass pyrolysis are indicated (Stamm and Harris, 1953; Shafizadeh et al., 1976). Accordingly, cellulose pyrolysis is endothermic (Brown et al., 1952; Roberts and Clough, 1963; Demirbas and Kucuk, 1994), but lignin pyrolysis is exothermic (Demirbas et al., 1996).

In fact, the pyrolysis of biomass is more complex than originally believed and the general changes that occur during biomass pyrolysis are (Babu and Chaurasia, 2003; Mohan et al., 2006):

1. Heat transfer from a heat source, to increase the temperature inside the fuel.
2. The initiation of primary pyrolysis reactions at this higher temperature releases volatiles and forms char.
3. The flow of hot volatiles toward cooler solids results in heat transfer between hot volatiles and cooler unpyrolyzed fuel.
4. Condensation of some of the volatiles in the cooler parts of the fuel, followed by secondary reactions, can produce tar.
5. Autocatalytic secondary pyrolysis reactions proceed while primary pyrolytic reactions simultaneously occur in competition.
6. Further thermal decomposition, reforming, water gas shift reactions, radicals recombination, and dehydrations can also occur, which are a function of the process's residence time/temperature/pressure profile.

A process has been investigated for the saccharification of wood, involving prehydrolysis, lignocellulose pyrolysis, and tar hydrolysis. In this process, ground wood was first prehydrolyzed to remove the more readily hydrolyzable hemicelluloses. The residual lignocellulose was then pyrolyzed rapidly to provide a tar-containing levoglucosan and its condensation products. The destructive reaction of cellulose commences at temperatures on the order 52°C and is characterized by a decreasing polymerization. Thermal degradation of cellulose proceeds through two types of reaction: a gradual degradation, decomposition and charring on heating at lower temperatures, and a rapid volatilization accompanied by the formation of levoglucosan on pyrolysis at higher temperatures. The glucose chains in cellulose are first cleaved to glucose and from this, in a second stage, glucosan is formed by the splitting off of one molecule of water. Since cellulose and levoglucosan have the same elementary formula, $C_6H_{10}O_5$, a yield of 100 percent of the latter might be expected. The initial degradation reactions include depolymerization, hydrolysis, oxidation, dehydration, and decarboxylation (Shafizadeh and Stevenson, 1982).

Hemicellulose species react more readily than cellulose during heating (Cohen et al., 2002). The thermal degradation of hemicellulose begins above 100°C during heating for 48 hours; hemicelluloses and lignin are depolymerized by steaming at high temperature for a short time (Shafizadeh et al., 1976).

Lignin is broken down by extensive cleavage of b-aryl ether linkages during steaming of wood under 215°C. It has been found that on analysis of the metoxyl groups after isothermal heating of dry distilled wood, lignin decomposition begins at approximately 280°C with a maximum rate occurring between 350 and 450°C and the completion of the reaction occurs at 450 and 500°C (Sandermann and Augustin, 1963).

The formation of char from lignin under mild reaction conditions is a result of the breaking of the relatively weak bonds, like the alkyl-aryl ether bond(s), and the consequent formation of more resistant condensed structures, as has already been noted (Domburg et al., 1974). One additional parameter which may also have an effect on the char formation is the moisture content of the kraft lignin used.

Water is formed by dehydration. In the pyrolysis reactions, methanol arises from the breakdown of methyl esters and/or ethers from decomposition of pectin-like plant materials (Goldstein, 1981). Methanol also arises from methoxyl groups of uronic acid (Demirbas and Gullu, 1998). Acetic acid is formed in the thermal decomposition of all three main components of wood. When the yield of acetic acid originating from the cellulose, hemicelluloses, and lignin is taken into account, the total is considerably less than the yield from the wood itself (Wenzl et al., 1970). Acetic acid comes from the elimination of acetyl groups, originally linked to the xylose unit.

If wood is completely pyrolyzed, resulting products are about what would be expected by pyrolyzing the three major components separately. The hemicelluloses would break down first, at temperatures of 200 to 250°C. Cellulose follows in the temperature range 240 to 350°C, with lignin being the last component to pyrolyze at temperatures of 280 to 500°C. A wide variety of organic compounds occur in the pyrolysis liquid fractions given in the literature (Beaumont, 1985). Degradation of xylan yields eight main products: water, methanol, formic, acetic and propionic acids, 1-hydroxy-2-propanone, 1-hydroxy-2-butanone, and 2-furfuraldeyde. The methoxy phenol concentration decreases with increasing temperature, while phenols and alkylated phenols increases. The formation of both methoxy phenol and acetic acid was possibly as a result of the Diels-Alder cycloaddition of a conjugated diene and unsaturated furanone or butyrolactone.

The chemical structure of the xylan as the 4-methyl-3-acetylglucoronoxylan has been described (Timell, 1967). Furthermore, it has been reporte that the pyrolysis of the pyroligneous acid produces 50% methanol, 18% acetone, 7% esters, 6% aldehydes, 0.5% ethyl alcohol, 18.5% water, and small amounts of furfural (Demirbas, 2000). The composition of the water soluble products was not ascertained but it has been reported to be composed of hydrolysis and oxidation products of glucose such as acetic acid, acetone, simple alcohols, aldehydes, and sugars (Sasaki et al., 1998).

Biophotolysis. The photosynthetic production of gas (e.g., hydrogen) employs microorganisms such as Cyanobacteria, which have been genetically modified to produce pure hydrogen rather than the metabolically relevant substances (notably $NADPH_2$). The conversion efficiency from sunlight to hydrogen is very small, usually under 0.1 percent, indicating the need for very large collection areas.

The current thinking favors ocean locations of the bioreactors. They have to float on the surface (due to rapidly decreasing solar radiation as function of depth) and they have to be closed entities with a transparent surface (e.g., glass), in order that the hydrogen produced is retained and in order for sunlight to reach the bacteria. Because hydrogen buildup hinders further production, there further has to be a continuous removal of the hydrogen produced, by pipelines to, for example, a shore location, where gas treatment and purification can take place. These requirements make it little likely that equipment cost can be kept so low that the very low efficiency can be tolerated.

A further problem is that if the bacteria are modified to produce maximum hydrogen, their own growth and reproduction is quenched. Presumably, there has to be made a compromise between the requirements of the organism and the amount of hydrogen produced for export, so that replacement of organisms (produced at some central biofactory) does not have to be made at frequent intervals. The implication of this is probably an overall efficiency lower than 0.05 percent.

In a life cycle assessment of biohydrogen produced by photosynthesis, the impacts from equipment manufacture are likely substantial. To this, one should add the risks involved in production of large amounts of genetically modified organisms. In conventional agriculture, it is claimed that such negative impacts can be limited, because of slow spreading of genetically modified organisms to new locations (by wind or by vectors such as insects, birds, or other animals).

In the case of ocean biohydrogen farming, the unavoidable breaking of some of the glass- or transparent-plastic-covered panels will allow the genetically modified organisms to spread over the ocean involved and ultimately the entire biosphere. A quantitative discussion of such risks is difficult, but the negative cost prospects of the biohydrogen scheme probably rule out any practical use anyway.

Biogas contains methane and can be recovered in industrial anaerobic digesters and mechanical-biological treatment systems. Landfill gas is a less clean form of biogas which is produced in landfills through naturally occurring anaerobic digestion. Paradoxically if this gas is allowed to escape into the atmosphere it is a potent greenhouse gas.

When biomass is heated with no oxygen or only about one-third the oxygen needed for efficient combustion (amount of oxygen and other conditions determine if biomass gasifies or pyrolyzes), it gasifies to a mixture of carbon monoxide and hydrogen (synthesis gas, syngas).

Combustion is a function of the mixture of oxygen with the hydrocarbon fuel. Gaseous fuels mix with oxygen more easily than liquid fuels, which in turn mix more easily than solid fuels. Syngas therefore inherently burns more efficiently and cleanly than the solid biomass from which it was made. Biomass gasification can thus improve the efficiency of large-scale biomass power facilities such as those for forest industry residues and specialized facilities such as black liquor recovery boilers of the pulp and paper industry, both major sources of biomass power. Like natural gas, syngas can also be burned in gas turbines, a more efficient electricity generation technology than steam boilers to which solid biomass and fossil fuels are limited.

Most electrical generation systems are relatively inefficient, losing half to two-thirds of the energy as waste heat. If that heat can be used for an industrial process, space heating, or another purpose, efficiency can be greatly increased. Small modular biopower systems are more easily used for such "cogeneration" than most large-scale electrical generation.

Just as syngas mixes more readily with oxygen for combustion, it also mixes more readily with chemical catalysts than solid fuels do, greatly enhancing its ability to be converted to other valuable fuels, chemicals, and materials. The Fischer-Tropsch process converts syngas to liquid fuels needed for transportation (Chap. 7). The water-gas shift process converts syngas to more concentrated hydrogen for fuel cells. A variety of other catalytic processes can turn syngas into a myriad of chemicals or other potential fuels or products.

8.3.2 Environmental Issues

There is a consensus amongst scientists that biomass fuels used in a sustainable manner result in no net increase in atmospheric carbon dioxide (CO_2). Some would even go as far as to declare that sustainable use of biomass will result in a net decrease in atmospheric CO_2. This is based on the assumption that all the CO_2 given off by the use of biomass fuels was recently taken in from the atmosphere by photosynthesis. Increased substitution of fossil fuels with biomass-based fuels would therefore help reduce the potential for global warming, caused by increased atmospheric concentrations of carbon dioxide.

Unfortunately, things may not be as simple as has been assumed above. Currently, biomass is being used all over the world in a very unsustainable manner, and the long-term effects of biomass energy plantations have not been proven. As well, the natural humus and dead organic matter in the forest soils is a large reservoir of carbon. Conversion of natural ecosystems to managed energy plantations could result in a release of carbon from the soil as a result of the accelerated decay of organic matter.

An ever increasing number of people on this planet are faced with hunger and starvation. It has been argued that the use of land to grow fuel crops will increase this problem. Hunger in developing countries, however, is more complex than just a lack of agricultural land. Many countries in the world today, such as the United States, have food surpluses. Much fertile agricultural land is also used to grow tobacco, flowers, food for domestic pets, and other "luxury" items, rather than staple foods. Similarly, a significant proportion of agricultural land is used to grow feed for animals to support the highly wasteful, meat-centered diet of the industrialized world. By feeding grain to livestock we end up with only about 10 percent of the caloric content of the grain. When looked at in this light, it does not seem to be so unreasonable to use some fertile land to grow fuel. Marginal land and underutilized agricultural land can also be used to grow biomass for fuel.

Acid rain, which can damage lakes and forests, is a by-product of the combustion of fossil fuels, particularly coal and oil. The high-sulfur content of these fuels together with hot combustion temperatures result in the formation of sulfur dioxide (SO_2) and nitrous oxides (NO_x), when they are burned to provide energy. The replacement of fossil fuels with biomass can reduce the potential for acid rain. Biomass generally contains less than 0.1 percent sulfur by weight compared to low-sulfur coal with 0.5 to 4 percent sulfur. Lower combustion temperatures and pollution control devices such as wet scrubbers and electrostatic precipitators can also keep emissions of NO_x to a minimum when biomass is burned to produce energy.

The final major environmental impact of biomass energy may be that of loss of biodiversity. Transforming natural ecosystems into energy plantations with a very small number of crops, as few as one, can drastically reduce the biodiversity of a region. Such "monocultures" lack the balance achieved by a diverse ecosystem, and are susceptible to widespread damage by pests or disease.

8.4 FUELS FROM BIOMASS

The production of biofuels to replace oil and natural gas is in active development, focusing on the use of cheap organic matter (usually cellulose, agricultural, and sewage waste) in the efficient production of liquid and gas biofuels which yield high net energy gain. The carbon in biofuels was recently extracted from atmospheric carbon dioxide by growing plants, so burning it does not result in a net increase of carbon dioxide in the Earth's atmosphere. As a result, biofuels are seen by many as a way to reduce the amount of carbon dioxide released into the atmosphere by using them to replace nonrenewable sources of energy.

Gasoline is a blend of hydrocarbons with some contaminants, including sulfur, nitrogen, oxygen, and certain metals. Ethanol and methanol are biofuels that provide alternative to gasoline (Table 8.1). Bioethanol is a fuel produced by processing familiar and renewable crops such as cereals, sugar beet, and maize using natural fermentation. Blended with petrol at 10 percent, bioethanol can be used in vehicles without the need to change fuel or engine specifications.

Biofuels are important because they replace petroleum fuels and can be used to fuel vehicles, but can also fuel engines or fuel cells for electricity generation (Kavalov and

TABLE 8.1 Current and Alternate Motor Fuels

Fuel type	Available motor fuel
Traditional fuels	Diesel and gasoline
Oxygenated fuels	Ethanol, methanol, methyl tertiary butyl ether (MTBE), ethyl tertiary butyl ether (ETBE), tertiary butyl alcohol (TBA), and tertiary amyl methyl ether (TAME)
Alternative fuels	Liquefied petroleum gases (LPG), ethanol, 85% (E85); ethanol, 95% (E95); methanol, 85% (M85); methanol, neat (M100); compressed natural gas (CNG); liquefied natural gas (LNG); biodiesel (BD); hydrogen; and electricity

Source: AFDC, 1997.

Peteves, 2005). Biofuels are generally more environmentally benign than traditional fuels and are further defined as being renewable, meaning that the feedstock used to make a particular biofuel can be replenished at a rate equal to or faster than the rate at which the biofuel is consumed.

8.4.1 Gaseous Fuels

Gaseous fuels are those fuels that are in the gaseous state under ambient conditions. In some circumstances, the definition of gaseous fuels may also include the low-boiling hydrocarbons such as pentane but, for the purpose of this text, such fuels are considered to be liquid fuels.

Biogas is a clean and renewable form of energy and the most important biogas components are methane (CH_4), carbon dioxide (CO_2), and sulfuric components (H_2S) (Coelho et al., 2006). The gas generally composes of methane (55–65 percent), carbon dioxide (35–45 percent), nitrogen (0–3 percent), hydrogen (0-1 percent), and hydrogen sulfide (0–1 percent) (Table 8.2) (Anunputtikul and Rodtong, 2004).

TABLE 8.2 Composition of Biogas

	% v/v
Methane (CH_4)	55–65
Carbon dioxide (CO_2)	35–45
Hydrogen sulphide (H_2S)	0–1
Nitrogen (N_2)	0–3
Hydrogen (H_2)	0–1
Oxygen (O_2)	0–2
Ammonia (NH_3)	0–1

Biogas could very well substitute for conventional sources of energy (i.e., fossil fuels) which are causing ecological-environmental problems and at the same time depleting at a faster rate (Santosh et al., 2004). Due to its elevated methane content, resultant of the organic degradation in the absence of molecular oxygen, biogas is an attractive source of energy. The physical, chemical, and biological characteristics of the manure are related to diet composition, which can influence the biogas composition (Mogami et al., 2006). Natural gas is about 90 to 95 percent methane, but biogas is about 55 to 65 percent methane. So biogas is basically low-grade natural gas (Speight, 2007). The biogas composition is an essential parameter, because it allows identifying the appropriate purification system, which aims to remove sulfuric gases and decrease the water volume, contributing to improve the combustion fuel conditions (Coelho et al., 2006). Biogas has a heat value of approximately 5.0 to 7.5 kWh/m^3 (Table 8.3).

TABLE 8.3 Typical Properties of Biogas

Density (dry basis)	1.2 kg/m^3
Heat value	5.0–7.5 kWh/m^3
Ignition point	700°C
Ignition concentration gas content	6–12%

Biogas production has usually been applied for waste treatment, mainly sewage sludge, agricultural waste (manure), and industrial organic waste streams (Hartmann and Ahring, 2005). The primary source, which delivers the necessary microorganisms for biomass biodegradation and as well, one of the largest single source of biomass from food/feed industry, is manure from animal production, mainly from cow and pig farms (Nielsen et al., 2007).

Anaerobic digestion of organic fraction municipal solid waste has been studied in recent decades, trying to develop a technology that offers waste stabilization with resources recovery (Nguyen et al., 2007). The anaerobic digestion of municipal solid waste is a process that has become a major focus of interest in waste management throughout the world. In India, the amounts of municipal solid waste generated in urban areas ranges from 350 to 600 gm/capita/day (Elango et al., 2006). The municipal solid waste stream in Asian cities is composed of high fraction of organic material of more than 50 percent with high moisture content (Juanga et al., 2005).

Currently, biogas production is mainly based on the anaerobic digestion of single energy crops. Maize, sunflower, grass, and sudan grass are the most commonly used energy crops. In the future, biogas production from energy crops will increase and requires to be based on a wide range of energy crops that are grown in versatile, sustainable crop rotations (Bauer et al., 2007).

A specific source of biogas is landfills. In a typical landfill, the continuous deposition of solid waste results in high densities and the organic content of the solid waste undergoes microbial decomposition. The production of methane-rich landfill gas from landfill sites makes a significant contribution to atmospheric methane emissions. In many situations the collection of landfill gas and production of electricity by converting this gas in gas engines is profitable and the application of such systems has become widespread. The benefits are obvious: useful energy carriers are produced from gas that would otherwise contribute to a buildup of methane greenhouse gas (GHG) in the atmosphere, which has stronger greenhouse gas impact than the carbon dioxide emitted from the power plant. This makes landfill gas utilization in general a very attractive greenhouse gas mitigation option, which is being increasingly deployed in world regions (Faaij, 2006).

In summary, biogas is most commonly produced by using animal manure mixed with water which is stirred and warned inside an airtight container, known as a digester.

Anaerobic processes could either occur naturally or in a controlled environment such as a biogas plant. In the complex process of anaerobic digestion, hydrolysis/acidification and methanogenesis are considered as rate-limiting steps.

Most biomass materials are easier to gasify than coal because they are more reactive with higher ignition stability. This characteristic also makes them easier to process thermochemically into higher-value fuels such as methanol or hydrogen. Ash content is typically lower than for most coals, and sulfur content is much lower than for many fossil fuels. Unlike coal ash, which may contain toxic metals and other trace contaminants, biomass ash may be used as a soil amendment to help replenish nutrients removed by harvest. A few biomass feedstocks stand out for their peculiar properties, such as high silicon or alkali metal contents—these may require special precautions for harvesting, processing, and combustion equipment. Note also that mineral content can vary as a function of soil type and the timing of feedstock harvest. In contrast to their fairly uniform physical properties, biomass fuels are rather heterogeneous with respect to their chemical elemental composition.

A number of processes allow biomass to be transformed into gaseous fuels such as methane or hydrogen (Sørensen et al., 2006). One pathway uses algae and bacteria that have been genetically modified to produce hydrogen directly instead of the conventional biological energy carriers. Problems are intermittent production, low efficiency, and difficulty in constructing hydrogen collection and transport channels of low cost. A second pathway uses plant material such as agricultural residues in a fermentation process leading to biogas from which the desired fuels can be isolated. This technology is established and in widespread use for waste treatment, but often with the energy produced only for on-site use, which often implies less than maximum energy yields. Finally, high-temperature gasification supplies a crude gas, which may be transformed into hydrogen by a second reaction step. In addition to biogas, there is also the possibility of using the solid by-product as a biofuel.

The technologies for gas production from biomass include: (a) fermentation, (b) gasification, and (c) direct biophotolysis.

8.4.2 Liquid Fuels

Generally, liquid fuels are those fuels which flow readily under ambient conditions. However, for the present purpose, liquid fuels also include those fuels that flow with difficulty under ambient conditions but will flow to the fuel chamber in heated pipes.

Alcohols. Alcohols are oxygenate fuels insofar as the alcohol molecule has one or more oxygen, which decreases to the combustion heat (Minteer, 2006, Chap. 1). Practically, any of the organic molecules of the alcohol family can be used as a fuel. The alcohols which can be used for motor fuels are methanol (CH_3OH), ethanol (C_2H_5OH), propanol (C_3H_7OH), and butanol (C_4H_9OH). However, only methanol and ethanol fuels are technically and economically suitable for internal combustion engines (Bala, 2005).

Ethanol (ethyl alcohol, CH_3CH_2OH), also referred to as *bioethanol*, is a clear, colorless liquid with a characteristic, agreeable odor. Currently, the production of ethanol by fermentation of corn-derived carbohydrates is the main technology used to produce liquid fuels from biomass resources (McNeil Technologies Inc., 2005). Furthermore, amongst different biofuels, suitable for application in transport, bioethanol and biodiesel seem to be the most feasible ones at present. The key advantage of bioethanol and biodiesel is that they can be mixed with conventional petrol and diesel respectively, which allows using the same handling and distribution infrastructure. Another important strong point of bioethanol and biodiesel is that when they are mixed at low concentrations (= 10 percent bioethanol in petrol and = 20 percent biodiesel in diesel), no engine modifications are necessary.

Ethanol can be blended with gasoline to create E85, a blend of 85 percent ethanol and 15 percent gasoline. E85 and blends with even higher concentrations of ethanol, E95; pure bioethanol (E100-fuel) has been used mainly in Brazil (Davis, 2006; Minteer, 2006, Chap. 7). More widespread practice has been to add up to 20 percent to gasoline (E20-fuel or gasohol) to avoid engine changes.

Ethanol has a higher octane number (108), broader flammability limits, higher flame speeds, and higher heats of vaporization than gasoline. These properties allow for a higher compression ratio, shorter burn time, and leaner burn engine, which lead to theoretical efficiency advantages over gasoline in an internal combustion engine.

On the other hand, the disadvantages of ethanol include its lower energy density than gasoline, its corrosiveness, low flame luminosity, lower vapor pressure, miscibility with water, and toxicity to ecosystems.

Ethanol from cellulosic biomass materials (such as agricultural residues, trees, and grasses) is made by first using pretreatment and hydrolysis processes to extract sugars, followed by fermentation of the sugars. Although producing ethanol from cellulosic biomass is currently more costly than producing ethanol from starch crops, several countries (including the United States) have launched biofuels initiatives with the objective of the economic production of ethanol from biosources. Researchers are working to improve the efficiency and economics of the cellulosic bioethanol production process.

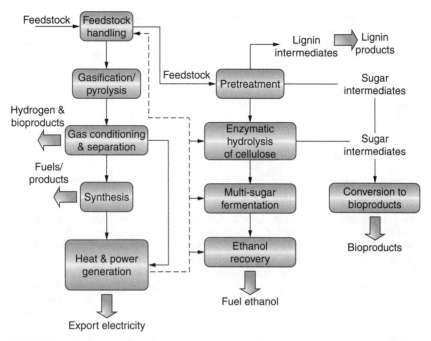

FIGURE 8.6 Schematic of a biorefinery. (*Source: Ruth, M.: "Development of a Biorefinery Optimization Model," Renewable Energy Modeling Series Forecasting the Growth of Wind and Biomass, National Bioenergy Centre, National Renewable Energy Laboratory, Golden, Colo., 2004, http://www.epa.gov/cleanenergy/pdf/ruth2_apr20.pdf.*)

In the process (Fig. 8.6), the carbohydrate from the biomass is converted into sugar; this is further converted to ethanol in a fermentation process that is similar to the process for brewing beer. Typical feedstocks for the process are sugar beets and from molasses and yield are on the order of 72.5 L of ethanol per ton of sugar cane. Modern crops yield 60 tons of sugar cane per hectare of land. Production of ethanol from biomass is one way to reduce both the consumption of crude oil and environmental pollution (Lang et al., 2001). Domestic production and use of ethanol for fuel can decrease dependence on foreign oil, reduce trade deficits, create jobs in rural areas, reduce air pollution, and reduce global climate change carbon dioxide buildup (Demirbas, 2005).

Ethanol can also be produced by synthesis from the chemical compound ethylene, which is derived from crude oil or natural gas, or by the fermentation of carbohydrates.

Methanol (methyl alcohol, wood alcohol, CH_3OH) is mainly manufactured from natural gas, but biomass can also be gasified to methanol. Methanol can be made with any renewable resource containing carbon such as seaweed, waste wood, and garbage. Methanol is stored and handled like gasoline because it is produced as a liquid. Methanol is currently made from natural gas, but it can also be made from a wide range of renewable biomass sources, such as wood or waste paper.

In comparison to gasoline, ethanol contains 35 percent oxygen by weight; gasoline contains none. The presence of the oxygen promotes more complete combustion which results in fewer tailpipe emissions. Compared to the combustion of gasoline, the combustion of ethanol substantially reduces the emission of carbon monoxide, volatile organic compounds, particulate matter, and greenhouse gases.

TABLE 8.4 Properties of Fuels Including Liquid Fuels from Biomass (BTL: Biomass-to-Liquids)

	Chemical formulae	Energy content, MJ/l	Density, kg/L	Octane number	Cetane number	Chemical feedstock
Oil petrol	C4–C12	31.2–32.2	0.72–0.77	90–95	–	No
Oil diesel	C15–C20	35.5–36.0	0.82–0.84	–	45–53	No
Oil naphtha	C5–C9	31.5	0.72	50	–	Yes
BTL naphtha	C5–C9	31.5	0.72	40	–	Yes
BTL diesel	C12–C20	33.1–34.3	0.77–0.78	–	70–80	No
Methanol	CH_3OH	15.4–15.6	0.79	110–112	5	Yes
Di-methyl-ether	CH_3OCH_3	18.2–19.3	0.66–0.67	–	55–60	Yes
Hydrogen	H_2	8.9	0.074	106	–	Yes

However, a unit of ethanol contains about 32 percent less energy than a liter of gasoline. One of the best qualities of ethanol is its octane rating.

Methanol can be used as one possible replacement for conventional motor fuels and has favorable properties (Table 8.4). Methanol has been seen as a possible large volume motor fuel substitute at various times during gasoline shortages. It was often used in the early part of the century to power automobiles before inexpensive gasoline was widely introduced. Methanol is poisonous and burns with an invisible flame. Methanol has, just like ethyl alcohol, a high octane rating and hence an Otto engine is preferable. If an ignition booster is used, methanol can be used in a diesel engine.

Methanol also offers important emissions benefits compared with gasoline. It can reduce hydrocarbon emissions by 30 to 40 percent with M85 and up to 80 percent with M100 fuels. Methanol costs less than gasoline, but has lower energy content. Taking this into account, costs for methanol in a conventional vehicle are slightly higher than those for gasoline. When used in fuel cells, which are considerably more efficient, fuel costs will be lower. Three important points of comparison are emissions, fuel economy, and octane quality.

Methanol and methanol blends have higher octane ratings than gasoline, which reduces engine knock and can produce in higher engine efficiency. The higher octane also gives methanol-fueled vehicles more power and quicker acceleration. A higher octane rating allows certain engine design parameters, such as compression ratio, and valve timing, to be altered in such ways that fuel economy and power are increased.

E100- and M100-fueled vehicles have difficulty starting in cold weather, but this is not a problem for E85 and M85 vehicles because of the presence of gasoline.

P-series fuels are a unique blend of natural gas liquids, ethanol, and hydrocarbons methyltetrahydrofuran ($CH_3C_4H_7O$) and is a biomass-derived product. These fuels are clear, colorless liquid blends (octane number: 89–93) that are formulated to be used in flexible fuel vehicles. Like gasoline, low vapor pressure formulations are produced to prevent excessive evaporation during summer and high vapor pressure formulations are used for easy starting in the winter. Also, these fuels are at least 60 percent nonpetroleum and the majority of the components that make up P-series fuels come from domestically produced renewable resources.

The P-series fuels provide significant emissions benefits over reformulated gasoline. Each unit of P-series fuel emits approximately 50 percent less carbon dioxide, 35 percent less hydrocarbons, and 15 percent less carbon monoxide than gasoline. The fuel also has 40 percent less ozone-forming potential. The other gasoline-subside ether, MTBE (methanol tertiary butyl ether) is a petroleum-derived product and is made from isobutene [$(CH3)_2CH=CH_2$] and methanol. The P-series fuel emissions are generally below those for reformulated gasoline and are well below federal emissions standards. P-series fuels join the list of alternatives to gasoline that includes ethanol (E85), methanol (M85), natural gas, propane, and electricity.

Biodiesel. Biodiesel is the generic name for fuels obtained by esterification of vegetable oil (Knothe et al., 2005; Bockey, 2006). The esterification can be done either by methanol or by ethanol. Biodiesel can be used in a diesel engine without modification and is a clean burning alternative fuel produced from domestic, renewable resources. The fuel is a mixture of fatty acid alkyl esters made from vegetable oils, animal fats, or recycled greases. Where available, biodiesel can be used in compression-ignition (diesel) engines in its pure form with little or no modifications.

Biodiesel is biodegradable, nontoxic, and essentially free of sulfur and aromatics. It is usually used as a petroleum diesel additive to reduce levels of particulates, carbon monoxide, hydrocarbons, and air toxics from diesel-powered vehicles. When used as an additive, the resulting diesel fuel may be called B5, B10, or B20, representing the percentage of the biodiesel that is blended with petroleum diesel.

Biodiesel is produced through a process in which organically derived oils are combined with alcohol (ethanol or methanol) in the presence of a catalyst to form the ethyl or methyl ester (Fig. 8.7). The biomass-derived ethyl or methyl esters can be blended with conventional diesel fuel or used as a neat fuel (100 percent biodiesel). Biodiesel can be made from any vegetable oil, animal fats, waste vegetable oils, or microalgae oils. Soybeans and Canola (rapeseed) oils are the most common vegetable oils used today.

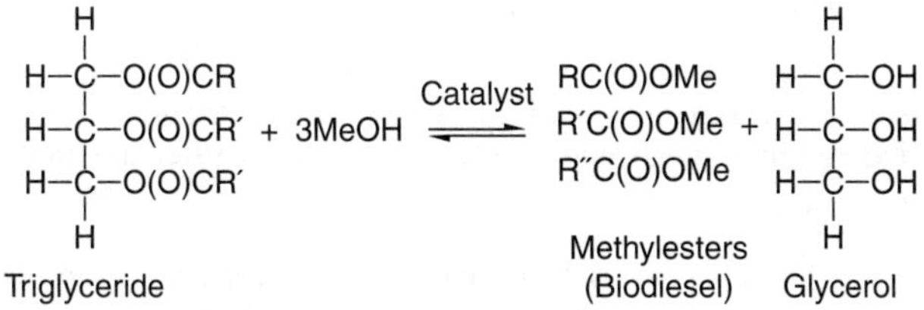

$$\begin{array}{c} H \\ | \\ H-C-O(O)CR \\ | \\ H-C-O(O)CR' \\ | \\ H-C-O(O)CR'' \\ | \\ H \end{array} + 3MeOH \xrightleftharpoons{\text{Catalyst}} \begin{array}{c} RC(O)OMe \\ R'C(O)OMe \\ R''C(O)OMe \end{array} + \begin{array}{c} H \\ | \\ H-C-OH \\ | \\ H-C-OH \\ | \\ H-C-OH \\ | \\ H \end{array}$$

Triglyceride   Methylesters   Glycerol
                     (Biodiesel)

FIGURE 8.7 The chemistry of biodiesel production.

Biodiesel is usually made from soybean oil or recycled cooking oils (Chap. 9). Animal fats, other vegetable oils, and other recycled oils can also be used to produce biodiesel, depending on their availability.

The production of biodiesel from vegetable oil represents another means of producing liquid fuels from biomass, and one which is growing rapidly in commercial importance. Commercially, biodiesel is produced from vegetable oils, including rapeseed, sunflower, and soybean oil, and animal fats (Fig. 8.8) (McNeil Technologies Inc. 2005). These oils and fats are typically composed of C_{14} to C_{20} fatty acid triglycerides (constituting approximately 90–95 percent by weight of the oil). In order to produce a fuel that is suitable for use in diesel engines, these triglycerides are converted to the respective alkyl esters and glycerol by base-catalyzed transesterification with short-chain alcohols (generally methanol). Thus, for every 10 lb of biodiesel produced, approximately 1 lb of glycerol is formed. Glycerol finds application in a wide range of industries (cosmetics, pharmaceuticals, as a plasticizer, etc.), although as biodiesel production grows, new uses will have to be developed to avoid a surplus of glycerol.

Fuel-grade biodiesel must be produced to strict industry specifications (ASTM D6751) in order to ensure proper performance. Biodiesel is the only alternative fuel to have fully completed the health effects testing requirements of the 1990 Clean Air Act Amendments;

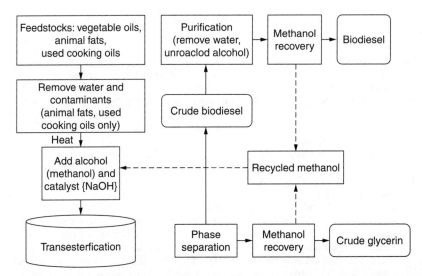

FIGURE 8.8 Production of biodiesel. *(Source: McNeil Technologies Inc.: "Colorado Agriculture IOF: Technology Assessments Liquid Fuels," Prepared Under State of Colorado Purchase Order # 01-336. Governor's Office of Energy Conservation and Management, Denver, Colo., 2005.)*

and biodiesel that meets ASTM D6751 and is legally registered with the Environmental Protection Agency (EPA) is a legal motor fuel for sale and distribution. Raw vegetable oil cannot meet biodiesel fuel specifications, therefore it is not registered with the EPA and it is not a legal motor fuel.

Bio-Oil. A totally different process than that used for biodiesel production can be used to convert biomass into a type of fuel similar to diesel which is known as *bio-oil*. The process (*fast pyrolysis, flash pyrolysis*) occurs when solid fuels are heated at temperatures between 350 and 500°C for a very short period of time (<2 seconds). The bio-oils currently produced are suitable for use in boilers for electricity generation.

In another process, the feedstock is fed into a fluidized bed (at 450–500°C) and the feedstock flashes and vaporizes. The resulting vapors pass into a cyclone where solid particles, char, are extracted. The gas from the cyclone enters a quench tower where they are quickly cooled by heat transfer using bio-oil already made in the process. The bio-oil condenses into a product receiver and any noncondensable gases are returned to the reactor to maintain process heating. The entire reaction from injection to quenching takes only 2 seconds.

8.4.3 Solid Fuels

For the present text, solid fuels are those fuels that are solid under ambient conditions and remain solid under mild heating. Thus, examples of solid fuels from biofuel feedstocks include wood and wood-derived charcoal (Chap. 9) and dried dung, particularly cow dung.

One widespread use of such fuels is in domestic cooking and heating. The biofuel may be burned on an open fireplace or in a special stove. The efficiency of this process may vary widely, from 10 percent for a well made fire (even less if the fire is not made carefully) up to 40 percent for a custom designed charcoal stove. Inefficient use of fuel is a cause of deforestation (though this is negligible compared to deliberate destruction to clear land for agricultural use) but more importantly it means that more work has to be put into gathering fuel, thus the quality of cooking stoves has a direct influence on the viability of biofuels.

8.4.4 Biofuels from Synthesis Gas

Alternatively, biomass can be converted into fuels and chemicals indirectly (by gasification to syngas followed by catalytic conversion to liquid fuels) or directly to a liquid product by thermochemical means. The process yields synthesis gas (syngas) composed primarily of hydrogen and carbon monoxide (Chap. 7), also called *biosyngas* (Cobb, 2007)

The production of high-quality syngas from biomass, which is later used as a feedstock for biomass-to-liquids (BTL) production, requires particular attention. This is due to the fact that the production of synthesis gas from biomass is indeed the novel component in the gas-to-liquids concept—obtaining syngas from fossil raw materials (natural gas and coal) is a relatively mature technology.

Gasification (qv) is actually *thermal degradation* of the feedstock in the presence of an externally supplied oxidizing (oxygen-containing) agent, for example, air, steam, and oxygen. Various gasification concepts have been developed over the years, mainly for the purposes of power generation. However, efficient biomass-to-liquids production imposes completely different requirements for the composition of the gas. The reason is that in power generation the gas is used as a fuel, while in biomass-to-liquids processing it is used as a chemical feedstock to obtain other products. This difference has implications with respect to the purity and composition of the gas.

In contrast, for biomass-to-liquids production only the amount of carbon monoxide and hydrogen is important (the larger the amount, the better), while the calorific value is irrelevant. The presence of other hydrocarbons and inert components should be avoided or at least kept as low as possible. This can be achieved via the following ways:

1. The amount of components other than CO and H_2 (primarily hydrocarbons) can be reduced via their further transformation into CO and H_2. This is however rather energy intensive and costly (two processes—gasification and transformation). As a result, the overall energy efficiency of syngas production and of biomass-to-liquids processing is also reduced, leading to higher production costs.
2. The amount of various components can be minimized via a more complete decomposition of biomass, thereby preventing the formation of undesirable components at the gasification step. This approach seems to be more appropriate from energy efficiency and cost point of view. The minimization of the content of various hydrocarbons is achieved by increasing temperatures in the gasifier, along with shortening the residence time of feedstocks inside the reactor. Because of this short residence time, the particle size of feedstocks should be small enough (in any case—smaller than in gasification for power generation) in order that complete and efficient gasification can occur.
3. In gasification for power generation, typically air is employed as oxidizing agent, as it is indeed the cheapest amongst all possible oxidizing agents. However, the application of air results in large amounts of nitrogen in the product gas, since nitrogen is the main constituent of air. The presence of such large quantities of nitrogen in the product gas does not hamper (very much) power generation, but it does hamper biomass-to-liquids production. Removing this nitrogen via liquefaction under cryogenic temperatures is extremely energy intensive, reduces substantially the overall biomass-to-liquids energy efficiency, and increases costs. Amongst other potential options (steam, CO_2, O_2), from a technical and economic point of view oxygen appears to be the most suitable oxidizing agent for biomass-to-liquids manufacturing. It is true that the oxygen-blown gasification

implies additional costs compared to the air-blown gasification, because of the oxygen production. Nevertheless, the energy and financial cost of producing oxygen seems to be far lower than the alternative energy and financial cost of removing nitrogen from the product gas from air-blown gasification. This is partly due to the fact that the production of high-purity oxygen (above 95 percent O_2) is a mature technology.

In principle, the larger the carbon and hydrogen content in raw materials, employed in gas-to-liquids processing, is, the easier and more efficient the carbon monoxide and hydrogen. Hence, the natural gas pathway is the most convenient one, since natural gas is gaseous and contains virtually carbon and hydrogen only. Solid raw materials (biomass, coal) involve more processing, because first they have to be gasified and then the obtained product gas should be cleaned up from other components such as nitrogen oxides (NO_x), sulfur oxides (SO_x), and particulate matter to the extent of getting as high as possible purity of syngas. Two basic types of biomass raw material are distinguished, namely, woody material and herbaceous material. Currently woody material accounts for about 50 percent of total world bioenergy potential. Another 20 percent is straw-like feedstock, obtained as a by-product from agriculture. The dedicated cultivation of straw-like energy crops could increase the herbaceous share up to 40 percent (Boerrigter and van der Drift, 2004; van der Drift et al., 2004).

8.5 USES

Biomass currently supplies 14 percent of the world's energy needs, but has the theoretical potential to supply 100 percent. Most present day production and use of biomass for energy is carried out in a very unsustainable manner with a great many negative environment consequences. If biomass is to supply a greater proportion of the world's energy needs in the future, the challenge will be to produce biomass and to convert and use it without harming the natural environment. Technologies and processes exist today which, if used properly, make biomass-based fuels less harmful to the environment than fossil fuels. Applying these technologies and processes on a site specific basis in order to minimize negative environment impacts is a prerequisite for sustainable use of biomass energy in the future.

Biodiesel and bioethanol are widely used in automobiles and freight vehicles. For example, in Germany most diesel fuel on sale at gas stations contains a few percent biodiesel, and many gas stations also sell 100 percent biodiesel. Some supermarket chains in the United Kingdom have switched to running their freight fleets on 50 percent biodiesel, and often include biofuels in the vehicle fuels they sell to consumers, and an increasing number of service stations are selling biodiesel blends (typically with 5 percent biodiesel).

In Europe, research is being undertaken into the use of biodiesel as domestic heating oil. A blend of 20 percent biodiesel with 80 percent kerosene (B20) has been tested successfully to power modern high-efficiency condensing oil boilers. Boilers needed a preheat burner to prevent nozzle blockages and maintain clean combustion. Blends with a higher proportion of biodiesel were found to be less satisfactory, owing to the greater viscosity of biodiesel than conventional fuels when stored in fuel tanks outside the building at typical U.K. winter temperatures.

Different combustion engines are being produced for very low prices lately. They allow the private house-owner to utilize low amounts of weak compression of methane to generate electrical and thermal power (almost) sufficient for a well insulated residential home.

Direct biofuels are biofuels that can be used in existing unmodified petroleum engines. Because engine technology changes all the time, exactly what a direct biofuel is can be hard to define; a fuel that works without problem in one unmodified engine may not work in another engine. In general, newer engines are more sensitive to fuel than older engines, but new engines are also likely to be designed with some amount of biofuel in mind.

Straight vegetable oil can be used in some (older) diesel engines. Only in the warmest climates can it be used without engine modifications, so it is of limited use in colder climates. Most commonly it is turned into biodiesel. No engine manufacturer explicitly allows any use of vegetable oil in their engines.

Biodiesel can be a direct biofuel. In some countries manufacturers cover many of their diesel engines under warranty for 100 percent biodiesel use. Many people have run thousands of miles on biodiesel without problem, and many studies have been made on 100 percent biodiesel. In many European countries, 100 percent biodiesel is widely used and is available at thousands of gas stations.

Ethanol is the most common biofuel, and over the years many engines have been designed to run on it. Many of these could not run on regular gasoline. It is open to debate if ethanol is a direct replacement in these engines though—they cannot run on anything else. In the late 1990s engines started appearing that by design can use either fuel. Ethanol is a direct replacement in these engines, but it is debatable if these engines are unmodified, or factory modified for ethanol.

Butanol is often claimed as a direct replacement for gasoline. It is not in wide spread production at this time, and engine manufacturers have not made statements about its use. While it appears that butanol has sufficiently similar characteristics with gasoline such that it should work without problem in any gasoline engine, no widespread experience exists.

There is some concern about the energy efficiency of biofuel production. Production of biofuels from raw materials requires energy (for farming, transport, and conversion to final product), and it is not clear what the overall efficiency of the process is. For some biofuels the energy balance may even be negative.

Since vast amounts of raw material are needed for biofuel production, monocultures and intensive farming may become more popular, which may cause environmental damages and undo some of the progress made toward sustainable agriculture.

8.6 A BIOREFINERY

Plants are very effective chemical mini-factories or refineries insofar as they produce chemicals by specific pathways. The chemicals they produce are usually essential manufacture (called metabolites) include sugars and amino acids that are essential for the growth of the plant, as well as more complex compounds.

Biorefining offers a key method to accessing the integrated production of chemicals, materials, and fuels. The biorefinery concept is analogous to that of an oil refinery (Chap. 3).

In a manner similar to the petroleum refinery, a biorefinery would integrate a variety of conversion processes to produce multiple product streams such as motor fuels and other chemicals from biomass. In short, a biorefinery would combine the essential technologies to transform biological raw materials into a range of industrially useful intermediates. However, the type of biorefinery would have to be differentiated by the character of the feedstock. For example, the *crop biorefinery* would use raw materials such as cereals or maize and the *lignocellulose biorefinery* would use raw material with high cellulose content, such as straw, wood, and paper waste.

In addition, a variety of methods techniques can be employed to obtain different product portfolios of bulk chemicals, fuels, and materials. Biotechnology-based conversion processes can be used to ferment the biomass carbohydrate content into sugars that can then be further processed. As one example, the fermentation path to lactic acid shows promise as a route to biodegradable plastics. An alternative is to employ thermochemical conversion processes which use pyrolysis or gasification of biomass (Chaps. 9 and 10) to produce a hydrogen-rich synthesis gas which can be used in a wide range of chemical processes (Chap. 7).

Thus, a biorefinery is a facility that integrates biomass conversion processes and equipment to produce fuels, power, and chemicals from biomass (Fig. 8.6) (Ruth, 2004). The biorefinery concept is analogous to the petroleum refinery, which produce multiple fuels and products from petroleum (Chap. 3).

A biorefinery can have two or more options for the production of biofuels from wood and other biomass materials (Mabee and Saddler, 2006). There is the (a) bioconversion, (b) thermal conversion, and (c) thermochemical conversion. Each of these options has merits but is selected depending on the feedstock and the desired product slate.

8.6.1 Bioconversion

The bioconversion option uses biologic agents to carry out a structured deconstruction of lignocellulose components. This platform combines process elements of pretreatment with enzymatic hydrolysis to release carbohydrates and lignin from the wood (Fig. 8.9).

The first step is a pretreatment stage which is based on existing pulping processes, however, traditional pulping parameters are defined by resulting paper properties and desired yields, while optimum bioconversion pretreatment is defined by the accessibility of the resulting pulp to enzymatic hydrolysis. This function of this step is to optimize the biomass feedstock for further processing and is designed to expose cellulose and hemicellulose for subsequent enzymatic hydrolysis, increasing the surface area of the substrate for enzymatic action to take place. The lignin is either softened or removed, and individual cellulosic fibers are released creating pulp.

In order to improve the ability of the pretreatment stage to optimize biomass for enzymatic hydrolysis, a number of nontraditional pulping techniques have been suggested (Mabee and Saddler, 2006) and include: (a) water-based systems, such as steam-explosion pulping, (b) acid treatment using concentrated or dilute sulfuric acid, (c) alkali treatment using recirculated ammonia, and (d) organic solvent pulping systems using acetic acid or ethanol. As with traditional pulping, pretreatment tends to work best with a homogenous batch of wood chips but the pretreatment option may have to be selected according to the type of lignocellulosic feedstock (Mabee et al., 2006a).

Once pretreated, the cellulose and hemicellulose components of wood can be hydrolyzed (in this option) using enzymes to facilitate bioconversion of the wood. Enzymatic hydrolysis of lignocellulose materials uses cellulase enzymes to break down the cellulosic microfibril structure into the various carbohydrate components.

The enzymatic hydrolysis step may be completely separated from the other stages of the bioconversion process, or it may be combined with the fermentation of carbohydrate intermediates to end products. Separate hydrolysis and fermentation (SHF) stages may offer this option more flexibility insofar as process adaptation to feedstock type and product slate is available. Simultaneous saccharification and fermentation (SSF) has been found to be highly effective in the production of specific end products, such as bioethanol (Mabee et al., 2006a).

The benefit of the bioconversion platform is that it provides a range of intermediate products, including glucose, galactose, mannose, xylose, and arabinose, which can be relatively easily processed into value-added bioproducts. The process also generates a quantity of lignin or lignin components; depending upon the pretreatment, lignin components may be found in the hydrolysate after enzymatic hydrolysis, or in the wash from the pretreatment stage. The chemical characteristics of the lignin are therefore heavily influenced by

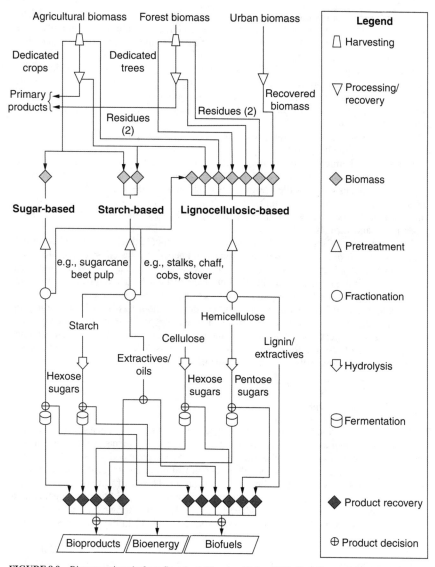

FIGURE 8.9 Bioconversion platform flowchart. *(Source: Mabee, W. E., D. J. Gregg, C. Arato, A. Berlin, R. Bura, N. Gilkes, et al.: "Update on Softwood-to-Ethanol Process Development," Appl. Biochem. Biotechnol., 2006a, 129–132:55–70. http://aic.uwex.edu/resources/documents/Comparingbiofueltechnology.pdf)*

the type of pretreatment that is employed. Finally, a relatively small amount of extractives may be retrieved from the process. These extractives are highly variable depending upon the feedstock employed, but may include resins, terpenes, or fatty acids.

Once hydrolyzed, six-carbon sugars can be fermented to ethanol using yeast-based processes. Five-carbon sugars, however, are more difficult to ferment and lack the efficiency of six-carbon sugar conversion. Bacterial fermentation under aerobic and anaerobic conditions is also an option to expand the variety of other products.

A large number of options on the various aspects of bioconversion are available (Mabee et al., 2004). The environmental performance of bioethanol, including air quality (NO_x, PM, SO_x, etc.) is also well documented as are the mass-balance and energy-balance of the bioconversion process and economic analyses.

However, the most fundamental issues for the bioconversion option include improving the effectiveness of the pretreatment stage, decreasing the cost of the enzymatic hydrolysis stage, and improving overall process efficiencies by capitalizing on synergies between various process stages. In fact, overall process economics needs improvement and this can be achieved by creating added-value coproducts.

In fact, one of the major advantages of the bioconversion option is the implication that the process can produce added-value products.

In the past, chemical products were a major part of the forest industry. A number of chemical forest products such as pitch (partially dried resins), pine tar (liquefied resins), turpentine (terpenes from distilled resins), rosin (the nonvolatile residues from resin distillation), and tall oil (obtained from alkaline pulping liquors) were the basis of the industry. These products were widely used in wooden shipbuilding and in the manufacture of soap, paper, paint, and varnishes.

The resurgence of interest in the production of chemicals (including fuels) from biosources is a means of searching to reduce reliance on petroleum-based products. Moreover, the products that advanced manufacturing processes may be generated from biological sources. There is the hope that the chemical products that can be derived from the biorefinery have the potential to become a significant part of a country's future economy.

To date, and in many countries, biochemical development has been based largely on sugars. In fact, sugars are one of the main intermediate products of the bioconversion option for a biorefinery. The issue, only in the context of the current text, then becomes the potential for the conversion of sugar products into fuels. Whether or not this is feasible technically and economically remains to be seen.

8.6.2 Thermal Conversion

The bioconversion option for a biorefinery uses biologic agents to carry out a structured deconstruction of lignocellulose components. This process combines elements of pretreatment with enzymatic hydrolysis to release carbohydrates and lignin from the wood (Fig. 8.10).

The thermal conversion option uses thermochemical processes to gasify wood, producing synthesis gases (sometimes called producer gases). This platform combines process elements of pretreatment, pyrolysis, gasification, cleanup, and conditioning to generate a mixture of hydrogen, carbon monoxide, carbon dioxide, and other gases. The products of this platform may be viewed as intermediate products, which can then be assembled into chemical building blocks and eventually end products.

In this process option, the only pretreatment required involves drying, grinding, and screening the material in order to create a feedstock suitable for the reaction chamber. The technology required for this stage is already available on a commercial basis, and is often associated with primary or secondary wood processing, or agricultural residue collection and distribution.

In the primary processing stage, the volatile components of biomass are subjected to pyrolysis, or combustion in the absence of oxygen, at temperatures ranging from 450 to 600°C and, depending on the residence time in the chamber, a variety of products can be achieved. If pyrolysis is rapid (short residence time), gaseous products, condensable liquids, and char are produced and overall yield of bio-oil can, under ideal conditions make up 60 to 75 percent of the original fuel mass. The oil produced can be used as a biofuel or as a feedstock for value-added chemical products. If the pyrolysis is slow (long residence time), the

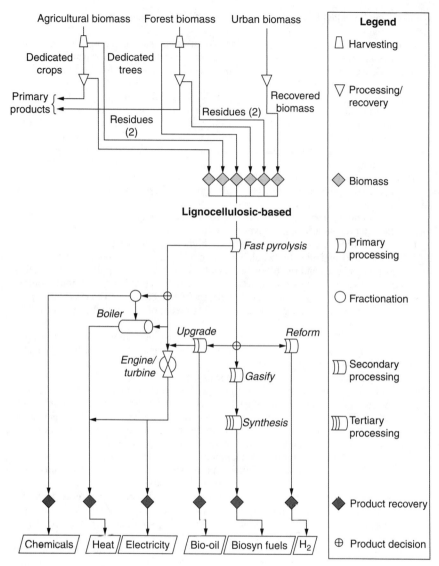

FIGURE 8.10 Thermochemical platform flowchart.

products are more likely to be gaseous and consist of carbon monoxide, hydrogen, methane, carbon dioxide, and water as well as volatile tar.

Slow pyrolysis, like fast pyrolysis, leaves behind a solid residue of char (or charcoal) which comprise approximately 10 to 25 percent by weight of the original feedstock. The char can be used as a fuel source to drive the pyrolysis process. If the pyrolysis is carried out at a higher temperature range (550–600°C), the gaseous products consist of carbon monoxide, hydrogen, methane, volatile tar, carbon dioxide, and water. Any char produced can be used as a fuel source to drive the pyrolysis process or can be gasified to produce synthesis gas, so-called because of the presence of carbon monoxide and hydrogen in the product stream.

After the production of syngas, a number of pathways may be followed to create biofuels. Proven catalytic processes for syngas conversion to fuels and chemicals exist, using syngas produced commercially from natural gas and coal. These proven technologies can be applied to biomass-derived syngas.

Methanol is one potential biofuel that can be generated through catalysis. The majority of methanol produced today is being derived from natural gas, however. Methanol has a high octane number (129) but relatively low energy (about 14.6 MJ/L) compared to gasoline (91–98 octane, 35 MJ/L). Because methanol has a favorable hydrogen/carbon ratio (4:1), it is often touted as a potential hydrogen source for future transportation systems.

Another potential biofuel that can be produced through the thermochemical platform is Fischer-Tropsch diesel (or biodiesel). This fuel was first discovered in 1923 and is commercially based on syngas made from coal, although the process could be applied to biomass-derived syngas. The process of converting carbon monoxide (CO) and hydrogen (H_2) mixtures to liquid hydrocarbons over a transition metal catalyst has become well established although process efficiency leaves much to be desired.

It is also possible to convert synthesis gas to higher-molecular-weigh products, including ethanol. Ethanol and other higher alcohols form as by-products of both Fischer-Tropsch and methanol synthesis, and modified catalysts have been shown to provide better yields. The thermal conversion option provides the opportunity for a number of additional coproducts, as well as energy in the form of heat or electricity and biofuels. Each component (i.e., carbon monoxide, carbon dioxide, methane, and hydrogen) of the gaseous products may be recovered, separated, and utilized.

Pyrolysis/gasification systems have been reported to be much more efficient for energy recovery, in terms of electricity generation, than traditional combustion. It has been estimated that typical biomass steam-generation plants have efficiencies in the low 20 percent range, compared to gasification systems with efficiencies that reach 60 percent (DOE, 2006). High efficiencies have been noted for both co-firing systems (where biomass is gasified together with a fossil fuel such as coal or natural gas) and in dedicated biomass gasification processes (Gielen et al., 2001). Because the potential for energy recovery is so much higher, gasification systems without any downstream catalysis may be able to increase bioenergy production with minimal impact on existing product streams in sawmilling or pulping operations.

Gasification technologies for the production of fuels from biomass are available but are often bypassed in favor of fossil fuels although this may change with rising fuel costs and uncertainty about the security of fossil reserves (Faaij, 2006).

Another issue is the quality of bio-based synthesis gas which are often more heterogeneous than natural gas-based synthesis gas. While technical approaches are well documented for the production of hydrogen, methanol, and Fischer-Tropsch liquids from synthesis gas, the input gases must be relatively clean in order for these processes to function in a commercially viable sense. Therefore, before catalysis, raw synthesis gas must be cleaned up in order to remove inhibitory substances that would deactivate the catalyst. These include sulfur, nitrogen, and chlorine compounds, as well as any remaining volatile tar.

The ratio of hydrogen to carbon monoxide may need to be adjusted and the carbon dioxide by-product may also need to be removed. One major problem with methanol synthesis is that biomass-based syngas tends to be hydrogen-poor compared to natural gas syngas. Methanol synthesis requires a ratio of 2:1 hydrogen/carbon monoxide to be cost-effective.

Common problems associated particularly with Fisher-Tropsch synthesis (Chap. 7) are low product selectivity (the unavoidable production of perhaps unwanted coproducts, including olefins, paraffins, and oxygenated products), and the sensitivity of the catalyst to contamination in the syngas that inhibit the catalytic reaction. With biomass-based syngas, this problem is amplified due to the heterogeneous nature of the syngas.

A final issue, perhaps of greater concern to policymakers, is that deployment on a large scale is required to gain necessary economies of scale for most of these processes, where the cost of syngas production can easily be more than 50 percent of the total process cost. This requirement for large facilities raises the level of capital required for infrastructure development, increasing risk to the investor; it also increases the amount of biomass required for operation, which makes it more difficult to supply the facility over the course of its operational lifetime.

One of the major implications of the thermal-based process option scenarios is the ability to generate excess heat and power. Self-generation of heat and power by the combustion of a portion of biomass feedstock can offset fossil fuel requirements, displacing the load on utilities and improve the environmental performance of the facility. Especially, since the cost of buying natural gas to generate heat and power internally has risen dramatically.

Other options for improved energy production include co-firing or cogeneration (i.e., combining biomass with fossil fuels in combustion). It is estimated that gasification technology has the potential to generate twice as much electricity per ton of black liquor as a conventional recovery boiler. This additional power can reduce the need to purchase natural gas, coal, oil, and electricity for everyday operations, increasing the economic performance of the facility.

Excess heat and power can be utilized for additional value-added processing, or can be distributed through a local network to provide district heating of nearby businesses and residences. The potential appeal of bioenergy as a product may be limited, however, by the regulatory regime which governs electricity generation, transmission, and sales.

8.6.3 Greenhouse Gas Production

Greenhouse gas production associated with lignocellulosic-based feedstocks is anticipated to be much lower than with conventional fuels. The environmental performance depends very much on the specific life cycle of the fuel, including the country in which the life cycle assessment was conducted, the feedstock on which the fuel is based, the vehicle used, the propulsion system, and the overall state of technology (Quirin et al., 2004; VIEWLS, 2005).

In general, fuels (and chemicals) made from lignocellulosic materials are characterized by reduced carbon dioxide emissions when compared to similar products derived from petroleum. Most biofuels can reduce greenhouse gas emissions significantly and substituting emissions by utilizing bio-based energy can create an overall negative emission for the fuel (VIEWLS, 2005).

Fischer-Tropsch fuels based on bio-residues are likely to have the lowest possible emissions; this is typical of diesel propulsion systems that have better energy recovery. If energy crops are utilized as a feedstock, the overall emissions rise slightly, because the benefit of residue disposal is lost. Ethanol from residues or from energy crops also have relatively low emissions, particularly compared to conventional fuels including gasoline and diesel fuel.

For ethanol from lignocellulose, there is a potential to reduce greenhouse has emissions with improved technology. This reflects the close-to-commercial status of the technology today, and the anticipated improvements that will be seen as this technology improves. For Fischer-Tropsch fuels, it is anticipated that commercial status will not be achieved until post-2010 (Mabee and Saddler, 2006).

8.6.4 Other Aspects

For example, a biorefinery using lignin as a feedstock would produce a range of valuable organic chemicals and liquid fuels that, at the present time, could supplement or even replace equivalent or identical products currently obtained from crude oil, coal, or gas. Thus, the biorefinery is analogous to an oil refinery in which crude oil is separated into a series of products, such as gasoline, heating oil, jet fuel, and petrochemicals.

By producing multiple products, a biorefinery can take advantage of the differences in biomass components and intermediates and maximize the value derived from the biomass feedstock. A biorefinery might, for example, produce one or several low-volume, but high-value, chemical products and a low-value, but high-volume liquid transportation fuel, while generating electricity and process heat for its own use and perhaps enough for sale of electricity. The high-value products enhance profitability, the high-volume fuel helps meet national energy needs, and the power production reduces costs and avoids greenhouse gas emissions.

As a feedstock, biomass can be converted by thermal or biological routes to a wide range of useful forms of energy including process heat, steam, electricity, as well as liquid fuels, chemicals, and synthesis gas. As a raw material, biomass is a nearly universal feedstock due to its versatility, domestic availability, and renewable character. At the same time, it also has its limitations. For example, the energy density of biomass is low compared to that of coal, liquid petroleum, or petroleum-derived fuels. The heat content of biomass, on a dry basis (7000–9000 Btu/lb) is at best comparable with that of a low-rank coal or lignite, and substantially (50–100 percent) lower than that of anthracite, most bituminous coals, and petroleum. Most biomass, as received, has a high burden of physically adsorbed moisture, up to 50 percent by weight. Thus, without substantial drying, the energy content of a biomass feed per unit mass is even less.

These inherent characteristics and limitations of biomass feedstocks have focused the development of efficient methods of chemically transforming and upgrading biomass feedstocks in a refinery. The refinery would be based on two "platforms" to promote different product slates.

The sugar-base involves breakdown of biomass into raw component sugars using chemical and biological means. The raw fuels may then be upgraded to produce fuels and chemicals that are interchangeable with existing commodities such as transportation fuels, oils, and hydrogen.

Although a number of new bioprocesses have been commercialized it is clear that economic and technical barriers still exist before the full potential of this area can be realized. One concept gaining considerable momentum is the biorefinery which could significantly reduce production costs of plant-based chemicals and facilitate their substitution into existing markets. This concept is analogous to that of a modern oil refinery in that the biorefinery is a highly integrated complex that will efficiently separate biomass raw materials into individual components and convert these into marketable products such as energy, fuels, and chemicals.

By analogy with crude oil, every element of the plant feedstock will be utilized including the low value lignin components. However, the different compositional nature of the biomass feedstock, compared to crude oil, will require the application of a wider variety of processing tools in the biorefinery. Processing of the individual components will utilize conventional thermochemical operations and state-of-the-art bioprocessing techniques. The production of biofuels in the biorefinery complex will service existing high volume markets, providing economy-of-scale benefits and large volumes of by-product streams at minimal cost for upgrading to valuable chemicals. A pertinent example of this is the glycerol by-product produced in biodiesel plants. Glycerol has high functionality and is a potential platform chemical for conversion into a range of higher value chemicals. The high-volume product streams in a biorefinery need not necessarily be a fuel but could also be a large-volume chemical intermediate such as ethylene or lactic acid.

A key requirement for delivery of the biorefinery concept is the ability to develop process technology that can economically access and convert the five- and six- membered ring sugars present in the cellulose and hemicellulose fractions of the lignocellulosic feedstock. Although engineering technology exists to effectively separate the sugar containing fractions from the lignocellulose, the enzyme technology to economically convert the five-ring sugars to useful products requires further development.

The construction of both large biofuel and renewable chemical production facilities coupled with the pace at which bioscience is being both developed and applied demonstrates that the utilization of nonfood crops will become more significant in the near term (Bourne, 2007). The biorefinery concept provides a means to significantly reduce production costs such that a substantial substitution of petrochemicals by renewable chemicals becomes possible. However, significant technical challenges remain before the biorefinery concept can be realized.

8.7 THE FUTURE

To secure a quality life for current and future generations, sufficient land, water, and energy must be available (Pimentel and Pimentel, 2006). By 2030, the world is projected to consume two-thirds more energy than today, with developing countries replacing the industrialized world as the largest group of energy consumers (Dorian et al., 2006). Energy consumption clearly is an important factor in future energy planning. In this century, green energy consumption may become an important parameter for indicating social, industrial, economical, and technological development (Ermis et al., 2007).

Therefore, the issue is not whether renewable biofuels will play a role in providing energy for transportation but to what extent and the implications of their use for the economy, for the environment, and for global security.

The rapidly growing interest in biofuels is being *fueled* by the realization that biofuels represent the only large near-term substitute for the petroleum-based fuels. As a result, biofuels are poised to be the potential solution to some very pertinent issues, such as rising oil prices, increasing national and global insecurity, climate instability, and local as well as global pollution levels.

The method chosen for biofuel production will be determined in part by the characteristics of the biomass available for processing. The majority of terrestrial biomass available is typically derived from agricultural plants and from wood grown in forests, as well as from waste residues generated in the processing or use of these resources. The primary barrier to utilizing this biomass is generally recognized to be the lack of low-cost processing options capable of converting these polymers into recoverable base chemical components (Lynd et al., 1999).

Currently, in the United States, much of the biomass being used for biofuel production includes agricultural crops that are rich in sugars and starch. Because of the prevalence of these feedstocks, the majority of U.S. activity toward developing new products has focused on bioconversion (BRDTAC, 2002). Bioconversion isolates sugars from biomass, which can then be processed into value-added products. Native sugars found in sugarcane and sugar beet can be easily derived from these plants, and refined in facilities that require the lowest level of capital input. Starch, a storage molecule which is a dominant component of cereal crops such as corn and wheat, is comprised wholly of glucose. Starch may be subjected to an additional processing in the form of an acid- or enzyme-catalyzed hydrolysis step to liberate glucose using a single family of enzymes, the amylases, which makes bioconversion relatively simple. Downstream processing of sugars includes traditional fermentation, which uses yeast to produce ethanol; other types of fermentation, including bacterial fermentation under aerobic and anaerobic conditions, can produce a variety of other products from the sugar stream.

Forest biomass or agricultural residues are almost completely comprised of lignocellulosic molecules (wood), a structural matrix that gives the tree or plant strength and form. This type of biomass is a prime feedstock for combustion, and indeed remains a major source of energy for the world today (FAO, 2005). The thermal conversion method utilizes pyrolysis and gasification processes to recover heat energy as well as the gaseous components of wood, known as synthesis which can then be refined into synthetic fuels (Chap. 7).

Lignocellulose is a complex matrix combining cellulose, hemicellulose, and lignin, along with a variable level of extractives. Cellulose is comprised of glucose, a six-carbon sugar, while hemicellulose contains both five- and six-carbon sugars, including glucose, galactose, mannose, arabinose, and xylose. The presence of cellulose and hemicellulose therefore makes lignocellulose a potential candidate for bioconversion. The ability of the bioconversion platform to isolate these components was initially limited, as the wood matrix is naturally resistant to decomposition. Recent advances, however, have made this process more commercially viable and there is added potential for value-added products that can utilize the lignin component of the wood.

The most fundamental issues for the bioconversion platform include improving the effectiveness of the pretreatment stage, decreasing the cost of the enzymatic hydrolysis stage, and improving overall process efficiencies by capitalizing on synergies between various process stages. There is also a need to improve process economics by creating coproducts that can add revenue to the process.

This type of application is a logical step on the path toward greater process efficiencies and increased energy self-generation. These types of systems could also provide surplus bioenergy, becoming an additional revenue stream.

Greenhouse gas production associated with lignocellulosic-based feedstocks is anticipated to be much lower than with conventional fuels. The environmental performance depends very much on the specific life cycle of the fuel, including the feedstock on which the fuel is based and the technology employed (VIEWLS, 2005).

The recent proliferation of global biofuel programs is due to several factors, not the least of which is high oil prices. Other factors, such as concern about (a) political instability in oil-exporting countries, (b) various countries seeking to bolster their agricultural industries, (c) climate-altering greenhouse gas emissions, and (d) urban air pollution are of equal importance depending upon the country under study. Continuing developments in biorefining technology have also brought greater attention to biofuels as a potentially large-scale and environmentally sustainable fuel.

However, the potential benefits of biofuels will only be realized if environmentally sustainable technologies are employed. Under the correct stewardship, the technologies described above will make it possible to produce biofuels from agricultural and forestry wastes, as well as from nonfood crops such as switchgrass that can be grown on degraded lands (Bourne, 2007).

Another potential benefit of biofuels is the role they could play in reducing the threat of global climate change. The transportation sector is responsible for about one-quarter of global energy-related greenhouse gas emissions, and that share is rising. Biofuels offer an option for reducing the demand for oil and associated transport-related warming emissions. However, the overall climate impacts of biofuels will depend upon several factors, the most important being changes in land use, choice of feedstock, and the various management practices.

Nevertheless, the greatest potential for reducing greenhouse gas emissions lies in the development of next-generation biofuel feedstocks and the associated technologies from conversion of these feedstocks to energy (Worldwatch Institute, 2006; Bourne, 2007).

8.8 REFERENCES

Adjaye, J. D., R. K. Sharma, and N. N. Bakhshi: "Characterization and Stability Analysis of Wood-Derived Bio-Oil," *Fuel Proc. Technol.*, 31,1992, pp. 241–256.

AFDC: "Replacement Fuel and Alternative Fuel Vehicle Technical and Policy Analysis: An Overview and Summary," Alternative Fuel Data Center, United States Department of Energy, Washington, D.C., 1997.

Albertazzi, S., F. Basile, and F. Trifirò: "Gasification of Biomass to Produce Hydrogen," in *Renewable Resources and Renewable energy: A Global Challenge*, M. Graziana and P. Fornaserio (eds.), CRC Press-Taylor and Francis Group, Boca Raton, Fla., 2007, chap. 6.

Anunputtikul, W. and S. Rodtong: "Laboratory Scale Experiments for Biogas Production from Cassava Tubers," The Joint International Conference on Sustainable Energy and Environment (SEE), Hua Hin, Thailand, Dec. 1–3, 2004.

Babu, B. V., and A. S. Chaurasia: "Modeling for Pyrolysis of Solid Particle: Kinetics and Heat Transfer Effects," *Energy Convers. Mgmt.*, 44, 2003, pp. 2251–2275.

Bala, B. K.: "Studies on Biodiesel from the Transformation of Vegetable Oils," *Energy Education Sci. Technol.*, 15, 2005, pp. 1–45.

Bauer, A., R. Hrbek, B. Amon, V. Kryvoruchko, A. Machmüller, K. Hopfner-Sixt, et al.: "Potential of Biogas Production in sustainable biorefinery concepts," 5th Research and Development Conference of Central- and Eastern European Institutes of Agricultural Engineering, Kiev, Jun. 20–24, 2007.

Beaumont, O.: "Flash Pyrolysis Products from Beech Wood," *Wood Fiber Sci.*, 17, 1985, pp. 228–239.

Bockey, D.: "Potentials for Raw Materials for the Production of Biodiesel: An Analysis," Union zur Förderung von Oel- und Proteinpflanzen e. V., Berlin Claire-Waldoff-Strasse 7, 10117 Berlin, Germany, 2006.

Boerrigter H. and A. van Der Drift: Biosyngas: "Description of R&D Trajectory Necessary to Reach Large-Scale Implementation of Renewable Syngas from Biomass," Energy Research Center of the Netherlands, Petten, Netherlands, 2004.

Bourne, J. K.: "Biofuels: Boon or Boondoggle," *National Geographic Magazine*, 212(4), 2007, p. 38–59.

BRDTAC: "Roadmap for Biomass Technologies in the United States," Biomass Research and Development Technical Advisory Committee, United States Department of Agriculture, Washington, D.C., 2002.

Brown, H. P., A. J. Panshin, and C. C. Forsaith: *Textbook of Wood Technology*, Volume 2, McGraw-Hill, New York, 1952.

Brown, R. C.: *Biorenewable Resources: Engineering New Products from Agriculture*, Iowa State Press, Ames, Iowa, 2003.

Cobb, J. T. Jr.: "Production of Synthesis Gas by Biomass Gasification," Proceedings, Spring National Meeting AIChE, Houston, Tex., Apr., 22–26, 2007.

Cohen, R., K. A. Jensen Jr., C. J. Houtman, and K. E. Hamel: Significant Levels of Extracellular Reactive Oxygen Species Produced by Brown rot Basidiomycetes on Cellulose, *FEBS Letters*, 531, 2002, pp. 483–488.

Coelho, S. T., S. M. S. G. Velazquez, V. Pecora, and F. C. Abreu: *Energy Generation with Landfill Biogas,* Book of proceedings of RIO6, World Climate & Energy Event, Rio de Janeiro, Brazil, Nov. 17–18, 2006.

Crocker, M. and C. Crofcheck: "Reducing national dependence on imported oil," *CAER Energeia*, 17(6), 2006.

Davis, G. W.: "Using E85 in Vehicles," in *Alcoholic Fuels,* S. Minteer (ed.), CRC-Taylor and Francis, Boca Raton, Fla., 2006, chap. 8.

Demirbas A. and M. M. Kucuk: "Kinetic Study on the Pyrolysis of Hazelnut Shell," *Cellulose Chem. Technol.*, 28, 1994, pp. 85–94.

Demirbas A., F. Akdeniz, Y. Erdogan, and V. Pamuk: "Kinetic for Fast Pyrolysis of Hazelnut Shell," *Fuel Sci. Technol. Int.*, 14, 1996, pp. 405–417.

Demirbas, A. and D. Güllü: "Acetic Acid, Methanol and Acetone from Lignocellosics by Pyrolysis," *Energy Edu Sci Technol.*, 1, 1998, pp. 111–115.

Demirbas, A.: "Biomass Resources for Energy and Chemical Industry," *Energy Educ. Sci. Technol.*, 5, 2000, pp. 21–45.

Demirbas, A.: "Bioethanol from Cellulosic Materials: A Renewable Motor Fuel from Biomass," *Energy Sources*, 27, 2005, pp. 327–337.

Demirbas, A.: "The Influence of Temperature on the Yields of Compounds Existing in Bio-oils Obtained from Biomass via Pyrolysis" *Fuel Proc Technol.*, 88, 2007, pp. 591–597.

Domburg G, G. Rossinskaya, and V. Sergseva: "Study of thermal stability of b-ether bonds in lignin and its models," *Therm. Anal. Proc. Int. Conf.*, 4th, Budapest, 2, 1974, p. 221.

DOE: *Electricity from Biomass*, United States Department of Energy, Washington, D.C., 2006.

Dorian, J. P., H. T. Franssen, and D. R. Simbeck: "Global Challenges in Energy," *Energy Policy*, 34, 2006, pp. 1984–1991.

Elango, D., M. Pulikesi, P. Baskaralingam, V. Ramamurthi, and S. Sivanesan: "Production of Biogas from Municipal Solid Waste with Domestic Sewage," *Energy Sources*, part A, 28, 2006, pp. 1127–1134.

Ermis, K., A. Midilli, I. Dincer, and M. A. Rosen: "Artificial Neural Network Analysis of World Green Energy Use," *Energy Policy*, 35, 2007, pp. 1731–1743.

Faaij, A.: "Modern Biomass Conversion Technologies," *Mitigation and Adaptation Strategies for Global Change*, 11, 2006, pp. 335–367.

FAO: "State of the World's Forests," Food and Agriculture Organization of the United Nations, Rome, Italy, 2005

Freeman, S. D.: *Winning Our Energy Independence: An Insider Show How*, Gibbs Smith, Salt Lake City, Utah, 2007.

Gielen, D.J., M. A. de Feber, A. J. Bos, and T. Gerlagh: "Biomass for Energy or Materials? A Western European Systems Engineering Perspective," *Energy Policy*, 29(4), 2001, 291–302.

Goldstein, I. S.: *Organic Chemical from Biomass*, CRC Press, Boca Raton, Fla., 1981, p. 13.

Juanga, J. P., P. Kuruparan, and C. Visvanathan: "Optimizing Combined Anaerobic Digestion Process of Organic Fraction of Municipal Solid Waste," International Conference on Integrated Solid Waste Management in Southeast Asian Cities, Siem Reap, Cambodia, July 5–7, 2005., pp.155–192.

Kavalov, B. and S. D. Peteves: "Status and Perspectives of Biomass-to-Liquid Fuels in The European Union," European Commission, Directorate General Joint Research Centre (DG JRC) Institute for Energy, Petten, Netherlands, 2005.

Knothe, G., J. Krahl, and J. van Gerpen: *The Biodiesel Handbook*, AOCS Press, Champaign, Ill., 2005.

Hartmann, H. and B. K. Ahring: "The future of biogas production," Risø International Energy Conference on "Technologies for Sustainable Energy Development in the Long Term," Riso-R-1517(EN), May 23–25, 2005, pp. 163–172.

Hocevar, S.: "Hydrogen Production and Cleaning from Renewable GFeedstock," in *Renewable Resources and Renewable Energy: A Global Challenge*, M. Graziana and P. Fornaserio (eds.). CRC Press-Taylor and Francis Group, Boca Raton, Fla., 2007, chap. 6.

Kavalov, B., and S. D. Peteves: "Status and Perspectives of Biomass-to-Liquid Fuels in the European Union," European Commission. Directorate General Joint Research Centre (DG JRC). Institute for Energy, Petten, Netherlands, 2005.

Lang, X., D.G. Macdonald, and G. A. Hill: "Recycle Bioreactor for Bioethanol Production from Wheat Starch II. Fermentation and Economics," *Energy Sources*, 23, 2001, pp. 427–436.

Lucia, L. A., D. S. Argyropoulos, L. Adamopoulos, and A. R. Gaspar: "Chemicals and energy from biomass," *Can. J. Chem.* 84, 2006, p. 960–970.

Lynd, L. R., C. E. Wyman, and T. U. Gerngross: "Biocommodity Engineering," *Biotechnol. Prog.*, 15(5), 1999, pp. 777–793.

Mabee, W.E. and J. N. Saddler: "Ethanol from Lignocellulose: Comparing Biofuel Technology Options," IEA Task 39 Report T39-P4, 2006, http://aic.uwex.edu/resources/documents/Comparingbiofueltechnology.pdf.

Mabee, W. E., D. J. Gregg, C. Arato, A. Berlin, R. Bura, N. Gilkes, et al.: "Update on Softwood-to-Ethanol Process Development," *Appl. Biochem. Biotechnol.*, 2006a, 129–132:55–70.

Mabee, W. E., E. D. G, Fraser, P. N, McFarlane, and J. N, Saddler: "Canadian Biomass Reserves for Biorefining," *Appl. Biochem. Biotechnol.,* 2006b, 22–40.

Mabee, W. E., D. J. Gregg, and J. N. Saddler: "Ethanol from Lignocellulose: Views to Implementation," Vancouver, BC, IEA Bioenergy Task , 2004, pp. 39–90.

McNeil Technologies Inc.: "Colorado Agriculture IOF: Technology Assessments Liquid Fuels," Prepared Under State of Colorado Purchase Order # 01-336. Governor's Office of Energy Conservation and Management, Denver, Colo., 2005.

Minteer, S.: "Alcoholic Fuels: An Overview," in *Alcoholic Fuels,* S. Minteer (ed.), CRC-Taylor and Francis, Boca Raton, Fla., 2006, chap. 1.

Mogami, C. A., C. F. Souza, V. T. Paim, I. F. Tinoco, Baeta, and R. S. Gates: "Methane Concentration in Biogas Produced from Dejections of Milk Goats Fed with Different Diets," ASABE Annual Meeting, Paper number 064068, Portland, Oreg., July 9–12, 2006.

Mohan, D., C. U. Pittman Jr., and P. H. Steele: "Pyrolysis of wood/biomass for bio-oil: A Critical Review," *Energy Fuels,* 20, 2006, pp. 848–889.

Mokhatab, S., W. A. Poe, and J. G. Speight: *Handbook of Natural gas Transmission and Processing.* CRC Press, Taylor and Francis Group, Boca Raton, Florida, 2007.

Narayan, R.: "Rationale, Drivers, Standards, and Technology for Biobased Materials," in *Renewable Resources and Renewable Energy: A Global Challenge,* M. Graziana and P. Fornaserio (eds.), CRC Press-Taylor and Francis Group, Boca Raton, Fla., 2007, chap. 1.

Nersesian, R. L.: *Energy for the 21st Century: A Comprehensive Guide to Conventional and Alternative Fuel Sources,* M.E. Sharpe Publishers, Armonk, N.Y., 2007.

Nguyen, P. H. L., P. Kuruparan, and C. Visvanathan.. "Anaerobic Digestion of Municipal Solid Waste as a Treatment Prior to Landfill," *Bioresource Technology,* 98, 2007, pp. 380–387.

Nielsen, J. B. H., P. Oleskowicz-Popiel, and T. Al Seadi: "Energy Crops Potentials for Bioenergy in EU-27," 15th European Biomass Conference & Exhibition, From Research to Market Deployment, Berlin, Germany, May 7–11, 2007.

NREL: "Dollars from Sense," National Renewable Energy Laboratory, Golden, Colo., 2003, http://www.nrel.gov/docs/legosti/fy97/20505.pdf.

NSCA: "The Potential Environmental and Rural Impacts of Biofuels Production in the UK—A Report of a Stakeholder Consultation Process," Cleaner Transport Forum and Institute for European Environmental Policy. NSCA, Brighton, U.K., 2004.

Pimentel, D. and M. Pimentel: "Global Environmental Resources Versus World Population Growth," *Ecological Economics,* 59, 2006, pp. 195–198.

Prins, M. J.: "Thermodynamic Analysis of Biomass Gasification and Torrefaction," Ph.D thesis, Technische Universiteit Eindhoven, 2005, ISBN: 90-386-2886-2.

Quirin, M., S.O, Gartner, M, Pehnt, and G.A, Reinhardt: 2004, 7(4): pp. 117–146.

Roberts, A. F. and G. Clough: "Thermal Degradation of Wood in an inert Atmosphere," Proceedings, 9th International Symposium Combustion, Pittsburgh, Pa., 1963.

Ruth, M.: "Development of a Biorefinery Optimization Model," Renewable Energy Modeling Series Forecasting the Growth of Wind and Biomass, National Bioenergy Centre, National Renewable Energy Laboratory, Golden, Colo., 2004.

Sandermann, W. and H. Augustin: "Chemical Investigations on the Thermal Decomposition of Wood," *Holz. Roh-Werkst.,* 12, 1963, pp. 256–265.

Santosh, Y., T. R. Sreekrishnan, S. Kohli, and V. Rana:[66] Enhancement of Biogas Production from Solid Substrates Using Different Techniques—A Review," *Bioresource Technology,* 95, 2004, pp. 1–10.

Sasaki, M., B. M. Kabyemela, R. M. Malaluan, S. Hirose, N. Takeda, T. Adschiri, and K. Arai: "Cellulose Hydrolysis in Subcritical and Supercritical Water," *J. Supercritical Fluids,* 13, 1998, pp. 261–268.

Shafizadeh, V., K. V. Sarkanen, and D. A. Tillman: *Thermal Uses and Properties of Carbohydrates and Lignins,* Academic Press, New York, 1976.

Shafizadeh, F. and T. T. Stevenson: "Saccharification of Douglas-Fir Wood by a Combination of Prehydrolysis and Pyrolysis," *J. Appl. Polym. Sci.,* 27, 1982, pp. 4577–4585.

Smith, I. M.: *Management of FGD Residues,* Center for Applied Energy Research, University of Kentucky, Lexington, Ky., 2006.

Sørensen, B. E., S. Njakou, and D. Blumberga: "Gaseous Fuels Biomass," Proceedings, World Renewable Energy Congress IX, WREN, London, U.K., 2006.

Speight, J. G.: *Natural Gas: A Basic Handbook,* Gulf Publishing Company, Houston, Tex., 2007.

Stamm, A. J. and E. Haris: *Chemical Processing of Wood,* Chemical Publishing Co., New York, 1953.

Timell, T. E.: "Recent Progress in the Chemistry of Wood Hemicelluloses," *Wood Sci. Technol.,* 1, 1967, pp. 45–70.

Tsai, W. T., M. K. Lee, and Y. M. Chang: "Fast Pyrolysis of Rice\Husk: Product Yields and Compositions," *Bioresource Technology,* 98(1), 2007, pp. 22–28.

van Der Drift A., R. van Ree, H. Boerrigter, and K. Hemmes: "Bio-Syngas: Key Intermediate for Large Scale Production of Green Fuels and Chemicals," Proceedings, 2d World Conference and Technology Exhibition "Biomass for Energy, Industry and Climate Protection," Rome, Italy, 2004. http://www.conference-biomass.com/Biomass2004/conference_Welcome.asp

van Loo, S. and J. Koppejan: *Handbook of Biomass Combustion and Co-Firing—International Energy Agency,* Prepared for the International Energy Agency, Task 32 of the Implementing Agreement on Bioenergy, Twente University Press, Netherlands, 2003.

VIEWLS: "Shift Gear to Biofuels,", SenterNovem, Utrecht, Netherlands, 2005, p. 60, http://www.tmleuven.be/project/viewls/home.htm.

Wenzl, H. F. J., F. E. Brauns, and D. A. Brauns: *The Chemical Technology of Wood,* Academic Press Inc., New York, 1970.

Worldwatch Institute: "Biofuls for Transport: Global Potential and Implications for Energy and Agriculture," Prepared by the Worldwatch Institute for the German ministry of Food, Agriculture, and Consumer Protection (BMELV) in Coordiantion with the German Agency for Technical Cooperation (GTZ) and the German Agency of Renewable Resources (FNR), Earthscan, London, UK, 2006.

Wright, L., R. Boundy, R. Perlack, S. Davis, and B. Saulsbury: *Biomass Energy Data Book*: Office of Planning, Budget and Analysis, Energy Efficiency and Renewable Energy, United States Department of Energy, Contract No. DE-AC05-00OR22725, Oak Ridge National Laboratory, Oak Ridge, Tenn., 2006.

Xiaohua, W. and F. Zhenmin: "Biofuel Use and Its Emission of Noxious Gases in Rural China," *Renew. Sustain Energy Review,* 8, 2004, pp. 183–192.

CHAPTER 9
FUELS FROM CROPS

Burning coal, oil, and gas inevitably produces carbon dioxide, a key greenhouse gas, as well as other pollutants, including acid-rain-related gases. By contrast, growing and burning energy crops is greenhouse gas neutral, as long as the regrowth rate balances the use rate, so that as much carbon dioxide is absorbed as is produced by combustion.

Crops have essentially many functions and benefit particularly for the *human animal*! Their products are not only a primary source of human food and animal feed, but also as source of timber, fiber, and biomass energy (BioCap Canada, 2004). In addition, crops have also an essential function to maintain ecological systems and natural environment.

Grain crops include corn, wheat, rice, barley, and other cereals. The seeds of these plants are typified by their high starch content that can be hydrolyzed to fermentable sugars for ethanol production. The *sugar crops,* including sugar cane, beet, and sweet sorghum, are preferable to the starch crops to the extent that sucrose is more readily hydrolyzed to fermentable sugars. However, sugar cane production is restricted to warm climates, and requires both high-quality land and irrigation. Sweet sorghum has considerable advantages over sugar cane as an energy crop. It can grow in a variety of soils and climatic conditions and its production costs have been estimated to be just over half those for sugar cane.

Most of crop production is used as foods and the concern is that biofuels can consume these crops. However, more crops are being cultivated for nonfood use such as pharmaceutical and nutritional products, chemical derivative products such as adhesives, paints, polymers, plastics, and industrial oils in forms of biodiesel, ethanol, two-cycle oils, transmission fluids, and lubricants.

Many efforts have been carried out to grow crops for food production in order to strengthen food security, and to alleviate the hunger and poverty particularly in developing countries. However, attention should also be directed toward cultivation of crops for nonfood use. Since, in some cases, cultivating for nonfood products may provide a higher value added product, a higher profit, and a higher income, that is, an increase in farm income. The hunger and poverty problems do not essentially rely on the incapability of individual to produce foods by themselves, but depend on capability of individual to access foods by their income.

Energy crops are plants grown specifically for use as a fuel. Although growing crops for fuel dates from medieval times, in their modern form energy crops are the most recent and innovative renewable energy option. Energy crops are important as a renewable energy technology because their use will produce a variety of economical, environmental, and energy benefits.

Wood is the oldest fuel known to man. Burning wood rather than fossil fuels can reduce the carbon dioxide emissions responsible for global climate changes. Wood fuel is carbon dioxide (CO_2) neutral. It gives off only as much CO_2 when burnt as it stores during its lifetime. In addition, wood fuel has very low levels of sulfur, a chemical that contributes to acid rain.

Biomass like wood can be used to produce electricity by direct combustion or gasification. It includes short rotation coppice and forestry waste. Biomass has some attractions as a fuel over some other renewables since it is not an intermittent resource—it can be supplied on a continuous basis to fuel base load plants.

Regularly coppiced plantations will actually absorb more carbon dioxide than mature trees—since carbon dioxide absorption slows once a tree has grown. Growing crops for fuel, particularly wood coppice, offers very promising developments for the future. Short rotation arable coppicing, using fast growing willows, is currently seen as an important source of fuel for electricity generation. The overall process involves several stages—growing over 2 or 3 years, cutting and converting to wood chip, storage and drying, transport to a power plant for combustion. And the combustion process can be very efficient, given the development of advanced cogeneration techniques.

Energy crop fuel contains almost no sulfur and has significantly less nitrogen than fossil fuels; therefore reductions in pollutants causing acid rain (SO_2) and smog (NO_x) may be realized. For example, the use of energy crops will greatly reduce greenhouse gas emissions. Burning fossil fuels removes carbon that is stored underground and transfers it to the atmosphere. Burning energy crops, on the other hand, releases carbon dioxide but as their growth requires carbon dioxide there is no net release of carbon into the atmosphere, that is, it creates a closed carbon cycle. Furthermore, where energy crops are gasified there is a net *reduction* of carbon dioxide. In addition, substantial quantities of carbon can be captured in the soil through energy crop root structures, creating a net carbon sink.

An additional environmental benefit is in water quality, as energy crop fuel contains less mercury than coal. Also, energy crop farms using environmentally proactive designs will create water quality filtration zones, uptaking and sequestering pollutants such as phosphorus from soils that leach into water bodies.

Also, growing energy crops on agricultural land that might otherwise be converted to residential or industrial use will reduce erosion/chemical runoff and enhance wildlife habitat. This will give energy producers and a renewable energy option with uniquely desirable characteristics. For example, energy crops differ from other sources of renewable energy in virtue of the fact that they can be grown to meet the needs of the market whereas other renewable resources (e.g., wind and wave power) must be harnessed where and when they occur.

Moreover, the security of energy supply will be significantly enhanced for many crop-growing countries. Energy crop production and use could reduce dependency on imported oil, although compete energy independence will remain a possible future event.

9.1 ENERGY CROPS

Markets for biomass crops have been slow to develop. The concept of co-firing biomass in existing power plants continues to show promise. Electric utilities are interested in co-firing biomass as a way to meet the renewable energy mandate, should it become law. In addition, there are air quality benefits from biomass fuels. Since biomass fuels are low in sulfur and nitrogen, they lower the smokestack emissions of sulfur oxides and nitrogen oxides—SO_x causes acid rain and NO_x contributes to smog—when burned with coal.

There are many opportunities for producing fuel from agricultural crops and crop residues. Possible end products include ethanol, vegetable oil, and solid cellulosic fuels. Ethanol can be produced from any grain, root, fruit, or juice crop containing fermentable carbohydrates. It also can be made from crops, residues, or woods that contain cellulose or other long-chain carbohydrates which can be hydrolyzed to fermentable sugars.

Vegetable oils are produced from numerous oil seed crops. Some of these oils have been evaluated in other laboratories as substitutes for diesel fuel. While all vegetable oils have high-energy content, most require some processing to assure safe use in internal combustion engines.

The simplest form of agricultural biomass energy use involves direct combustion of cellulosic crops or residues, such as hay, straw, or corn fodder, to heat space or produce steam. Such fuels are useful for heating farm buildings and small commercial buildings in rural areas and for drying crops. Ideally, energy crops should be produced on land not needed for food production. This use should not increase the erosion hazard or cause other environmental damage.

On the other hand, a variety of crops can be grown specifically to provide sources of energy and, once established, a stand of perennial biomass/energy crop is expected to remain productive for a period of 6 years or more.

Thus, perennial crops that regenerate annually from buds at the base of the plant offer the greatest potential for energy-efficient production. These include: (a) short rotation coppice, (b) miscanthus, (c) reed canary grass, (d) cordgrass, (e) switchgrass, (f) reed plants, (g) Jerusalem artichoke, and (h) sorghum.

9.1.1 Short Rotation Coppice

Short rotation coppice (or SRC) refers to fast-growing deciduous trees which are grown as energy crops, such as willow and poplar trees. The species of short rotation coppice that are most suitable, and therefore most popular, for use as energy crops are poplar and willow (and possibly also birch) because they both require deep, moisture-retentive soils for proper growth. Willow, in particular, is able to endure periods of water logging and is therefore better suited to wetter soils.

Short rotation coppice is harvested on a 2- to 5-year cycle, although commonly every 3 years. Willow and poplar are well adapted to cool climates and tolerate temporary wet conditions better than most of other species.

The harvesting of the wood takes place at the vegetative rest during the winter months. At this time of the year the wood has a water content of around 50 percent. Harvesting can be carried out using three main methods: (a) the coppice is cut and bundled, (b) the coppice is cut and chipped in a single operation, then blown into a trailer, and (c) an intermediate system where the coppice is cut into billets and blown into a trailer. After harvesting, the product may be stored for a few weeks in order to reduce its moisture content to a satisfactory level for use in energy production. Dry short rotation coppice can then be burnt under controlled conditions to produce other fuels, gas, or liquid, which are then used for electricity generation.

A plantation could be viable for up to 30 years before replanting becomes necessary, although the yield decreases after 5 to 6 rotations.

9.1.2 Miscanthus

Miscanthus is a hardy perennial rhizomatous grass that produces a crop of bamboo-like cane up to 4 m tall. Rhizomatous implies that it spreads naturally by means of underground storage organs (rhizomes). Miscanthus can grow up to 10 ft tall and theoretically can give an annual harvest of up to 12 tons per acre. Like other bioenergy crops, the harvested stems of miscanthus may be used as fuel for production of heat and electric power, or for conversion to other useful products such as ethanol. Miscanthus is high in lignin and lignocellulose fiber and uses the C4 pathway. It can be grown in a cool climate and on many types of arable land. Miscanthus does not require a big input of fertilizers due to its capability to recycle large amounts of nutrients.

Miscanthus is also well equipped for high productivity under relatively cool temperatures and may require substantial amounts of water for maximal growth (its growth could therefore also have valuable environmental benefits by acting as absorbing disposal areas for waste water and some industrial effluents). Furthermore, Miscanthus seems to grow well in most soil conditions (bar thin droughty soils) but appears to thrive within areas which are currently best-suited to maize production. The advantages of Miscanthus as an energy crop are that it multiplies very rapidly, has a high yield which is relatively dry and can be harvested annually (from its second season onward) compared with every 2 to 4 years for short rotation coppice.

Further advantages are that miscanthus can be grown and harvested with existing farm machinery; it requires little or no pesticide/fertilizer input after establishment and the harvest can use the same infrastructure for storage and transport as short rotation coppice. Finally, Miscanthus has a similar calorific value per unit weight as wood and therefore could possibly be used in the same power plant or those designed for agricultural residues.

9.1.3 Reed Canary Grass

Reed canary grass is a robust C3 perennial grass, widely distributed across temperate regions of Europe, Asia, and also North America. It usually grows in damp areas.

This rhizonous grass can grow up to 6 ft in height and usually flowers in June and July. Reed canary grass is usually harvested once per year either by mowing or baling using a high density baler. Crop duration of 10 years or more may be possible. Due to its low moisture content, canary grass (those harvested in spring) can be easily converted to pellets, briquettes, and powder.

The advantages of reed canary grass as an energy crop are its good adaptation to cool temperate climates and poor wet soil conditions and, conversely, its ability to withstand drought. Crucially, for the purposes of biomass production, reed canary grass is also able to attain high dry matter content earlier than Miscanthus. The crop responds well to nitrogen and phosphate and it may be used in a bed system to remove nutrients from waste water, as well as to stabilize areas at risk of soil erosion.

9.1.4 Cordgrass

Cordgrasses are perennial grasses from Western Europe, Africa, and North America. They are pioneer colonists of muddy coastal salt flats, spreading by means of rhizomes to form clumps and mats which help to reduce erosion and reclaim land from the sea. Cordgrasses are expected to be able to produce good yields of biomass in poor soil conditions.

9.1.5 Switchgrass

The utilization of energy crops such as *switchgrass* (*Panicum virgatum*, L., Poaceae) is a concept with great relevance to current ecologic and economic issues on a global scale. Development of a significant national capacity to utilize perennial forage crops, such as switchgrass as biofuels, could provide an important new source of energy from perennial cropping systems, which are compatible with conventional farming practices, would help reduce degradation of agricultural soils, lower national dependence on foreign oil supplies, and reduce emissions of greenhouse gases and toxic pollutants to the atmosphere (McLaughlin, 1997).

Switchgrass is a perennial sod-forming grass with thick strong stems. The advantages of switchgrass as an energy crop are that it is fast-growing, remarkably adaptable, and high-yielding. Further advantages of switchgrass are that it can be harvested, using conventional equipment, either annually or semiannually for 10 years or more before replanting is needed and that it is able to reach deep into the soil for water and use water very efficiently.

Switchgrass is a sod-forming, warm season grass, which combines good forage attributes and soil conservation benefits typical of perennial grasses (Moser and Vogel, 1995). Switchgrass was an important part of the native, highly productive North American tallgrass prairie (Weaver, 1968; Risser et al., 1981). While the original tallgrass prairies have been severely reduced by cultivation of prairie soils, remnant populations of switchgrass are still widely distributed geographically within North America. Switchgrass tolerates diverse growing conditions, ranging from arid sites in the shortgrass prairie to brackish marshes and open woods. The range of switchgrass extends from Quebec to Central America. Two major ecotypes of switchgrass occur, a thicker stemmed lowland type better adapted to warmer, moister habitats of its southern range, and a finer stemmed upland type, more typical of mid to northern areas (Vogel et al., 1985). The ecologic diversity of switchgrass can be attributed to three principal characteristics, genetic diversity associated with its open pollinated reproductive mode, a very deep, well-developed rooting system, and efficient physiologic metabolism. In the southern range, switchgrass can grow to more than 3 m in height, but what is most distinctive is the deep, vigorous root system, which may extend to depths of more than 3.5 m (Weaver, 1968). It reproduces both by seeds and vegetatively and, with its perennial life form, a stand can last indefinitely once established. Standing biomass in root systems may exceed that found aboveground (Shiflet and Darby, 1985), giving perennial grasses such as switchgrass, an advantage in water and nutrient aquisition even under stressful growing conditions.

Physiologically, switchgrass, like maize, is a C4 species, fixing carbon by multiple metabolic pathways with a high water use efficiency (Moss et al., 1969; Koshi et al., 1982). In general C4 plants such as grasses will produce 30 percent more food per unit of water than C3 species such as trees and broadleaved crops and grasses and are well adapted to the more arid production areas of the mid-western United States where growth is more limited by moisture supply.

Besides showing great promise as an energy crop for energy production, switchgrass also restores vital organic nutrients to farmed-out soils and with its extensive network of stems and roots (the plants extend nearly as far below ground as above), it is also a valuable soil stabilization plant.

9.1.6 Reed Plants

Reed plants are a potentially prolific producer of biomass, capable of yielding 20 to 25 t/ha of dry matter annually for a number of years. They can grow up to 6 m, are spread by means of stout rhizomes and stolons, and are commonly found in swampy ground and shallow water throughout temperate and subtropical areas.

9.1.7 Jerusalem Artichoke

The high-fructose syrups that can be derived from the tubers produced by the *Jerusalem artichoke* may be used for the production of ethanol and other industrial raw materials. Jerusalem artichokes also produce a large amount of top growth which may also prove to be a useful source of biomass for energy purposes.

9.1.8 Sorghum

Sorghum is an annual tropical grass with large genetic variation that is a crop with the potential for energy production. Sweet sorghum has been selected for its sugar content and is normally grown for molasses production. Forage sorghum has been selected for high yields of reasonably good quality animal feed. Sorghum varieties producing tall plants with large stems make the best candidates for biomass production. Both sweet and forage sorghum have a high potential for lodging. Lodging can result in harvest problems with ensuing loss of yield from both initial and ratoon crops.

Sweet sorghums produce sugar syrups which could form the basis of fermentation processes for methane or ethanol production and some of the forage types of the plant may be suitable for biomass production.

9.1.9 Hemp

Hemp is an annual short day, C3 plant with a high cellulose and lignin content in its stems and a high fat and protein content in its seeds. The entire plant consisting of bast fibers, leaves, seeds, and processable remains can be used as a solid fuel when compacted. The stalk contains a very strong and durable fiber. The average height is 7 ft but it can reach up to 12 ft. For the energetic use the whole hemp crop is harvested.

Hemp can be easily planted and cultivated but its growing is prohibited in many countries as it can be used as a drug. The plants vegetative period is about 100 days, with the main growth period in summer months followed by flowering in toward the end of summer. The yields are high but the high plant moisture content can create storage problems. Two different harvesting technologies are available for the energy use of the whole hemp crop.

The first is the whole-fiber-technology where, with a modified chopping technology, the hemp culm is cut into 24- to 24-in lengths. The straw stays for 4 weeks on the field for drying before the bales are pressed. Another technology is wet harvesting. This technology includes the chopping of the crops followed by silaging. For combustion it is necessary to press the hemp silage into bales to reduce the water content of the fuel.

9.2 PROCESSES

There are four basic types of processes used to generate energy from crops: direct combustion, gasification, pyrolysis, and anaerobic digestion.

9.2.1 Direct Combustion

Direct combustion involves burning the energy crop and then using the resulting hot combustion gases to raise steam. The steam is, in turn, used to drive a steam turbine which drives a generator to produce electricity. The conversion efficiency from energy crop to energy is fairly low, especially for small systems, but this is balanced by the relatively low capital cost of direct combustion systems and the fact that the technology is tried and tested. Furthermore, using the waste heat produces much better efficiencies and economics.

Combustion facilities can burn many types of biomass fuel, including wood, agricultural residues, wood pulping liquor, municipal solid waste (MSW), and refuse-derived fuel. Combustion technologies convert biomass fuels into several forms of useful energy for commercial or industrial uses: hot air, hot water, steam, and electricity.

A furnace is the simplest combustion technology. In a furnace, biomass fuel burns in a combustion chamber, converting biomass into heat energy. As the biomass burns, hot gases are released. These hot gases contain about 85 percent of the fuel's potential energy. Commercial and industrial facilities use furnaces for heat either directly or indirectly through a heat exchanger in the form of hot air or water.

A biomass-fired boiler is a more adaptable direct combustion technology because a boiler transfers the heat of combustion into steam. Steam can be used for electricity, mechanical energy or heat. Biomass boilers supply energy at low cost for many industrial and commercial uses.

Pile burners consist of cells, each having an upper and a lower combustion chamber. Biomass fuel burns on a grate in the lower chamber, releasing volatile gases. The gases burn in the upper (secondary) combustion chamber and pile burners must be shut down periodically to remove ash.

Although capable of handling high-moisture fuels and fuels mixed with dirt, pile burners have become obsolete with the development of more efficient combustion designs with automated ash-removal systems.

In a stationary or traveling grate combustor, an automatic feeder distributes the fuel onto a grate, where the fuel burns. Combustion air enters from below the grate. In the stationary grate design, ashes fall into a pit for collection. In contrast, a traveling grate system has a moving grate that drops the ash into a hopper.

Fluidized bed combustors burn biomass fuel in a hot bed of granular material, such as sand. Injection of air into the bed creates turbulence resembling a boiling liquid. The turbulence distributes and suspends the fuel. This design increases heat transfer and allows for operating temperatures below 972°C (1781°F), reducing nitrogen oxide (NO_x) emissions. Fluidized bed combustors can handle high-ash fuels and agricultural biomass residue.

Conventional combustion equipment is not designed for burning agricultural residues. Straw and grass contain alkali (potassium and sodium) compounds, which are also present in all annual crops and crop residues and in the annual growth of trees and plants. During combustion, alkali combines with silica, which is also present in agricultural residues. This reaction causes slagging and fouling problems in conventional combustion equipment designed for burning wood at higher temperatures.

Volatile alkali lowers the fusion temperature of ash. In conventional combustion equipment having furnace gas exit temperatures above 1450°F, combustion of agricultural residue causes slagging and deposits on heat transfer surfaces. Specially designed boilers with lower furnace exit temperatures could reduce slagging and fouling from combustion of these fuels. Low-temperature gasification may be another method of using these fuels for efficient energy production while avoiding the slagging and fouling problems encountered in direct combustion.

Combustion facilities that produce electricity from steam-driven turbine-generators have a conversion efficiency of 17 to 25 percent. Using a boiler to produce both heat and electricity (cogeneration) improves overall system efficiency to as much as 85 percent. That is, cogeneration converts 85 percent of the fuel's potential energy into useful energy in two forms: electricity and steam heat.

Two cogeneration arrangements, or cycles, are possible for combining electric power generation with industrial steam production. Steam can be used in an industrial process first and then routed through a turbine to generate electricity. This arrangement is called a bottoming cycle. In the alternate arrangement, steam from the boiler passes first through a turbine to produce electric power. The steam exhaust from the turbine is then used for industrial processes or for space and water heating. This arrangement is called a topping cycle.

The direct-fired gas turbine is another combustion technology for converting biomass to electricity. In this technology, fuel pretreatment reduces biomass to a particle size of less than 2 mm and a moisture content of less than 25 percent. Then the fuel is burned with compressed air. Cleanup of the combustion gas reduces particulate matter before the gas expands through the turbine stage. The turbine drives a generator to produce electricity.

Co-firing biomass as a secondary fuel in a coal-burning power plant using high-sulfur coal could help reduce sulfur dioxide and nitrogen oxide emissions. Also, co-firing decreases net carbon dioxide emissions from the power plant (if the biomass fuel comes from a sustainable source). Co-firing may require wood fuel preparation or boiler modifications to maintain boiler efficiency.

9.2.2 Gasification

Gasification is a high temperature process which produces gas that can then be used in an internal combustion engine or fuel cell.

Briefly, gasification was used as long ago as the early 1800s. The process was rather crude and the fuel was most often coal. The gas product (town gas, Chap. 5) was used for heating and lighting. For example, the gas was piped to street lights as early as 1846 in England.

The gasification process produces a fuel gas from crops by heating them under carefully controlled temperature, pressure, and atmospheric conditions. A key to gasification is using less air or oxygen than is usually found in combustor. The product (biogas or fuel gas), like natural gas, can be burned in high-efficiency gas turbines.

The gasifier is the heart of the gasification process (Chaps. 5 and 7). Gasifiers are designed to process the fuel in a variety of ways consistent with the type of fuel, the end use of the gas, the size of the process, and the source of oxygen. The oxygen may be introduced as a pure gas or may come from air or steam. Some gasifiers operate under pressure, others do not.

The simplest type of gasifier is the *fixed bed counter current gasifier*. The biomass is fed at the top of the reactor and moves downward as a result of the conversion of the biomass and the removal of ashes. The air intake is at the bottom and the gas leaves at the top. The biomass moves in counter current to the gas flow, and passes through the drying zone, the distillation zone, reduction zone, and the oxidation zone.

The major advantages of this type of gasifier are its simplicity, high charcoal burn-out, and internal heat exchange leading to low gas exit temperatures and high gasification efficiency. In this way also fuels with high moisture content (up to 50 percent by weight) can be used. Major drawbacks are the high amounts of tar and pyrolysis products, because the pyrolysis gas is not lead through the oxidation zone. This is of minor importance if the gas is used for direct heat applications, in which the tars are simply burnt. In case the gas is used for engines, gas cleaning is required, resulting in problems of tar-containing condensates.

In the conventional *downdraft gasifier* (sometimes called the *co-flow* gasifier), biomass is fed at the top of the reactor and air intake is at the top or from the sides. The gas leaves at the bottom of the reactor, so the fuel and the gas move in the same direction. The pyrolysis gasses are lead through the oxidation zone (with high temperatures) and or more or less burnt or cracked. Therefore the producer gas has low tar content and is suitable for engine applications. In practice however, a tar-free gas is seldom if ever achieved over the whole operating range of the equipment. Because of the lower level of organic components in the condensate, downdraft gasifiers suffer less from environmental objections than updraft gasifiers.

Successful operation of a downdraft gasifier requires drying the biomass fuel to a moisture content of less than 20 percent. The advantage of the downdraft design is the very low tar content of the producer gas. However, disadvantages of the downdraft gasifier are: (a) the high amounts of ash and dust particles in the gas, (b) the inability to operate on a number of unprocessed fuels, often pelletization or briquetting of the biomass is necessary, (c) the outlet gas has a high temperature leading to a lower gasification efficiency, and (d) the moisture content of the biomass must be less than 25 percent by weight.

A more recent development is the *open core* gasifier design for gasification of small sized biomass with high-ash content. The producer gas is not tar-free; it contains approximately 0.05 kg tar per kilogram of gas. In the open core gasifier the air is sucked over the whole cross section from the top of the bed. This facilitates better oxygen distribution since the oxygen will be consumed over the whole cross section, so that the solid bed temperature will not reach the local extremes (hot spots) observed in the oxidation zone of conventional gasifiers due to poor heat transfer. Moreover, the air nozzles in conventional gasifiers generate caves and create obstacles that may obstruct solid flow especially for solids of low bulk, like rice husk. On the other hand, the entry of air through the top of the bed creates a downward flow of the pyrolysis gases, and transports the tars products to the combustion zone. Thus, flow problems due to the caking of rice husk caused by back mixing of tar are avoided.

The gasification process is amenable to a variety of biomass feedstocks such as waste rice hulls, wood waste, grass, and the dedicated energy crops. Gasification is a clean process with few air emissions and, when crops are used as the feedstock, little or no ash results.

In the gasifier, biomass is converted into a gaseous mixture of hydrogen, carbon monoxide, carbon dioxide, and other compounds by applying heat under pressure in the presence of steam and a controlled amount of oxygen. The biomass produces synthesis gas (Chap. 7).

$$C_6H_{12}O_6 + O_2 + H_2O \rightarrow CO + CO_2 + H_2 + \text{other products}$$

The above reaction uses glucose as a surrogate for cellulose. Biomass has highly variable composition and complexity, with cellulose as one major component.

The gasification process could play a significant role in meeting the goal of greenhouse gas mitigation. It is likely that both in the transition phase to a hydrogen economy and in the steady state, a significant fraction of hydrogen might be derived from domestically abundant crops. In addition, the co-firing applications of crops biomass (and other biomass) with coal (Chap. 5), biomass can provide up to 15 percent of the total energy input of the fuel mixture. Such concepts address greenhouse gas mitigation by co-firing biomass and coal to offset the losses of carbon dioxide to the atmosphere that are inherent in coal combustion processes (even with the best-engineered capture and storage of carbon). Since growth of biomass fixes atmospheric carbon, its combustion leads to no net addition of atmospheric carbon dioxide even if vented. Thus, co-firing of crops or crop residues (or other biomass with coal) in an efficient coal gasification process, affording the opportunity for capture and storage of carbon dioxide, could lead to a net reduction of atmospheric carbon dioxide. Cheaper, though less plentiful, biomass residue could supplant crops as gasifier feedstock leading to a less significant impact on the environment than would farming bioenergy crops.

Along similar lines, willow biomass crops have been shown to be a good fuel for farm-based power production using advanced gasification technology (Pian et al., 2006). The fuel gas can be used for generating electricity, using micro-turbines modified to operate on low-Btu gas, or for other farm energy needs. Willow biomass was found to make a fuel for ash-rejection gasifiers with a predicted net gasification efficiency of about 85 percent. Analysis showed that developing a method to co-gasify willow with various amounts of low-cost wastes, such as dairy farm animal waste, can be an excellent way to reduce the fuel cost, to increase the overall fuel availability and help work around problems resulting from seasonal availability of bioenergy crops. Co-gasification of dairy farm wastes along with willow offers an economical way to dispose the wastes and manage nutrient flows on a dairy farm.

Agricultural residues can be divided into two groups: (a) crop residues and (b) agro-industrial residues. Crop residues are plant materials left behind in the farm after removal of the main crop produce. The remaining materials could be of different sizes, shapes, forms, and densities like straw, stalks, sticks, leaves, haulms, fibrous materials, roots, branches, and twigs.

Due to high-energy content, straw is one of the best crop residues for solid biofuels. However, straw has several disadvantages—it has a higher-ash content, which results in lower-calorific value. In order to improve its bulk density, the straw is generally baled before transportation. Straw burning requires a specific technology. There are four basic types of straw burners: (a) those that accept shredded, loose straw; (b) burners that use densified straw products such as pellets, briquettes, or cubes, and straw logs; (c) small, square bale burners and; (d) round bale burners. To be suitable for heat and electricity production straw should not have a large content of moisture, preferably not more than 20 percent as the moisture reduces the boiler efficiency. Also straw color as well as straw chemistry should be considered before burning as it indicates the quality of the straw.

The agro-industrial residues are by-products of the postharvest processes of crops such as cleaning, threshing, sieving, and crushing. These could be in the form of husk, dust, and straw. Furthermore, the quantity of agricultural residues produced differs from crop to crop and is affected by soil type and irrigation conditions. Production of agricultural residues is directly related to the corresponding crop production and ratio between the main crop produce and the residues, which varies from crop to crop and, at times, with the variety of the seeds in one crop itself. Thus, for known amounts of crop production, it may be possible to estimate the amounts of agricultural residues produced using the residue-to-crop ratio.

Most crop or agricultural residues are not found throughout the year but are available only at the time of harvest. The amount available depends upon the harvesting time, storage-related characteristics, and the storage facility.

9.2.3 Pyrolysis

Pyrolysis is a medium temperature method which produces gas, oil, and char from crops which can then be further processed into useful fuels or feedstock (Boateng et al., 2007).

Pyrolysis is often considered to be the gasification of biomass in the absence of oxygen. However, the chemistry of each process may differ significantly. In general, biomass does not gasify as easily as coal, and it produces other hydrocarbon compounds in the gas mixture exiting the gasifier; this is especially true when no oxygen is used. As a result, typically an extra step must be taken to reform these hydrocarbons with a catalyst to yield a clean syngas mixture of hydrogen, carbon monoxide, and carbon dioxide.

Fast pyrolysis is a thermal decomposition process that occurs at moderate temperatures with a high heat transfer rate to the biomass particles and a short hot vapor residence time in the reaction zone. Several reactor configurations have been shown to assure this condition and to achieve yields of liquid product as high as 75 percent based on the starting dry biomass weight. They include bubbling fluid beds, circulating and transported beds, cyclonic reactors, and ablative reactors.

Fast pyrolysis of biomass produces a liquid product (Fig. 9.1), pyrolysis oil or bio-oil that can be readily stored and transported. Pyrolysis oil is a renewable liquid fuel and can also be used for production of chemicals. Fast pyrolysis has now achieved a commercial

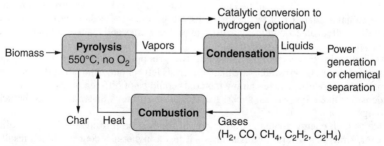

FIGURE 9.1 Biomass liquefaction by pyrolysis.

success for production of chemicals and is being actively developed for producing liquid fuels. Pyrolysis oil has been successfully tested in engines, turbines, and boilers, and has been upgraded to high quality hydrocarbon fuels.

In the 1990s several fast pyrolysis technologies reached near-commercial status and the yields and properties of the generated liquid product, bio-oil, depend on the feedstock, the process type and conditions, and the product collection efficiency (http://www1.eere.energy.gov/biomass/pyrolysis.html).

Direct hydrothermal liquefaction involves converting biomass to an oily liquid by contacting the biomass with water at elevated temperatures (300–350°C) with sufficient pressure to maintain the water primarily in the liquid phase (12–20 MPa) for residence times up to 30 minutes. Alkali may be added to promote organic conversion. The primary product is an organic liquid with reduced oxygen content (about 10 percent) and the primary by-product is water containing soluble organic compounds. (http://www1.eere.energy.gov/biomass/pyrolysis.html).

The importance of the provisions for the supply of feedstocks as crops and other biomass are often underestimated since it is assumed that the supplies are inexhaustible. While this may be true over the long-term, short-term supply of feedstocks can be as much a risk as any venture.

9.2.4 Anaerobic Digestion

Anaerobic digestion is a natural process and is the microbiologic conversion of organic matter to methane in the absence of oxygen (Fig. 9.2). The decomposition is caused by natural bacterial action in various stages and occurs in a variety of natural anaerobic environments, including water sediment, water-logged soils, natural hot springs, ocean thermal vents, and the stomach of various animals (e.g., cows). The digested organic matter resulting from the anaerobic digestion process is usually called *digestate*.

Symbiotic groups of bacteria perform different functions at different stages of the digestion process. There are four basic types of microorganisms involved. Hydrolytic bacteria break down complex organic wastes into sugars and amino acids. Fermentative bacteria then convert those products into organic acids. Acidogenic microorganisms convert the

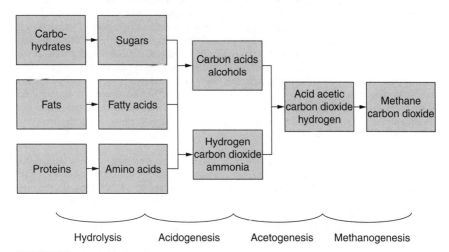

FIGURE 9.2 Anaerobic digestion.

acids into hydrogen, carbon dioxide, and acetate. Finally, the methanogenic bacteria produce biogas from acetic acid, hydrogen, and carbon dioxide.

The process of anaerobic digestion occurs in a sequence of stages involving distinct types of bacteria. Hydrolytic and fermentative bacteria first break down the carbohydrates, proteins, and fats present in biomass feedstock into fatty acids, alcohol, carbon dioxide, hydrogen, ammonia, and sulfides. This stage is called hydrolysis (or liquefaction).

Next, acetogenic (acid-forming) bacteria further digest the products of hydrolysis into acetic acid, hydrogen, and carbon dioxide. Methanogenic (methane-forming) bacteria then convert these products into biogas.

The combustion of digester gas can supply useful energy in the form of hot air, hot water, or steam. After filtering and drying, digester gas is suitable as fuel for an internal combustion engine, which, combined with a generator, can produce electricity. Future applications of digester gas may include electric power production from gas turbines or fuel cells. Digester gas can substitute for natural gas or propane in space heaters, refrigeration equipment, cooking stoves, or other equipment. Compressed digester gas can be used as an alternative transportation fuel.

Thus, there are three principal byproducts of anaerobic digestion: (a) biogas, (b) acidogenic digestate, and (c) methanogenic digestate.

Biogas is a gaseous mixture comprising mostly of methane and carbon dioxide, but also containing a small amount of hydrogen and occasionally trace levels of hydrogen sulfide. Biogas can be burned to produce electricity, usually with a reciprocating engine or microturbine. The gas is often used in a cogeneration arrangement, to generate electricity and use waste heat to warm the digesters or to heat buildings.

Since the gas is not released directly into the atmosphere and the carbon dioxide comes from an organic source with a short carbon cycle, biogas does not contribute to increasing atmospheric carbon dioxide concentrations; because of this, it is considered to be an environment friendly energy source. The production of biogas is not a steady stream; it is highest during the middle of the reaction. In the early stages of the reaction, little gas is produced because the number of bacteria is still small. Toward the end of the reaction, only the hardest to digest materials remain, leading to a decrease in the amount of biogas produced.

The second by-product (*acidogenic digestate*) is a stable organic material comprised largely of lignin and chitin, but also of a variety of mineral components in a matrix of dead bacterial cells; some plastic may be present. This resembles domestic compost and can be used as compost or to make low-grade building products such as fibreboard.

The third by-product is a liquid (*methanogenic digestate*) that is rich in nutrients and can be an excellent fertilizer dependent on the quality of the material being digested. If the digested materials include low-levels of toxic heavy metals or synthetic organic materials such as pesticides or polychlorobiphenyls, the effect of digestion is to significantly concentrate such materials in the digester liquor. In such cases further treatment will be required in order to dispose of this liquid properly. In extreme cases, the disposal costs and the environmental risks posed by such materials can offset any environmental gains provided by the use of biogas. This is a significant risk when treating sewage from industrialised catchments.

Nearly all digestion plants have ancillary processes to treat and manage all of the byproducts. The gas stream is dried and sometimes sweetened before storage and use. The sludge liquor mixture has to be separated by one of a variety of ways, the most common of which is filtration. Excess water is also sometimes treated in sequencing batch reactors (SBR) for discharge into sewers or for irrigation.

Digestion can be either *wet* or *dry*. Dry digestion refers to mixtures which have a solid content of 30 percent or greater, whereas wet digestion refers to mixtures of 15 percent or less.

In recent years, increasing awareness that anaerobic digesters can help control the disposal and odor of animal waste has stimulated renewed interest in the technology. New digesters now are being built because they effectively eliminate the environmental hazards of dairy farms and other animal feedlots.

Anaerobic digester systems can reduce fecal coliform bacteria in manure by more than 99 percent, virtually eliminating a major source of water pollution. Separation of the solids during the digester process removes about 25 percent of the nutrients from manure, and the solids can be sold out of the drainage basin where nutrient loading may be a problem. In addition, the digester's ability to produce and capture methane from the manure reduces the amount of methane that otherwise would enter the atmosphere. Scientists have targeted methane gas in the atmosphere as a contributor to global climate change.

Controlled anaerobic digestion requires an airtight chamber, called a digester. To promote bacterial activity, the digester must maintain a temperature of at least 68°F. Using higher temperatures, up to 150°F, shortens processing time and reduces the required volume of the tank by 25 to 40 percent. However, there are more species of anaerobic bacteria that thrive in the temperature range of a standard design (mesophillic bacteria) than there are species that thrive at higher temperatures (thermophillic bacteria). High-temperature digesters also are more prone to upset because of temperature fluctuations and their successful operation requires close monitoring and diligent maintenance.

The biogas produced in a digester (*digester gas*) is actually a mixture of gases, with methane and carbon dioxide making up more than 90 percent of the total. Biogas typically contains smaller amounts of hydrogen sulfide, nitrogen, hydrogen, methyl mercaptans, and oxygen.

Methane is a combustible gas. The energy content of digester gas depends on the amount of methane it contains. Methane content varies from about 55 to 80 percent. Typical digester gas, with a methane concentration of 65 percent, contains about 600 Btu/ft^3 of energy.

There are three basic digester designs and all of them can trap methane and reduce fecal coliform bacteria, but they differ in cost, climate suitability, and the concentration of manure solids they can digest.

A *covered lagoon digester*, as the name suggests, consists of a manure storage lagoon with a cover. The cover traps gas produced during decomposition of the manure. This type of digester is the least expensive of the three.

Covering a manure storage lagoon is a simple form of digester technology suitable for liquid manure with less than 3 percent solids. For this type of digester, an impermeable floating cover of industrial fabric covers all or part of the lagoon. A concrete footing along the edge of the lagoon holds the cover in place with an airtight seal. Methane produced in the lagoon collects under the cover. A suction pipe extracts the gas for use. Covered lagoon digesters require large lagoon volumes and a warm climate. Covered lagoons have low capital cost, but these systems are not suitable for locations in cooler climates or locations where a high water table exists.

A *complete mix digester* converts organic waste to biogas in a heated tank above or below ground. A mechanical or gas mixer keeps the solids in suspension. Complete mix digesters are expensive to construct and cost more than plug-flow digesters to operate and maintain.

Complete mix digesters are suitable for larger manure volumes having solids concentration of 3 to 10 percent. The reactor is a circular steel or poured concrete container. During the digestion process, the manure slurry is continuously mixed to keep the solids in suspension. Biogas accumulates at the top of the digester. The biogas can be used as fuel for an engine-generator to produce electricity or as boiler fuel to produce steam. Using waste heat from the engine or boiler to warm the slurry in the digester reduces retention time to less than 20 days.

Plug-flow digesters are suitable for ruminant animal manure that has a solids concentration of 11 to 13 percent. A typical design for a plug-flow system includes a manure collection system, a mixing pit, and the digester itself. In the mixing pit, the addition of water adjusts the proportion of solids in the manure slurry to the optimal consistency. The digester is a long, rectangular container, usually built below-grade, with an airtight, expandable cover.

New material added to the tank at one end pushes older material to the opposite end. Coarse solids in ruminant manure form a viscous material as they are digested, limiting solids separation in the digester tank. As a result, the material flows through the tank in a *plug*. Average retention time (the time a manure plug remains in the digester) is 20 to 30 days. Anaerobic digestion of the manure slurry releases biogas as the material flows through the digester. A flexible, impermeable cover on the digester traps the gas. Pipes beneath the cover carry the biogas from the digester to an engine-generator set.

A plug-flow digester requires minimal maintenance. Waste heat from the engine-generator can be used to heat the digester. Inside the digester, suspended heating pipes allow hot water to circulate. The hot water heats the digester to keep the slurry at 25 to 40°C (77 to 104°F), a temperature range suitable for methane-producing bacteria. The hot water can come from recovered waste heat from an engine generator fueled with digester gas or from burning digester gas directly in a boiler.

9.3 ETHANOL

Ethanol is the predominant fuel produced form crops and has been used as fuel in the United States since at least 1908. Although early efforts to sustain an ethanol program failed, oil supply disruptions in the Middle East and environment concerns over the use of lead as a gasoline octane booster renewed interest in ethanol in the late 1970s. At present, extending the volume of conventional gasoline is a significant end use for ethanol, as is its use as an oxygenate. To succeed in these markets, the cost of ethanol must be close to the wholesale price of gasoline, currently made possible by the federal ethanol subsidy. However, in order for ethanol to compete on its own merits the cost of producing it must be reduced substantially.

The production of ethanol from corn is a mature technology that holds much potential (Nichols et al., 2006). Substantial cost reductions may be possible, however, if cellulose-based feedstocks are used instead of corn. Producers are experimenting with units equipped to convert cellulose-based feedstocks, using sulfuric acid to break down cellulose and hemicellulose into fermentable sugar. Although the process is expensive at present, advances in biotechnology could decrease conversion costs substantially. The feed for all ethanol fermentations is sugar—traditionally a hexose (a six-carbon or "C6" sugar) such as those present naturally in sugar cane, sugar beet, and molasses. Sugar for fermentation can also be recovered from starch, which is actually a polymer of hexose sugars (*polysaccharide*).

Biomass (Chap. 8), in the form of wood and agricultural residues such as wheat straw, is viewed as a low cost alternative feed to sugar and starch. It is also potentially available in far greater quantities than sugar and starch feeds. As such it receives significant attention as a feed material for ethanol production. Like starch, wood and agricultural residues contain polysaccharides. However, unlike starch, while the cellulose fraction of biomass is principally a polymer of easily fermented six-carbon sugars, the hemicellulose fraction is principally a polymer of five-carbon sugars, with quite different characteristics for recovery and fermentation the cellulose and hemicellulose in biomass are bound together in a complex framework of crystalline organic material known as lignin.

These differences mean that recovery of these biomass sugars is more complex than recovery of sugars from a starch feed. Once recovered, fermentation is also more complex than a simple fermentation of six-carbon sugars. The current focus is on the issues of releasing the sugars (hydrolysis) and then fermenting as much of the six- and five-carbon sugars as possible to produce ethanol.

There are several different methods of hydrolysis: (a) concentrated sulfuric acid, (b) dilute sulphuric acid, (c) nitric acid, and (d) acid pretreatment followed by enzymatic hydrolysis.

9.3.1 Ethanol Production

Ethanol is produced from the fermentation of sugar by enzymes produced from specific varieties of yeast. The five major sugars are the five-carbon xylose and arabinose and the six-carbon glucose, galactose, and mannose. Traditional fermentation processes rely on yeasts that convert six-carbon sugars to ethanol. Glucose, the preferred form of sugar for fermentation, is contained in both carbohydrates and cellulose. Because carbohydrates are easier than cellulose to convert to glucose, the majority of ethanol currently produced in the United States is made from corn, which produces large quantities of carbohydrates. Also, the organisms and enzymes for carbohydrate conversion and glucose fermentation on a commercial scale are readily available.

The conversion of cellulosic biomass to ethanol parallels the corn conversion process. The cellulose must first be converted to sugars by hydrolysis and then fermented to produce ethanol. Cellulosic feedstocks (composed of cellulose and hemicellulose) are more difficult to convert to sugar than are carbohydrates. Two common methods for converting cellulose to sugar are dilute acid hydrolysis and concentrated acid hydrolysis, both of which use sulfuric acid.

Dilute acid hydrolysis occurs in two stages to take advantage of the differences between hemicellulose and cellulose. The first stage is performed at low temperature to maximize the yield from the hemicellulose, and the second, higher temperature stage is optimized for hydrolysis of the cellulose portion of the feedstock. Concentrated acid hydrolysis uses a dilute acid pretreatment to separate the hemicellulose and cellulose. The biomass is then dried before the addition of the concentrated sulfuric acid. Water is added to dilute the acid and then heated to release the sugars, producing a gel that can be separated from residual solids. Column chromatography is used to separate the acid from the sugars.

Both the dilute and concentrated acid processes have several drawbacks. Dilute acid hydrolysis of cellulose tends to yield a large amount of by-products. Concentrated acid hydrolysis forms fewer by-products, but for economic reasons the acid must be recycled. The separation and concentration of the sulfuric acid adds more complexity to the process. The concentrated and dilute sulfuric acid processes are performed at high temperatures (100 and 220°C) which can degrade the sugars, reducing the carbon source and ultimately lowering the ethanol yield. Thus, the concentrated acid process has a smaller potential for cost reductions from process improvements such as acid recovery and sugar yield for the concentrated acid process could provide higher efficiency for both technologies.

Another approach involves countercurrent hydrolysis, a two-stage process. In the first stage, cellulose feedstock is introduced to a horizontal cocurrent reactor with a conveyor. Steam is added to raise the temperature to 180°C (no acid is added at this point). After a residence time of about 8 minutes, during which some 60 percent of the hemicellulose is hydrolyzed, the feed exits the reactor. It then enters the second stage through a vertical reactor operated at 225°C. Very dilute sulfuric acid is added to the feed at this stage, where virtually all of the remaining hemicellulose and, depending on the residence time, anywhere from 60 percent to all of the cellulose is hydrolyzed. The countercurrent hydrolysis process offers higher efficiency (and, therefore, cost reductions) than the dilute sulfuric acid process. This process may allow an increase in glucose yields to 84 percent, an increase in fermentation temperature to 55°C, and an increase in fermentation yield of ethanol to 95 percent.

The greatest potential for ethanol production from biomass, however, lies in enzymatic hydrolysis of cellulose. The enzyme cellulase, now used in the textile industry to stone wash denim and in detergents, simply replaces the sulfuric acid in the hydrolysis step. The cellulase can be used at lower temperatures, 30 to 50°C, which reduces the degradation of the sugar. In addition, process improvements now allow simultaneous saccharification and fermentation (SSF). In the SSF process, cellulase and fermenting yeast are combined, so that as sugars are produced, the fermentative organisms convert them to ethanol in the same step.

Once the hydrolysis of the cellulose is achieved, the resulting sugars must be fermented to produce ethanol. In addition to glucose, hydrolysis produces other six-carbon sugars from cellulose and five-carbon sugars from hemicellulose that are not readily fermented to ethanol by naturally occurring organisms. They can be converted to ethanol by genetically engineered yeasts that are currently available, but the ethanol yields are not sufficient to make the process economically attractive. It also remains to be seen whether the yeasts can be made hardy enough for production of ethanol on a commercial scale.

A large variety of feedstocks is currently available for producing ethanol from cellulosic biomass. The materials being considered can be categorized as agricultural waste, forest residue, and energy crops. Agricultural waste available for ethanol conversion includes crop residues such as wheat straw, corn stover (leaves, stalks, and cobs), rice straw, and bagasse (sugar cane waste). Forestry waste includes underutilized wood and logging residues; rough, rotten, and salvable dead wood; and excess saplings and small trees. Energy crops, developed and grown specifically for fuel, include fast-growing trees, shrubs, and grasses such as hybrid poplars, willows, and switchgrass.

Although the choice of feedstock for ethanol conversion is largely a cost issue, feedstock selection has also focused on environmental issues. Materials normally targeted for disposal include forest thinnings collected as part of an effort to improve forest health, and certain agricultural residues, such as rice straw. Although forest residues are not large in volume, they represent an opportunity to decrease the fire hazard associated with the dead wood present in many forests. Small quantities of forest thinnings can be collected at relatively low cost, but collection costs rise rapidly as quantities increase.

Agricultural residues, in particular corn stover, represent a tremendous resource base for biomass ethanol production. Agricultural residues, in the long-term, would be the sources of biomass that could support substantial growth of the ethanol industry. At conversion yields of around 60 to 100 gal/dry ton, the available corn stover inventory would be sufficient to support 7 to 12 billion gallons of ethanol production per year.

The cost of agricultural residues is not nearly as sensitive to supply as is the cost of forest residues, although the availability of corn stover could be affected by a poor crop year. The relatively low rise in cost as a function of feedstock use is due to the relatively high density of material available that does not involve competition for farmland. In addition, the feedstock is located in the corn-processing belt, an area that has an established infrastructure for collecting and transporting agricultural materials. It is also located near existing grain ethanol plants, which could be expanded to produce ethanol from stover. Initially, locally available labor and residue collection equipment might have to be supplemented with labor and equipment brought in from other locations for residue harvesting and storage operations, if the plants involved are of sufficient scale. Eventually, however, when the local collection infrastructure has been built up, costs would come down.

Dedicated energy crops such as switchgrass, hybrid willow, and hybrid poplar are another long-term feedstock option. Switchgrass is grown on a 10-year crop rotation basis, and harvest can begin in the first year in some locations and the second year in others. Willows require a 22-year rotation, with the first harvest in the fourth year and subsequent harvests every 3 years thereafter. Hybrid poplar requires 6 years to reach harvest age in the Pacific Northwest, 8 years in the Southeast, Southern Plains, and South Central regions, and 10 years in the Corn Belt, Lake States, Northeast, and Northern Plains regions. Thus, if it were planted in the spring of 2000, switchgrass could be harvested in 2000 or 2001, willow could be harvested in 2004, and poplars could be harvested in 2006, 2008, or 2010, depending on the region.

The use of cellulosic biomass in the production of ethanol also has environmental benefits. Converting cellulose to ethanol increases the net energy balance of ethanol compared to converting corn to ethanol. The net energy balance is calculated by subtracting the energy required to produce a gallon of ethanol from the energy contained in a gallon of ethanol (approximately 76,000 Btu). Corn-based ethanol has a net energy balance of 20,000 to 25,000 Btu/gal, whereas cellulosic ethanol has a net energy balance of more than 60,000 Btu/gal. In addition, cellulosic ethanol use can reduce greenhouse gas emissions. Cellulosic ethanol can produce an 8 to 10 percent reduction in greenhouse gas emissions when used in E10 and a 68 to 91 percent reduction when used in E85.

9.4 OTHER ALCOHOLS

9.4.1 Methanol

Methanol is a colorless, odorless, and nearly tasteless alcohol and is also produced from crops and is also used as a fuel. Methanol, like ethanol, burns more completely but releases as much or more carbon dioxide than its gasoline counterpart. The balance is often seen as the various biprocesses that draw carbon dioxide from the atmosphere so there is no net modern release, as there is for fossil fuels.

Methanol and other chemicals were routinely extracted from wood in the nineteenth and early twentieth centuries. However, the original route for methanol recovery from biomass was quite different to current routes. Methanol was originally recovered from wood as a by-product of charcoal manufacture, and was often called "wood alcohol." Pyrolysis (heating wood in the absence of air) to above 270°C in a retort causes thermal cracking or breakdown of the wood and allows much of the wood to be recovered as charcoal. The watery condensate leaving the retort contained methanol, amongst other compounds.

In 1923, commercial production of methanol from synthesis gas by a catalytic process was commenced. Now almost all of the methanol used worldwide comes from the processing of natural gas.

In general, methanol production from natural gas feed consists of three steps: (a) synthesis gas (syngas) generation—in the case of natural gas feed, syngas production consists of converting methane (CH_4) into carbon monoxide (CO) and hydrogen (H_2) via steam reforming; (b) synthesis gas upgrading—primarily removal of CO_2, plus any contaminants such as sulfur; and (c) methanol synthesis and purification—reacting the carbon monoxide, hydrogen, and steam over a catalyst in the presence of a small amount of CO_2 and at elevated temperature and pressure. The methanol synthesis is an equilibrium reaction and excess reactants must be recycled to optimize yields.

Modern methods proposed for the production of methanol from biomass involve the conversion of the biomass to a suitable synthesis gas, after which processing steps are very similar to those developed for methanol production from natural gas. However, the gasification techniques proposed are still at a relatively early stage of development using biomass feed and the methods are based on similar techniques used widely already with natural gas as feed.

Before biomass can be gasified it must be pretreated to meet the processing constraints of the gasifier. This typically involves size reduction, and drying to keep moisture contents below specific levels. Thereafter, biomass gasification involves heating biomass in the presence of low levels of oxygen (i.e., less than required for complete combustion to carbon dioxide and water). Above certain temperatures the biomass will break down into a gas stream and a solid residue. The composition of the gas stream is influenced by the operating conditions for the gasifier, with some gasification processes more suited than others to producing a gas for methanol production. In particular, simple gasification with air creates a synthesis gas stream that is diluted with large quantities of nitrogen. This nitrogen is detrimental to subsequent processing to methanol and so techniques using indirect gasification or an oxygen feed are preferred. For large-scale gasification, pressurized systems are considered to be more economic than atmospheric systems.

Once the economic optimum synthesis gas is available the methanol synthesis takes place. This typically uses a copper-zinc catalyst at temperatures of 200 to 280°C and pressures of 50 to 100 atm.

The crude methanol from the synthesis loop contains water produced during synthesis as well as other minor by-products. Purification is achieved in multistage distillation, with the complexity of distillation dictated by the final methanol purity required.

9.4.2 Propanol and Butanol

Propanol and butanol are considerably less toxic and less volatile than methanol. In particular, butanol has a high flash point of 35°C (95°F), which is a benefit for fire safety, but may be a difficulty for starting engines in cold weather.

The fermentation processes to produce propanol and butanol from cellulose are fairly tricky to execute, and the *Clostridium acetobutylicum* currently used to perform these conversions produces an extremely unpleasant smell, and this must be taken into consideration when designing and locating a fermentation plant. This organism also dies when the butanol content of whatever it is fermenting rises to 7 percent. For comparison, yeast dies when the ethanol content of its feedstock hits 14 percent. Specialized strains can tolerate even greater ethanol concentrations—so-called turbo yeast can withstand up to 16 percent ethanol. However, if ordinary *Saccharomyces* yeast can be modified to improve its ethanol resistance, scientists may yet one day produce a strain of the Weizmann organism with a butanol resistance higher than the natural boundary of 7 percent. This would be useful because butanol has a higher energy density than ethanol, and because waste fiber left over from sugar crops used to make ethanol could be made into butanol, raising the alcohol yield of fuel crops without there being a need for more crops to be planted.

9.5 BIODIESEL

Biodiesel is a diesel-equivalent fuel derived from biologic sources (such as, vegetable oils) which can be used in unmodified diesel-engine vehicles. It is thus distinguished from the straight vegetable oils or waste vegetable oils used as fuels in some diesel vehicles. In the current context, biodiesel refers to alkyl esters made from the transesterification of vegetable oils or animal fats.

Biodiesel fuel is made from the oil of certain oilseed crops such as soybean, canola, palm kernel, coconut, sunflower, safflower, corn and hundreds of other oil producing crops. The oil is extracted by the use of a press and then mixed in specific proportions with other agents which cause a chemical reaction. The results of this reaction are two products, biodiesel and soap. After a final filtration, the biodiesel is ready for use. After curing, the glycerin soap which is produced as a by-product can be used as is, or can have scented oils added before use.

Biodiesel is made through a chemical process (*transesterification*) whereby the glycerin is separated from the fat or vegetable oil. The process leaves behind two products: (a) methyl esters (the chemical name for biodiesel) and (b) glycerin (a valuable by-product usually sold to be used in soaps and other products).

Biodiesel (fatty acid methyl esters; FAME) is a notable alternative to the widely used petroleum derived diesel fuel since it can be generated by domestic natural sources such as soybeans, rapeseeds, coconuts, and even recycled cooking oil, and thus reduces dependence on diminishing petroleum fuel from foreign sources. In addition, because biodiesel is largely made from vegetable oils, it reduces life cycle greenhouse gas emissions by as much as 78 percent (Ban-Weiss et al., 2007).

Vegetable oils and animal fats belong to an extensive family of chemicals called lipids. Lipids are bioproducts from the metabolism of living creatures. As a result, they can be found widely distributed in nature. Their biofunctions are diverse, but they are most known for their energy storage capacity. Most lipids can easily dissolve in common organic solvents, meaning that they are hydrophobic. If a lipid is a solid at 25°C, it is classified as a fat; otherwise, it is oil. Typically, fats are produced by animals and oils by plants, but both are mainly made of triglyceride (TG) molecules, which are triesters of glycerol (a triol) and free fatty acids (long alkyl chain carboxylic acids). Other glyceride species, such as diglycerides and monoglycerides, are obtained from triglycerides by the substitution of one and two fatty acid moieties, respectively, with hydroxyl groups (Lotero et al., 2006).

Biodiesel production is a very modern and technological area for researchers due to the relevance that it is winning everyday because of the increase in the petroleum price and the environmental advantages (Marchetti et al., 2005). The successful introduction and commercialization of biodiesel in many countries around the world has been accompanied by the development of standards to ensure high product quality and user confidence (Knothe, 2005). In general, biodiesel compares well to petroleum-based diesel (Lotero et al., 2006). Pure biodiesel fuel (100 percent esters of fatty acids) is called B100. When blended with diesel fuel the designation indicates the amount of B100 in the blend, for example, B20 is 20 percent B100 and 80 percent diesel, and B5 used in Europe contains 5 percent B100 in diesel (Pinto et al., 2005).

Biodiesel is biodegradable and nontoxic, and typically produces about 60 percent less net carbon dioxide emissions than petroleum-based diesel, as it is itself produced from atmospheric carbon dioxide via photosynthesis in plants. This figure can actually differ widely between fuels depending upon production and processing methods employed in their creation. Pure biodiesel is available at many gas stations in Germany.

9.5.1 Feedstocks for Biodiesel

Soybeans are widely used as a source of biodiesel. However, a variety of oils can be used to produce biodiesel. These include:

1. Virgin oil feedstock; rapeseed, and soybean oils are most commonly used, soybean oil alone accounting for about 90 percent of all fuel stocks. It also can be obtained from field pennycress and *Jatropha* other crops such as mustard, flax, sunflower, canola, palm oil, hemp, and even algae show promise.
2. Waste vegetable oil (WVO).
3. Animal fats including tallow, lard, yellow grease, chicken fat, and the by-products of the production of omega-3 fatty acids from fish oil.

Worldwide production of vegetable oil and animal fat is not yet sufficient to replace liquid fossil fuel use. Furthermore, there are objections to the vast amount of farming and the resulting over-fertilization, pesticide use, and land use conversion that they say would be needed to produce the additional vegetable oil. Many advocates suggest that waste vegetable oil is the best source of oil to produce biodiesel. However, the available supply is drastically less than the amount of petroleum-based fuel that is burned for transportation and domestic heating in the world.

9.5.2 Transesterification (Alcoholysis)

Transesterification is the conversion of triacylglycerol lipids by alcohols to alkyl esters without first isolating the free fatty acids (May, 2004).

Transesterification of a vegetable oil was conducted as early as 1853 many years before the first diesel engine became functional. Rudolf Diesel's prime model, a single 10-ft(3-m)-iron cylinder with a flywheel at its base, ran on its own power for the first time in Augsburg, Germany, on August 10, 1893. Diesel later demonstrated his engine and received the *Grand Prix* (highest prize) at the World Fair in Paris, France in 1900. This engine stood as an example of Diesel's vision because it was powered by peanut oil, which is a biofuel, though not *biodiesel*, since it was not transesterified. He believed that the utilization of biomass fuel was the real future of his engine and that the use of vegetable oils for engine fuels would become as important as petroleum and the coal-tar products of the present time.

The purpose of transesterification of vegetable oils to their methyl esters (biodiesel) process is to lower the viscosity of the oil. The transesterification reaction is affected by alcohol type, molar ratio of glycerides to alcohol, type and amount of catalyst, reaction temperature, reaction time, and free fatty acids and water content of vegetable oils or animal fats. The transesterification reaction proceeds with or without a catalyst by using primary or secondary monohydric aliphatic alcohols having one- to eight-carbon atoms as follows (Demirbas, 2006a; 2006b):

$$\text{Triglycerides} + \text{Monohydric alcohol} \leftrightarrow \text{Glycerin} + \text{Mono-alkyl esters}$$

Generally, the reaction temperature near the boiling point of the alcohol is recommended (Çanakçi and Özsezen, 2005). The reactions take place at low temperatures (approximately 65°C) and at modest pressures (2 atm, 1 atm = 14.7 psi = 101.325 kPa). Biodiesel is further purified by washing and evaporation to remove any remaining methanol. The oil (87 percent), alcohol (9 percent), and catalyst (1 percent) are the inputs in the production of biodiesel (86 percent), the main output (Lucia et al., 2006). Pretreatment is not required if the reaction is carried out under high pressure (9000 kPa) and high temperature (240°C), where simultaneous esterification and transesterification take place with maximum yield obtained at temperatures ranging from 60 to 80°C at a molar ratio of 6:1 (Barnwal and Sharma, 2005). The alcohols employed in the transesterification are generally short chain alcohols such as methanol, ethanol, propanol, and butanol. It was reported that when transesterification of soybean oil using methanol, ethanol, and butanol was performed, 96 to 98 percent of ester could be obtained after 1 hour (Dmytryshyn et al., 2004).

9.5.3 Catalytic Transesterification

Transesterification reactions can be catalyzed by *alkalis* (Fig. 9.3) (Korytkowska et al., 2001; Dmytryshyn et al., 2004; Stavarache et al., 2005; Varghaa and Truterb, 2005; Meher et al., 2006a), *acids* (Lee et al., 2000; Goff et al., 2004; Lopez et al., 2005; Liu et al., 2006), or *enzymes* (Watanabe et al., 2000, 2002; Ghanem, 2003; Reyes-Duarte et al., 2005; Royon et al., 2007; Shah and Gupta, 2007; Bernardes et al., 2007). The catalytic transesterification

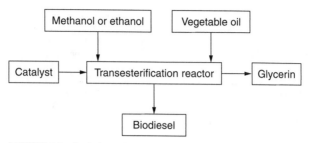

FIGURE 9.3 Catalytic production of biodiesel.

of vegetable oils with methanol is an important industrial method used in biodiesel synthesis. Also known as methanolysis, this reaction is well studied and established using acids or alkalis, such as sulfuric acid or sodium hydroxide as catalysts. However, these catalytic systems are less active or completely inactive for long chain alcohols. Usually, industries use sodium or potassium hydroxide or sodium or potassium methoxide as catalyst, since they are relatively cheap and quite active for this reaction (Macedo et al., 2006). Enzyme-catalyzed procedures, using lipase as catalyst, do not produce side reactions, but the lipases are very expensive for industrial scale production and a three-step process was required to achieve a 95 percent conversion. The acid-catalyzed process is useful when a high amount of free acids are present in the vegetable oil, but the reaction time is very long (48–96 hours), even at the boiling point of the alcohol, and a high molar ratio of alcohol was needed (20:1 wt/wt to the oil) (Stavarache et al., 2005).

The transesterification process is catalyzed by alkaline metal alkoxides, and hydroxides, as well as sodium or potassium carbonates. Alkali-catalyzed transesterification with short-chain alcohols, for example, generates high yields of methyl esters in short reaction times (Jeong and Park, 1996). The alkaline catalysts show high performance for obtaining vegetable oils with high quality, but a question often arises, that is, the oils contain significant amounts of free fatty acids which cannot be converted into biodiesels but to a lot of soap (Furuta et al., 2004). These free fatty acids react with the alkaline catalyst to produce soaps that inhibit the separation of the biodiesel, glycerin, and wash water (Çanakci and van Gerpen, 2003). Triglycerides are readily transesterified in a batch operation in the presence of alkaline catalyst at atmospheric pressure and at a temperature of approximately 60 to 70°C with an excess of methanol (Srivastava and Prasad, 2000). It often takes at least several hours to ensure the alkali (NaOH or KOH) catalytic transesterification reaction is complete. Moreover, removal of these catalysts is technically difficult and brings extra cost to the final product (Demirbas, 2002, 2003). Nevertheless, they are a good alternative since they can give the same high conversions of vegetable oils just by increasing the catalyst concentration to 1 or 2 mole percent. Alkaline metal alkoxides (as CH_3ONa for the methanolysis) are the most active catalysts, since they give very high yields (>98 percent) in short reaction times (30 minutes) even if they are applied at low molar concentrations (0.5 mole percent) (Schuchardta et al., 1998).

The transesterification process is catalyzed by sulfuric, hydrochloric, and organic sulfonic acids. In general, acid-catalyzed reactions are performed at high alcohol-to-oil molar ratios, low-to-moderate temperatures and pressures, and high acid-catalyst concentrations (Lotero et al., 2006). These catalysts give very high yields in alkyl esters but these reactions are slow, requiring typically temperature above 100°C and more than 3 hours to complete the conversion (Meher et al., 2006b). Studies of the acid-catalyzed system have been very limited in number. No commercial biodiesel plants to date have been reported to use the acid-catalyzed process. Despite its relatively slow reaction rate, the acid-catalyzed process offers benefits with respect to its independence from free fatty acid content and the consequent absence of a pretreatment step. These advantages favor the use of the acid-catalyzed process when using waste cooking oil as the raw material (Zhang et al., 2003).

Enzyme (such as lipase)-catalyzed reactions have advantages over traditional chemical-catalyzed reactions: the generation of no by-products, easy product recovery, mild reaction conditions, and catalyst recycling. Also, enzymatic reactions are insensitive to free fatty acids and water content in waste cooking oil (Kulkarni and Dalai, 2006). As for the enzyme-catalyzed system, it requires a much longer reaction time than the other two systems (Zhang et al., 2003). The enzyme reactions are highly specific and chemically clean. Because the alcohol can be inhibitory to the enzyme, a typical strategy is to feed the alcohol into the reactor in three steps of 1:1 mole ratio each. The reactions are very slow, with a three step sequence requiring from 4 to 40 hours, or more. The reaction conditions are modest, from 35 to 45°C (van Gerpen et al., 2004). The main problem of the enzyme-catalyzed process is the high cost of the lipases used as catalyst (Royon et al., 2007).

Synthesis of biodiesel using enzymes such as *Candida antarctica*, *Candida rugasa*, *Pseudomonas cepacia*, *immobilized lipase (Lipozyme RMIM)*, *Pseudomonas spp.*, and *Rhizomucor miehei* is well reported in the literature. In the previously mentioned work of Shah and Gupta (2007), the best yield 98 percent (w/w) was obtained by using *P. cepacia* lipase immobilized on celite at 50 C in the presence of 4 to 5 percent (w/w) water in 8 hours.

9.5.4 Noncatalytic Supercritical Methanol Transesterification

The transesterfication of triglycerides by supercritical methanol (SCM), ethanol, propanol, and butanol has proved to be the most promising process. Recently, a catalysts-free method was developed for biodiesel production by employing supercritical methanol (Fig. 9.4) (Saka and Kusdiana, 2001). The supercritical treatment at 350°C, 43 MPa, and 240 seconds with a molar ratio of 42 in methanol is the optimum condition for transesterification of rapeseed oil to biodiesel fuel (Kusdiana and Saka, 2004).

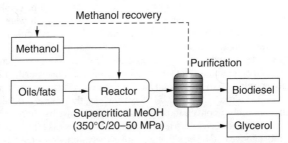

FIGURE 9.4 Production of biodiesel by the one-stage transesterification of oils/fats using supercritical methanol.

To achieve more moderate reaction conditions, further effort was made through the two-step preparation (Fig. 9.5). In this method, oils/fats are, first, treated in subcritical water for hydrolysis reaction to produce fatty acids. After hydrolysis, the reaction mixture is separated into oil phase and water phase by decantation. The oil phase (upper portion) is mainly fatty acids, while the water phase (lower portion) contains glycerol in water. The separated oil phase is then mixed with methanol and treated at supercritical condition to produce biodiesel thorough methyl esterification. After removing unreacted methanol and water produced in reaction, biodiesel can be obtained as biodiesel. Therefore, in this

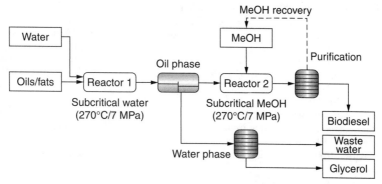

FIGURE 9.5 Production of biodiesel by the two-stage transesterification of oils/fats using supercritical methanol.

process, methyl esterification is the main reaction for biodiesel formation, while in the one-step method, transesterification is the major one (Saka and Minami, 2006).

Reaction by supercritical methanol has some advantages: (a) the glycerides and free fatty acids are reacted with equivalent rates, (b) the homogeneous phase eliminates diffusive problems, (c) the process tolerates great percentages of water in the feedstock catalytic process require the periodical removal of water in the feedstock or in intermediate stage to prevent catalyst deactivation, (d) the catalyst removal step is eliminated, and (e) if high methanol-to-oil ratios are used, total conversion of the oil can be achieved in a few minutes (Vera et al., 2005). Some disadvantages of the one-stage supercritical method are clear: (a) it operates at very high pressures (25–40 MPa), (b) the high temperatures bring along proportionally high heating and cooling costs, (c) high methanol-to-oil ratios (usually set at 42) involve high costs for the evaporation of the unreacted methanol, and (d) the process as posed to date does not explain how to reduce free glycerol to less than 0.02 percent as established in the ASTM D6584 or other equivalent international standards.

9.5.5 Effect of Reaction Parameters on Conversion Yield

The main factors affecting transesterification are the molar ratio of glycerides to alcohol, catalyst, reaction temperature and pressure, reaction time, and the contents of free fatty acids and water in oils.

Effect of Free Fatty Acids and Moisture. The free fatty acids and moisture content are key parameters for determining the viability of the vegetable oil transesterification process (Meher et al., 2006b). In the transesterification, free fatty acids and water always produce negative effects, since the presence of free fatty acids and water causes soap formation, consumes catalyst, and reduces catalyst effectiveness, all of which result in a low conversion (Demirbas and Karslıoglu, 2007). These free fatty acids react with the alkaline catalyst to produce soaps that inhibit the separation of the biodiesel, glycerin, and wash water (Çanakçi and van Gerpen, 2003). To carry the base catalyzed reaction to completion; a free fatty acid value lower than 3 percent is needed (Meher et al., 2006b).

The presence of water has a greater negative effect on transesterification than that of the free fatty acids. In the transesterification of beef tallow catalyzed by sodium hydroxide (NaOH) in presence of free fatty acids and water, the water and free fatty acid contents must be maintained at specified levels (Ma et al., 1998; Kusdiana and Saka, 2004a; Bala, 2005).

The Effect of Reaction Temperature and Time. Transesterification can occur in different temperatures depending on the type of oil employed (Ma and Hanna, 1999). A few works reported the reaction at room temperature (Marinetti, 1962, 1966; Graboski and McCormick, 1998; Encinar et al., 2002).

The effect of reaction temperature on production of propyl oleate was examined at the temperature range from 40 to 70°C with free *P. fluorescens* lipase (Iso et al., 2001). The conversion ratio to propyl oleate was observed highest at 60°C, whereas the activity highly decreased at 70°C.

The conversion rate increases with reaction time. The transesterification of rice bran oil with methanol was studied at molar ratios of 4:1, 5:1, and 6:1 (Gupta et al., 2007). At molar ratios of 4:1 and 5:1, there was significant increase in yield when the reaction time was increased from 4 to 6 hours. Among the three molar ratios studied, ratio 6:1 gave the best results.

The Effect of Molar Ratio and Alcohol Type. One of the most important factors that affect the yield of ester is the molar ratio of alcohol to triglyceride. Although the stoichiometric molar ratio of methanol to triglyceride for transesterification is 3:1, higher molar ratios are used to enhance the solubility and to increase the contact between the triglyceride and alcohol molecules (Noureddini et al., 1998). In addition, investigation of the effect of molar ratio on the transesterification of sunflower oil with methanol showed that when the molar ratio varied from 6:1 to 1:1 and concluded that 98 percent conversion to ester was obtained at a molar ratio of 6:1 (Freedman et al., 1986).

Another important variable affecting the yield of methyl ester is the type of alcohol to triglyceride. In general, short-chain alcohols such as methanol, ethanol, propanol, and butanol can be used in the transesterification reaction to obtain high methyl ester yields. Çanakci and van Gerpen (1999) investigated the effect of different alcohol types on acid-catalyzed transesterification of pure soybean oil. They obtained yields from 87.8 to 95.8 percent after 48 and 96 hours of reaction.

The Effect of Catalyst. Catalysts used for the transesterification of triglycerides are classified as alkali, acid, and enzyme. Alkali-catalyzed transesterification is much faster than acid-catalyzed transesterification and is most often used commercially (Ma and Hanna, 1999) and, quite often, for the base-catalyzed transesterification the best yields were obtained when the catalyst was used in small concentration, that is, 0.5 percent wt/wt of oil (Stavarache et al., 2005). On the other hand, data show that during the production of free and bound ethyl ester (FAEE) from castor oil, hydrochloric acid is much more effective than sodium hydroxide at higher reaction temperatures (Meneghetti et al., 2006).

9.5.6 Properties

Biodiesel is a liquid which varies in color between golden and dark brown depending on the feedstock from which it is produced. It is practically immiscible with water, has a high boiling point and low vapor pressure. Typical methyl ester biodiesel has a flash point of approximately 150°C (302°F), making it rather nonflammable. Biodiesel has a density of approximately 0.88 g/cm^3, less than that of water. Biodiesel uncontaminated with starting material can be regarded as nontoxic but it is recommended that no one drink any!

Biodiesel has a viscosity similar to diesel produced from petroleum (*petrodiesel*). It can be used as an additive in formulations of diesel to increase the lubricity of pure ultra-low sulfur diesel (ULSD) fuel, which is advantageous because it has virtually no sulfur content. Much of the world uses a system known as the "B" factor to state the amount of biodiesel in any fuel mix, in contrast to the "BA" or "E" system used for ethanol mixes. For example, fuel containing 20 percent biodiesel is labeled B20. Pure biodiesel is referred to as B100. Blends of 20 percent biodiesel with 80 percent petroleum diesel (B20) can generally be used in unmodified diesel engines. Biodiesel can also be used in its pure form (B100), but may require certain engine modifications to avoid maintenance and performance problems. Biodiesel has about 5 to 8 percent less energy density, but better lubricity and more complete combustion can make the energy output of a diesel engine only 2 percent less per volume when compared to petrodiesel—or about 35 MJ/L.

9.5.7 Technical Standards

The common international standard for biodiesel is EN 14214 while ASTM D6751 is the most common standard referenced in the United States and Canada. In Germany, the requirements for biodiesel are fixed in the DIN EN 14214 standard and in the UK the requirements for biodiesel is fixed in the BS EN 14214 standard, although these last two standards are essentially the same as EN 14214 and are just prefixed with the respective national standards institution codes.

There are standards for three different varieties of biodiesel, which are made of different oils: (a) RME, rapeseed methyl ester, DIN E51606, (b) PME, vegetable methyl ester, purely vegetable products, DIN E51606), and (c) FME, fat methyl ester, vegetable and animal products, according to DIN V51606. The standards ensure that the following important factors in the fuel production process are satisfied: (a) complete reaction, (b) removal of glycerin, (c) removal of catalyst, (d) removal of alcohol, (e) absence of free fatty acids, and (f) low sulfur content.

Basic industrial tests to determine whether the products conform to the standards typically include gas chromatography, a test that verifies only the more important of the variables above. Tests that are more complete are more expensive. Fuel meeting the quality standards is very nontoxic, with a toxicity rating (LD_{50}) of greater than 50 mL/kg.

9.5.8 Uses

Biodiesel can be used in pure form (B100) or may be blended with petroleum diesel at any concentration in most modern diesel engines. Biodiesel will degrade natural rubber gaskets and hoses in vehicles (mostly found in vehicles manufactured before 1992), although these tend to wear out naturally and most likely will have already been replaced to gaskets that are nonreactive to biodiesel. The higher lubricity index of biodiesel compared to petrodiesel is an advantage and can contribute to longer fuel injector life. However, biodiesel is a better solvent than petrodiesel, and has been known to break down deposits of residue in the fuel lines of vehicles that have previously been run on petrodiesel. As a result, fuel filters and injectors may become clogged with particulates if a quick transition to pure biodiesel is made, as biodiesel cleans the engine in the process.

Pure, nonblended biodiesel can be poured straight into the tank of any diesel vehicle. As with normal diesel, low-temperature biodiesel is sold during winter months to prevent viscosity problems. Some older diesel engines still have natural rubber parts which will be affected by biodiesel.

The temperature at which pure (B100) biodiesel starts to gel varies significantly and depends upon the mix of esters and therefore the feedstock oil used to produce the biodiesel. For example, biodiesel produced from low erucic acid varieties of canola seed (RME) starts to gel at approximately −10°C. Biodiesel produced from tallow tends to gel at around +16°C. As of 2006, there are a very limited number of products that will significantly lower the gel point of straight biodiesel. Winter operation is possible with biodiesel blended with other fuel oils including #2 low sulfur diesel fuel and #1 diesel/kerosene but the exact blend depends on the operating environment.

Biodiesel may contain small but problematic quantities of water. Although it is hydrophobic (nonmiscible with water molecules), there are indications that biodiesel, it is said to be, at the same time, is hygroscopic to the point of attracting water molecules from atmospheric moisture. In addition, there may be water that is residual to processing or resulting from storage tank condensation. The presence of water is a problem because: (a) water reduces the heat of combustion of the bulk fuel which means more smoke, harder starting, less power, (b) water causes corrosion of vital fuel system components: fuel pumps, injector pumps, and fuel lines, (c) water freezes to form ice crystals at 0°C (32°F) and the crystals provide sites for nucleation and accelerate the gelling of the residual fuel, and (d) water accelerates the growth of microbe colonies, which can plug up a fuel system so biodiesel users who have heated fuel tanks therefore face a year-round microbe problem.

Chemically, transesterified biodiesel comprises a mix of mono-alkyl esters of long-chain fatty acids. The most common form uses methanol to produce methyl esters as it is the cheapest alcohol available, though ethanol can be used to produce an ethyl ester biodiesel and higher alcohols such as isopropanol and butanol have also been used. Using alcohols of higher molecular weights improves the cold flow properties of the resulting ester, at the cost of a less efficient transesterification reaction. A lipid transesterification production process is used to convert the base oil to the desired esters. Any free fatty acids in the base oil are either converted to soap and removed from the process, or they are esterified (yielding more biodiesel) using an acidic catalyst. After this processing, unlike straight vegetable oil, biodiesel has combustion properties very similar to those of petroleum diesel, and can replace it in most current uses.

A by-product of the transesterification process is the production of glycerol. For every unit of biodiesel that is manufactured, 0.1 unit of glycerol is produced. Originally, there was a valuable market for the glycerol, which assisted the economics of the process as a whole. However, with the increase in global biodiesel production, the market price for this crude glycerol (containing 20 percent water and catalyst residues) is lower and affords an operational challenge. Usually the crude glycerol has to be purified, typically by performing vacuum distillation after which the refined glycerol (of more than 98 percent purity) can then be utilized directly, or converted into other products.

The extra lubrication provided by biodiesel fuel helps improve the longevity of your engine, as well as boosting engine performance, also helping eliminate engine knocks and noise. In addition, biodiesel fuel can be stored in any type of tank and has a much higher flash point (approximately 300°C) compared to petrodiesel approximately (150°C).

9.6 HYDROCARBONS

Biodiesel fuel can be made from renewable resources, such as, vegetable oils and animal fats. For example, there are certain species of flowering plants belonging to different families which convert a substantial amount of photosynthetic products into latex. The latex of such plants contains liquid hydrocarbons of high molecular weight (10,000). These hydrocarbons can be converted into high grade transportation fuel (i.e., petroleum). Therefore, hydrocarbon producing plants are called petroleum plants or petroplants and their crop as petrocrop. Natural gas is also one of the products obtained from hydrocarbons. Thus, petroleum plants can be an alternative source for obtaining petroleum to be used in diesel engines. Normally, some of the latex-producing plants of families Euphorbiaceae, Apocynaceae, Asclepiadaceae, Sapotaceae, Moraceae, Dipterocarpaceae, and others are petroplants. Similarly, sunflower (family Composiae), *Hardwickia pinnata* (family Leguminosae) are also petroplants. Some algae also produce hydrocarbons.

Different species of *Euphorbia* of family Euphorbiaceae serve as the petroplants that have the potential to be a renewable substitute for the conventional petroleum sources. Latex of *Euphorbia lathyrus* contains fairly high percentage of terpenoids. These can be converted into high grade transportation fuel. Similarly the carbohydrates (hexoses) from such plants can be used for ethanol formation.

Sugarcane (*Saccharum officinarum*, family Gramineae) is the main source of raw material for sugar industry. The wastes from sugar industry include bagasse, molasses, and press mud. After extracting the cane juice for sugar production, the cellulosic fibrous residue that remains is called bagasse. It is used as the raw material (biomass) and processed variously for the production of fuel, alcohols, single cell protein as well as in paper mills. Molasses is an important by-product of sugar mills and contains 50 to 55 percent fermentable sugars. One ton of molasses can produce about 280 L of ethanol. Molasses is used for the production of animal feed, liquid fuel, and alcoholic beverages.

Sugar beet (*Beta vulgaris*, family Chenopodiaceae) is yet another plant which contains a high percentage of sugars stored in fleshy storage roots. It is also an important source for production of sugar as well as ethanol.

Raising crops like sugarcane, sugar beet, tapioca, potato, maize etc. purely for production of ethanol is described as energy cropping, and the crops are called energy crops. Cultivation of plants (trees) for obtaining fuel/firewood is described as energy plantation.

Other plants, such as Euphorbiaceae plants like *Euphorbia lathyris*, milkweed (*Asclepia speciosa*), and the tree legume (*Copaifera multijuga*), produce hydrocarbons which can be converted into and used as diesel, called biodiesel (Table 9.1). In addition, some freshwater and marine algae are also known to accumulate hydrocarbons. Some algae like *Chlamydomonas* and anaerobic bacteria like *Clostridium* produce hydrogen gas, which can be used as a pollution free fuel. Long-term research efforts are needed to develop these organisms as sources of bioenergy.

The euphorbeans and milkweeds can be grown in relatively dry environments on lands not suited for crop production; this makes them highly attractive sources of biofuels. The Euphorbias are relatives of plants used to produce rubber; they produce a latex, which has about 30 percent hydrocarbons emulsified in water.

However, hydrocarbons, as such, are not usually produced from crops, there being insufficient amount of the hydrocarbons present in the plant tissue to make the process economic. However, biodiesel is produced from crops thereby offering an excellent renewable fuel for diesel engines.

TABLE 9.1 Typical Plants Used as a Source of Energy

Nature of biomass	Plant species	Estimated annual production (tons)	Predominant mode of energy use
Wood	*Butea monosperma, Casuarina equisetifolia, Eucalyptus globulus, Leucaena leucocephala, Melia azadirachta, Tamarix dioica*	1.3×10^{10}	Firewood (ca. 50% of harvest)
Starch	Cereals, millets, root, and tuber crops, e.g., potato	1.9×10^{9}	Bioethanol
Sugar	Sugarcane, sugarbeet	1.2×10^{8}	Bioethanol, e.g., in Brazil
Hydrocarbons	*Euphorbia lathyris, Aslepia speciosa, Copaifera multijuga,* alage	—	Biodiesel
Wastes	Crop residues, animal/human refuge, sewage, etc.	—	Biogas

9.7 REFERENCES

Bala, B. K.: "Studies on Biodiesels from Transformation of Vegetable Oils for Diesel Engines," *Energy Edu. Sci. Technol.*, 15, 2005, pp. 1–43.

Ban-Weiss, G. A., J. Y. Chen, B. A. Buchholz, and R. W. Dibble: "A Numerical Investigation into the Anomalous Slight NO_x Increase When Burning Biodiesel; A New (Old) Theory," *Fuel Processing Technology,* 88, 2007, pp. 659–667.

Barnwal, B. K. and M. P. Sharma: "Prospects of Biodiesel Production from Vegetable Oils in India," *Renew. Sust. Energy Rev.*, 9, 2005, pp. 363–378.

Bernardes, O. L., J. V. Bevilaqua, M. C. M. R. Leal, D. M. G. Freire, and M. A. P. Langone: "Biodiesel Fuel Production by the Transesterification Reaction of Soybean Oil Using Immobilized Lipase," *Applied Biochemistry and Biotechnology,* 2007, pp. 137–140, 105–114.

BioCap Canada: "An Assessment of the Opportunities and Challenges of a Bio-Based Economy for Agriculture and Food Research in Canada," Canadian Agri-Food Research Council, Ottawa Ontario, Canada, 2004.

Boateng, A. A., D. E. Daugaard, N. M. Goldberg, and K. B. Hicks: *Ind. Eng. Chem. Res.*, 46, 2007, pp. 1891–1897.

Çanakçi, M., and J. van Gerpen: "Biodiesel Production via Acid Catalysis," *Trans. ASAE,* 42, 1999, pp. 1203–1210.

Çanakçi, M., and J. van Gerpen: "A Pilot Plant to Produce Biodiesel from High Free Fatty Acid Feedstocks," *Trans. ASAE,* 46, 2003, pp. 945–955.

Çanakçi, M. and A. N. Özsezen: "Evaluating Waste Cooking Oils as Alternative Diesel Fuel," *G. U. Journal of Science,* 18, 2005, pp. 81–91.

Demirbas, A.: "Biodiesel From Vegetable Oils via Transesterification in Supercritical Methanol," *Energy Convers. Mgmt.,* 43, 2002, pp. 2349–2356.

Demirbas, A.: Biodiesel Fuels from Vegetable Oils via Catalytic and Non-Catalytic Supercritical Alcohol Transesterifications and Other Methods: A Survey," *Energy Convers. Mgmt.,* 44, 2003, pp. 2093–2109.

Demirbas, A.: "Biodiesel from Sunflower Oil in Supercritical Methanol with Calcium Oxide," *Energy Convers. Mgmt.,* 48, 2006a, pp. 937–941.

Demirbas, A.: "Biodiesel Production via Non-Catalytic SCF Method and Biodiesel Fuel Characteristics," *Energy Convers. Mgmt.,* 47, 2006b, pp. 2271–2282.

Demirbas, A. and S. Karslıoglu: "Biodiesel Production Facilities from Vegetable Oils and Animal Fats," *Energy Sources,* part A 29, 2007, pp. 133–141.

Dmytryshyn, S. L., A.K. Dalai, S. T. Chaudhari, H. K. Mishra, and M. J. Reaney: "Synthesis and Characterization of Vegetable Oil Derived Esters: Evaluation for Their Diesel Additive Properties," *Bioresource Technology,* 92, 2004, pp. 55–64.

Encinar, J. M., J. F. Gonzalez, J. J. Rodriguez, and A. Tejedor: Biodiesel Fuels from Vegetable Oils: Transesterification of Cynara Cardunculus L. Oils with Ethanol," *Energy and Fuel,* 16, 2002, pp. 443–450.

Freedman, B., R. O. Butterfield, and E. H. Pryde: "Transesterification Kinetics of Soybean Oil," *J. Am. Oil. Chem. Soc.,* 63, 1986, pp. 1375–1380.

Furuta, S., H. Matsuhasbi, and K. Arata: "Biodiesel Fuel Production with Solid Superacid Catalysis in Fixed Bed Reactor Under Atmospheric Pressure," *Catalysis Communications,* 5, 2004, pp. 721–723.

Ghanem, A.: The Utility of Cyclodextrins in Lipase-Catalyzed Transesterification in Organic Solvents: Enhanced Reaction Rate and Enantioselectivity," *Org. Biomol. Chem.,* 1, 2003, pp. 1282–1291.

Goff, M.J., N.S. Bauer, S. Lopes, W.R. Sutterlin, and G. J. Suppes: "Acid-Catalyzed Alcoholysis of Soybean Oil," *J. Am. Oil. Chem. Soc.,* 81, 2004, pp. 415–420.

Graboski, M. S. and R. L. McCormick: "Combustion of Fat and Vegetable Oil Derived Fuels in Diesel Engines," *Prog. Energy Combust. Sci.,* 24, 1998, pp. 125–164.

Gupta, P. K., R. Kumar, P. S. Panesar, and V. K. Thapar: Parametric Studies on Bio-diesel prepared from Rice Bran Oil, Agricultural Engineering International: the CIGR Ejournal, Manuscript EE 06 007, vol. 9, Apr., 2007.

Jeong, G. T., and D. H. Park: "Batch (One- and Two-Stage) Production of Biodiesel Fuel From Rapeseed Oil," *Applied Biochemistry and Biotechnology,* 131, 1996, pp. 668–679.

Iso, M., B. Chen, M. Eguchi, T. Kudo, and S. Shrestha: "Production of Biodiesel Fuel from Triglycerides and Alcohol Using Immobilized Lipase," *Journal of Molecular Catalysis B: Enzymatic,* 16, 2001, pp. 53–58.

Knothe, G.: "Dependence of Biodiesel Fuel Properties on the Structure of Fatty Acid Alkyl Esters," *Fuel Processing Technology,* 86, 2005, pp. 1059–1070.

Korytkowska, A., I. Barszczewska-Rybarek, and M. Gibas: "Side-Reactions in the Transesterification of Oligoethylene Glycols by Methacrylates," *Designed Monomers and Polymers,* 4, 2001, pp. 27–37.

Koshi, P. T., J. Stubbendieck, H. V. Eck, and W. G. McCully: "Switchgrass: Forage Yield, Forage Quality, and Water Use Efficiency," *J. Range Mgt.,* 35, 1982, pp. 623–627.

Kulkarni, M. G. and A. K. Dalai: "Waste Cooking Oils An Economical Source for Biodiesel: A Review," *Ind. Eng. Chem. Res.,* 45, 2006, pp. 2901–2913.

Kusdiana, D. and S. Saka: "Effects of Water on Biodiesel Fuel Production by Supercritical Methanol Treatment," *Bioresource Technology,* 91, 2004, pp. 289–295.

Lee, Y., S. H. Park, I. T. Lim, K. Han, and S. Y. Lee: "Preparation of Alkyl (R)-(2)-3-Hydroxybutyrate by Acidic Alcoholysis of Poly-(R)-(2)-3-Hydroxybutyrate," *Enzyme and Microbial. Technol.,* 27, 2000, pp. 33–36.

Liu, Y., E. Lotero, and J.G. Goodwin Jr.: "Effect of Water on Sulfuric Acid Catalyzed Esterification," *Journal of Molecular Catalysis A: Chemical,* 245, 2006, pp. 132–140.

Lopez, D. E., J. G. Goodwin Jr., D. A. Bruce, and E. Lotero: "Transesterification of Triacetin with Methanol on Solid Acid and Base Catalysts," *Applied Catalysis A: General,* 295, 2005, pp. 97–105.

Lotero, E., J. G. Goodwin Jr., D. A. Bruce, K. Suwannakarn, Y. Liu, and D. E. Lopez: "The Catalysis of Biodiesel Synthesis," *Catalysis,* 19, 2006, pp. 41–83.

Lucia, L. A., D. S. Argyropoulos, L. Adamopoulos, and A. R. Gaspar: "Chemicals and Energy from Biomass," *Can. J. Chem.,* 8, 2006, pp. 960–970.

Ma, F., L. D. Clements, and M. A. Hanna: "The Effect of Catalyst, Free Fatty Acids, and Water on Transesterification of Beef Tallow," *Trans. ASAE,* 41, 1998, pp. 1261–1264.

Ma, F., and M. A. Hanna: "Biodiesel Production: A Review," *Bioresource Technology,* 70, 1999, pp. 1–15.

Macedo, C.C.S., F. R. Abreu, A. P. Tavares, M. P. Alves, L. F. Zara, J. C. Rubim, et al.: "New Heterogeneous Metal-Oxides Based Catalyst for Vegetable Oil Transesterification," *J. Braz. Chem. Soc.,* 17, 2006, pp. 1291–1296.

Marchetti, J. M., V. U. Miguel, and A. F. Errazu: "Possible Methods for Biodiesel Production," *Renew. Sust. Energy Rev.,* 11, 2005, pp. 1300–1311.

Marinetti, G. V.: "Hydrolysis of Lecithin with Sodium Methoxide," *Biochemistry,* 1, 1962, pp. 350–353.

Marinetti, G. V.: "Low Temperature Partial Alcoholysis of Triglycerides," *J. Lipid Res.,* 7, 1966, pp. 786–788.

May, C. Y.: "Transesterification of Palm Oil. Effect of Reaction Parameters," *Journal of Oil Palm Research,* 16, 2004, pp. 1–11.

McLaughlin, S. B. "Forage crops as Bioenergy Fuels: Evaluating the Status and Potential," Proceedings, VXIII International Grassland Congress, Winnipeg, Manitoba, Canada, Jun. 8–10, 1997.

Meher, L.C., M. G. Kulkarni, A. K. Dalai, and S. N. Naik: "Transesterification of Karanja (Pongamia Pinnata) Oil by Solid Basic Catalysts," *European Journal of Lipid Science and Technology,* 108, 2006a, pp. 389–397.

Meher, L. C., D. V. Sagar, and S. N. Naik: "Technical Aspects of Biodiesel Production by Transesterification— A Review," *Renewable and Sustainable Energy Reviews,* 10, 2006b, pp. 248–268.

Meneghetti, P. S. M., M. R. Meneghetti, C. R. Wolf, E. C. Silva, G. E. S. Lima, D. A. Coimbra, et al.: "Ethanolysis of Castor and Cottonseed Oil: A Systematic Study Using Classical Catalysts," *JAOCS,* 83, 2006, pp. 819–822.

Moser, L. E. and K. P. Vogel: "Switchgrass, Big Bluestem, and Indiangrass," in *Forages: An Introduction to Grassland Agriculture,* M. E. Heath, D. A. Miller, and C. J. Nelson (eds.), Iowa State University Press, Ames, Iowa, 1995, pp. 409–420.

Moss, D. N., E. G. Krenzer, and W. A. Brun: "Carbon Dioxide Compensation Points in Related Plant Species," *Science*, 164, 1969, pp. 187–188.

Nichols, N. N., B. S. Dien, R. J. Bothast, and M. A. Cotta: "The Corn Ethanol Industry," in *Alcoholic Fuels*, S. Minteer (ed.), CRC-Taylor & Francis, Boca Raton, Fla., 2006, chap. 4.

Noureddini, H., D. Harkey, and V. Medikonduru: "A Continuous Process for the Conversion of Vegetable Oils into Methyl Esters of Fatty Acids," *JAOCS*, 75, 1998, pp. 1775–1783.

Pian, C. C. P., T. A. Volk, L. P.Abrahamson, E. H. White, and J. Jarnefeld: "Biomass Gasification for Farm-Based Power Generation Applications," Wessex Institute of Technology Press, *Transactions on Ecology and the Environment*, 92, 2006, pp. 267.

Pinto, A. C., L. N. N. Guarieiro, M. J. C. Rezende, N. M. Ribeiro, E. A. Torres, W. A. Lopes, et al.: "Biodiesel: An Overview," *J. Braz. Chem. Soc.*, 16, 2005, pp. 1313–1330.

Reyes-Duarte, D., N. Lopez-Cortes, M. Ferrer, F. Plou, and A. Ballesteros: "Parameters Affecting Productivity in the Lipase-Catalysed Synthesis of Sucrose Palmitate," *Biocatalysis and Biotransformation*, 23, 2005, pp. 19–27.

Risser, P.G., E. C. Birney, H. D. Blocker, S. W. May, W. J. Parton, and J. A. Wiens: *The True Prairie Ecosystem*, US/IBP Synthesis Series 16, Hutchinson Ross Publishers, 1981.

Royon, D., M. Daz, G. Ellenrieder, and S. Locatelli: "Enzymatic Production of Biodiesel from Cotton Seed Oil Using T-Butanol as a Solvent," *Bioresource Technology*, 98, 2007, pp. 648–653.

Saka, S. and D. Kusdiana: "Biodiesel Fuel from Rapeseed Oil as Prepared in Supercritical Methanol," *Fuel*, 80, 2001, pp. 225–231.

Saka, S. and E. Minami: A Novel Non-catalytic Biodiesel Production Process by Supercritical Methanol as NEDO "High Efficiency Bioenergy Conversion Project," The 2nd Joint International Conference on "Sustainable Energy and Environment (SEE 2006)", Bangkok, Thailand, Nov. 21–23, 2006.

Schuchardta, U., R. Serchelia, and R. M. Vargas: "Transesterification of Vegetable Oils: a Review," *J. Braz. Chem. Soc.*, 9, 1998, pp. 199–210.

Shah, S., and M. N. Gupta: "Lipase Catalyzed Preparation of Biodiesel from Jatropha Oil in a Solvent Free System," *Process Biochemistry*, 42, 2007, pp. 409–414.

Shiflet, T. N. and G. M. Darby: "Forages and Soil Conservation," in *Forages: The Science of Grassland Culture*, M. E. Heath, R. F. Barnes, and D. S. Metcalf (eds.), Iowa State University Press, Ames, Iowa, 1985, pp. 21–32.

Srivastava. A. and R. Prasad: "Triglycerides-Based Diesel Fuels," *Renew. Sust. Energy Rev.*, 4, 2000, pp. 111–133.

Stavarache, C., M. Vinatoru, R. Nishimura, and Y. Maed: "Fatty Acids Methyl Esters from Vegetable Oil by Means of Ultrasonic Energy," *Ultrasonics Sonochemistry*, 12, 2005, pp. 367–372.

van Gerpen, J., B. Shanks, R. Pruszko, D. Clements, and G. Knothe: "Biodiesel Analytical Methods: August 2002–January 2004," National Renewable Energy Laboratory, NREL/SR-510-36240, Colo., July, 2004.

Varghaa, V. and P. Truterb: "Biodegradable Polymers by Reactive Blending Transesterification of Thermoplastic Starch with Poly(Vinyl Acetate) and Poly(Vinyl Acetate-Co-Butyl Acrylate)," *European Polymer Journal*, 41, 2005, pp. 715–726.

Vera, C. R., S. A. D'Ippolito, C. L. Pieck, and J. M. Parera: "Production of Biodiesel by a Two-Step Supercritical Reaction Process with Adsorption Refining," in 2d Mercosur Congress on Chemical Engineering and 4th Mercosur Congress on Process Systems Engineering (ENPROMER-2005), Rio de Janeiro, Brazil, Aug. 14–18, 2005.

Vogel, K. P., C. I. Dewald, H. J. Gorz, and F. A. Haskins: "Development of Switchgrass, Indiangrass, and Eastern Gamagrass: Current Status and Future," Range Improvement in Western North America, Proceedings. Society of Range Management Salt Lake City, Utah, Feb. 14, 1985, pp. 51–62.

Watanabe, Y., Y. Shimada, and A. Sugihara: "Continuous Production of Biodiesel Fuel from Vegetable Oil Using Immobilized Candida Antarctica Lipase," *J. Am. Oil. Chem. Soc.*, 77, 2000, pp. 355–360.

Watanabe, Y., Y. Shimada, A. Sugihara, and T. Tominaga: "Conversion of Degummed Soybean Oil to Biodiesel Fuel with Immobilized Candida Antarctica Lipase," *J. Mol. Catal. B: Enzym.*, 17, 2002, pp. 151–155.

Weaver, J. E.: *Prairie Plants and Their Environment,* University of Nebraska Press, Lincoln, Neb., 1968.

Wierzbicka, A., L. Lillieblad, J. Pagels, M. Strand, A. Gudmundsson, A. Gharibi, et al.: "Particle Emissions from District Heating Units Operating on Three Commonly Used Biofuels," *Atmospheric Environment,* 39, 2005, pp. 139–150.

Zhang, Y., M. A. Dube, D. D. McLean, and M. Kates: "Biodiesel Production from Waste Cooking oil: 1. Process Design and Technological Assessment," *Bioresource Technology,* 89, 2003, pp. 1–16.

CHAPTER 10
FUELS FROM WOOD

Biofuels are presently the only renewable source of liquid transportation fuels and offer many potential environment and economic benefits. The production of the raw biomass material and its subsequent conversion to fuels creates local jobs, provides regional economic development, and can increase farm and forestry incomes. Biofuels also offer many environment benefits including reduction of carbon dioxide emissions associated with global climate change and improved waste utilization. The chemical composition of many biofuels also leads to improved engine performance and reduces unwanted pollutants such as carbon monoxide and unburned hydrocarbons. Billions of liters of ethanol are used annually for transportation fuels, and biodiesel is gaining popularity in some regions.

Biomass (Chap. 8) is a catch-all term that also includes wood and is generally made up of woody plant residue and complex starches. The largest percentage of biomass used to create energy is wood, but other bioproducts, such as fast-growing switchgrass (Chap. 9), are being investigated as sources of energy. The three largest sources of biomass used for fuel are cellulose, hemicellulose, and lignin. Biomass processing results in the end-products, biochemical products, biofuels, and biopower, all of which can be used as fuel sources. The production of biochemical products involves converting biomass into chemicals to produce electricity; biofuels are biomass converted into liquids for transportation; and biopower is made by either burning biomass directly (as with a wood-burning stove) or converting it into a gaseous fuel to generate electric power. Currently, production of electricity from biomass constitutes 3.3 percent of the United States' energy supply.

In fact, it is confidently predicted that wood will remain a major renewable resource for man's future (Youngs, 1982).

10.1 HISTORY

Of all of the alternate energy sources (relative to fossil fuels) the use of wood predates the others and dates from prebiblical times (pre-4000 B.C.). In peace and in war, in the Old and the New World, man first turned to wood for his basic needs and later learned to use advanced science to employ wood as his most sophisticated raw material, being infinitely versatile and an easily renewable source.

The manner in which wood was used by early cultures is difficult to determine, as wood artifacts have largely disappeared. Certainly the use of wood for fire is one of the first and most significant contributions of this resource to the development of society. No doubt man built early pole structures from the small trees growing along the rivers and later he would build more solid structures from planks, turf, mud, and adobe. The Scandinavians developed the basic principles of timber framing which were probably known in Europe in the Bronze Age and framing eventually became the preeminent method of wood building in the Western World, reflecting developments in structural engineering that had been worked out with wood mostly through trial and error.

One of the first uses of wood for water transport was probably a hollowed-out log. Around 4000 B.C., the Egyptians were making ships from bundles of reeds and their earliest wooden boats copied the hull frame of the reed boats. For larger vessels, the Egyptians imported cedar from Lebanon. One reason for the northward expansion of Egypt's influence was to ensure its cedar supply. Records show that the Egyptian shipbuilder could use wood on a grand scale. Queen Hatshepsut's barge, built in 1500 B.C. to transport granite obelisks from Aswan to Thebes, had a displacement of some 7500 tons, and 30 oar-powered tugs were needed to tow it.

According to Theophrastus, a pupil of Aristotle, we know what was available for shipbuilding in Ancient Greece and the ship-building woods are silver fir, fir, and cedar. Silver fir is used for lightness; for merchant ships, fir is used because of its resistance to rot. In Syria and Phoenicia, cedar is used because of the lack of fir.

Technologic improvement in land transport was slower than that of water transport. From 7000 B.C. onward, wood sledges were used for heavy loads such as stones, and archeologists reason that the massive stones in the great monument at Stonehenge on Salisbury Plain, England, must have been moved on sledges placed on rollers, which may have inspired the discovery of the wheel. But we still have no record of when and where the wheel was invented, though surely the first axle was made of wood.

Another significant contribution of wood to the ancient world was for war devices. Examples include the catapult, which enabled a man to attack his enemy from a safe distance, the battering ram and scaling ladder, the tortoise, and the siege tower. Although the choice of materials for these purposes was quite limited, the properties of wood made it eminently suitable. High strength and low weight were highly valued characteristics of wood then, just as they are today. These siege engines were integral to the expansion of both Greek and Roman civilizations and of the science, technology, and philosophy that developed under the tutelage of the great thinkers and teachers of the times.

Ancient man was using wood to conquer his world as well as build it and explore it. Then some unknown woodman in Ancient Greece invented a primitive wooden lathe, and man found himself on the threshhold of the age of machines. When he entered that age, he would find ways to make wood work for him to unprecedented degrees. From the basic concept of the lathe and the ability to shape wood to circular symmetry developed new concepts of both materials use and machine development.

In Europe the water-and-wood phase reached a high plateau around the sixteenth century with the work of Leonardo da Vinci and his talented contemporaries. At about this time, the availability of timber diminished, particularly in the United Kingdom. The scarcity was caused by the expansion of agriculture, the increasing use of wood as a structural material and fuel, and from growing demands of the smelting furnaces. To smelt one cannon took several tons of wood. By the seventeenth century, Europeans were turning to coal for the domestic hearth, and when the secret of smelting metal with coal was discovered, coal became the unique basis for industrial technology until late in the nineteenth century.

In early nineteenth century America, a seemingly inexhaustible supply of timber existed. The technology here was geared to exploiting the use of all natural resources to make up for the scarcity in capital and labor. But the technologic advances of the nineteenth century, along with the increasing population, would have a major impact on American forests. Railroads, telegraph lines, charcoal-fueled steel mills, and other industries were consuming immense quantities of wood. The Civil War made a heavy demand, too. One gun factory alone used 28,000 walnut trees for gunstocks. During the latter half of the nineteenth century, the volume of lumber produced each year rose from 4 thousand million board feet to about 35 thousand million.

As with many other industries of this time, lumbering was a highly competitive business. Quick profits were the name of the game. This encouraged careless and extravagantly wasteful harvesting and manufacturing methods. The visible devastation that resulted encouraged a new concern for America's forests. Theories were published that purported to prove that the fall of ancient empires, radical changes of climate, and the spread of epidemics could be attributed to deforestation.

But America's wood-and-water phase reached its own plateau around 1850 and about 200 years after that phase had peaked in Europe. Our heads were turned by European technology that was now based on the coal-and-iron complex. Some of our traditional uses of wood—for fuel, pavement, sailing ships, charcoal, and iron smelting—were taken over by coal, steel, and stone. However, demand for timber was maintained as many new uses of wood, for paper, plywood, telegraph and telephone poles, railroad ties, and chemicals entered the picture. The selection from among competing materials was based partly on cost and availability and partly on properties and performance. It is also noteworthy that such a range of choices coincided with the rapid mechanization and increasing technical complexity of our society. Nevertheless, in the late nineteenth century the use of wood products had begun to level off. For the time being, most of the country stopped worrying about a timber scarcity. Coal was abundant and iron and steel could be manufactured.

Up to the latter part of the nineteenth century no appreciable systematic research on wood occurred—no research of the type we now call wood science. Wood had been used by early experimenters to make instruments and other research equipment, and early engineers had used it as a construction material and a material with which to work out engineering problems and designs. Methods for pulping wood to make paper had been worked out by the paper industry, too. Further, both cotton and wood had been used by chemists as a source of cellulose for man-made fibers. This led to work on cellulose acetate reactions with solvents that led to the ability to produce that compound as both film and fiber. These advances provided a base for the subsequent technology of nylon and established the principles by which countless numbers and kinds of linear high polymers can be synthesized.

The carriage business provided an early milestone for a new era of wood research. In 1889 the Carriage Builders Association was concerned about the scarcity of northern oak, a species long preferred for their craft. The builders wondered if southern oak, in plentiful supply, possessed the same desirable characteristics as the northern species. The Division of Forestry of the U.S. Department of Agriculture stepped in to help solve the problem. Its research confirmed that suitable material could be obtained from the south as well as the north. This incident was an important step toward comprehensive wood research as we know it today. From 1890 to 1910, small amounts of money were appropriated by the Division of Forestry to universities for wood research. Studies of the mechanical properties of wood were begun, along with wood preservation and wood drying studies.

In 1910 the Division of Forestry, in cooperation with the University of Wisconsin, established the world's first comprehensive forest products-laboratory in Madison, Wisconsin, to centralize the federally sponsored wood science efforts in the country. The birth of a full-fledged wood research laboratory could not have happened much earlier. The leaps and bounds science had taken in the nineteenth century provided the necessary foundation for such a laboratory. Each of the major branches of experimental science made such great progress then that in retrospect its earlier state seemed rudimentary. Scientists would call this century the Golden Age.

During World War II, wood research covered the whole gamut of possible wartime uses of wood but after the war the importance of timber products declined, on a relative scale, as the importance of minerals increased, due in part to abundant low-cost energy in the form of coal and then petroleum. It is worth noting, however, that metric tons of timber products produced in the United States then exceeded that of all metals and plastics combined, just as it does today. So, while timber declined in relative importance and public awareness, it remained the major product of American manufacture.

Today low cost and accessible energy can no longer be taken for granted. We are back to a point where many people, including materials scientists and engineers, are beginning to appreciate the need for renewable resources like wood. This appreciation is heightened and fed by the fact that the United States finds itself blessed with a timber inventory that is increasing each year. Unfortunately, much of this is not of the large clear sixes and high quality to which we are accustomed.

On the other hand, the past abundance of timber and the dispersion of the industry have worked against advances in technology for the efficient production, conversion, and use of wood products. Fortunately, and despite its relatively recent origin as a recognized field of study, wood science has had an appreciable effect on wood technology as well as science in general. The study of wood chemistry has contributed to our understanding of the principal components of wood—cellulose and lignin—and their reactions. Early research on hydrolysis of cellulose was prompted by fuel needs in World War I, but contributed much to our knowledge of this form of chemical reaction. Similarly, research on nitrocellulose was prompted by the needs for explosives. Accompanying studies of saccharification and fermentation are contributing much to our scientific knowledge in those areas. Engineering studies of wood as an orthotropic material contributed strongly to the concept of sandwich construction, now commonly used in aircraft design, as well as to the early development of glass-fiber-reinforced plastics in the 1950s and 1960s.

Another research focus is on use of wood for fuel, which still plays a big part in man's existence. Today about half of the world's annual wood harvest is burned for those same products primitive man valued from his wood fire—heat and light. But much of this is in the less developed countries. In most developed countries, use of wood for fuel peaked in the last century. But with the energy situation as it is today, even developed countries are turning to wood for fuel. It is renewable, relatively cheap, low in ash content, and negligible in sulfur content.

On the other hand, wood is bulky, has less than half the heat of combustion of fuel oil, and in its green state is heavy to ship. Furthermore the cost of a wood-burning system may be three to four times that of a gas-burning installation because of fuel storage, handling, and air quality control systems. These drawbacks have kindled interest in production of liquid and gaseous fuels from wood. Much research is devoted to improving existing technology and devising new approaches, but such fuels are still expensive compared with petroleum-based fuels.

Finally, closely related to the conversion of wood to liquid or gaseous fuel is the use of the chemical storehouse that is wood to produce a wide range of *silvichemicals*. Research has shown how to produce useful products from cellulosic polymers, wood and bark extractives, oleoresins, and pulping liquors. Many processes of these types already form the basis of chemical production on a commercial scale. But the potential to use wood as a chemical feedstock is much greater than has so far been realized. Whole wood can be gasified, liquefied, or pyrolized in ways comparable with those used for coal to yield a wide variety of chemicals. Cellulose, as a glucose polymer, can be hydrolyzed to the glucose monomer by acid or enzymes, and the glucose then fermented to ethanol. The ethanol can be used as a fuel or as a source of other important chemicals such as ethylene or butadiene.

As an alternative, use of glucose as substrate for fermentation would make possible production of antibiotics, vitamins, and enzymes. Hemicelluloses can easily be converted to simple sugars which can be used to produce ethanol or furfural, a potential raw material for nylon or other synthetics.

Lignin can be pyrolyzed, hydrogenated, and hydrolyzed to yield phenols, which can be further processed to benzene. Once the technology and economics are feasible, future plants will manufacture a variety of these very significant chemicals from wood, now derived from petroleum or other resources.

10.2 COMPOSITION AND PROPERTIES

The cellulose content of wood varies between species in the range of 40 to 50 percent. Some lignocellulosic materials can have more cellulose than wood. Cellulose is an organic polymer, consisting solely of units of glucose held together in a straight chain macromolecule. These glucose units are bound together by β-(1,4)-glycosidic linkages which establish as the repeat unit for cellulose chains; cellulose must be hydrolyzed to glucose before fermentation to ethanol.

Lignocellulose is a complex matrix combining cellulose, hemicellulose, and lignin, along with a variable level of extractives. Cellulose is comprised of glucose, a six-carbon sugar, while hemicellulose contains both five- and six-carbon sugars, including glucose, galactose, mannose, arabinose, and xylose. The presence of cellulose and hemicellulose therefore makes lignocellulose a potential candidate for bioconversion. The ability of the bioconversion platform to isolate these components was initially limited, as the wood matrix is naturally resistant to decomposition. Recent advances, however, have made this process more commercially viable. Costs remain higher than for starch-based bioconversion, but there is added potential for value-added products that can utilize the lignin component of the wood.

By forming intramolecular and intermolecular hydrogen bonds between hydroxyl groups within the same cellulose chain and the surrounding cellulose chains, the chains tend to be arranged parallel and form a crystalline macromolecular structure. Bundles of linear cellulose chains (in the longitudinal direction) form a microfibril which is oriented in the cell wall structure (Hashem et al., 2007). Cellulose is insoluble in most solvents and has a low accessibility to acid and enzymatic hydrolysis.

Unlike cellulose, hemicelluloses consist of different monosaccharide units. In addition, the polymer chains of hemicelluloses have short branches and are amorphous. Because of the amorphous morphology, hemicelluloses are partially soluble in water. Hemicelluloses are related to plant gums in composition, and occur in much shorter molecule chains than cellulose. The hemicelluloses, which are present in deciduous woods chiefly as pentosans and in coniferous woods almost entirely as hexosans, undergo thermal decomposition very readily. Hemicelluloses are derived mainly from chains of pentose sugars, and act as the cement material holding together the cellulose micelles and fiber (Theander, 1985).

The backbone of the chains of hemicelluloses can be a homopolymer (generally consisting of single sugar repeat unit) or a heteropolymer (mixture of different sugars). Among the most important sugar of the hemicelluloses component is xylose. In hardwood xylan, the backbone chain consists of xylose units which are linked by β-(1,4)-glycosidic bonds and branched by α-(1,2)-glycosidic bonds with 4-O-methylglucuronic acid groups (Hashem et al., 2007). In addition, O-acetyl groups sometime replace the OH groups in position C_2 and C_3. For softwood xylan, the acetyl groups are fewer in the backbone chain but softwood xylan has additional branches consisting of arabinofuranose units linked by α-(1,3)-glycosidic bonds to the backbone. Hemicelluloses are largely soluble in alkali and, as such, are more easily hydrolyzed.

Lignins are polymers of aromatic compounds the function of which is to provide structural strength, provide sealing of water-conducting system that links roots with leaves, and protect plants against degradation. Lignin is a macromolecule, which consists of alkylphenols and has a complex three-dimensional structure. Lignin is covalently linked with xylans in the case of hardwoods and with galactose-glucose-mannose units in softwoods. Lignin is a natural polymer which together with hemicelluloses acts as a cementing agent matrix of cellulose fibers in the woody structures of plants.

Most lignin applications are based on technical lignins which are separated during pulping processes, and hydrolysis of lignin, which is obtained during wood acidic hydrolysis.

The basic chemical phenylpropane units of lignin are bonded together by a set of linkages to form a very complex matrix. This matrix comprises a variety of functional groups (such as hydroxyl, methoxyl, and carbonyl groups) which impart a high polarity to the lignin macromolecule (Hashem et al., 2007).

The amount of extractable constituents is an important parameter which directly affects the heating value of wood (or, for that matter, any biomass). A high content of extractable constituents makes it desirable as fuel. The extractable constituents usually have low molecular weight and are soluble in neutral solvents. For example, terpenes, lignans and other aromatics, fats, waxes, fatty acids and alcohols, turpentines, tannins, and flavonoids are categorized as extractable constituents. The contents of extractives vary among wood species, geographical site, and season and usually represent from between 3 and 10 percent by weight of wood.

Wood cut in the spring and summer contains more water than that cut in the early part of the winter. A cord (8 ft long, 4 ft wide, and 4 ft high) of hard wood, such as ash or maple, is about equal in heating value to 1 ton of bituminous coal; soft woods, such as pine and poplar, have less than half this amount. Wood burns with a long flame and makes comparatively little smoke; but its calorific intensity is low, averaging from 3000 to 4000 cal/kg of airdried wood. It is, however, easily kindled, the fire quickly reaches its maximum intensity, and a relatively small quantity of ash is formed. Wood is too expensive for industrial use, except in a few special cases, where freedom from dirt and smoke is necessary.

Of other cellulose materials, shavings, sawdust, and straw are used for fuel in some places. They are bulky and difficult to handle, while their heat value, which depends on the amount of moisture they contain, is seldom more than from one-third to one-half that of subbituminous coal. Waste matter such as spent tan-bark and *bagasse* (crushed sugar cane), and the pulp from sugar beets is sometimes used for fuel for evaporation for steam, but owing to the large amount of moisture they contain, the heat value is very low.

When considering the type of wood for use as firewood, several characteristics are important. These include heat value, ease of splitting, weight per unit volume, ease of starting, amount of smoking, and coaling qualities. Moisture content of the wood, number of knots and pitch content affect these characteristics of the more common woods used as firewood.

All woods dried to the same moisture content contain approximately the same heat value per pound—from 8000 to 9500 Btu for fully dried wood and 5500 to 8500 Btu for air-seasoned wood. However, the heat content of any fire depends on wood density, resin, ash, and moisture. A general rule for estimating heat value of firewood is: *one cord of wellseasoned hardwood (weighing approximately 2 tons) burned in an airtight, draft-controlled wood stove with a 55 to 65 percent efficiency is equivalent to approximately 175 gal of #2 fuel oil or 225 therms of natural gas consumed in normal furnaces having 65 to 75 percent efficiencies.*

Forest biomass or agricultural residues are almost completely comprised of lignocellulosic molecules (wood), a structural matrix that gives the tree or plant strength and form. This type of biomass is a prime feedstock for combustion, and indeed remains a major source of energy for the world today (FAO, 2005). The thermochemical platform utilizes pyrolysis and gasification processes to recover heat energy as well as the gaseous components of wood, known as *synthesis gas* or *syngas*. Syngas can then be refined into synthetic fuels, including Fischer-Tropsch, methanol, and ethanol, through the process of catalytic conversion.

The wood biomass comprises stem wood from ordinary forestry, dedicated (short-rotation) forestry, as well as various residues and wood wastes. However, two core types of biomass raw material are distinguished: (a) woody and (b) herbaceous (Table 10.1). Currently woody material accounts for about 50 percent of total world bioenergy potential. Another 20 percent is straw-like feedstock, obtained as a by-product from agriculture.

TABLE 10.1 Properties of Various Woody and Herbaceous Feedstocks Compared to Coal and Natural Gas

	Bituminous Coal	Natural gas	Wood	Bark	Willow	Forest residues	Wood chips	Wood pellets	Cereal straw	Dedicated energy crops
Ash, d%	8.5–10.9	0	0.4–0.5	3.5–8	1.1–4.0	1–3	0.8–1.4	0.4–1.5	3–10	6.2–7.5
Moisture, w%	5–10	0	5–60	45–65	50–60	50–60	20–50	7–12	14–25	15–20
NCV, MJ/kg	26–28.3	48	18.5–20	18.0–23	18.4–19.2	18.5–20	19.2–19.4	16.2–19	16.5–17.4	17.1–17.5
Density, kg/m³	1100–1500	NA	390–640	320	120	NA	250–350, 320–450	500–780	100–170	200
Volatile matter, w%	25–40	100	>70	69.6–77.2	>70	>70	76–86	>70	70–81	>70
Ash melting point, T°C	1100–1400	—	1400–1700	1300–1700	NA	NA	1000–1400	>1120	700–1000	700–1200
C, d%	76–87	75	48–52	48–52	47–51	48–52	47–52	48–52	45–48	45.5–46.1
H, d%	3.5–5	24	6.2–6.4	4.6–6.8	5.8–6.7	6.0–6.2	6.1–6.3	6.0–6.4	5.0–6.0	5.7–5.8
N, d%	0.8–1.5	0.9	0.1–0.5	0.3–0.8	0.2–0.8	0.3–0.5	<0.3	0.27–0.9	0.4–0.6	0.50–1.0
O, d%	2.8–11.3	0.9	38–42	24.3–42.4	40–46	40–44	38–45	≈40	36–48	41–44
S, d%	0.5–3.1	0	<0.05	<0.05	0.02–0.10	<0.05	<0.05	0.04–0.08	0.05–0.2	0.08–0.13
Cl, d%	<0.1	—	0.01–0.03	0.01–0.03	0.02–0.05	0.01–0.04	0.02	0.02–0.04	0.14–0.97	0.09
K, d%	0.003	—	0.02–0.05	0.1–0.4	0.2–0.05	0.1–0.4	≈0.02	NA	0.69–1.3	0.3–0.5
Ca, d%	4–12	—	0.1–1.5	0.02–0.08	0.2–0.7	0.2–0.9	≈0.04	NA	0.1–0.6	9

Source: Kavalov, B. and S. D. Peteves: "Status and Perspectives of Biomass-to-Liquid Fuels in the European Union," European Commission. Directorate General Joint Research Centre (DGJRC), Institute for Energy, Petten, The Netherlands, 2005.

Residual wood is the material that is a refuse without objective value within a specific context, otherwise it constitutes a material at the end of its usefulness. Thus, a number of woody materials can be included in the group of residues and waste: thinning and logging residues from forest industry (tops, branches, and small-size stems), demolition wood and railway sleepers, fiberboard residues, cutter shavings, and plywood residues. Residual woody material is believed to be a very promising bioenergy resource, since it is available at much lower or negligible cost compared to wood logs and short-rotation forestry.

However, the availability of residual woody biomass depends on the primary wood yield and typically accounts for 25 to 45 percent of all harvested wood on average. The heterogeneous composition of the residual and waste woody biomass (content of moisture, impurities, etc.) might sometimes preclude its application for biomass-to-liquids production. Hence, preliminary treatment of the residual and waste woody biomass may be necessary, in order to make it appropriate for processing to liquid fuels (Calis et al., 2002; van Loo and Koppejan, 2003; US DOE, 1996).

Fireplaces and wood stoves, popular aesthetic accessories of the recent past, are rapidly gaining prominence as primary or supplemental heat sources for homes. The rising costs, and in some instances, actual shortages of conventional domestic-heating energies have led to greatly increased utilization of wood as a heating fuel.

10.3 ENERGY FROM WOOD

Combustion remains the most common way of converting biomass into energy. It is well understood, relatively straightforward and commercially available, and can be regarded as a proven technology. However, the desire to burn uncommon fuels, improve efficiencies, cut costs, and decrease emission levels results in new technologies being continuously developed.

On the other hand, wood as a feedstock for, one of nature's most common methods of storing solar energy, is a renewable energy source. It is a relatively clean, efficient, safe energy source having low sulfur content and is generally found throughout the country. Its primary products of combustion are carbon dioxide, water vapor, and ash. The ash content is low—only 1 to 2 percent by weight—and that which does remain can be used as a worthwhile soil conditioner.

A wood fire is easy to start and produces a large quantity of heat in a short time as well as adding a cheerful atmosphere to the home. An ample air supply to the wood fire is important to ensure complete burning or combustible gases. Wood fires are ideal where heat is required only occasionally, for warming a living area on cool days or for supplying extra heat in extremely cold weather. When considering wood as a primary heat source, several factors must be carefully weighed to ensure satisfactory results and acceptable deficiencies.

Generally, hardwoods which provide long-burning fires contain the greatest total heating value per unit of volume. Softwoods which give a fast burning, cracking blaze are less dense and contain less total heating value per unit of volume. However, the amount and types of wood fuel used vary considerably between regions, mainly due to different local situations and conditions.

Charcoal continues to be used as an important industrial source of energy. For example, in Brazil, some 6 million metric tons of charcoal is produced every year for use in heavy industry, such as steel and alloy production. The industrial demand for charcoal in the last few years has led to new, more efficient, and large-scale technologies, mainly aimed at improving charcoal yield and quality.

On the other hand, the production and consumption of black liquor (Table 10.2), which is a by-product of pulp and paper production, are concentrated in developed countries with large paper industries. In the pulp and paper industry, black liquors are widely used for heat

TABLE 10.2 Estimated World Total Wood Fuels Consumption, 2002

FUELWOOD	Million tonnes	Million m³	PJ	Million toe	%
Africa	441	609	6088	145.4	31.3
North and Central America	121	167	1673	40.0	8.6
South America	111	153	1528	36.5	7.9
Asia	671	925	9254	221.0	47.6
Europe	58	81	806	19.2	4.1
Oceania	6	9	86	2.0	0.4
Total World	1411	1946	19458	464.7	100.0

CHARCOAL	Million tonnes of charcoal	Million m³ wood*	PJ	Million toe	%
Africa	15	89	453	10.8	50.5
North and Central America	2	12	64	1.5	7.1
South America	7	42	211	5.1	23.6
Asia	5	29	145	3.5	16.2
Europe	1	4	23	0.5	2.5
Oceania	0	0	1	0.0	0.1
Total World	29	176	897	21.4	100.0

* Estimated amount of wood used for charcoal production.

BLACK LIQUOR	Million m³ wood equivalent	PJ	Million toe	%
Africa	No data	No data	No data	No data
North and Central America	160	1599	38.2	49.4
South America	60	601	14.4	18.6
Asia	41	414	9.9	12.8
Europe	59	592	14.1	18.3
Oceania	3	29	0.7	0.9
Total World	323	3234	77.3	100.0

TOTAL WOOD FUELS	Million tonnes	Million m³	PJ	Million toe	%
Africa		698	6541	156.2	27.7
North and Central America		340	3335	79.7	14.1
South America		255	2341	56.0	9.9
Asia		995	9812	234.4	41.6
Europe		144	1420	33.8	6.0
Oceania		12	115	2.7	0.5
Total World	1771	2443	23589	563.4	100.0

Note: Total World figures include small amounts for Middle East.

and power production. Almost all this industry's energy needs are met by black liquors, with surplus electricity being sold to the public grid in some cases. About 50 percent of black liquor consumption takes place in North America, followed by Europe and South America, both with about 19 percent, and Asia with 13 percent.

Some industry associations are also making special voluntary contributions to reduce the consumption of fossil fuels and increase the use of wood-based energy. For instance, the European paper industry aims to achieve an approximately 25 percent increase in the amount of wood fuel used for on-site heat and power production by 2010, and increase the share of wood in its on-site total primary energy consumption from 49 to 56 percent. In other words, energy will be the main product of the forests, and energy and environmental policies which have been enacted present new opportunities for its further development. A combination of factors such as higher oil prices and technological developments in wood fuel production, transportation, and combustion is also making wood fuels more attractive.

The dynamics of wood fuel flows are complex and very site-specific. The development of sustainable wood energy systems remains one of the most critical issues to be addressed by policy makers and community planners. With society giving increasing attention to sustainability issues, in the case of wood energy in both developing and developed countries, economical, environmental, and social issues deserve particular attention.

Most of the uses of wood are accounted for by combustion in intermediate or large-scale units outside the forest industries (e.g., in schools, hospitals, barracks, or district-heating plants), with minor volumes going to the production of charcoal. Very small volumes were used in a few European countries to generate electricity or to manufacture solid fuels (e.g., briquettes). No wood is used at present in the region to make synthetic liquid or gaseous fuels. Use of energy wood by the forest industries and users has grown faster than use by households.

The technical platform chosen for biofuel production is determined in part by the characteristics of the biomass available for processing. The majority of terrestrial biomass available is typically derived from agricultural plants and from wood grown in forests, as well as from waste residues generated in the processing or use of these resources. Currently, the primary barrier to utilizing this biomass is generally recognized to be the lack of low-cost processing options capable of converting these polymers into recoverable base chemical components (Lynd et al., 1999).

In the United States, much of the biomass being used for first-generation biofuel production includes agricultural crops that are rich in sugars and starch. Because of the prevalence of these feedstocks, the majority of activity toward developing new products has focused on the bioconversion platform (BRDTAC, 2002a). Bioconversion isolates sugars from biomass, which can then be processed into value-added products. Native sugars found in sugarcane and sugar beet can be easily derived from these plants, and refined in facilities that require the lowest level of capital input. Starch, a storage molecule which is a dominant component of cereal crops such as corn and wheat, is comprised wholly of glucose. Starch may be subjected to an additional processing in the form of an acid- or enzyme-catalyzed hydrolysis step to liberate glucose using a single family of enzymes, the amylases, which makes bioconversion relatively simple. Downstream processing of sugars includes traditional fermentation, which uses yeast to produce ethanol; other types of fermentation, including bacterial fermentation under aerobic and anaerobic conditions, can produce a variety of other products from the sugar stream.

In order to incorporate all aspects of biofuel production, including the value of coproducts and the potential of the industry to diversify their product offering, we employ the biorefinery concept. The biorefinery concept is important because it offers many potential environment-, economy-, and security-related benefits to our society. Biorefineries provide the option of coproducing high-value, low-volume products for niche markets together with lower-value commodity products, such as industrial platform chemicals, fuels, or energy, which offsets the higher costs that are associated with processing lignocellulose (Keller 1996, BRDTAC 2002b).

10.4 FUELS FROM WOOD

Energy from wood can come in several forms. There are three options currently available for producing heat from wood: (a) logs, (b) woodchips, and (c) reconstituted fuels such as pellets and briquettes. Other options which are at the research stage include liquid fuels produced from wood and bales of compressed forest residues.

Logs and wood chips can come from forest thinnings, coppicing, tree-surgery, and pruning operations—even whole trees can be chipped to provide fuel. Woodchips can also be produced from fast-growing varieties of willow and poplar which are cultivated by short rotation coppicing specifically to produce energy. On the other hand, wood pellets are made from highly compressed dry wood shavings and sawdust, usually from sawmilling and joinery operations.

After drying and resizing (in which the wood is cut and split in sizes that are easy to transport), the main processes for the use of wood as fuel are:

1. *Carbonization*: the process of burning wood or biomass in the absence of air breaks it down into liquids, gases, and charcoal.
2. *Gasification*: this is the process in which solid biomass fuels (e.g., wood, charcoal) are broken down by the use of heat to produce a combustible gas, known as producer gas.
3. *Densification*: to overcome the bulky nature, low thermal efficiency, and smoke emission of wood and agriculture residues, these can be processed to produce smokeless briquettes with or without binders.
4. *Liquid fuel production*: for example ethanol production by alcohol by hydrolysis or anaerobic digestion; liquid fuels can be produced from several biomass types such as wood pulp residues (black liquor), sugarcane, and cassava.
5. *Combustion*: biomass fuels can also be used for power generation and/or heat production by direct combustion, either in its primary form or after one or more of the above mentioned transformation processes.

Generally, biomass fuels such as wood are consumed by direct combustion in their primary form after drying and resizing, or in the form of charcoal. New conversion technologies such as gasification and anaerobic digestion have been developed to provide alternatives and to match biomass resources with modern end-use devices. However, as of yet these technologies are not widely used, so also in the near future wood energy conversion will consist mainly of resizing, drying, and charcoal production.

Nevertheless, in recent years there has been increasing use of wood and other biomass for energy conversion in modern applications, such as combined heat and power generation (CHP) and cogeneration.

Wood fuels consist of three main commodities: fuel wood, charcoal, and black liquor. Fuel wood and charcoal are traditional forest products derived from the forest, trees outside forests, wood-processing industries, and recycled wooden products from society. Black liquors are by-products of the pulp and paper industry (Fig. 10.1).

10.4.1 Gaseous Fuels

Wood can be used to make both liquid and gaseous fuels. When wood is heated in the absence of air, or with a reduced air supply it is possible to produce a liquid fuel which can be used in a similar way to conventional oil fuels. It can be used to run internal combustion engines in vehicles or generators. The gas produced from wood is a mixture of hydrogen and carbon monoxide which is similar to the coal gas which was made before the arrival of natural gas from the North Sea. This wood gas can be used in internal combustion engines or in gas turbines which can be used to power generators. Although the liquid fuels are rarely produced from wood at present, wood gas is important in other countries for producing electricity in more remote areas.

Thus, gasification technology is an attractive route for the production of fuel gases from biomass. By gasification, solid biomass is converted into a combustible gas mixture

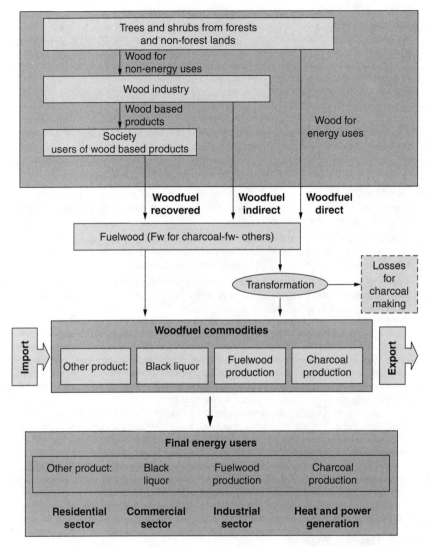

FIGURE 10.1 Wood fuel balance scheme from supply source to end user.

normally called *producer gas* consisting primarily of hydrogen (H_2) and carbon monoxide (CO), with lesser amounts of carbon dioxide (CO_2), water (H_2O), methane (CH_4), and higher hydrocarbons (C_xH_y), as well as nitrogen (N_2) and particulates.

The gasification is carried out at elevated temperatures, 500°C and 1500°C and at atmospheric or elevated pressures. The process involves conversion of biomass, which is carried out in absence of air or with less air than the stoichiometric requirement of air for complete combustion. Partial combustion produces carbon monoxide as well as hydrogen which are both combustible gases. Solid biomass fuels, which are usually inconvenient and have low efficiency of utilization, can be converted into gaseous fuel. The energy in producer gas is 70 to 80 percent of the energy originally stored in the biomass. The producer gas can serve in different ways: it can be burned directly to produce heat or can be used as a fuel for gas engines and gas turbines to generate electricity; in addition, it can also be used as a feedstock (syngas) in the production of chemicals, for example, methanol. The diversified applications of the producer gas make the gasification technology very attractive.

A variety of biomass gasifiers have been developed and can be grouped into four major classes: (a) fixed bed updraft or countercurrent gasifier, (b) fixed bed downdraft or cocurrent gasifier, (c) bubbling fluidized bed gasifier, and (d) circulating fluidized bed gasifier. Differentiation is based on the means of supporting the biomass in the reactor vessel, the direction of flow of both the biomass and oxidant, and the way heat is supplied to the reactor. The processes occurring in any gasifier include drying, pyrolysis, reduction, and oxidation. The unique feature of the updraft gasifier is the sequential occurrence of the chemical processes: they are separated in relative position in the gasifier and therefore by time.

As an illustration, using the updraft gasifier (Fig. 10.2), biomass and air are fed in an opposite direction. In the highest zone, biomass is heated up and releases its moisture. In the pyrolysis zone, biomass undergoes a further increase in temperature and decomposes into hydrocarbons, gas products, and char in the temperature range of 150 to 500°C. The major reactions are:

$$\text{Biomass} \rightarrow C_xH_y + C_xH_yO_z + H_2O + CO_2 + CO + H_2$$

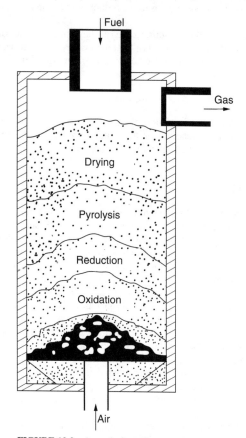

FIGURE 10.2 An updraft gasifier.

The hydrocarbon fraction consists of methane to high-boiling tar. The composition of this fraction can be influenced by many parameters, such as particle size of the biomass, temperature, pressure, heating rate, residence time, and catalysts. The operative reactions are:

$$C + CO_2 \rightarrow 2CO$$

$$C + H_2O \rightarrow H_2 + CO$$

$$2C + 2H_2O \rightarrow 2H_2 + CO_2 + CO$$

$$C + 2H_2 \rightarrow CH_4$$

$$CO + H_2O \rightarrow CO_2 + H_2$$

$$CO + 3H_2 \rightarrow CH_4 + H_2O$$

The composition of the producer gases varies widely with the properties of the biomass, the gasifying agent, and the process conditions. Depending on the nature of the raw solid feedstock and the process conditions, the char formed from pyrolysis contains 20 to 60 percent of the energy input. Therefore the gasification of char is an important step for the complete conversion of the solid biomass into gaseous products and for an efficient utilization of the energy in the biomass.

The producer gases from the reduction zone rise beyond the reduction zone. When they come into contact with the cooler biomass, the temperature drops down and the aforementioned reactions are frozen. The unreacted char further undergoes the oxidation with air in the lowest zone:

$$C + O_2 \rightarrow CO_2$$

As a result, ash is left at the bottom of the reactor. The produced carbon dioxide flows upward and is involved in the reactions in the reduction zone. The heat released in the oxidation zone drives both the reduction and pyrolysis processes.

10.4.2 Liquid Fuels

As already noted (Sec. 10.4.1), wood can be used to make both liquid and gaseous fuels. When wood is heated in the absence of air, or with a reduced air supply it is possible to produce a liquid fuel which can be used in a similar way to conventional oil fuels. It can be used to run internal combustion engines in vehicles or generators. The gas produced from wood is a mixture of hydrogen and carbon monoxide (synthesis gas) which is similar to the coal gas which was made before the arrival of natural gas from the North Sea. This wood gas can be used in internal combustion engines or in gas turbines which can be used to power generators. Although the liquid fuels are rarely produced from wood at present, wood gas is important in other countries for producing electricity in more remote areas.

Processes for making liquid fuels from wood have been understood and available for longer than such fuels have been used in vehicles for transportation. However, the economics of liquid fuels from wood as compared to liquid fuels from fossil fuels have been unfavorable. Nevertheless, the recent phenomenal increases in the cost of gasoline and diesel fuel, that seem to bear little relationship to the increases in petroleum process, has caused a renewed interest in liquid fuels from sources other than petroleum. Wood is a fuel source of interest!

The three approaches that are most promising for making liquid fuels from wood are methanol, ethanol, and diesel fuel, but other liquid fuels from wood are possible

(Enecon, 2002). Methanol was the first fuel from wood and is often called *wood alcohol*. Ethanol has been the focus of research at the Forest Products Laboratory. There has been little attention to diesel fuel from wood, although there has been some research on production from synthesis gas and through utilization of extractable materials from wood.

The United States accounts for about 23 percent of the world's emissions of carbon dioxide. Of the United States sources of carbon dioxide, electric power accounted for 35 percent, transportation 30 percent, industry 24 percent, and residences 11 percent. Obviously, if we're to do our share in reducing carbon dioxide emissions, we should consider making a change in using more non-fossil fuels. The transportation industry is based almost totally on the use of liquid fossil fuels and measures are under consideration to reduce this consumption.

Liquid fuels that could be suitable for use in transportation vehicles have been made from wood for a long time. Methanol was commonly called wood alcohol, and this term is still used. Cellulose which is the largest wood component could be dissolved in concentrated acid solutions and converted to sugar, a precursor for making ethanol. A dilute sulfuric acid hydrolysis process was used to make ethanol during World War I and wood hydrolysis received considerable attention in Europe during the period between the World Wars I and II. Wood hydrolysis plants continue to operate in Russia.

However, methanol and ethanol are not the only transportation fuels that might be made from wood. A number of possibilities exist for producing alternatives. The most promising biomass fuels, and closest to being competitive in current markets without subsidy, are (a) ethanol, (b) methanol, (c) ethyl-t-butyl ether, and (d) methyl-t-butyl ether. Other candidates include isopropyl alcohol, sec-butyl alcohol, t-butyl alcohol, mixed alcohols, and t-amyl methyl ether.

Ethanol or grain alcohol is not restricted to grain as a feedstock. It can be produced from other agricultural crops and ligno-cellulose compounds such as wood. It has often been advocated as a motor fuel, and has been used frequently in times of gasoline scarcity. Today Brazil is the only country that uses large quantities of ethanol as a motor fuel, but even in the United States we use close to a billion gallons per year. In Brazil, 95 percent alcohol is used as a neat fuel or anhydrous ethanol is used in admixture with gasoline. In the United States we use anhydrous ethanol in mixtures of 10 percent ethanol with 90 percent gasoline. The high cost of ethanol production in comparison to gasoline is a major disadvantage, and in the United States only subsidies for biomass ethanol make it competitive. However, because of the perceived ability of ethanol and other oxygenated fuels including alcohols and ethers to reduce air pollution in 90 percent carbon monoxide and ozone nonattainment areas in the United States, the cost disadvantage may become secondary, at least in these areas. Other reasons for considering fuels alternative to petroleum include energy security within national borders, balance of trade, and tax policies.

Another possibility for oxygenated fuels is methanol. Methanol could conceivably be made from grain, but its most common source is natural gas. Use of natural gas is better for reducing carbon dioxide production in comparison to other fossil fuels, but use of renewable fuels instead of natural gas would be still better. It can be made from coal or wood with more difficulty and lower efficiency than from natural gas.

Methanol has long been used as the fuel for race cars at Indianapolis and some other race tracks, not only because of its clean-burning characteristics, but also because of its efficiency, low tire hazard, and high octane rating.

High octane rating is characteristic of all oxygenated fuels, including ethanol, methanol, ethyl-t-butyl ether, and methyl-t-butyl ether. A large part of the success of ethanol from grain in the current United States mix of motor fuels is due to its ability to raise octane rating in a 10 percent mixture of ethanol with 90 percent gasoline. However, it is the recent phenomenal growth in the use of methyl-t-butyl ether (MTBE) as an octane enhancer that has captured worldwide attention. Methyl-t-butyl ether is made by reacting isobutylene with methanol. Ethyl-t-butyl ether (ETBE) is made by using ethanol instead of methanol. Thus either ethanol or methanol from either grain or wood could be a factor in making t-butyl ether octane enhancers. The characteristics of ethers are generally closer to those of gasoline than those of alcohols.

Ethers are benign in their effect on fuel system materials and are miscible in gasoline; therefore, they are not subject to phase separation in the presence of water, as are methanol and ethanol. Ethers are nonpolar. They are of low volatility and thus give low evaporative emissions.

Alternative fuels from wood, as well as grain, have a potential for being competitive with gasoline and diesel motor fuels from petroleum, even without subsidization. Today, ethanol from grain and a slight amount from wood are competing, but only with a large Federal and some State subsidies. However, environmental and octane-enhancing benefits of ethanol and other oxygenated fuels that may be produced from grain and wood may make them worth more than comparisons on fuel value alone would indicate.

Diesel fuel or gasoline from wood is a possibility through a number of approaches. The one that appears simplest is to use an exudation or gum from a tropical wood species, *Copaifera,* which is said to be directly combustible in a diesel engine. The Fischer-Tropsch pyrolysis process, used successfully for converting coal to synthesis gas in South Africa could also be used to make synthesis gas from wood (Fig. 10.3). Synthesis gas could then be used to make gasoline or diesel fuel. Or methanol could be produced from wood and then, by a catalyzed reaction known as the Mobil process, be transformed to gasoline.

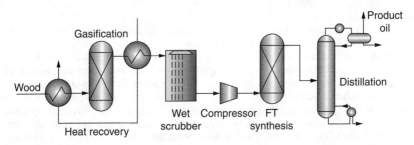

FIGURE 10.3 Diesel from wood by the Fischer-Tropsch process.

Although, ideally, there should be additional pilot testing for any process to produce ethanol or methanol from wood commercially, technology for ethanol production has been developed and subjected to some pilot testing. The technology is available for fairly rapid implementation, should the need for alternative fuels become pressing as the result of another global petroleum emergency. Depending on feedstock costs and other variables, ethanol from wood might or might not be able to compete with ethanol from corn. Another important consideration is production and marketing of by-products such as high fructose corn syrup and distillers dry grains from corn and molasses and/or furfural from wood. The two-stage, dilute sulfuric acid hydrolysis process is a possibility for commercial application in producing ethanol from low-grade hardwoods.

In the two-stage hydrolysis process, for every 100 kg of oven-dry wood feedstock about 20 kg carbohydrates suitable for processing to ethanol are obtained from the second stage. There are more carbohydrates derived from the first stage, about 24.9 kg, but many of these first stage carbohydrates are not necessarily fermentable to ethanol.

Ethanol is a possibility if xylose can be fermented to ethanol economically. Fermentation of the xylose and glucose from the first stage could result in almost doubling the ethanol production as compared to only fermenting the glucose from the second stage. Other possible products from the first stage carbohydrates are single-cell protein, furfural, and feed molasses.

Methanol was once produced from wood as a by-product of charcoal manufacture, but overall yields were low. To produce methanol from wood with a significantly higher yield would require production of synthesis gas in a process similar to that used for production of methanol from coal. Such processes for gasifying wood are less fully developed than the two-stage hydrolysis process for production of ethanol. Another consideration in producing liquid fuels from wood is the amount of wood available to manufacture the fuels. For converting wood to liquid fuel, the most optimistic assumption normally used is that wood could be converted to liquid fuel in the ratio of Btu in liquid fuel/Btu in wood = 0.5. At least in the short run it would be difficult to find more than 100 million dry tons of wood per year (1.7 Quads equivalent) for this purpose. This would calculate to be a maximum of 13 billion gallons per year if the output was methanol and the energy content of a ton of dry wood is assumed to be 17 million Btu. On the order of 11.5 billion gallons per year would be needed if methanol were added to gasoline at the rate of 10 percent methanol to 90 percent gasoline.

Projections of wood use for energy to 2010 are modest. Currently, 2.7 Quads of energy are produced from wood. Projections of the total for 2010 are about 4.0 Quads. This does not provide for a total growth of even 1.7 Quads. However, if we really want to get serious about doing something to deter atmospheric CO_2 accumulation the availability of wood for energy, including solid as well as liquid fuel, could be increased; up to 10 Quads per year might be very realistic. This runs counter to decreased usage of wood for all purposes, particularly from the national forests, to provide more wilderness, habitat for threatened and endangered species, clean water, and other environment considerations. However it must be remembered that, in some cases, more harvesting and cleanup of residues is needed to increase the vigor of forest growth, to protect the forest against wildfire, and to prepare the soil for new growth. In many cases open broadcast burning of logging slash is being outlawed, and harvest of this material for fuel instead of open burning is a viable option for better forest management as well as for profit.

10.4.3 Solid Fuels

Burning wood as a fuel is the largest current use of biomass derived energy. Wood can be used in many forms as a solid fuel for cooking or heating, occasionally for steam engines, and steam turbines that generate electricity. The particular form of wood fuel used depends upon (among other things) its source, quantity, and quality. Available forms include logs, bolts, blocks, firewood, stove-wood (often from split blocks), charcoal, chips, sheets, pellets, and sawdust. Sawmill waste and construction industry by-products also include various forms of lumber tailings.

The use of wood as a fuel source for home heat is as old as civilization itself. Historically, it was limited in use only by the distribution of technology required to make a spark. Wood heat is still common throughout much of the world, although it has been mainly replaced with coal, oil, or natural gas heating. Wood heating has been singled out as a serious health hazard in many regions of the world.

Early examples include the use of wood heat in tents. Fires were constructed on the ground, and a smoke hole in the top of the tent allowed the smoke to escape by convection. In permanent structures, hearths were constructed—surfaces of stone or another noncombustible material upon which a fire could be built. Smoke escaped through a smoke hole in the roof. The development of the chimney and the fireplace allowed for more effective exhaustion of the smoke. Masonry heaters or stoves went a step further by capturing much of the heat of the fire and exhaust in a large thermal mass, becoming much more efficient than a fireplace alone.

The metal stove was a technologic development concurrent with the industrial revolution. Stoves were manufactured or constructed pieces of equipment that contained the fire on all sides and provided a means for controlling the draft—the amount of air allowed to reach the fire. Stoves have been made of a variety of materials. Cast iron is among the more common. Soapstone (talc), tile, and steel have all been used. Metal stoves are often lined with refractory materials such as firebrick, since the hottest part of a wood-burning fire will burn away steel over the course of several years' use.

The Franklin stove was developed in the United States by Benjamin Franklin. More a manufactured fireplace than a stove, it had an open front and a heat exchanger in the back that was designed to draw air from the cellar and heat it before releasing it out the sides. The heat exchanger was never a popular feature and was omitted in later versions. So-called "Franklin" stoves today are made in a great variety of styles, though none resembles the original design.

The airtight stove, originally made of steel, allowed greater control of combustion, being more tightly fitted than other stoves of the day. Airtight stoves became common in the nineteenth century.

Fuel Wood. Fuel wood (i.e., wood itself) is the most common solid fuel and continues to be widely used as a major source of energy for households, especially in developing countries. Charcoal is also increasingly used in many African countries by urban dwellers, as a result of a relentless process of migration of people from rural areas toward urban centers. The major energy end-use in households is cooking: about 86 percent of fuel wood consumed in urban households in India is for this purpose, while the rest is mostly used for water heating. In Africa, more than 86 percent of total wood fuels consumption was attributed to the household sector in 1994. Dependence on wood fuels to meet household energy needs is especially high in most of sub-Saharan Africa, where 90 to 98 percent of residential energy consumption is met from this source.

In short, the use of fuel wood and charcoal remains the dominant source of energy for most developing countries. It is estimated that over 2 billion poor people depend on fuel wood and/or charcoal for meeting their basic daily energy needs for cooking and heating. For them, wood fuels are not only vital to the nutritional stability of rural and urban households, but are also often essential in food-processing industries for baking, brewing, smoking, curing, and electricity production.

In places with high fuel wood and charcoal consumption (due to high population density with low income and/or severe climatic conditions) and weak supply sources, strong pressures are put on existing tree resources, and deforestation and devegetation problems remain a matter of great concern. In addition, urbanization and economic development are bringing about changes in consumption patterns in developing countries, which in turn are leading to major changes in the household energy sector. A pronounced shift from fuel wood to charcoal, especially in Africa, is observed. This issue has raised concerns among environmentalists and those responsible for forest development and management because these charcoal-making activities are, in most cases, carried out illegally. Moreover, they put a much higher pressure on natural forests than the extraction of fuel wood, which is often produced from trees outside forests or from other sources not involving the destruction of forests.

In developed countries, heat production by households also remains the major use of fuel wood. For instance, in the European Union wood fuels account for around 60 percent of the total wood energy consumed, although their utilization as an industrial energy source for electricity and heat generation is increasing, as a result of new energy policies enacted in most countries to comply with climate change mitigation programs.

When used as a solid fuel, some fuel wood (firewood) is harvested in *woodlots* managed for that purpose, but in heavily wooded areas it is more usually harvested as a by-product of natural forests. Deadfall that has not started to rot is preferred, since it is already partly seasoned. Standing dead timber is considered better still, as it is both seasoned, and has less rot. Harvesting this form of timber reduces the speed and intensity of bushfires. Harvesting timber for firewood is normally carried out by hand with chainsaws. Thus, longer pieces—requiring less manual labor, and less chainsaw fuel—are less expensive (but the user must ensure that the lengths will fit in the firebox!) Prices also vary considerably with the distance from wood lots, and quality of the wood.

Firewood usually relates to timber or trees unsuitable for building or construction. Firewood is a renewable resource provided the consumption rate is controlled to sustainable levels. The shortage of suitable firewood in some places has seen local populations damaging huge tracts of bush thus leading to further desertification.

As with any fire, burning wood fuel creates numerous by-products, some of which may be useful (heat and steam), and others that are undesirable, irritating, or dangerous. Thus, before wood is used there are several processes that are necessary for application:

1. *Wood selection:* Firewood can be sold by weight. Freshly cut (*green*) wood can contain from 40 to 60 percent moisture by weight, whereas properly seasoned wood contains only 15 to 20 percent. Select the driest wood when buying by weight. Green wood will shrink approximately 8 percent in volume (i.e., approximately a cubit feet per cord) when properly seasoned.

2. *Preparation and drying:* Wood should be dried as much as possible before burning. Properly seasoned wood has about 7700 Btu maximum usable energy per pound versus only about 5000 Btu available from green wood. For best results, season or air-dry wood for at least 6 to 8 months after cutting. This should bring the moisture content down to 15 to 20 percent by weight. The best time to cut wood is during the winter or early spring before the sap runs. If the tree is felled when fully leafed out, let it lie until leaves have become crisp to allow leaves to draw out as much moisture as possible from tree before further cutting. Drying time is greatly reduced if wood is cut into firewood length and split, especially pieces larger than 8 in in diameter. Splitting is easiest when wood is frozen or green and should be done before wood is stacked. Wood must be properly stacked for satisfactory drying. The greater the surface area exposed to air, the more rapid the drying. Therefore, stack wood loosely and keep it off moist ground. The stack should be located in an open area for good air circulation—avoid stacking in wood lots for seasoning.

3. *Storage of wood:* Store firewood outdoors, under partial or full protection from the elements, and no closer than 25 ft from the house. Keep area around wood clear of weeds, leaves, debris, and the likes, to discourage rodents, snakes, insects, and other unwanted pests from making their home in the stacked wood. Avoid storing large quantities in the house, warm garage, or basement because the heat will activate insect and fungi or spore activity and bring about hatching of any insect eggs in or on the wood.

4. *Building a fire*: Before lighting a fire, make sure the thermostat is turned down so air heated by the central furnace will not go up the chimney. The easiest and best fire for either a stove or fireplace is achieved with a mixture of softwoods for easy igniting with hardwoods for longer burning and good coaling qualities. A cardinal rule of fireplace management is to keep a thick bed for glowing coals that drop through. The coals yield a steady heat and aid in igniting fresh fuel as it is added. Keep the fire burning by adding small amounts of wood at regular intervals. A small, hot fire is much better than a large, roaring blaze because it burns more completely and produces less creosote.

5. *Precautions*: Coal should never be burned in a stove or heater designed for wood. Artificial or manufactured logs, which are composites of sawdust, chips, colorful

chemicals, starch binders, and wax should be burned only in open brick fireplaces. The wax burns at too hot a temperature for metal stoves and chimneys. When using manufactured logs in fireplaces, never crumble the burning log with tongs or poker. Avoid using wood salvaged from poles, posts, and lumber that has been treated with wood preservatives such as creosote or pentachlorophenol. These chemical compounds may vaporize upon combustion and cause respiratory problems for those breathing the fumes. Wet, green wood, or highly resinous wood should not be burned because of the large amounts of wood tars, creosote, and wood extractives given off which can coat chimney flues and cause serious chimney fires if ignited.

One by-product of wood burning is wood ash, which in moderate amounts is a fertilizer (mainly potash), contributing minerals, but is strongly alkaline as it contains sodium hydroxide (lye). Wood ash can also be used to manufacture soap.

Smoke, containing water vapor, carbon dioxide, and other chemicals and aerosol particulates can be an irritating (and potentially dangerous) by-product of partially burnt wood fuel. A major component of wood smoke is fine particles that may account for a large portion of particulate air pollution in some regions. During cooler months, wood heating accounts for as much as 60 percent of fine particles in Melbourne, Australia.

Slow combustion stoves increase efficiency of wood heaters burning logs, but also increase particulate production. Low pollution/slow combustion stoves are a current area of research. An alternative approach is to use pyrolysis to produce several useful biochemical by-products, and clean burning charcoal, or to burn fuel extremely quickly inside a large thermal mass, such as a masonry heater. This has the effect of allowing the fuel to burn completely without producing particulates while maintaining the efficiency of the system.

For example, wood-derived pyrolysis oil contains specific oxygenated compounds in relatively large amounts (Phillips et al., 1990). A current comprehensive review focuses on the recent developments in the wood/biomass pyrolysis and reports the characteristics of the resulting bio-oils, which are the main products of fast wood pyrolysis. Sufficient hydrogen added to the synthesis gas to convert all of the biomass carbon into methanol carbon would more than double the methanol produced from the same biomass base (Phillips et al., 1990).

Rapid heating and rapid quenching produce the intermediate pyrolysis liquid products, which condense before further reactions break down higher-molecular-weight species into gaseous products. High reaction rates minimize char formation. Under some conditions, no char is formed. At higher fast pyrolysis temperatures, the major product is gas. Many researchers have attempted to exploit the complex degradation mechanisms by conducting pyrolysis in unusual environments.

Pyrolysis is the simplest and almost certainly the oldest method of processing a fuel in order to produce a better one. Pyrolysis can also be carried out in the presence of a small quantity of oxygen (gasification), water (steam gasification) or hydrogen (hydrogenation). One of the most useful products is methane, which is a suitable fuel for electricity generation using high-efficiency gas turbines.

Cellulose and hemicelluloses form mainly volatile products on heating due to the thermal cleavage of the sugar units. The lignin forms mainly char since it is not readily cleaved to lower molecular weight fragments. The progressive increase in the pyrolysis temperature of the wood led to the release of the volatiles thus forming a solid residue that is different chemically from the original starting material. Cellulose and hemicelluloses initially break into compounds of lower molecular weight. This forms an "activated cellulose" which decomposes by two competitive reactions: one forming volatiles (anhydrosugars) and the other forming char and gases. The thermal degradation of the activated cellulose and hemicelluloses to form volatiles and char can be divided into categories depending on the reaction temperature. Within a fire all these reactions take place concurrently and consecutively. Gaseous emissions are predominantly a product of pyrolitic cracking of the fuel. If flames are present, fire temperatures are high, and more oxygen is available from thermally induced convection.

In some of the most efficient burners, the temperature of the smoke is raised to a much higher temperature where the smoke will itself burn. This may result in significant reduction of smoke hazards while also providing additional heat from the process. By using a catalytic converter, the temperature for obtaining cleaner smoke can be reduced. Some U.S. jurisdictions prohibit sale or installation of stoves that do not incorporate catalytic converters.

Most of the nonhousehold fuel wood consumption in developing countries is in commercial and industrial activities such as crop drying, tea processing, and tobacco curing, as well as the brick and ceramic industries. Fuel wood consumption by these sectors is smaller than that in households; nevertheless, it cannot be overlooked, as it can constitute 10 to 20 percent of fuel wood use, as seen in some Asian countries. In Africa, it is estimated that consumption of wood fuels in industry accounted for about 9.5 percent of the total in 1994. In developed countries, fuel wood uses for electricity and heat generation at industrial sites or in municipal district heating facilities are rapidly rising as a substitute for fossil fuels.

The most commonly known solid fuel produced from wood is *charcoal*, but there are other sources such as coconut shells and crop residues.

Charcoal is produced in kilns by a process called pyrolysis, that is, breaking down the chemical structure of wood under high temperature in the absence of air. During the process, first the water is driven from the wood (drying), and then the pyrolysis starts when the temperature in the kiln is high enough. When the pyrolysis is complete, the kiln gradually cools down, after which the charcoal can be removed from the kiln. Because some of the wood is burned to drive off the water, dry wood produces better charcoal at a higher efficiency. Typically, approximately two-thirds of the energy is lost in the process, but charcoal has advantages over fuel wood like stoves with higher efficiency, higher convenience, and easier distribution.

The oldest and probably still the most widely used method for charcoal production is the earth kiln. Two varieties exist, the earth pit kiln and the earth mound kiln. An earth pit kiln is constructed by first digging a small pit in the ground. Then the wood is placed in the pit and lit from the bottom, after which the pit is first covered with green leaves or metal sheets and then with earth to prevent complete burning of the wood. The earth mound kiln is built by covering a mound or pile of wood on the ground with earth. The mound is preferred over the pit where the soil is rocky, hard or shallow, or the water table is close to the surface. Mounds can also be built over a long period, by stacking gathered wood in position and allowing it to dry before covering and burning.

Earth kilns can be made at minimal cost, and are often used near wood resources, since they can be made entirely from local materials. Earth kilns can be made in any size, with the duration of the process ranging from 3 days to 2 months. Gross variations in the quality of the charcoal can occur, because in one batch some of the wood is burned and some of the wood is only partly carbonized. Efficiencies are generally low, around 10 to 20 percent by weight and 20 to 40 percent in energy terms. The efficiency and the quality varies depending on the construction of the kiln (e.g., walls can be lined with rocks or bricks and external chimneys can be used), and the monitoring of the carbonization process.

Several other types of charcoal kilns have been developed, which generally have higher efficiencies but also require higher investments than the earth kiln. Two often-used types are fixed kilns made of mud, clay, bricks, and portable steel kilns. Fixed kilns usually have a beehive shape. Smaller beehives are usually made of mud and are not very durable. Larger beehives are made of bricks and have external chimneys. Beehive kilns have an opening for loading the wood and unloading the charcoal, which is closed after loading.

Portable steel kilns can be made from oil drums, and can be used both in horizontal and vertical position. They generally have a short lifetime. When used in the horizontal position, an opening is made in the side, through which the wood is loaded. For the vertical kiln the top is cut out and used as a lid.

The production of *briquettes* (*briquetting* or *densification*) is used to improve characteristics of materials for transport and use as energy source. Raw materials include sawdust, loose crop residues, and charcoal fines. The material is compacted under pressure, and depending on the material, the pressure, and the speed of densification, additional binders may be needed to bind the material. The two main briquetting technologies are the piston press and the screw press. In the piston press the material is punched into a die by a ram with a high pressure. In the screw press, the material is compacted continuously by a screw. With the screw press, generally briquettes of higher quality can be produced.

Logs and Wood Chips. Log-fired heating conjures up images of open fires and log stoves for many people, but there are also sophisticated, controllable log-boilers available which can provide central heating and hot water. Logs are readily available in many areas where there are existing markets for domestic fuel wood. Logs should be well seasoned before burning to ensure the most efficient combustion, and should ideally be stored under cover in a well ventilated log store for a year or more.

Logs are ideal for providing heat for domestic buildings and are suitable for heating loads of up to around 50 kW. Log-heating systems do require manual stoking once a day, which makes logs less convenient than more automated-heating systems such as woodchips and pellets. Log-fired systems are available with outputs greater than 50 kW to heat larger building such as village halls; however, these larger systems will require more frequent fuelling.

Wood chips are made from whole trees, branch wood, or coppice products which have been mechanically shredded by a chipping machine. For some types of boilers, the wood needs to have been air-dried before chipping, or the chips dried before burning. Wood chips are a bulky fuel and sufficient storage and delivery access needs to be considered when designing a heating system. Transport costs can be high, but if wood chip is sourced within 20 miles it can be a very cost-effective fuel.

The potential for woodchip heating in many countries is high. The use of a timber resource for local woodchip heating would provide a valuable economic return and stimulate the rural economy. Woodchip systems can provide automated, clean, and convenient heating for larger domestic properties with outputs of 20 to 30 kW, up to large-scale systems for hospitals, factories, schools, and district-heating schemes with heating loads in the megawatt range. Woodchips can also be used to fuel combine heat and power plants in which the heat produced during electricity generation is used to provide hot water, and is not lost as in conventional power stations.

Pellets. Pellets made of compressed sawdust or wood shavings have been available in many parts of the world for more than several decades. Because they are compressed, pellets offer a more concentrated form of fuel than wood chips. Consequently they need less storage space and are easier to handle. The manufacture of wood pellets requires more energy than woodchips and the capital cost for production plant is high; pellets are therefore more expensive than chips.

Pellets can be used to fuel a variety of appliances and heating systems. The smallest are pellet stoves with outputs of up to around 9 kW which are suitable for heating individual rooms. These stoves are electronically controlled and can deliver regulated heat output and only need fuelling once every few days. Pellet boilers are available in a wide range of outputs from small domestic scale to large industrial scale to heat schools or hospitals.

For the domestic user, pellets offer the most user-friendly form of wood heating. In Scandinavia, wood pellets are delivered by tanker and are pumped into storage silos, which feed, automatically into the boiler.

Charcoal. Charcoal is the blackish residue consisting of impure carbon obtained by removing water and other volatile constituents from animal and vegetation substances. Charcoal is usually produced by heating wood, sugar, bone char, or other substances in the absence of oxygen. The soft, brittle, lightweight, black, porous material resembles coal and is 85 to 98 percent carbon with the remainder consisting of volatile chemicals and ash (Table 10.3).

TABLE 10.3 Characteristics of Charcoal Briquettes

	Lump charcoal	Briquettes	
		Without filler	With filler
Ash, % w/w	3–4	8	25
Moisture, % w/w	5	5	5
Carbon, % w/w	Balance	Balance	Balance
Volatiles, % w/w	10–15	10–15	10–15
Binder, % w/w	—	10	10
Calorific value, kJ/Kg	28,000	25,000	22,000
Btu/lb	12,050	10,750	9,500

The first part of the word is of obscure origin, but the first use of the term "coal" in English was as a reference to charcoal. In this compound term, the prefix *chare* meant *turn* with the literal meaning being *to turn to coal*. The independent use of *char*, meaning to scorch, to reduce to carbon, is comparatively recent and must be a back-formation from the earlier charcoal. It may be a use of the word *charren* or *churn*, meaning to turn, that is, wood changed or turned to coal, or it may be from the French *charbon*. A person who manufactured charcoal was formerly known as a collier (also as a wood collier). The word *collier* was also used for those who mined or dealt in coal, and for the ships that transported it.

Historically, production of wood charcoal in districts where there is an abundance of wood dates back to a very remote period, and generally consists of piling billets of wood on their ends so as to form a conical pile, openings being left at the bottom to admit air, with a central shaft to serve as a flue. The whole pile is covered with turf or moistened clay. The firing is begun at the bottom of the flue, and gradually spreads outward and upward. The success of the operation depends upon the rate of the combustion. Generally, 100 parts of wood yield about 25 parts by weight (60 parts by volume) of charcoal but this yield is very dependent on the type of wood.

The production of charcoal (at its height employing hundreds of thousands, mainly in Alpine and neighboring forests) was a major cause of deforestation, especially in Central Europe. In England, many woods were managed as coppices, which were cut and to regrow cyclically, so that a steady supply of charcoal would be available (in principle) forever; complaints (as early as in seventeenth century England) about shortages may relate to the results of temporary over-exploitation or the impossibility of increasing production. In fact, the increasing scarcity of easily harvestable wood was a major factor that led the switch to the fossil fuel equivalents, mainly coal and brown coal.

The modern process of carbonizing wood, either in small pieces or as sawdust in cast iron retorts, is extensively practiced where wood is scarce, and also for the recovery of valuable by-products (wood spirit, pyroligneous acid, wood tar), which the process permits. The question of the temperature of the carbonization is important; charcoal produced made at 300°C is brown, soft and friable, and readily inflames at 380°C while charcoal, made at higher temperatures it is hard and brittle, and does not fire until heated to about 700°C.

Commercial charcoal is found in either lump, briquette, or extruded forms: (a) *lump charcoal*, which is made directly from hardwood material and usually produces far less ash than briquettes, (b) *briquettes*, which are produced by compressing charcoal, typically made from sawdust and other wood by-products, with a binder (usually starch) and other additives, (c) *extruded charcoal*, which is made by extruding either raw ground wood or carbonized wood into logs without the need for a binder since the heat and pressure of the extruding process hold the charcoal together.

Animal charcoal or *bone black is* the carbonaceous residue obtained by the dry distillation of bones; it contains only about 10 percent carbon, the remainder being calcium and magnesium phosphates (80 percent) and other inorganic material originally present in the bones. It is generally manufactured from the residues obtained in the glue and gelatin industries.

Charcoal fines have a much lower purity than lump charcoal and, in addition to charcoal, contain fragments, mineral sand, and clay picked up from the earth and the surface of the wood and its bark. The fine powdered charcoal (produced from bark, twigs, and leaves) has a higher ash content than normal wood charcoal. Most of this undesired high-ash material can be separated by screening the fines and rejecting undersize material passing, say, a 2- to 4-mm screen.

Fines cannot be burned by the usual simple charcoal burning methods but if fines could be fully used, overall charcoal production would rise by 10 to 20 percent. Briquetting—turning fines into lumps of charcoal—seems an obvious answer.

Briquetting requires a binder to be mixed with the charcoal fines, a press to form the mixture into a cake or briquette which is then passed through a drying oven to cure or set it by drying out the water so that the briquette is strong enough to be used in the same burning apparatus as normal lump charcoal.

Charcoal needs addition of a sticking or agglomerating material to enable a briquette to be formed. The binder should preferably be combustible, though a noncombustible binder effective at low concentrations can be suitable. Starch is preferred as a binder though it is usually expensive. Tar and pitch from coal distillation or from charcoal retorts have been used for special purpose briquettes but they must be carbonized again before use to form a properly bonded briquette.

The binders which have been tried are many but, as stated, the most common effective binder is starch. About 4 to 8 percent of starch made into paste with hot water is adequate. First, the fines are dried and screened. Undersized fines are rejected and oversized hammer-milled. This powder is blended with the starch paste and fed to the briquetting press and the resulting briquettes are dried in a continuous oven at about 80°C. The starch sets through loss of water, binding the charcoal into a briquette which can be handled and burned like ordinary lump charcoal in domestic stoves and grates. Briquettes bonded with tar or pitch and subsequently carbonized in charcoal furnaces to produce a metallurgic charcoal briquette of adequate crushing strength are needed.

Charcoal fines when available in large quantities do have industrial uses such as in metallurgic and calcining operations. For example, in charcoal iron making fine charcoal can be injected at the base of the blast furnace with the air blast. Fine charcoal is excellent for producing sinter, partially reduced iron ore, to provide a high grade feed to the blast furnace. This is one of the best ways to use charcoal fines as the amount which can be used is not limited to a percentage of the total as is the case of injection into the base of the blast furnace. Pulverized fine and lump charcoal can be burned in rotary furnaces producing cement clinker and calcium bauxite.

10.5 REFERENCES

BRDTAC: "Roadmap for Biomass Technologies in the United States," Biomass Research and Development Technical Advisory Committee, Washington, D.C., 2002a.

BRDTAC: "Vision for Bioenergy and Bio-Based Products in the United States," Biomass Research and Development Technical Advisory Committee, Washington, D.C., 2002b.

Calis, H., J. Haan, H. Boerrigter, A. van Der Drift, G. Peppink, R. van Den Broek, et al.: "Preliminary Techno-Economic Analysis of Large-Scale Synthesis Gas Manufacturing from Imported Biomass," Proceedings, Expert Meeting on Pyrolysis and Gasification of Biomass and Waste, Strasbourg. TU Delft Publication, Technical University of Delft, Delft, Netherlands, 2002

Enecon: "Wood for Alcohol Fuels: Status of Technology and Cost/Benefit Analysis of Farm Forestry for Bioenergy," A Report for the RIRDC/Land & Water Australia/FWPRDC/MDBC, Joint Venture Agroforestry Program by Enecon Pty Ltd. in association with Centre for International Economics, Stephen Schuck & Associates Pty Ltd. RIRDC Publication No 02/141, RIRDC Project No EPL-2A. Nov., 2002.

FAO: "State of the World's Forests," Food and Agriculture Organization of the United Nations, Rome, Italy, 2005.

Hashem, A., R. A. Akasha, A. Ghith, and D. A. Hussein: "Adsorbent-Based on Agricultural Wastes for Heavy Metal and Dye Removal: A Review," *Energy Edu. Sci. Technol.,* 19, 2007, pp. 69–86.

Kavalov, B. and S. D. Peteves: "Status and Perspectives of Biomass-to-Liquid Fuels in the European Union," European Commission. Directorate General Joint Research Centre (DG JRC)," Institute for Energy, Petten, The Netherlands, 2005.

Keller, F. A.: "Integrated Bioprocess Development for Bioethanol Production," in *Handbook on Bioethanol: Production and Utilization,* C. E. Wyman (ed.), Taylor & Francis, Washington, D.C., 1996, pp. 351–79.

Lynd, L. R., C. E. Wyman, and T. U. Gerngross: "Biocommodity Engineering," *Biotechnol. Prog.,* 15(5), 1999, pp. 777–93.

Phillips, V. D., C. M. Kinoshita, D. R. Neill, and P. K. Takahashi: "Thermochemical Production of Methanol from Biomass in Hawaii," *Appl. Energy,* 35, 1990, 167–75.

Theander, O.: In *Fundamentals of Thermochemical Biomass Conversion,* R. P. Overand, T. A. Mile, and L. K. Mudge (eds.), Elsevier Applied Science Publisher, New York, 1985.

US DOE: "Biomass Power Program," Report No. DOE/GO-10096-345 DE97000081, National Renewable Energy Laboratory, United States Department of Energy, Washington D.C., 1996.

van Loo, S. and J. Koppejan: *Handbook of Biomass Combustion and Co-Firing. International Energy Agency,* Twente University Press, University of Twente, Enschede, Netherlands, 2003.

Youngs, R. L.: *Interdisciplinary Science Reviews,* 7(3), 1982, p. 211.

CHAPTER 11
FUELS FROM DOMESTIC AND INDUSTRIAL WASTE

Waste is the remains or by-product from a process for which no use is planned or foreseen. The words *domestic* (or *municipal*) and *industrial* are qualifiers of the source of the waste and, to some extent, also descriptive of the contents of the waste. Once a material has been designated as waste, it remains waste until it has been fully recovered and no longer poses a potential threat to the environment.

Thus, *domestic waste* (also known as *rubbish, garbage, trash,* or *junk*) is unwanted or undesired material (although the old adage *one man's waste is another man's treasure* sometimes applies). *Waste* is the general term; though the other terms are used loosely as synonyms, they have more specific meanings: rubbish or trash are mixed household waste including paper and packaging; food waste or garbage (North America) is kitchen and table waste; and junk or scrap is metallic or industrial material (Table 11.1). There are other categories of waste as well: sewage, ash, manure, and plant materials from garden operations, including grass cuttings, fallen leaves, and pruned branches.

On the other hand, *industrial waste* is waste type produced by industrial factories, mills, and mines. It has existed since the outset of the industrial revolution. *Toxic waste* and *chemical waste* are two more specific designations of industrial waste.

Municipal solid waste (MSW) is a waste type that includes predominantly household waste (domestic waste) with sometimes the addition of commercial wastes collected by a municipality within a given area. They are in either solid or semisolid form and generally exclude industrial hazardous wastes. The term *residual waste* relates to waste left from household sources containing materials that have not been separated out or sent for reprocessing.

There are five broad categories of municipal solid waste: (a) biodegradable waste such as food and kitchen waste, green waste, paper—which can also be recycled; (b) recyclable material such as paper, glass, bottles, cans, metals, and certain plastics; (c) inert waste such as construction and demolition waste, dirt, rocks, debris; (d) composite wastes which include waste clothing and waste plastics such as toys; and (e) domestic hazardous waste (also called *household hazardous waste*) and toxic waste such as discarded medications, paints, chemicals, light bulbs, fluorescent tubes, spray cans, fertilizer and pesticide containers, batteries, and shoe polish.

Some components of waste can be recycled once recovered from the waste stream, for example, plastic bottles, metals, glass, or paper. The biodegradable component of wastes (e.g., paper and food waste) can be composted or anaerobically digested (Chap. 9) to produce soil improvers and renewable fuels. If it is not dealt with in a sustainable manner biodegradable waste can contribute to greenhouse gas emissions and by implication climate change.

On the other hand, there is also electronic waste which is a waste consisting of any broken or unwanted electrical or electronic appliance. While there is no generally accepted definition of electronic waste, in most cases electronic waste consists of electronic products that were used for data processing, telecommunications, or entertainment in private households and businesses that are now considered obsolete, broken, irreparable, or of no

TABLE 11.1 Constituents of Domestic/Municipal Waste

Component	Description
Food wastes	The animal, fruit, or vegetable residues (also called garbage) resulting from the handling, preparation, cooking, and eating of foods. Because food wastes are putrescible, they will decompose rapidly, especially in warm weather.
Rubbish	Combustible and non-combustible solid wastes, excluding food wastes or putrescible materials. Typically combustible rubbish consists of materials such as paper, cardboard, plastics, textiles, rubber, leather, wood, furniture, and garden trimmings. Non-combustible rubbish consists of items such as glass, crockery, tin cans, aluminium cans, ferrous and non-ferrous metals, dirt, and construction material.
Ashes and residues	Materials remaining from the burning wood, coal, coke, and other combustible wastes. Ashes and residues are normally composed of fine, powdery materials, cinders, clinkers, and small amounts of burned and partially burned materials.
Demolition and construction wastes	Wastes from razed buildings and other structures are classified as demolition wastes. Wastes from the construction, remodelling, and repairing of residential, commercial, and industrial buildings and similar structures are classified as construction wastes. These wastes may include dirt, stones, concrete, bricks, plaster, lumber, shingles, and plumbing, heating, and electrical parts. They are usually of an inert nature. The main exception is asbestos, where special disposal is required.
Special wastes	Wastes such as street sweepings, roadside litter, catch-basin debris, dead animals, trash like abandoned vehicles, electrical appliances are classified as special wastes.
Treatment plant wastes and dredged soil	The solid and semisolid wastes from water, sewage and industrial waste water treatment facilities are included in this classification. Sewage sludge is a slurry of fine organic-rich particles with a highly variable chemical composition depending on the sources of the effluent and the type and efficiency of the treatment processes. Sewage sludges tend to concentrate heavy metals and water-soluble synthetic organic compounds, but they may also contain greases, oils, and bacteria. Dredged materials are excavated from river estuaries, harbours, and other waterways to aid navigation. It is estimated that 10 percent of dredged materials are contaminated by oil, heavy, metals, nutrients, and organochlorine compounds.

further use due to planned obsolescence. Despite its common classification as a waste, disposed electronics are a considerable category of secondary resource due to their significant suitability for direct reuse (e.g., many fully functional computers and components are discarded during upgrades), refurbishing, and material recycling of its constituents.

It is a point of concern considering that many components of such equipments are considered toxic and are not biodegradable but they are not precursors to fuels and other than recognition though the above paragraph will not be considered in the context of the present text.

In nature there is no waste but, on the other hand, waste is a by-product of the human economic system and the technology that drives it. Since the industrial revolution, human society has developed economies that are largely unrelated from nature and the natural order of events. They have created public and private wealth, material well-being, a plentiful food supply, comfortable living conditions, personal freedoms, and enhanced culture, but at a price.

Our economic system is based on the idea that by applying our technology to the natural resources we extract from the environment, we can create endless wealth through the provision of continuously improved goods and services that people are prepared to work for to obtain. The downside of this process has been explosive population growth, depletion of finite natural resources, environmental damage (including the potential for climate change), and the loss of biodiversity, as well as pollution and waste.

In modern society, waste comes in a variety of forms and must be dealt with on a day-to-day basis. Industry produces huge amounts of *industrial waste* and *domestic waste* makes a large contribution to the general waste problem. In spite of the recognition many insidious waste products escape (inadvertently or deliberately) into the surrounding environment. Thus, there are numerous pollution incidents. On the other hand, there are technologies available for the treatment of most of the wastes we produce. The level of treatment is largely a matter of cost but conversion of waste to new products is a concept that has long been ready to hatch. Now is the time.

11.1   INDUSTRIAL AND DOMESTIC WASTE

Industrial waste is waste type produced by industrial factories, mills, and mines. It has existed since the outset of the industrial revolution. Toxic waste and chemical waste are two designations of industrial waste.

Domestic waste is waste from householders which cannot be recycled, composted, reused, or disposed of by other means.

Much of what human society discards contains usable material, much of it in the form of recoverable energy. Paper, wood, cloth, food waste, and plastics are the main potential energy sources in waste. The remainder of the waste consists of glass, metals, and miscellaneous rubble. Domestic waste is typically disposed of by tipping it into large holes in the ground—landfill sites. Sometimes the waste is incinerated first and only the remaining ash and noncombustible material is sent to a landfill. Increasingly, a proportion of the waste is being separated for recycling at some stage along the way.

It is possible to recover energy from landfill sites and from waste incinerators. Domestic waste could also provide feedstock for a number of other conversion systems, all of which could recover useful energy while reducing the requirement for landfill sites. However, whatever the energy technology, domestic waste is a low-grade fuel. Its consistency is variable and not well suited to mechanical handling systems; the proportions of the various constituents will vary from load to load; the moisture content and heating value will vary; and the proportion of noncombustible material will keep the heating value low. All of this can lead to inefficient combustion if the process is not well controlled, making it more difficult to control toxic emissions from plastics and other materials.

There is also a potential conflict between the recycling of materials and the recovery of energy from those materials. The main benefit of domestic waste as a fuel source is that, as with most other waste streams, energy technology can reduce the waste disposal problem.

Recently interest has grown in the burning of garbage/domestic waste to produce electricity. This is not a new idea although in the past when waste was burned it created pollution that could even be toxic. Today, the technology exists to remove almost all the pollutants from the fumes produced during the energy production cycle. Special filters remove dangerous chemicals and particles that would normally be found in the fumes. The domestic waste is sorted usually by hand to remove materials that can be recycled.

Steel is removed using electromagnets and this is stored until there is enough quantity for recycling to be economically viable. Aluminum, in the form of cans is removed by hand. The waste is then fed into the hopper of a furnace. When the doors slide open it falls into the burning chamber. Gas is normally used to start the fire which burns at a high temperature, destroying the domestic waste. Whilst the waste burns it heats a water tank, in turn, producing steam. The steam is used to turn turbines, producing electricity. Once steam has been produced, the production of electrical power is no different than that used in any other power station. The high pressure steam is used to turn electrical turbines which produce electricity. The advantage of this way of producing electricity is that the domestic waste that would normally be buried in landfill sites or even dumped far out at sea is burned. This means that vast areas of land that would have to be used for landfill are free for agriculture or for building.

Finally, although not presented in the current context, electronic waste is worthy of mention.

Electronic waste (*e-waste* or WEEE—waste electrical and electronic equipment) is a waste type consisting of any broken or unwanted electrical or electronic appliance. It is a point of concern considering that many components of such equipment are considered toxic and are not biodegradable.

Electronic waste includes computers, entertainment electronics, mobile phones, and other items that have been discarded by their original users. While there is no generally accepted definition of electronic waste, in most cases electronic waste consists of electronic products that were used for data processing, telecommunications, or entertainment in private households and businesses that are now considered obsolete, broken, or irreparable. Despite its common classification as a waste, disposed electronics are a considerable category of secondary resource due to their significant suitability for direct reuse (e.g., many fully functional computers and components are discarded during upgrades), refurbishing, and material recycling of its constituent raw materials. Reconceptualization of electronic waste as a resource thus preempts its potentially hazardous qualities.

However, electronic waste is a valuable source for secondary raw materials, if treated properly, however if not treated properly it is major source of toxins. Rapid technology change, low initial cost, and even planned obsolescence have resulted in a fast growing problem around the globe. Technical solutions are available but in most cases a legal framework, a collection system, logistics, and other services need to be implemented before a technical solution can be applied.

11.2 EFFECTS OF WASTE

Wastes generated from the domestic and industrial sources increase continuously with rising population. In general, the lack of facilities for disposal of waste caused overall of landfill sites resulting in hazards for the environment and for public health. These effects include: (a) air pollution, (b) pollution of surface waters, (c) changes in soil fertility, and (d) changes in the landscape and visual discomfort.

Air pollution by unpleasant odors and wind-carried suspensions is obvious in areas located nearby urban waste landfills, since these are currently run in complete ignorance of practices such as cell operation and inert matter lagging. The leakage on the slopes of landfills situated nearby surface water bodies adds to the pollution of such waters by organic substances and suspensions. The urban waste landfills that are not waterproof often represent sources of groundwater pollution by nitrates and nitrites, as well as other pollutants. Water leakage adversely affects the quality of the adjoining soils, which brings along consequential effects to their utilization.

Withdrawing land from the natural or economic circuit in order to set up waste landfills is a process which may be regarded as temporary, yet, in terms of sustainable development, it extends over at least two generations when adding up the time periods required for planning works (1–3 years), running (15–30 years), and ecological reconstruction and subsequent monitoring (15–20 years).

In terms of biodiversity, any domestic or industrial waste landfill means the subsequent elimination of a number of 30 to 300 species (microbiologic population of the soil not included) on each hectare of the area intended to host a landfill. Moreover, changes are likely to occur in areas close to the landfill, such as: (a) ruderal species, that is, plant species that are first to colonize disturbed lands that are specific to polluted areas, would become dominant in the vegetal assemblies, and (b) some mammals, birds, insects would desert the area, in favor of species that feed on refuse (rats, crows).

Although the effects on flora and fauna are, in theory, limited to the time period over which the landfill is operated, the ecologic reconstruction performed after the area has been relieved from its technologic use will not be able to retrieve the initial biologic balance, as the evolution of that biosystem has been irreversibly modified. The practices used currently in the collection, transport, and storage of urban waste facilitate the multiplication and dissemination of the pathogenic agents and their accompanying breed: insects, rats, crows, stray dogs.

Thus, waste in any form (domestic or industrial) represents a health hazard, due to its disposal and composition. In terms of composition, the focus is often on toxic substances such as heavy metals (lead, cadmium), pesticides, solvents, and used oils that form an integral part of the waste.

Indeed, the challenge for waste disposal arises from the joint storage of hazardous materials (including toxic sludge, oil products, dyeing residues, metallurgic slag) and solid domestic waste. This situation is likely to generate inflammable, explosive, or corrosive mixtures and combinations thereof. On the other hand, the presence of easily degradable household may facilitate the decomposition of complex hazardous components, and thus diminish environmental pollution.

Another negative aspect is the fact that several recyclable and useful materials are stored in the same place as materials that cannot be recycled; consequently, these materials blend together and become chemically and biologically contaminated, which renders their retrieval rather difficult.

Thus, the problems faced by waste management activities may be summarized as follows: (a) storage in open grounds is the most used method to remove waste ultimately, (b) existing landfills may be located in sensitive places (i.e., in the close proximity to lodgings, surface or ground water, leisure areas), (c) existing waste landfills may be improperly designed from an environment protection point of view, thus allowing for water and soil pollution in those areas, (d) currently waste landfills may require a review of waste-handling practices insofar as waste layers are not compacted and there is no strict control of the quality and quantity of waste that is dumped on the landfill leading to the potential for fire and/or the emanation of unpleasant odors.

All of the above lead to the conclusion that specific measures need to be taken with regard to waste management, which would be adequate in each phase of the waste dumping process. Environment-monitoring activities should comprise the observance of these measures. However, one answer to these issues is to convert the waste to usable products either through (a) the production of gaseous fuels, or (b) the production of liquid fuels, or (c) the production of solid fuels.

Such efforts may not only solve the depletion of fuels from fossil sources but also assist in the disposal of waste materials and the ensuing environment issues. However, before entering upon the process descriptions for waste conversion, it is necessary to understand the composition of domestic and industrial waste.

The term *waste* incorporates a number of streams in a modern society. Relevant to gasification, pyrolysis, incineration are streams such as chemical waste, medical waste, waste from auto-shredder, paper, plastics, textiles, metals, glass, and sundry other components which may form part of the industrial waste and domestic waste. Like feedstocks for any process, these raw materials require different processes for optimal operation.

Generating waste at current levels is incompatible with a sustainable future. While the problem of waste is serious, a variety of initiatives are being taken to address the various threats. These include moves toward waste minimization, waste segregation and recycling, cleaner production with regular waste audits, green chemistry, renewable energy and energy efficiency, and developing the concept of industrial ecosystems. The issues involved are much more than technical problem.

The historic approach to solid waste has been to bury it in landfill. This is becoming increasingly problematic because the diminishing availability of suitable landfill sites and the increasingly stringent conditions being applied to landfills mean that charges have increased and will continue to increase. Major problems associated with landfills are the leachate containing toxic heavy metals and the methane gas that is produced.

This is a major issue because of the variability of industrial solid, liquid, and gaseous wastes, and the capacity of modern processing industries to produce huge quantities of waste. Fortunately regulatory processes are now such that the numerous disasters caused previously should not be repeated. However there are remaining problem sites that constitute long-term hazards.

Thus, in order to reduce the amount of landfill, the amount of waste must be reduced. This involves either (a) cutting back on the use of many materials or (b) use of the waste by conversion to useful products. Either option will reduce the amount if waste sent to landfill sites. The first option certainly reduces the amount of landfill material but is often more difficult to achieve. However, the second option offers the attractive proposition of the production of fuel products. Thus, waste conversion becomes an attractive option to landfill disposal and the result is the generation of a usable product in the form of a gaseous, liquid, or solid fuel.

However, ancillary of the second option is the heterogeneity of waste material. In fact, it is obvious that many waste streams are not subject to direct processing and will require special measures in the form of specific pretreatment of the waste prior to processing.

One form of pretreatment is separation and recycling of waste components thereby removing a portion of the waste stream for recycling and other uses. The result of this separation at the source is the remaining residual waste (i.e., the waste stream from which recyclable materials have been removed) that is not destined for any use other than landfill is sent to the conversion reactor.

11.3 FUELS FROM WASTE INCINERATION

A significant number and variety of organic wastes are combusted in energy recovery systems including municipal solid waste, various forms of refuse-derived fuel produced from municipal solid waste as well as a broad range of other specific specialty wastes. The practice of incinerating waste has become increasingly prevalent in order to accomplish disposal in a cost-effective and environmentally sensitive manner. Combustion of such wastes reduces the volume of material which must be sent to a landfill, reduces the airborne emissions resulting from plant operations and landfill operations, and permits some economic benefit through energy recovery.

The technologies used to combust wastes depend on the form and location of components to be burned. The issues associated with energy generation from waste include fuel composition characteristics, combustion characteristics, formation and control of airborne emissions including both criteria pollutants and air toxics, for example, trace metals, and the characteristics of bottom- and fly-ash generated from waste combustion.

The concept of the incineration of domestic and industrial waste to gaseous fuels is one of the first suggestions from the disposal of waste and the generation of energy. With this in mind, waste incineration has been, and continues to be practice in many pasts of the world. In fact, co-combustion of waste in (large and efficient) coal-fired power plants is an especially attractive option because of the high conversion efficiency of these plants. However, there are issues that need attention and while it is not intended here to use the following comments to be a critique of waste incineration, here are lessons to be learned and changes to be made in this manner of waste removal.

Although not in keeping with the main context of this text, that is, the production of fuels, the incineration of wastes is worthy of mention insofar as waste combustion (although the main goal of incineration is volume reduction with the sterilization of the waste as a significant side-effect) can be a source of energy. Thus, the incineration process may also be used to produce steam and electricity. However, as the moisture content of the waste increases a self-sustaining combustion process is not possible. Nevertheless, domestic waste and industrial waste are feedstocks for incinerators.

Nevertheless, incineration is recognized as a waste treatment technology for disposal of the organic constituents of the waste. Incineration and other high temperature waste treatment systems convert the waste into ash, flue gases, particulate, and heat, which can in turn be used to generate electricity. The flue gases are cleaned for pollutants before it is dispersed in the atmosphere. Furthermore, incineration with energy recovery is one of several waste-to-energy technologies such as gasification and anaerobic digestion.

Modern incinerators reduce the volume of the original waste by 95 to 96 percent, depending upon composition and degree of recovery of materials such as metals from the ash for recycling. While incineration does not replace landfilling, it does reduce the necessary volume significantly. In fact, incineration has particularly strong benefits for the treatment of certain waste types in niche areas such as clinical wastes and certain hazardous wastes where pathogens and toxins can be destroyed by high temperatures.

The incinerator is, simply, a furnace for burning refuse and modern incinerators include pollution mitigation equipment such as flue gas cleaning. There are various types of incinerator plant design: (a) simple incinerator, (b) fixed or moving grate incinerator, (c) rotary-kiln incinerator, and (d) the fluidized bed incinerator.

The *simple incinerator* is an older and simpler kind of incinerator that is, essentially, a brick-lined cell with a metal grate over a lower ash pit, with one opening at the top or side for loading and another opening at the side for removing incombustible solids called clinkers. Many small incinerators formerly found in apartment houses have now been replaced by waste compactors.

The *fixed* or *moving grate incinerator* is a large fixed hearth incinerator with a moving grate. The moving grate enables the movement of waste through the combustion chamber to be optimized to allow a more efficient and complete combustion. These incinerators are typically used for combustion of municipal wastes, and are thus referred to as MSWIs (municipal solid waste incinerators).

In the fixed or moving grate incinerator, the waste is introduced by a waste crane through the "throat" at one end of the grate, from where it moves down over the descending grate to the ash pit in the other end. Here the ash is removed through a water lock. Part of the combustion air (primary combustion air) is supplied through the grate from below. This air flow also has the purpose of cooling the grate itself. Cooling is important for the mechanical strength of the grate, and many moving grates are also water cooled internally. Secondary combustion air is supplied into the boiler at high speed through nozzles over the grate. It facilitates complete combustion of the flue gases by introducing turbulence for better mixing and by ensuring a surplus of oxygen.

The incinerator must be designed to ensure that the flue gases reach a temperature of at least 850°C in order to ensure proper breakdown of organic toxins. This includes backup auxiliary burners (often fueled by oil), which are fired into the boiler in case the heating value of the waste becomes too low to reach this temperature alone. The flue gases are then cooled by heat transfer and heat the steam to typically 400°C at a pressure of 550 to 600 psi bar for the electricity generation in the turbine. At this point, the flue gas has a temperature of around 200°C, and is passed to the flue-gas-cleaning system.

The *rotary-kiln incinerator* has a primary chamber and secondary chamber. The primary chamber consists of an inclined refractory lined cylindrical tube. Movement of the cylinder on its axis facilitates movement of waste. In the primary chamber, there is conversion of solid fraction to gases, through volatilization, destructive distillation, and partial combustion reactions. The secondary chamber is necessary to complete gas phase combustion reactions.

The clinker spills out at the end of the cylinder. A tall flue gas stack, fan, or steam jet supplies the needed draft. Ash drops through the grate, but many particles are carried along with the hot gases. The particles and any combustible gases may be combusted in an *afterburner*. To control air pollution, the combustion product gases are further treated with acid gas scrubbers to remove sulfuric acid and nitric acid emissions, and then routed through bag houses to remove particulates before the gases are released into the atmosphere.

The *fluidized bed incinerator* uses a strong airflow through a sand bed until a point is reached where the sand particles separate to let the air through and mixing and churning occurs, thus a fluidized bed is created and fuel and waste can now be introduced. The sand with the pretreated waste and/or fuel is kept suspended on pumped air currents and takes on a fluidlike character. The bed is thereby thoroughly mixed and agitated keeping small inert particles and air in a fluidlike state. This allows all of the mass of waste, fuel, and sand to be fully circulated through the furnace.

The heat produced by an incinerator can be used to generate steam which may then be used to drive a turbine in order to produce electricity. However, the amount of energy produced is very much dependant upon the type and composition of the waste used as the feedstock. Much of the current thinking involves coincineration of a specified amount of the waste (calculated on the basis of the carbon content of the waste) with, for example, coal to produce a consistent amount of energy.

Incineration has a number of outputs such as the ash and the emission to the atmosphere of flue gas. Before the flue gas cleaning, the flue gases may contain significant amounts of particulate matter, heavy metals, dioxins, furans, sulfur dioxide, hydrochloric acid, and polynuclear aromatic hydrocarbons (carcinogens).

The most publicized concerns about the incineration of municipal solid wastes involve the fear that it produces significant amounts of dioxin and furan emissions. Dioxins and furans are considered by many to be serious health hazards. Older generation incinerators that were not equipped with modern gas-cleaning technologies were indeed significant sources of dioxin emissions. However, modern incinerators (due to advances in emission control designs and stringent regulations) must limit and even mitigate such emissions.

The quantity of pollutants in the flue gas from incineration plants is reduced by several processes.

Particulate matter is collected by particle filtration, most often electrostatic precipitators and/or baghouse filters. Hydrogen chloride and sulfur dioxide are removed in scrubbers or as dry desulfurization by injection limestone slurry into the flue gas before the particle filtration. Waste water from scrubbers must subsequently pass through a waste-water treatment plant. Nitrogen oxide (NO_x) emissions are either reduced by catalytic reduction with ammonia in a catalytic converter (selective catalytic reduction) or by a high temperature reaction with ammonia in the furnace (selective noncatalytic reduction). Heavy metals are often adsorbed on injected active carbon powder, which is collected by the particle filtration.

Incineration produces fly ash and bottom ash just as is the case when coal is combusted. The total amount of ash produced by municipal solid waste incineration ranges from 15 to 20 percent by weight of the waste and the fly ash amounts to 0 to 10 percent of the total ash. The fly ash, by far, constitutes more of a potential health hazard than does the bottom ash because the fly ash often contains high concentrations of heavy metals such as lead, cadmium, copper, and zinc as well as small amounts of dioxins and furans. The bottom ash seldom contains significant levels of heavy metals.

While fly ash is always regarded as hazardous waste, bottom ash is generally considered safe for regular landfill after a certain level of testing defined by the local legislation. Ash, which is considered hazardous, may generally only be disposed of in landfills which are carefully designed to prevent pollutants in the ash from leaching into underground aquifers—or after chemical treatment to reduce its leaching characteristics.

In spite of the promise shown by the use of incinerators for waste disposal and, in some cases, energy production (in the form of fuel gases), the use of incinerators for waste management is controversial. On one hand, it is felt that:

- The concerns over the health effects of dioxin and furan emissions have been significantly lessened by advances in emission control designs and very stringent new government regulations that have resulted in large reductions in the amount of dioxins and furans emissions.
- Modern incineration plants generate electricity and heat that can be sold to the regional electric grid and can sell steam to district heating systems or industrial customers.
- The bottom ash residue remaining after combustion has been shown to be a non-hazardous solid waste that can be safely landfilled or possibly reused.
- In densely populated areas, finding space for additional landfills is becoming increasingly difficult.
- Incineration of municipal solid waste avoids the release of carbon dioxide and methane. Every ton of municipal solid waste incinerated prevents about 1 ton of carbon dioxide equivalents from being released to the atmosphere.
- Incineration of medical waste and sewage sludge produces an end product ash that is sterile and nonhazardous.

On the other hand, it is felt that:

- The end-product ash must still be safely disposed. Many countries deny the use of incineration as an option for the treatment of their municipal solid wastes because of the potential of environment pollution.
- There are still concerns by many about the health effects of dioxin and furan emissions into the atmosphere from old incinerators.
- Incinerators emit varying levels of heavy metals such as vanadium, manganese, chromium, nickel, arsenic, mercury, lead, and cadmium, which can be toxic at very minute levels.
- Other advanced alternative technologies are available such as biologic treatments combined with anaerobic digestion and gasification.
- Incinerators produce soot particles [$PM_{2.5}$, nanoparticles or particles with a size less than 2.5 μm (i.e., $<1 \times 10^{-6}$ m)] in the furnace. Even with modern particle filtering of the flue gases (e.g., baghouse filters) some of these small particles are emitted to the atmosphere.
- Although waste combustion can be used to generate power, a portion of that power is consumed by fans, pumps, and other electrically powered components.

However, with the ban on landfilling untreated waste in various countries, modern design waste-to-energy plants have been built in the last decade, with more under construction and there is renewed interest in waste-to-energy.

In summary, incineration is not a complete method of disposal, its main advantage is that it produces a residue that is substantially reduced in volume and may be relatively inert. In addition, the efficiency of an incinerator is, on occasion, due to excessive throughput or choice of operating parameters that are less that desired. For example, the operating temperature inside the plant should be at the specified temperature (usually about 850°C) to reduce the possible effects of the most dangerous air pollutants and the production of dioxins. A lower than specified combustion temperature implies incomplete incineration, leaving part of the waste in the ashes and other combustion residues. Furthermore, the exhaust gases may be unnecessarily polluting because of the low combustion temperature and the by-products of such combustion may be sent to an unprotected landfill for disposal.

11.4 GASEOUS FUELS

There are two ways of producing gaseous fuel from waste: (a) gasification of the waste at high temperatures by partial oxidation and then conversion of materials containing carbon into synthesis gas (mainly hydrogen and carbon monoxide) and (b) production of biogas, mainly methane, by the anaerobic digestion of waste.

Gasification is the general term used for processes where heat is used to transform a feedstock, usually a solid feedstock such as coal (Chap. 5) or biomass (Chap. 8) or wood (Chap. 10) into a gaseous fuel. Similarly, the gasification process is also applicable to the conversion of any solid carbonaceous waste to a clean burning gaseous fuel. Whether starting with coffee grounds, municipal trash, or junk tires, the end product is a flexible gaseous fuel and with the relevant process modifications production of liquid and solid fuels is also possible.

Biochemical conversion options can be divided into digestion (production of biogas, a mixture of mainly methane and carbon dioxide) and fermentation (production of ethanol). With respect to thermochemical conversion options, a distinction can be made between combustion, gasification, and pyrolysis.

Anaerobic digestion is a process whereby organic waste is broken down in a controlled, oxygen free environment by bacteria naturally occurring in the waste material. Methane rich biogas is produced thus facilitating renewable energy generation. As a result, materials that are currently going to landfill can be utilized; natural methane emissions are reduced and conventional generation with its associated carbon emissions is displaced. The residual nutrient rich liquor and digestate is suitable for use as fertilizer on the farmland surrounding such a plant, reducing the need for artificial fertilizer. In summary, anaerobic digestion is a form of composting and its use for the conversion of waste to gaseous fuel by anaerobic processes is covered elsewhere in this text (Chap. 12).

The concept of the production of fuel gas by gasification of waste is not new (Rensfelt and Östman, 1996). Whilst solid wastes were somewhat in focus for the gasification and pyrolysis activities, around 1970, biomass gasification took over the central role when the oil prices went up after 1973. This is reflected in availability of a wide range of processes that has been discussed elsewhere (Rensfelt and Östman, 1996) and will not be repeated here. Most of the processes were designed for wood or wood waste (Chap. 10) and, surprisingly, some of these processes have also been proposed for waste handling and are taking up the move to commercialization that was last evident two decades ago.

11.4.1 Chemistry

The principle behind waste gasification and the production of gaseous fuels is that that waste contains carbon and it is this carbon that is converted to gaseous products via the usual gasification chemistry. Thus, when fuel is fed to a gasifier, water and volatile matter are released fast and a char residue is left to react further. The char gasification is what mainly controls the conversion achieved in the process. From solid carbon, product gas is formed according to the following main reactions:

$$C + H_2O \rightarrow CO + H_2$$

$$C + CO_2 \rightarrow 2CO$$

$$C + 2H_2 \rightarrow CH_4$$

In addition to the reactions of solid carbon, the most important reaction is the water-gas shift reaction, which takes place in the gas phase:

$$CO + H_2O \leftrightarrow H_2 + CO_2$$

The product gas generally contains large amounts of hydrogen and carbon monoxide and a small amount of methane, as well as carbon dioxide and steam, and in air gasification nitrogen. In addition, a significant amount of other organic components in the gas, known as tar, is formed. Tar formation is a well-known phenomenon in the thermal reactions of coal (Speight, 1994 and references cited therein) and can also be anticipated to form in the gasification or pyrolysis of any complex carbonaceous material.

Biomass-based fuels (and many components of domestic and industrial waste can be included in this category) are non-fossil fuels. Vegetative biomass can be classified into the categories woody, herbaceous, agricultural by-products, energy crops, and black liquor (a by-product of the forest industry) (Grace et al. 1989). Because of the high concentration of organic materials in industrial waste and in domestic waste (including discarded food), many constituents of waste are actually biomass. But the issues arise from the other constituents of the waste that make the waste truly heterogeneous and, in many cases, of indeterminate composition.

11.4.2 Reactors

Most of the reactors were of the fixed bed type (updraft or downdraft) and are considered to be adequate processes for local use in terms of the alternate, that is, landfilling the waste. However, whether wastes should be handled locally or in large-scale, centralized units is another issue that is still under debate.

Thus, a process for waste gasification (Fig. 11.1) must be chosen to reflect the character of the waste relation to the main questions in this report as stated in the introduction.

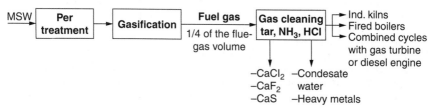

FIGURE 11.1 Process Flow for the Gasification of Municipal Solid Waste (MSW).

However, few of the original processes touted as potential methods for waste gasification may not be in current commercial use and reference must be made to gasification processes (for coal) that are in current commercial use (Chap. 5).

The reason for the coverall lack of commercial processes for waste gasification hinges on several areas.

For example, most of the processes proposed for waste gasification did not include a separation step. Feedstock homogeneity is a prerequisite for many gasifiers and feedstock heterogeneity and process scale-up can lead to intended to mechanical to a number of mechanical problems, shut-downs, sintering, and hot sots leading to corrosion and eventual failure of the reactor wall.

On the other hand, it is only recently that the fundamental knowledge about waste and gasification has been understood. It is not only conversion of the carbon in the waste to gaseous products but the heterogeneity of the noncarbon constituents of the waste and their influence on the process chemistry was difficult to estimate. In fact, the gasification process is a complex chemical process that was seriously underestimated, even misunderstood, by many engineers. Indeed, several of the processes were merely considered as *thermal process* without any reference to the details of the chemistry of the conversion.

In addition, most of these process efforts included use of a fixed bed reactor on the basis of applicability of coal gasification is also applicable to waste. The most common equipment was a shaft reactor with a bottom temperature of approximately 1000°C leading to transport problems within the reactor bed and in the shaft to the ash outlet, such as ash sintering. The character of coal ash and its effect on gasification is well understood. The effect of ash from waste is not yet fully understood and requires extra effort before a suitable reactor can be proposed for waste gasification.

Finally, most of the systems proposed for waste gasification produced tar or a mixture of tar and gas and very few of the processes included gas cleaning. Thus, the tar-rich gas caused problems on the gas side as well as tar condensation in the pipes to the combustor. On the other hand, waste that contained metals, glass, and inorganic materials which sinter and melt at higher temperatures also caused reactor problems.

Overall the heterogeneous composition of domestic and industrial waste has not been beneficial in terms of the gasification of the waste. Recycling programs in which the domestic waste is first separated into various components may be a necessary prelude to waste gasification. With energy prices (crude oil and natural gas) currently at an all-time high, and threatening to continue at high levels for the foreseeable future, the time is now for development of waste gasification processes.

11.5 LIQUID FUELS

Using ethanol as the example, numerous waste streams could be exploited for ethanol production. They are often inexpensive to obtain, and in many instances they have a negative value attributable to current disposal costs. Some principal waste streams currently under consideration include mixed paper from municipal solid waste, cellulosic fiber fines from recycled paper mills, as well as various biofeedstocks (Chap. 8). However, each waste stream has its own unique characteristics, and they generally vary from one source or time to another. Therefore, ethanol or, for that matter, the production of liquid fuels from waste streams is not only feedstock dependent but is also process dependent.

11.5.1 Hydrocarbon Fuels

Using the term loosely, the standard way of preparing liquid fuels from carbonaceous material is (a) gasify the feedstock to produce synthesis gas and (b) convert the synthesis gas by the Fischer-Tropsch reaction to hydrocarbons (Chap. 7).

Briefly, synthesis gas (*syngas*) (Chap. 7) is the name given to a gas mixture that contains varying amounts of carbon monoxide and hydrogen generated by the gasification of a carbonaceous fuel to a gaseous product with a heating value. Examples include steam reforming of natural gas or liquid hydrocarbons to produce hydrogen, the gasification of coal and in some types of waste-to-energy gasification facilities.

Synthesis gas consists primarily of carbon monoxide, carbon dioxide, and hydrogen, and has less than half the energy density of natural gas. Synthesis gas is combustible and often used as a fuel source or as an intermediate for the production of other chemicals. Synthesis gas for use as a fuel is most often produced by gasification of coal or municipal waste mainly by the following paths:

$$C + O_2 \rightarrow CO_2$$

$$CO_2 + C \rightarrow 2CO$$

$$C + H_2O \rightarrow CO + H_2$$

When used as an intermediate in the large-scale industrial synthesis of hydrogen and ammonia, it is also produced from natural gas (via the steam-reforming reaction) as follows:

$$CH_4 + H_2O \rightarrow CO + 3H_2$$

The Fischer-Tropsch synthesis is, in principle, a carbon-chain-building process, where methylene groups are attached to the carbon chain. The actual reactions that occur have been, and remain, a matter of controversy, as it has been the last century since 1930s.

Even though the overall Fischer-Tropsch process is described by the following chemical equation:

$$(2n+1)H_2 + nCO \rightarrow C_nH_{(2n+2)} + nH_2O$$

The initial reactants in the above reaction (i.e., CO and H_2) can be produced by other reactions such as the partial combustion of a hydrocarbon:

$$C_nH_{(2n+2)} + 1/2\ nO_2 \rightarrow (n+1)H_2 + nCO$$

for example (when n = 1), methane (in the case of gas-to-liquids applications):

$$2CH_4 + O_2 \rightarrow 4H_2 + 2CO$$

Or by the gasification of any carbonaceous source, such as biomass and, in the current context, carbonaceous waste:

$$C + H_2O \rightarrow H_2 + CO$$

The energy needed for this endothermic reaction is usually provided by (exothermic) combustion with air or oxygen:

$$2C + O_2 \rightarrow 2CO$$

A modification of this concept involves the proposed addition of hydrogen from a carbon-free energy source, such as solar or nuclear power, during the gasification step (Agrawal et al., 2007). The theoretical concept is a hybrid process insofar as liquid hydrocarbon fuels are proposed to be produced from biomass wherein biomass is the carbon source and hydrogen is supplied from carbon-free energy. Modeling of this biomass-to-liquids process indicates that the concept has the potential to provide the transportation sector for the foreseeable future, using the existing infrastructure.

In addition, although not mentioned because the concept is modeled on the complete gasification of the feedstock, the addition of hydrogen will not only suppress the formation of carbon dioxide but also the formation of other by-products thereby increasing the efficiency of the process. The presence of hydrogen will also reduce the amount of char and tarry by-products which is in complete agreement with the observed beneficial effect of hydroprocesses in petroleum refineries has been known for decades (Speight, 2007; Ancheyta and Speight, 2007). The hydrogen and energy balance may be even better if water (also produced during gasification) (Chaps. 5 and 8) could be used as the hydrogen source.

The pyrolysis of waste tires is also a source of liquid fuels; the process by which this is achieved is often called *thermal depolymerization* (TDP). This term, however, is more a term of convenience since it is rare that depolymerization, in the true scientific sense, actually occurs.

Thermal depolymerization is a term that is used to describe the reduction of complex organic materials into lower molecular weight products. Nevertheless, in this context, thermal cracking and catalytic cracking as used in the petroleum industry (Chap. 3) are examples of thermal depolymerization. The feedstocks for such processes may also include non-oil-based waste products, such as old tires, offal, wood, and plastic. However, the so-called thermal depolymerization process does not in any way mimic the natural geologic processes thought to be involved in the production of fossil fuels, such as petroleum and natural gas.

In the thermal depolymerization process, conversion efficiencies can be very high but is feedstock dependent. Quantifying the potential amount of product (e.g., biodiesel) can be done on a feedstock carbon versus product carbon basis but such estimates are only approximate.

Quantifying the amount of oil which could be produced from tire pyrolysis can be done in a similar manner to quantifying the amount of biodiesel produced from waste vegetable oil. It is simply a case of equating the amount of waste product available to the production yields. In the case of tire pyrolysis the number of used tires produced each year is well documented. What is less certain however is the amount of suitable road transport oil that can be yielded from the pyrolysis process itself.

11.5.2 Biodiesel

Another source of waste for the production of liquid fuels is waste cooking oil; usually biodiesel is the desired product (Zhang et al., 2003). In fact, waste cooking oil is a valuable asset for conversion to liquid fuels and the quantity of biodiesel which could be produced from waste cooking oil can be quantified in a manner similar to that of biodiesel produced direct from agriculture. The only difference is that rather than calculating crop yields it is necessary to quantify the amount of waste oil available.

Biodiesel is a light- to dark-yellow liquid that is immiscible with water and has a high boiling point and low vapor pressure. Typical methyl ester biodiesel has a flash point of approximately 150°C (302°F), making it nonflammable. Biodiesel has a density of 0.88 g/cm^3. Biodiesel has a viscosity similar to *petrodiesel* (the current term to differentiate the different forms of diesel) and can be used as an additive in formulations of diesel to increase the lubricity of pure ultra-low sulfur diesel (ULSD) fuel. Much of the world uses a system known as the "B" factor to state the amount of biodiesel in any fuel mix, in contrast to the "BA" or "E" system used for ethanol mixes. For example, fuel containing 20 percent biodiesel is labeled B20 whereas pure biodiesel is referred to as B100.

Biodiesel is a renewable fuel that can be manufactured from algae, vegetable oils, animal fats, or recycled restaurant greases; it can be produced locally in most countries. It is safe, biodegradable, and reduces air pollutants, such as particulates, carbon monoxide, and hydrocarbons. Blends of 20 percent biodiesel with 80 percent petroleum diesel (B20) can generally be used in unmodified diesel engines. Biodiesel can also be used in its pure form (B100), but may require certain engine modifications to avoid maintenance and performance problems.

Biodiesel is an alternative to petroleum-based diesel but it is produced from biodegradable materials and can be used as fuel in diesel engines. It can also be used in boilers or furnaces designed to use heating oils or in oil-fueled lighting equipment or used neat (100 percent biodiesel), or it can be blended with petroleum diesel.

Biodiesel is made by chemically reacting vegetable oil or animal fat (or combinations of oils and fats) with alcohol (usually nearly pure methanol or ethanol) and a catalyst (sodium hydroxide). The chemical reaction converts constituents of the vegetable oil into an ester (*biodiesel fuel*) and glycerol. Since the biodiesel is less dense than the glycerol, it floats on top of the glycerol and may be pumped off, or the glycerol can be drained off the bottom. The fuel can then be filtered and used in heating or lighting applications. It is possible to use the product in diesel engines without further processing, but caution is advised because of the potential for engine damage and it is recommended that impurities (unreacted alcohol and sodium hydroxide) are removed by a washing.

Almost any variety of oil or grease (from new food-grade vegetable oil to used cooking oil or trap grease to waste water treatment-plant grease) can be converted into biodiesel. However, the amounts of reactants (oil, methanol, and sodium hydroxide) vary to some degree, depending on what oil is used. The amount of methanol and sodium hydroxide must be sufficient to react with the vegetable oil, but excessive amounts of these reactants should not be used. Just as an engine requires excess air to be sure that all the fuel burns, it takes excess methanol to be sure all of the oil reacts.

The chemistry of biodiesel production is known as transesterification insofar as the process involves reaction of glyceride-containing plant oil with a short chain alcohol such as methanol or ethanol:

$$\begin{bmatrix} O-OCR_1 \\ O-OCR_2 \\ O-OCR_3 \end{bmatrix} \xrightarrow[\text{3 EtOH}]{\text{Cat. base}} R_1 COOEt + R_2 COOEt + R_3 COOEt + C_3H_5(OH)_3$$

Triglyceride / Ethyl esters of fatty acids / Glycerol

Animal and plant fats and oils are typically made of triglycerides which are esters of free fatty acids with the trihydric alcohol, glycerol. In the transesterification reaction, the alcohol is deprotonated with a base to make it a stronger nucleophile. Commonly, ethanol or methanol is used. Usually, this reaction will proceed either exceedingly slowly or not at all. Heat, as well as an acid or base are used to help the reaction proceed more quickly.

Almost all biodiesel is produced using the base-catalyzed technique as it is the most economical process requiring only low temperatures and pressures and producing over 98 percent conversion yield (provided the starting oil is low in moisture and free fatty acids).

The major steps required to synthesize biodiesel are:

Purification: If waste vegetable oil is used, it is filtered to remove dirt, charred food, and other nonoil material often found. Water is removed because its presence causes the triglycerides to hydrolyze to give salts of the fatty acids instead of undergoing transesterification to give biodiesel. The crude oil may be stirred with a drying agent such as magnesium sulfate to remove the water in the form of water of crystallization. The drying agent can be separated by decanting or by filtration but the viscosity of the oil may not allow the drying agent to mix thoroughly.

Neutralization of free fatty acids: A sample of the cleaned oil is titrated against a standard solution of base in order to determine the concentration of free fatty acids (RCOOH) present in the waste vegetable oil sample. The quantity (in moles) of base required to neutralize the acid is then calculated.

Transesterification: While adding the base, a slight excess is factored in to provide the catalyst for the transesterification. The calculated quantity of base (usually sodium hydroxide) is added slowly to the alcohol and it is stirred until it dissolves. Sufficient alcohol is added to make up three full equivalents of the triglyceride, and an excess is added to drive the reaction to completion. The solution of sodium hydroxide in the alcohol is then added to a warm solution of the waste oil, and the mixture is heated (typically 50°C) for several hours (4–8 hours typically) to allow the transesterification to proceed. A condenser may be used to prevent the evaporative losses of the alcohol. Care must be taken not to create a closed system which can explode.

Workup: Once the reaction is complete, the glycerol should sink. When ethanol is used, it is reported that an emulsion often forms. This emulsion can be broken by standing, centrifugation, or the addition of a low-boiling (easily removed) nonpolar solvent, decanting, and distilling. The top layer, a mixture of biodiesel and alcohol, is decanted. The excess alcohol can be distilled off, or it can be extracted with water. If the latter, the biodiesel should be dried by distillation or with a drying agent.

11.5.3 Ethanol

Ethanol is produced from organic feedstocks, such as biomass, by a fermentation process. However, recent efforts have focused on the production of ethanol from waste materials using a gasification process (http://www.syntecbiofuel.com/technology.html).

The overall process consists of a thermochemical conversion of synthesis gas which is then converted to higher molecular weight alcohols by a catalytic process after which high purity ethanol is separated by distillation.

The process is similar to modern day methanol and gas-to-liquids processes (Chaps. 2 and 7). The key differentiating factors are the catalysts and their operating parameters.

Most current research efforts have centered on bacteria, enzyme, and acid/solvent hydrolysis fermentation based ethanol production. However, the action of bacteria and enzymes are often very dependent on the feedstock. In addition, these processes generate a small quantity of ethanol over a period of days and the dilute aqueous ethanol product must be distilled to recover the ethanol. In the gasification-synthesis process, various carbonaceous feedstocks can be used, with appropriate modifications in the syngas production step.

A new system for converting trash into ethanol and methanol could help reduce the amount of waste piling up in landfills while displacing a large fraction of the fossil fuels used to power vehicles in the United States (Bullis, 2007). The technology developed does not incinerate refuse, so it does not produce the pollutants that have historically plagued efforts to convert waste into energy. Instead, the technology vaporizes organic materials to produce synthesis gas (Chap. 7) that can be used to synthesize a wide variety of fuels and chemicals. In addition to processing municipal waste, the technology can be used to create ethanol out of agricultural biomass waste, providing a potentially less expensive way to make ethanol than current corn-based plants.

The new system makes synthesis gas in two stages. In the first stage, waste is heated in a 1200°C chamber into which a controlled amount of oxygen is added to partially oxidize carbon and free hydrogen. Not all of the organic material is converted and some forms char which is then gasified when researchers pass it through arcs of plasma. The remaining inorganic materials, including toxic substances, are oxidized and incorporated into a pool of molten glass which hardens into a material that can be used for building roads or discarded as a safe material in landfills. The second stage is a catalyst-based process for converting synthesis gas into equal ethanol and methanol.

$$CO + 2H_2 \rightarrow CH_3OH$$

$$CO + 3H_2 \rightarrow C_2H_5OH$$

11.5.4 Other Alcohols

It is fairly common knowledge that the alcohols can be made from organic materials by fermentation. There is, however, a potential for the production of alcohols from organic waste.

Historically, the production of methanol, ethanol, and higher molecular alcohols from syngas has been known since the beginning of the last century. There are several processes that can be used to make mixed alcohols from synthesis gas including isosynthesis, variants of Fischer-Tropsch synthesis, oxosynthesis involving the hydroformylation of olefins, and homologation of methanol and lower molecular weight alcohols to make higher alcohols. However, in the context of the Fischer-Tropsch process, depending on the process and its operating conditions, the most abundant products are usually methanol and carbon dioxide but methanol can be recycled to produce higher molecular weight alcohols.

With the development of various gas-to-liquids processes (Chaps. 2 and 7) it was recognized that higher alcohols were by-products of these processes when catalysts or conditions were not optimized. Modified Fischer-Tropsch (or methanol synthesis) catalysts can be promoted with alkali metals to shift the products toward higher alcohols. Synthesis of higher molecular weight alcohols is optimal at higher temperatures and lower space velocities compared to methanol synthesis and with a ration of hydrogen/carbon monoxide ratio of approximately 1 rather than 2 or greater.

The first step in the synthesis to ethanol and higher alcohols is the formation of a carbon-carbon bond. Linear alcohols are then produced in stepwise fashion:

$$nCO + 2nH_2 \rightarrow C_nH_{(2n+1)}OH + (n-1)H_2O$$

Stoichiometry suggests that the carbon monoxide/hydrogen ratio is optimum at 2, but the simultaneous presence of water-gas shift leads to an optimum ratio closer to 1.

As in other synthesis gas conversion processes, the synthesis of higher molecular weight alcohols generates significant heat and an important aspect is choice of the proper reactor to maintain even temperature control which then maintains catalyst activity and selectivity. In fact the synthesis of higher molecular weight alcohols is carried out in reactors similar to those used in methanol and Fischer-Tropsch synthesis. These include shell and tube reactors with shell-side cooling, trickle bed, and slurry bed reactors.

Catalysts for the synthesis of higher molecular weight alcohols generally fall mainly into four groups: (a) modified high pressure methanol synthesis catalysts, such as alkali-doped ZnO/Cr_2O_3, (b) modified low pressure methanol catalysts, such as alkali-doped Cu/ZnO

and $Cu/ZnO/Al_2O_3$, (c) modified Fischer-Tropsch catalysts, such as alkali-doped $CuO/CoO/Al_2O_3$, and (d) alkali-doped sulfides, such as mainly molybdenum sulfide (MoS_2).

In the process, the feedstock enters the process (Fig. 11.2) and is converted to synthesis gas (Chap. 7) with the desired carbon monoxide/hydrogen ratio which is then reacted, in the presence of a catalyst, into methanol (CH_3OH), ethanol (CH_3CH_2OH), and higher molecular weight alcohols.

$$CO + 2H_2 \rightarrow CH_3OH$$

$$CO + 3H_2 \rightarrow C_2H_5OH$$

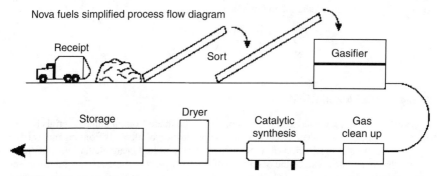

FIGURE 11.2 Alcohols from waste.

The catalytic synthesis process makes several different alcohols depending, in part, on residence time in the reactor and the nature of the catalyst. The alcohols can be separated by distillation and dried to remove water.

Another aspect of this concept is to produce the alcohols by a catalytic method in which the catalyst is in the form of unsupported nano-sized particles or supported on a high surface area support such as carbon, alumina, or silica (Mahajan and Jackson, 2004). In either arrangement the nano-sized catalyst is suspended in an inert solvent, such as a high molecular weight hydrocarbon solvent to form a slurry. The synthesis gas is then passed through the catalyst slurry to produce alcohols in the product stream. The operating temperature range is from 200 to 300°C at a pressure ranges from 500 to 3000 psi.

A further aspect of the waste-to-alcohols concept is the use of a plasma field in which temperatures are reputed (but not yet proved) to reach 30,000°C. The feedstock can be materials such as waste coal, used tires, wood wastes, raw sewage, municipal solid wastes, biomass, discarded roofing shingles, coal waste (*culm*), and discarded corn stalks. The plasma field breaks down the feedstock into their core elements in a clean and efficient manner.

11.6 SOLID FUELS

The use of organic waste as cooking fuel in both rural and urban areas is not new. In seventeenth-century England, the rural poor often burned dried cow dung because of the acute shortage of wood fuel due to widespread deforestation (Lardinois and Klundert, 1993). In some countries, cattle and buffalo dung is still used as relatively good cooking fuel. In the beginning of the nineteenth century, sawdust briquettes were made with binding materials such as tar, resins, and clay bind the small particles together. None of these processes attained great importance because of their relatively high costs compared to wood and conventional charcoal fuels (Lardinois and Klundert, 1993).

Fuel briquettes emerged as a significant business enterprise in the twinteeth century. In the 1950s, several economic methods were developed to make briquettes without a binder. A multitude of factories throughout the world produced literally tens of millions of tons of usable and economic material that met the household and industrial energy needs. During the two world wars, households in many European countries made their own briquettes from soaked waste paper and other combustible domestic waste using simple lever-operated presses (Lardinois and Klundert, 1993). Modern industrial briquetting machines, although much larger and more complex, operate on the same principle although the marketed briquettes are now sold at a premium for occasional backyard barbeques rather than for everyday use.

For over 100 years informal waste collectors in Cairo have separated and dried organic waste products for sale as solid fuel for domestic use. This process faded somewhat when fossil fuel sources became available. Switching to conventional fuels may prove advantageous to those who can afford them, but given the economic and energy conditions in many cities, urban and agricultural wastes remain a viable alternative fuel.

Briquetting is undergoing resurgence, principally due to the convergence of three critical factors. First, the recent developments in briquette processing and binding have dramatically changed the economics of using fuel briquettes as an energy resource. Secondly, a shortage of fuel wood has become increasingly severe in most of the developing countries. Finally, there has been a steady increase by environmental concerns to address the problem of domestic and urban waste disposal, a dilemma that briquetting can help remedy.

11.6.1 Briquette Manufacture

Generally, briquette manufacture (*briquetting*) involves the collection of combustible materials that are not usable as such because of their low density, and compressing them into a solid fuel product of any convenient shape that can be burned like, wood or charcoal. Thus, the material is compressed to form a product of higher bulk density, lower moisture content, and uniform size, shape, and material properties. Briquettes are easier to package and store, cheaper to transport, more convenient to use, and their burning characteristics are better than those of the original organic waste material.

The raw material of a briquette must bind during compression; otherwise, when the briquette is removed from the mould, it will crumble. Improved cohesion can be obtained with a binder but also without, since under high temperature and pressure, some materials such as wood bind naturally. A binder must not cause smoke or gummy deposits, while the creation of excess dust must also be avoided. Two different sorts of binders may be employed. Combustible binders are prepared from natural or synthetic resins, animal manure or treated, dewatered sewage sludge. Noncombustible binders include clay, cement, and other adhesive minerals. Although combustible binders are preferable, noncombustible binders may be suitable if used in sufficiently low concentrations. For example, if organic waste is mixed with too much clay, the briquettes will not easily ignite or burn uniformly. Suitable binders include starch (5–10 percent) or molasses (15–25 percen) although their use can prove expensive (Lardinois and Klundert, 1993). It is important to identify additional, inexpensive materials to serve as briquette binders in Kenya and their optimum concentrations. The exact method of preparation depends upon the material being briquetted as illustrated in the following three cases of compressing sugar bagasse, sawdust, and urban waste into cooking briquettes.

11.6.2 Bagasse Briquettes

Surplus bagasse presents a disposal problem for many sugar factories (Keya, 2000). Briquetting technology remains simple, and involves the following steps: (a) size reduction in which the bagasse is chopped, rolled, or hammered, (b) drying in which moisture is removed by open air drying or by using forced, heated air in a large rotating drum, (c) carbonization in which the bagasse is combusted in a limited supply of oxygen in a buried pit or trench until it carbonizes into charcoal, (d) feedstock preparation in which the carbonized bagasse is mixed with a binder such as clay or molasses, (e) compaction and extrusion in which the material is passed through a machine-operated or manually-operated extruder to form *rolls* of charcoal, (f) drying in which the rolls are air-dried for 1 to 3 days, causing them to break into chunks, and (g) packaging in which the briquettes are made ready for sales.

11.6.3 Sawdust Briquettes

Sawdust is waste material from all types of primary and secondary wood processing. Between 10 and 13 percent of a log is reduced to sawdust in milling operations. Sawdust is bulky, and is therefore expensive to store and transport. Also, the calorific value of sawdust is quite low, so that briquetting is an ideal way to reduce the bulk, to increase the density, and thus to increase the calorific value. The equipments required for producing sawdust briquettes consist of a drier, a press and an extruder with a tapered screw, and a large revolving disk.

Sawdust briquettes are formed under sufficiently high pressure to produce cohesion between wood particles. In the process, the lignin softens and binds the briquette, so no additional binder is required. The use of sawdust briquettes has several advantages, including (a) price which is usually about the same as fuel wood but is much more convenient to use as they do not require further cutting and chopping, (b) good burning characteristics in any kind of solid fuel stove and boiler, (c) quick ignition followed by clean burning and only leaving 1 to 6 percent by weight ash, (d) sulfur-free and burn without producing an odor, and (e) a heat content of approximately 18,000 kJ/kg which is almost equivalent to the heat content of medium quality coal.

11.6.4 Urban Waste Briquettes

Solid waste disposal is one of the most serious urban environmental problems in developing countries. Many municipal authorities collect and adequately dispose (in places other than landfill sites) less than half of these wastes. This failure is attributed to (a) an inadequate number of landfill sites, (b) a variety of environmental regulations, (c) the absence of sufficient capacity for waste processing and recycling, and last, but not least, (d) the planned obsolescence of packaging and many of the items that form the basis of the waste (Kibwage, 2002).

Open or crude dumping is the most common method used by municipal authorities. Waste poses a health hazard when it lies scattered in the streets and at the dumping sites. It is now an accepted environmental philosophy that wastes have value and should be utilized based on principles of *reduce*, *reuse*, *recover*, and *recycle*. Through recycling, urban wastes can be transformed into useful products. Waste paper and leaves, in particular, provide a potentially important, alternative source of cooking fuel.

11.7 THE FUTURE

Energy from waste provides a huge opportunity for large cities. If all of the city waste that typically goes to a landfill (where it emits greenhouse gases such as methane) was utilized as a fuel source, it could generate enough to provide electricity to up to a million or more homes and provide heat to 500,000 homes (depending, of course, upon the size of the city).

Many large cities would be well advised to generate a renewable gases and liquid fuels supply chain from waste using non-incineration technologies including anaerobic digestion of organic waste and/or sewage producing biogas, and pyrolysis and gasification of wood paper, plastic, light industrial waste, producing synthetic diesel or synthesis gas. Renewable gases and liquid fuels are also hydrogen-rich fuels, contributing to the renewable energy economy.

Typically, a new facility comprising commercially available anaerobic digestion plant and pyrolysis/gasification plant would be located at a waste collection site and supply renewable gases to local mixed development either by direct connection, where the development is in proximity to the waste collection site, or by renewable gas pipeline or transportation (in compressed or liquid form), where the development is not in proximity to the waste collection site.

The renewable gases and liquid fuels will provide low-carbon fuels for low and zero carbon developments and transport applications, and to supply non-potable water from waste dewatering (squeezing liquid out of waste) which is part of the process for these alternative fuel technologies. Local low and zero carbon developments will be supplied with heat and electricity from fuel cell or combined cooling, heat, and power (CCHP or trigeneration), distributed via a district energy network, with local heat fired absorption chillers for chilled water services, displacing electric air conditioning and refrigeration via the heat-to-cool process of heat-fired absorption chillers and thereby significantly reducing electricity consumption, particularly in summer.

Renewable fuels from waste could be the single largest form of indigenous renewable energy in cities where there are enough potential waste management sites to supply the needs for low and zero carbon fuel.

In addition, recycling organic combustible materials into fuel briquettes contributes to solving urban needs such as income-generation, insufficient land for waste disposal, and maintaining environmental quality. Since the earth's resources are finite, greater resource recovery and utilization are essential to achieve an acceptable level of organic waste management. Enhancing the recovery of organic waste can restore various natural cycles, thus preventing the loss of raw materials, energy, and nutrients.

On the other hand, the demand for energy in many countries is expected to add to the emission of greenhouse gas through burning of fossil fuels. There is urgent need to promote climate-friendly technologies and fuel briquetting appears to be one such technology that addresses the multiple needs of society and the environment.

Current needs are focused on finding better binders for bagasse briquettes, improved calorific values and combustion by producing higher density briquettes, introducing more efficient extrusion methods, and reducing production costs.

Technologies which enable the production of high quality fuels from a variety of non-food feedstock sources and waste streams could provide an important alternative to biofuel feedstock sources which compete with uses for food.

11.8 REFERENCES

Agrawal, Singh, N. R., Ribeiro, F. H., and W. N. Delgass: "Sustainable Fuel for the Transportation Sector," *Proceedings of the National Academy of Science,* 104(12), 2007, pp. 482–3.

Ancheyta, J. and J. G. Speight: *Hydroprocessing of Heavy Oils and Residua,* CRC Press, Taylor and Francis Group, Boca Raton, Fla., 2007.

Bullis, K.: "Ethanl from Trash," Technology Review, Massachusetts Institute of Technology, 2007. http://www.technologyreview.com/Energy/18084/.

Grace, T. M., E. W. Malcom, and M. J. Kocurek: *Pulp and Paper Manufacture,* vol. 5. Alkaline Pulping, The Joint Textbook Committee Of The Paper Industry, Atlanta: Tappi, Ga., 1989.

Keya, N.C.O. 2000. Nzoia Sugar Company Annual Reports 1980-2000. Nzoia Sugar Company. Bungoma, Kenya.

Kibwage, J.K. 2002. Integrating the Informal Recycling Sector into the Solid Waste Management Planning in Nairobi City. Ph.D. Thesis, Maseno University, Kenya.

Lardinois, I and A. Klundert: "Organic Waste: Options for Small-Scale Resource Recovery," WASTE Consultants, The Netherlands, 1993.

Mahajan, D. and G. R. Jackson: "Method for the Production of Mixed Alcohols from Synthesis Gas," United States Patent 6753353, Jun. 22, 2004.

Rensfelt, E., and A. Östman:. "IEA Biomass Agreement," Task X, Biomass Utilization. Biomass Thermal Gasification and Gas Turbines Activity, Sub-task 6—Gasification of Waste, Report No. TPS 96/19, 1996.

Speight, J. G.: *The Chemistry and Technology of Coal,* 2nd ed., Marcel Dekker inc., New York, 1994.

Speight, J. G.: *The Chemistry and Technology of Petroleum,* 4th ed., CRC Press, Taylor and Francis Group, Boca Raton, Fla., 2007.

Zhang Y., M. A. Dubé, D. D. McLean, and M. Kates: *Bioresource Technology,* 89(1), 2003, 1–16.

CHAPTER 12
LANDFILL GAS

Landfill gas is produced by the microbial decomposition of landfilled waste in an oxygen free (anaerobic) atmosphere. The gas is approximately 65 percent methane and 35 percent carbon dioxide plus traces of other organic compounds and the composition varies with age and the type of waste (Table 12.1). The odor of landfill gas is associated with trace compounds such as hydrogen sulfide (H_2S), mercaptans (R-SH), and ethylene ($CH_2=CH_2$).

The rate and duration of gas production depends on the nature of the waste and oxygen availability. The gas can move or migrate in any direction within the site, depending on the permeability of the waste layers. It can also migrate off-site, if measures are not taken to prevent it. One indication that gas is escaping from the site and venting through the ground outside is the presence of yellow and dying vegetation.

Landfill gas, being predominantly methane, is flammable. (Methane has a flammable range of between 5 and 15 percent by volume in air.) As the gas migrates, it can accumulate in voids, underground services, buildings, and other such places resulting in a risk of fire and explosion.

A landfill site is an area of land that has been specifically engineered to allow for the deposition of waste on to and into it. Municipal solid waste landfill sites are a large source of human-related methane emissions and, in some countries such as the United States, can account for up to 25 percent of these emissions. At the same time, methane emissions from landfills represent a lost opportunity to capture and use a significant energy resource.

Historically, a landfill has been called a *dump* or a *midden* and has been the most convenient method of waste disposal for millennia. Landfills may include internal waste disposal sites (where a producer of waste carries out their own waste disposal at the place of production) as well as sites used by many producers. Many landfills are also used for other waste management purposes, such as the temporary storage, consolidation, and transfer, or processing of waste material (sorting, treatment, or recycling). A landfill also may refer to ground that has been filled in with soil and rocks instead of waste materials, so that it can be used for a specific purpose, such as for building houses.

A main concern with landfill gas is the possibility of gas accumulation inside buildings constructed on or close to a landfill site. Protection of new buildings is usually achieved using a dual barrier approach: low permeability gas membrane to resist the passage of gas, and some form of ventilation/extraction system to disperse the gas safely to the atmosphere. The requirements for particular developments will be specified by the relevant planning authorities. Gas monitoring may also be a requirement and, if so, this indicates that gas detectors must be installed at the most effective location.

TABLE 12.1 Typical Constituents of Landfill Gas

Component	Percent by volume	Characteristics
Methane	45–60	Methane is a naturally occurring gas. It is colorless and odorless. Landfills are the single largest source of U.S. man-made methane emissions.
Carbon dioxide	40–60	Carbon dioxide is naturally found at small concentrations in the atmosphere (0.03%). It is colorless, odorless, and slightly acidic.
Nitrogen	2–5	Nitrogen comprises approximately 79% of the atmosphere. It is odorless, tasteless, and colorless.
Oxygen	0.1–1	Oxygen comprises approximately 21% of the atmosphere. It is odorless, tasteless, and colorless.
Ammonia	0.1–1	Ammonia is a colorless gas with a pungent odor.
NMOCs (non-methane organic compounds)	0.01–0.6	NMOCs are organic compounds (i.e., compounds that contain carbon). (Methane is an organic compound but is not considered an NMOC.) NMOCs may occur naturally or be formed by synthetic chemical processes. NMOCs most commonly found in landfills include acrylonitrile, benzene, 1,1-dichloroethane, 1,2-cis dichloroethylene, dichloromethane, carbonyl sulfide, ethylbenzene, hexane, methyl ethyl ketone, tetrachloroethylene, toluene, trichloroethylene, vinyl chloride, and xylenes.
Sulfides	0–1	Sulfides (e.g., hydrogen sulfide, dimethyl sulfide, mercaptans) are naturally occurring gases that give the landfill gas mixture its rotten-egg smell. Sulfides can cause unpleasant odors even at very low concentrations.
Hydrogen	0–0.2	Hydrogen is an odorless, colorless gas.
Carbon monoxide	0–0.2	Carbon monoxide is an odorless, colorless gas.

Source: Tchobanoglous G, H. Theisen, and S. Vigil: *Integrated Solid Waste Management, Engineering Principles and Management Issues,* McGraw-Hill, Inc. New York, 1993, pp. 381–417.
US EPA: *Compilation of Air Pollutant Emissions Factors, AP-42,* 5th ed., vol. 1, Stationary Point and Area Sources, U.S. Environmental Protection Agency, Washington, D.C., Jan., 1995, sec. 2.4—"Municipal Solid Waste Landfills," http://www.epa.gov/ttn/chief/ap42/ch02/ and http://www.atsdr.cdc.gov/HAC/landfill/PDFs/Landfill_2001_ch2mod.pdf.

12.1 LANDFILL CLASSIFICATION

Landfill sites are classified according to the type(s) of waste material contained therein, namely:

1. *Hazardous waste landfill site*: Waste disposal units constructed to specific design criteria and which receive wastes meeting the local definition of hazardous waste. These landfills are generally constructed to be secure repositories for material that presents a serious hazard to human health, such as high-level radioactive waste. They are restricted, by permit or law, to the types of waste that they may handle (chemical vs. radioactive, liquid vs. dry). Double liner systems are the norm for hazardous waste landfills.

2. *Sanitary landfill site*: They are also called modern, engineered, or secure landfills; these usually have physical barriers such as liners and leachate collection systems,

and procedures to protect the public from exposure to the disposed wastes. The term sanitary landfill normally refers to those where municipal solid waste is disposed of, as well as other wastes high in organic material. In some countries, all landfills are sanitary landfills.

3. *Inert waste landfill site*: Waste disposal units that receive wastes which are chemically and physically stable and do not undergo decomposition, such as sand, bricks, concrete, or gravel.
4. *Dump site*: They are also simply called landfills, dumps are landfills that are not engineered with the special protective measures required by sanitary landfills. They are most common in rural, remote, and developing areas. Many jurisdictions prohibit the use of nonsanitary landfills for the disposal of municipal solid waste. Other jurisdictions that do allow dumps may require them to be constructed according to some engineering standard to mitigate the risk for environmental contamination, such as by limiting the slope, requiring compaction, or ensuring that the cell is high enough above the groundwater table.

Whatever the classification, the landfill site should be designed in such fashion that there is minimal harm to the environment and that the land may be restored in some way after the period of waste disposal. Despite its connotations as an *out of sight, out of mind* solution (Hester and Harrison, 2002, p. 28), landfill is a well used method of waste management that is applied to the majority of solid wastes in many countries.

The production of alternate fuels from landfill gas is an emerging area. Landfill gas has been successfully delivered to the natural gas pipeline system as both a high-Btu and medium-Btu fuel. Landfill gas has also been converted to vehicle fuel in the form of compressed natural gas (CNG), with a number of liquefied natural gas (LNG) and methanol production projects in the planning stage (Chap. 7) (Wisbrock, 2006).

Most modern landfill sites are engineered to contain liquid leachate or landfill gas produced by decomposing organic waste, even to the point of requiring a minimum of one landfill liner. This consists of a layer of compacted clay with a minimum required thickness and a maximum allowable hydraulic conductivity. A deep geologic repository of high level radioactive waste is not generally classified as a landfill site.

Municipal solid waste contains significant portions of organic materials that produce a variety of gaseous products when dumped, compacted, and covered in landfills. Anaerobic bacteria thrive in the oxygen-free environment, resulting in the decomposition of the organic materials and the production of primarily carbon dioxide and methane. Carbon dioxide is likely to leach out of the landfill because it is soluble in water. Methane, on the other hand, which is less soluble in water and lighter than air, is likely to migrate out of the landfill. Landfill gas energy facilities capture the methane (the principal component of natural gas) and combust it for energy.

However, not all landfills are capable of producing landfill gas that is usable as an energy gas. Gas production is dependent upon the composition of the waste in the landfill and this, in turn, dictates the classification of the landfill.

12.2 LANDFILL GAS

Landfill gas is not the same as *natural gas* or *methane*; each has a separate identity and the names should not be used interchangeably. The term *landfill methane* is deceiving as it may seem to imply (incorrectly) that landfill gas is simply methane.

Instead of allowing landfill gas to escape into the air, it can be captured, converted, and used as an energy source. Using landfill gas helps to reduce odors and other hazards associated with landfill gas emissions, and it helps prevent methane from migrating into the atmosphere and contributing to local smog and global climate change.

Landfill gas is produced from organic waste disposed off in landfill (Barlaz and Ham, 1993). As the waste is covered and compressed mechanically and by the pressure of higher levels (perhaps analogous to the formation of petroleum from organic detritus; Chap. 3) and as conditions become anaerobic the organic waste is broken down by microorganisms and gas is produced. Landfill gas is hazardous because of: (a) the risk of explosion, (b) the potential global warming through methane as a greenhouse gas, and (c) the presence of volatile organic compounds (VOCs) that are precursors to photochemical smog.

The rate and duration of gas production depends on the nature of the waste and oxygen availability. The gas can move or *migrate* in any direction within the site, depending on the permeability of the waste layers. It can also migrate off-site, if measures are not taken to prevent it. One indication that gas is escaping from the site and venting through the ground outside is the presence of yellow and dying vegetation.

Gas-monitoring and control measures are the responsibility of the site operator and the aim of gas control is to prevent migration off-site. This can be achieved by a gas barrier (lining of the excavation) combined with stone filled vent trenches which can vent gas to atmosphere. Venting can also be carried out using borehole pipes spread throughout the site or at selected points where gas can be pumped to them through a gas drainage system inside the tip. If the quantity or quality (odor) of the gas is such that it requires destroying, a flare system may be adopted. Landfill gas can also be exploited for industrial use as fuel for heat or to produce electricity.

The end result of landfill is generally the restoration of the land to agriculture or similar purposes. This can be achieved by capping the area with an impermeable layer of clay followed by layers of subsoil and soil. Aftercare of the site may be necessary and will involve monitoring the land for gas.

Landfill gas is extracted from landfills using a series of wells and a blower/flare (or vacuum) system. This system directs the collected gas to a central point where it can be processed and treated depending upon the ultimate use for the gas. From this point, the gas can be simply flared or used to generate electricity, replace fossil fuels in industrial and manufacturing operations, fuel greenhouse operations, or be upgraded to pipeline quality gas.

As already noted, the composition of landfill gas is variable and the evolution rate of the gas from the landfill and the quantity of gas are dependent on a number of factors: (a) waste composition, (b) age of waste, (c) presence of oxygen, (d) moisture content, (e) temperature, (f) waste input rate, (g) ambient pH, (h) waste density (closely or loosely packed), and (i) site management strategy/strategies.

Perhaps most important (Crawford and Smith, 1985) are the first five, that is, waste composition, age of refuse, presence of oxygen in the landfill, moisture content, and temperature.

The *composition of the waste*, especially the amount of organic waste present in a landfill, determines the level of bacterial activity and the degree of bacterial decomposition. Some types of organic waste contain nutrients, such as sodium, potassium, calcium, and magnesium, which help bacteria thrive. When these nutrients are present, landfill gas production increases. Alternatively, some wastes contain compounds that harm bacteria, causing less gas to be produced. For example, methane-producing bacteria can be inhibited when waste has high salt concentrations.

The *presence of oxygen* in a landfill also dictates the mode of decomposing. Only when oxygen is used up will bacteria begin to produce methane. The more oxygen present in a landfill, the longer aerobic bacteria can decompose waste during the *first phase* of the decomposition. If waste is loosely buried or frequently disturbed, more oxygen is available, so that oxygen-dependent bacteria live longer and produce carbon dioxide and water for longer periods. If the waste is highly compacted, however, methane production will begin earlier as the aerobic bacteria are replaced by methane-producing anaerobic bacteria in the *third phase* of the decomposition. Methane gas starts to be produced by the anaerobic bacteria only when the oxygen in the landfill is used up by the aerobic bacteria; therefore, any oxygen remaining in the landfill will slow methane production. Barometric highs will tend to introduce atmospheric oxygen into surface soils in shallow portions of a landfill, possibly altering bacterial activity. In this scenario, waste in the *fourth phase* of the decomposition, for example, might briefly revert to first phase until all the oxygen is used up again.

The *presence of water* in a landfill increases gas production because moisture encourages bacterial growth and transports nutrients and bacteria to all areas within a landfill. A moisture content of 40 percent or higher, based on wet weight of waste, promotes maximum gas production (e.g., in a capped landfill).Waste compaction slows gas production because it increases the density of the landfill contents, decreasing the rate at which water can infiltrate the waste. The rate of gas production is higher if heavy rainfall and/or permeable landfill covers introduce additional water into a landfill.

Temperature effects can vary but, generally, warm temperatures increase bacterial activity, which in turn increases the rate of landfill gas production. Colder temperatures inhibit bacterial activity. Typically, bacterial activity drops off dramatically below 50°F. Weather changes have a far greater effect on gas production in shallow landfills because the bacteria are not as insulated against temperature changes as compared to deep landfills where a thick layer of soil covers the waste. A capped landfill usually maintains a stable temperature, maximizing gas production. Bacterial activity releases heat, stabilizing the temperature of a landfill between 77 and 113°F, although temperatures up to 158°F have been noted. Temperature increases also promote volatilization and chemical reactions. As a general rule, emissions of organic compounds other than methane double with every 18°F increase in temperature.

The *age of refuse* is also an important factor—recently buried waste will produce more gas than older waste. Landfills usually produce appreciable amounts of gas within 1 to 3 years. Peak gas production usually occurs 5 to 7 years after wastes are dumped. Almost all gas is produced within 20 years after waste is dumped; however, small quantities of gas may continue to be emitted from a landfill for 50 or more years. A low-methane yield scenario, however, estimates that slowly decomposing waste will produce methane after 5 years and continue emitting gas over a 40-year period. Different portions of the landfill might be in different phases of the decomposition process at the same time, depending on when the waste was originally placed in each area. The amount of organic material in the waste is an important factor in how long gas production lasts.

In addition, the *pressure* of the gas is also affected by (a) the rate of landfill gas evolution since a higher rate of gas evolution will result in an increase in the pressure of the landfill, (b) the permeability of the surrounding waste types and/or strata (rock type) since more permeable waste and/or rock types will result in lower pressures as they will allow the landfill gas to migrate further than less permeable types, and (c) variations in leachate levels since an increase in leachate levels lead to a decrease in volume and thus the pressure within the landfill increases.

Landfill gas can present a potential hazard, thus making correct landfill monitoring procedures a requirement. The potential effects, on a local scale, range from fire and/or explosion to toxicity, asphyxiation, and vegetation death. One of the problems is that landfill gas is hard to control as it is not a static medium. Depending on the surrounding pressure levels, landfill gas can move laterally (through lines of weakness in the waste), vertically, (around gas/leachate wells) and also through areas of weakness where the landfill waste borders the surrounding strata.

Aside from environmental concerns of methane emissions and unpleasant odors associated with landfill gas, uncontrolled landfill gas can present a serious explosion hazard. It has been known to migrate underground to adjoining property and be set off by such things as furnace pilot lights. For this reason, regulations require that the gas from large landfills be controlled, usually by the drilling of wells into the landfill and establishing a system of collection pipes to draw the gas out of the landfill. As a result of the slight vacuum used in these collection systems, air may be drawn in adding another contaminant to the methane, particularly in wells near the perimeter of the landfill. Most of this gas is flared, but over the last 20 years an increasing number of landfill operators have found ways to convert a liability into an asset by using landfill gas to produce energy for sale.

In addition, regulations implemented under the Clean Air Act require that an estimated many large landfills install gas collection and control systems, which means that at the very least they will have to collect and flare the gas. The EPA estimates that another 600 landfills have the potential to support gas to energy projects. This represents a good opportunity for the use of technology that will enable the economical enrichment of landfill gas to pipeline quality gas without reliance on an off-and-on a tax credit.

A landfill gas facility (Figs. 12.1 and 12.2) will typically consist of gas well fields, a gas-processing plant, and a gas delivery pipeline to the customer. This will allow (a) recovery of the crude gas being generated within the landfill, (b) the ability to process the gas to produce medium-Btu gas (about 500 Btu/ft^3), and (c) delivery to a pipeline of specification gas.

At the facility, the gas is drawn from the wells and sent to a main collection header and thence to the processing plant. Vapor-liquid separators are used to separate water and any liquid condensate from the gas. The condensate is usually a mixture of hydrocarbons that resembles contaminated gasoline or kerosene. Further vapor contaminant removal is accomplished by solvent scrubbing. The crude gas is compressed and then cooled and scrubbed by countercurrent solvent flow.

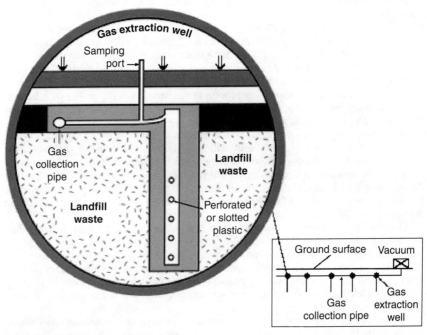

FIGURE 12.1 Landfill gas well.
http://www.epa.gov/landfill/over-photos.htm#1

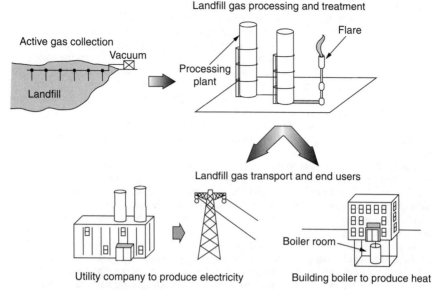

FIGURE 12.2 Landfill gas system.
http://www.epa.gov/landfill/over-photos.htm#4

12.3 BIOGAS AND OTHER GASES

12.3.1 Biogas

Biogas typically refers to a (biofuel) gas produced by the anaerobic digestion or fermentation of organic matter including manure, sewage sludge, municipal solid waste, biodegradable waste, or any other biodegradable feedstock, under anaerobic conditions. Biogas is comprised primarily of methane and carbon dioxide (Table 12.2). Depending on where it is produced, biogas is also called: *swamp gas*, *marsh gas*, *landfill gas*, and *digester gas*.

Biogas production by anaerobic digestion is popular for treating biodegradable waste because valuable fuel can be produced while destroying disease-causing pathogens and reducing the volume of disposed waste. The methane in biogas combusts more cleanly than coal, and produces more energy with less emissions of carbon dioxide. Harvesting biogas is an important role of waste management because methane is a greenhouse gas with a greater

TABLE 12.2 Composition of Biogas

Constituent	%
Methane, CH_4	50–75
Carbon dioxide, CO_2	25–50
Nitrogen, N_2	0–10
Hydrogen, H_2	0–1
Hydrogen sulfide, H_2S	0–3
Oxygen, O_2	0–2

global warming potential than carbon dioxide. The carbon in biogas was generally recently extracted from the atmosphere by photosynthetic plants, so releasing it back into the atmosphere adds less total atmospheric carbon than the burning of fossil fuels.

The composition of biogas varies depending upon the composition of the waste material in the landfill and the anaerobic digestion process. Landfill gas typically has methane concentrations around 50 percent. Advanced waste treatment technologies can produce biogas with 55 to 75 percent methane; often air is introduced (5 percent by volume) for microbiologic desulfurisation. For example, the constituents (by volume) of biogas generally are: methane (50–75 percent), carbon dioxide (25–50 percent), nitrogen (0–10 percent), hydrogen (0–1 percent), hydrogen sulfide (0–3 percent), and oxygen (0–2 percent).

If biogas is cleaned up sufficiently, biogas has the same characteristics as natural gas. In this instance the producer of the biogas can utilize the local gas distribution networks. The gas must be very clean to reach pipeline quality. Water (H_2O), hydrogen sulfide (H_2S), and particulates are removed if present at high levels or if the gas is to be completely cleaned. Carbon dioxide is less frequently removed, but it must also be separated to achieve pipeline quality gas. If the gas is to be used without extensively cleaning, it is sometimes cofired with natural gas to improve combustion. Biogas cleaned up to pipeline quality is called renewable natural gas.

In this form, the gas can be used in any application that natural gas is used for. Such applications include distribution via the natural gas grid, electricity production, space heating, water heating, and process heating. If compressed, it can replace compressed natural gas for use in vehicles, where it can fuel an internal combustion engine or fuel cells.

12.3.2 Renewable Natural Gas

Renewable natural gas is a biogas which has been upgraded to a quality similar to natural gas. A biogas is a gas obtained from biomass. By upgrading the quality to that of natural gas, it becomes possible to distribute the gas to customers via the existing gas grid, and burned within existing appliances. Renewable natural gas is also known as *sustainable natural gas*, and is a subset of *synthetic natural gas* or *substitute natural gas* (SNG).

Renewable natural gas can be produced economically, and distributed via the existing gas grid, making it an attractive means of supplying existing premises with renewable heat and renewable gas energy, while requiring no extra capital outlay of the customer. The existing gas network allows distribution of gas energy over vast distances at a minimal cost in energy. Existing networks would allow biogas to be sourced from remote markets that are rich in low-cost biomass (e.g., Russia or Scandinavia).

In terms of the present context, renewable natural gas can be manufactured through two main processes: (a) anaerobic digestion of organic (normally moist) material or (b) thermal gasification of organic (normally dry) material that has been mined from the landfill. In both cases the gas from primary production has to be upgraded in a secondary step to produce gas that is suitable for injection into the gas grid.

12.3.3 Gober Gas

Gober gas is a biogas generated out of cow dung. In India, gober gas is generated at the countless number of micro plants (an estimated more than 2 million) attached to households. The gober gas plant is basically an airtight circular pit made of concrete with a pipe connection. The manure is directed to the pit (usually from the cattle shed). The pit is then filled with a required quantity of water (usually waste water). The gas pipe is connected to the kitchen fire place through control valves. The flammable methane gas generated out of this is practically odorless and smokeless. The residue left after the extraction of the gas is used as biofertiliser. Owing to its simplicity in implementation and use of cheap raw materials in the villages, it is often quoted as one of the most environmentally sound energy source for the rural needs.

12.4 FORMATION OF LANDFILL GAS

Most landfill gas is produced by bacterial decomposition, which occurs when organic waste is broken down by bacteria naturally present in the waste and in the soil used to cover the landfill. Organic wastes include food, garden waste, street sweepings, textiles, and wood and paper products. Bacteria decompose organic waste in *four phases*, and the composition of the gas changes during each phase (Fig. 12.3).

During the *first phase* of decomposition, aerobic bacteria (bacteria that live only in the presence of oxygen) consume oxygen while breaking down the long molecular chains of complex carbohydrates, proteins, and lipids that comprise organic waste. The primary

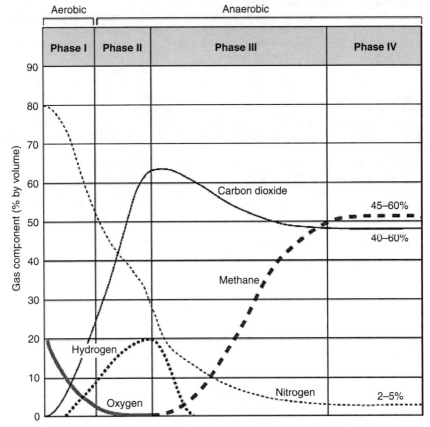

FIGURE 12.3 Variation of the composition of landfill gas with the onset of the various bacterial decomposition phases.

by-product of this process is carbon dioxide. Nitrogen content is high at the beginning of this phase, but declines as the landfill moves through the four phases. The chemistry of this first phase continues until the available oxygen is depleted. This phase can last for days or months, depending on how much oxygen is present when the waste is disposed in the landfill. Oxygen levels will vary according to factors such as how loose or compressed the waste was when it was buried.

The *second phase* starts after the oxygen in the landfill has been consumed. Using an anaerobic process (a process that does not require oxygen), bacteria convert compounds created by aerobic bacteria into acetic, lactic, and formic acids and alcohols such as methanol and ethanol. The landfill becomes highly acidic. As the acids mix with the moisture present in the landfill, they cause certain nutrients to dissolve, making nitrogen and phosphorus available to the increasingly diverse species of bacteria in the landfill. The gaseous by-products of these processes are carbon dioxide and hydrogen. If the landfill is disturbed or if oxygen is somehow introduced into the landfill, microbial processes return those chemical processes typical of the *first phase* of decomposition.

The *third phase* commences when certain kinds of anaerobic bacteria consume the organic acids produced in the *second phase* and form acetate, an organic acid. This process causes the landfill to become a more neutral environment in which methane-producing bacteria begin to establish themselves.

Methane- and acid-producing bacteria have a symbiotic, or mutually beneficial, relationship. Acid-producing bacteria create compounds for the methanogenic bacteria to consume. Methanogenic bacteria consume the carbon dioxide and acetate, too much of which would be toxic to the acid-producing bacteria.

The *fourth phase* of deocmpostion begins when both the composition and production rates of landfill gas remain relatively constant. The gas produced in this phase gas usually contains approximately 45 to 60 percent methane by volume, 40 to 60 percent carbon dioxide, and 2 to 9 percent other gases, such as sulfides. Gas is produced at a stable rate, typically for about 20 years. However, gas will continue to be emitted for 50 or more years after the waste is placed in the landfill (Crawford and Smith, 1985). Gas production might last longer, for example, if greater amounts of organics are present in the waste, such as at a landfill receiving higher than average amounts of domestic animal waste.

As might be anticipated from the above paragraphs, there are many reactions that occur amongst the array of waste that is sent to landfill. Of the wastes that are disposed to landfill, municipal wastes are likely to play the greatest role in chemical reactions, with perhaps some arising as a result of hazardous waste. However, as the name suggests, inert wastes are chemically unreactive.

The majority of reactions occur as a result of the biodegradation, or breakdown, of organic material via microbial activity, but chemical interactions can also be seen to exist in the case of inorganic materials.

Many of the biochemical reactions in landfill sites occur as a result of the presence of the reactive, organic wastes which account for more than half of the average household's waste (Hester and Harrison, 2002 p. 29). There are five major reaction processes involved, with different microorganisms involved at each stage.

Although much more complex, the chemistry occurring at landfill sites is generally represented in the simplest form as the decomposition of cellulose and hemicellulose.

For cellulose:

$$(C_6H_{10}O_5)_n + nH_2O \rightarrow 3nCO_2 + 3nCH_4$$

For hemicellulose:

$$2(C_5H_8O_4)_n + 2nH_2O \rightarrow 5nCO_2 + 5nCH_4$$

The reactions actually occur in stages and cellulose and hemicellulose are not the only starting materials.

12.4.1 Organic Reactions

Many of the biochemical reactions in landfill sites occur as a result of the presence of the reactive organic wastes (Hester and Harrison, 2002, p. 29). There are five major reaction processes involved (described in greater detail in the Web site), with different microorganisms involved at each stage.

Hydrolysis and Aerobic Degradation. The initial stage of organic decomposition occurs during the placement of the waste in the landfill and for the period of time after, when oxygen is available within the waste. These chemical processes are initiated and facilitated by the presence of aerobic microbiota which metabolize a fraction of the organic waste to produce simpler hydrocarbons, water, carbon dioxide, and, as this is an exothermic reaction, heat (the heat generated can raise the temperature of the waste to up to 70 to 90°C, however, compacted waste achieves lower temperatures due to the reduced availability of oxygen).

In these reactions, water and carbon dioxide are produced in the greatest concentrations. The carbon dioxide can dissolve in the water, forming a leachate that is rich in carbonic acid, which, in turn, lowers the pH of the surroundings. This stage generally lasts for a matter of days or weeks, depending on the amount of oxygen that is available within the waste.

Hydrolysis and Fermentation. The removal of oxygen during the *hydrolysis and aerobic degradation* facilitates a change in conditions from aerobic (oxygen present) to anaerobic (oxygen absent). Thus the majority of microbiota found within the waste change to anaerobic species. Carbohydrates are hydrolyzed (a chemical process in which a molecule is split into two parts by the addition of a water molecule) to sugars, which are then further decomposed to form carbon dioxide, hydrogen, ammonia, and organic acids. Proteins decompose via deaminization [the removal of an amino (NH_2) group] to form ammonia, carboxylic acids, and carbon dioxide. The leachate that is produced at this stage contains ammoniacal nitrogen in high concentration. Acetic acid is the main organic acid formed but propionic, butyric, lactic, and formic acids and acid derivative products are also produced, and their formation is affected by the composition of the initial waste material. The temperatures in the landfill drop to between 30 and 50°C. Gas composition may rise to levels of up to 80 percent carbon dioxide and 20 percent hydrogen.

Acetogenesis. In this stage, anaerobic conditions are still present and the organic acids that were formed in the hydrolysis and fermentation stage are now converted, via specific microorganisms to acetic acid, acetic acid derivatives, carbon dioxide, and hydrogen. Other microorganisms convert carbohydrates directly to acetic acid in the presence of carbon dioxide and nitrogen. Hydrogen and carbon dioxide levels begin to diminish toward the end of this stage, with the lower hydrogen concentrations promoting the methane-generating microorganisms (methanogens), which subsequently generate methane and carbon dioxide from the organic acids and their derivatives generated in the earlier stages.

Methanogenesis. This stage encompasses the main processes that lead to the production of landfill gas. At this point, the chemical processes involved are comparatively slow and can take many years to complete. Oxygen-depleted, anaerobic conditions still remain as in the previous two stages. Low levels of hydrogen are required to promote the methanogenic organisms, which generate carbon dioxide and methane from the organic acids and their derivatives such as acetates and formates formed in the earlier stages. Methane generation may also occur from the direct conversion of hydrogen and carbon dioxide (via microorganisms) into methane and water: Hydrogen concentrations, produced during stages 2 and 3, therefore fall to low levels during this stage 4.

Oxidation. Oxidation processes mark the final stage of the reactions involved in the biodegradation of waste. As the acids are used up in the production of landfill gas (as seen in stage 4), new aerobic microorganisms slowly replace the anaerobic forms and reintroduce oxygen to the region. Microorganisms that convert methane to carbon dioxide and water may also become established.

12.4.2 Inorganic Reactions

Hester and Harrison (2002) provide two examples of reactions involving inorganic waste: (a) sulfate-containing materials and (b) heavy metals.

In the former case, plaster board (in this case the sulfate is present in the form of gypsum) will be reduced under anaerobic, oxygen-deficient, conditions to produce a metal sulfide (via association with free metal ions). Under acidic conditions, the sulfate will be released as hydrogen sulfide (H_2S), which has the potential to present a serious hazard. The gas is toxic and can cause respiratory arrest if large concentrations are inhaled. Also, at concentrations above 100 ppm, a person's ability to detect the gas is affected by rapid temporary paralysis of the olfactory nerves in the nose, leading to a loss of the sense of smell. This unusual property makes it extremely dangerous to rely totally on the sense of smell to warn of the presence of hydrogen sulfide.

Reduced sulfate, in the form of sulfide, will react with heavy metals, (see above) producing a metal sulfide, which is comparatively immobile. Other insoluble salts such as metal carbonates may also be formed. Although these are potential pollutants, their immobility reduces their potential to migrate out of the landfill site boundary.

12.4.3 Landfill Leachate

In addition to landfill gas, the other product of landfill chemistry is *landfill leachate.*

Landfill leachate is the liquid formed within a landfill site that is comprised of the liquids that enter the site (including rainwater) and the material that is leached from the wastes as the infiltrating liquids percolate downwards through the waste (Hester and Harrison, 2002, p.45).

One of the most important problems with designing and maintaining a landfill is managing the leachate that is generated when water passes through the waste. The leachate consists of many different organic and inorganic compounds that may be either dissolved or suspended. Regardless of the nature of the compounds, they pose a potential pollution problem for local ground and surface waters (Barton et al., 2003).

Many factors influence the production and composition of leachate. One major factor is the climate of the landfill. For example, where the climate is prone to higher levels of precipitation there will be more water entering the landfill and therefore more leachate generated. Another factor is the site topography of the landfill that influences the runoff patterns and again the water balance within the site.

Current leachate treatment options include recycling and reinjection, on-site treatment, and discharge to a municipal water treatment facility or a combination. However, with stricter regulations regarding ground and surface water contamination, landfills have to find new treatment alternatives.

The vast majority of wastes will produce a leachate of sorts if water is allowed to percolate through it but the proportions of the leachate constituents are chiefly determined by the composition and solubility of the waste components. For example, if the waste is "changing in composition" (e.g., due to weathering or biodegradation), then leachate quality will change with time. This is particularly the case in landfills containing municipal waste.

At the start, the leachate produced is typically generated under the aerobic, oxygen-rich conditions, and this produces a complex solution with near neutral pH of 7.5. This process generally only lasts a few days or weeks and is relatively unimportant in terms of the quality of the leachate. However the heat produced during aerobic decomposition can sometimes rise to as high as 80 to 90°C, and if this heat is retained it can enhance the later stages of leachate production.

As waste decomposition progresses, the oxygen within the waste gradually becomes diminished, and the conditions become anaerobic. This initially creates high concentrations of soluble degradable organic compounds and an acidic pH. Ammonium and metal concentrations also rise during this phase.

After several months or years, the onset of methanogenic conditions changes the pH of the leachate, making it more neutral or even slightly alkaline. As biodegradation nears completion, aerobic conditions may return, and the leachate will eventually cease to be hazardous to the environment.

The character of the leachate can be determined by means of the *toxicity characteristic leaching procedure* (TCLP).

The TCLP is a soil extraction method for chemical analysis. An analytical method to simulate leaching through a landfill. The leachate is analysed for appropriate substances.

The toxicity characteristic leaching procedure comprises four fundamental procedures: (a) sample preparation for leaching, (b) sample leaching, (c) preparation of leachate for analysis, and (d) leachate analysis and is generally useful for classifying waste material for disposal options. Extremely contaminated material is expensive to dispose and grading is required to ensure safe disposal and to avoid paying for disposal of *clean fill*. The main problem is that the toxicity characteristic leaching procedure test is based on the assumption that the waste material will be buried in landfill along with organic material. Organic matter is not really buried with other waste anymore (composting usually applies) and other leachate techniques may be more appropriate.

As an example of the method, the pH of the sample material is first established, and then leached with an acetic acid/sodium hydroxide solution at a 1:20 mix of sample to solvent. The leachate solution is sealed in extraction vessel for general analytes, or possibly pressure sealed as in zero-headspace extractions (ZHE) for volatile organic compounds, cyanides, or sulfites, and tumbled for 18 hours to simulate an extended leaching time in the ground.

12.4.4 Landfill Gas Monitoring

The one foreseeable disadvantage is that, while the global and local environment effects of uncontrolled releases would be reduced, emissions such as CO_2, CO, NO_x, SO_x, and other components would increase leading to other detrimental effects. In a sense, the problem of landfill gas is simply being shifted rather than solved and monitoring is essential. Landfill gas monitoring is the process by which gases that are released from landfill are electronically monitored.

Landfill gas production results from chemical reactions and microbes acting upon the waste as they begin to break down the putrescible materials. In the landfill, due to the constant production of landfill gas, pressure increases within the landfill provoke its release into the atmosphere. Such emissions lead to important environment, hygiene, and security problems in the landfill. Due to the risk presented by landfill gas there is a clear need to monitor gas produced by landfills. Techniques for the monitoring of landfill gas are:

1. Surface monitoring: Used to check the integrity of caps on waste, check on borehole monitoring. May give preliminary indications of off-site gas migration. Typical regulatory limit is 500 ppm of methane (CH_4) by volume.
2. Subsurface monitoring (gas probes): Used to enable point source monitoring of gas concentrations in the local environment around the probe. These may contain a single probe or multiple probes (at different depths) at a single point. Probes typically form a ring around a landfill. The distance between probes vary (based upon several factors) and rarely exceeds 1000 ft. Also known as migration or perimeter probes.
3. Excavated trenches and pits backfilled around standpipes: Provides means of measuring gas concentrations in shallow sites.
4. Gas monitoring boreholes or wells: Preferred method for measuring gas concentrations and sometimes flow rates both on and off site.
5. Ambient air samplers: Used to sample the air up-wind and down-wind of a landfill, over the period of several hours. Samples are taken in special plastic bags or steel canisters and analyzed at a laboratory. Where a landfill is situated in a location where the wind direction changes from day to night (like along a coastline), a 24-hour sampling period may be broken into day and night components.
6. The use of leachate wells: May be used for gas monitoring but are not comparable with, or substitutes for specially designed gas monitoring points.

12.4.5 Landfill Mining

Landfill mining and reclamation (LFMR) is a process whereby solid wastes which have previously been landfilled are excavated and processed. The function of landfill mining is to reduce the amount of landfill mass encapsulated within the closed landfill and/or temporarily remove hazardous material to allow protective measures to be taken before the landfill mass is replaced. In the process mining recovers valuable recyclable materials, a combustible fraction that may serve as a precursor to fuel, soil, and landfill space. The aeration of the landfill soil is a secondary benefit regrading the landfills future use. The overall appearance of the landfill mining procedure is a sequence of processing machines laid out in a functional conveyor system. The operating principle is to excavate, sieve, and sort the landfill material.

The concept of landfill mining was introduced when it was recognized that waste contains many resources with high value—the most notable of which are nonferrous metals such as aluminum cans and metal scrap. The concentration of aluminum in many landfills is higher than the concentration of aluminum in bauxite from which the metal is derived. However, the potential to mine a fuel precursor such as wood or compressed organic material adds extra impetus to the concept of landfill mining.

Mining of landfill sites that were created for specific purposes is often the easiest form of landfill mining. However, mining municipal landfills has to be based on the expected content of the landfill. Older landfills in the United States were often capped and closed, essentially entombing the waste. This can be beneficial for waste recovery. It can also create a higher risk for toxic waste and leachate exposure as the landfill has not fully processed the stewing wastes. Mining of bioreactor landfills and properly stabilized modern sanitary landfills provides its own benefits. The biodegradable wastes are more easily sieved out, leaving the nonbiodegradable materials readily accessible. The quality of these materials for recycling and reprocessing purposes is not as high as initially recycled materials, however materials such as aluminum and steel are usually excluded from this.

Landfill mining is most useful as a method to remediate hazardous landfills. Landfills that were established before landfill liner technology was well established often leak their unprocessed leachate into underlying aquifers. This is both an environmental hazard and also a legal liability. Mining the landfill simply to lay a safe liner is a last, but sometimes necessary resort.

In the mining process, an excavator or front end loader uncovers the landfilled materials and places them on a moving floor conveyor belt to be taken to the sorting machinery. A trommel is used to separate materials by size. First, a large trommel separates materials like appliances and fabrics. A smaller trommel then allows the biodegraded soil fraction to pass through leaving nonbiodegradable, recyclable materials on the screen to be collected. An electromagnet is used to remove the ferrous material from the waste mass as it passes along the conveyor belt. A front end loader is used to move sorted materials to trucks for further processing. Odor control sprayers are wheeled tractors with a cab and movable spray arm mounted on a rotating platform. A large reservoir tank mounted behind the cab holds neutralizing agents, usually in liquid form, to reduce the smell of exposed wastes.

Excavators dig up waste mass and transport it, with the help of front end loaders, onto elevator and moving floor conveyor belts. The conveyor belts empty into a coarse, rotating trommel. The large holes in the screen allow most wastes to pass through, leaving behind the over-sized, nonprocessable materials. The over-sized wastes are removed from inside the screen. The coarse trommel empties into the fine rotating trommel. The fine rotating trommel allows the soil fraction to pass through, leaving mid-sized, nonbiodegradable, mostly recyclable materials. The materials are removed from the screen. These materials are put on a second conveyor belt where an electromagnet removes any metal debris. Depending on the level of resource recovery material can be put through an air classifier which separates light organic material from heavy organic material. The separate streams are then loaded, by front end loaders, onto trucks either for further processing or for sale. Further manual processing can be done on site if processing facilities are too far away to justify the transportation costs.

Landfill mining is also possible in countries where land is not available for new landfill sites. In this instance landfill space can be reclaimed by the extraction of biodegradable waste and other substances then refilled with wastes requiring disposal.

There are several proposals to use the mined landfill as feedstock to an anaerobic digester. However, there are some drawbacks to anaerobic digesters, mainly relating to the type of feedstock used. If the feedstock contains toxic chemicals (such as pesticides and wood preservatives), the digestate will be highly toxic through concentration of the toxic constituents. If the digestate cannot be used as a fertilizer, it becomes waste once more—but it is now, having been mined from an old site, it will, more than likely, be classified as a toxic waste.

12.5 GAS MIGRATION

Once gases are produced under the landfill surface, they generally move away from the landfill. Gases tend to expand and fill the available space, so that they move or *migrate* through the limited pore spaces within the refuse and soils covering of the landfill. The tendency of some landfill gases, such as methane, that is lighter than air, is to move upward, usually through the landfill surface. However, of the hydrocarbons, methane is the only one that is lighter than air. Higher molecular weight hydrocarbon such as ethane, propane, and butane are denser than air (Table 12.3) and may collect in pockets rather than migrate with

TABLE 12.3 Density and Vapor Density of the Lower Boiling Hydrocarbons

	Molecular weight	Specific gravity	Vapor density air = 1
Methane	16	0.553	0.56
Ethane	30	0.572	1.04
Propane	44	0.504	1.50
Butane	58	0.601	2.11
Pentane	72	0.626	2.48
Hexane	86	0.659	3.00

the methane. This is an error often made by many of the arm-waving specialists who do not check gas properties and such deficiencies of knowledge can cause problems when pockets are suddenly opened to the air (e.g., during landfill mining) and the flammability limit or flash point of the collected gas is reached.

In addition to the density phenomenon, upward movement of landfill gas can be inhibited by densely compacted waste or landfill cover material (e.g., by daily soil cover and caps). When upward movement is inhibited, the gas tends to migrate horizontally to other areas within the landfill or to areas outside the landfill, where it can resume its upward path. Basically, the gases follow the path of least resistance. Some gases, such as carbon dioxide, are denser than air and will collect in subsurface areas, such as utility corridors. Three main factors influence the migration of landfill gases: diffusion (uniform concentration), pressure, and permeability.

12.5.1 Diffusion (Uniform Concentration)

Diffusion is the natural tendency of a gas to reach a uniform concentration in a given space, whether it is a room or the earth's atmosphere. Gases in a landfill move from areas of high gas concentrations to areas with lower gas concentrations. Because gas concentrations are generally higher in the landfill than in the surrounding areas, landfill gases diffuse out of the landfill to the surrounding areas with lower gas concentrations.

12.5.2 Pressure

Gases accumulating in a landfill create areas of high pressure in which gas movement is restricted by compacted refuse or soil covers and areas of low pressure in which gas movement is unrestricted. The variation in pressure throughout the landfill results in gases moving from areas of high pressure to areas of low pressure. Movement of gases from areas of high pressure to areas of lower pressure is known as *convection*. As more gases are generated, the pressure in the landfill increases, usually causing subsurface pressures in the landfill to be higher than either the atmospheric pressure or indoor air pressure. When pressure in the landfill is higher, gases tend to move to ambient or indoor air.

12.5.3 Permeability

Gases will also migrate according to where the pathways of least resistance occur. Permeability is a measure of how well gases and liquids flow through connected spaces or pores in refuse and soils. Dry, sandy soils are highly permeable (many connected pore spaces), while moist clay tends to be much less permeable (fewer connected pore spaces). Gases tend to move through areas of high permeability (e.g., areas of sand or gravel) rather than through areas of low permeability (e.g., areas of clay or silt). Landfill covers are often made of low-permeability soils, such as clay. Gases in a covered landfill, therefore, may be more likely to move horizontally than vertically.

12.5.4 Other Factors Affecting Gas Migration

The direction, speed, and distance of landfill gas migration depend on a number of factors.

Landfill Cover. If the landfill cover consists of relatively permeable material, such as gravel or sand, then gas will likely migrate up through the landfill cover. If the landfill cover consists of silts and clays, it is not very permeable; gas will then tend to migrate horizontally underground. If one area of the landfill is more permeable than the rest, gas will migrate through that area.

Natural and Man-Made (Anthropogenic) Pathways. Drains, trenches, and buried utility corridors (such as tunnels and pipelines) can act as conduits for gas movement. The natural geology often provides underground pathways, such as fractured rock, porous soil, and buried stream channels, where the gas can migrate.

Wind Speed and Direction. Landfill gas naturally vented into the air at the landfill surface is carried by the wind. The wind dilutes the gas with fresh air as it moves it to areas beyond the landfill. Wind speed and direction determine the gas's concentration in the air, which can vary greatly from day to day, even hour by hour. In the early morning, for example, winds tend to be gentle and provide the least dilution and dispersion of the gas to other areas.

Moisture. Wet surface soil conditions may prevent landfill gas from migrating through the top of the landfill into the air above. Rain and moisture may also seep into the pore spaces in the landfill and "push out" gases in these spaces.

Groundwater Levels. Gas movement is influenced by variations in the groundwater table. If the water table is rising into an area, it will force the landfill gas upward.

Temperature. Increases in temperature stimulate gas particle movement, tending also to increase gas diffusion, so that landfill gas might spread more quickly in warmer conditions. The landfill generally maintains a stable temperature, freezing and thawing cycles can cause the soil surface to crack, causing landfill gas to migrate in an upward direction and/or in a horizontal direction. Frozen soil over the landfill may provide a physical barrier to upward landfill gas migration, causing the gas to migrate further from the landfill horizontally through soil.

Barometric and Soil Gas Pressure. The difference between the soil gas pressure and barometric pressure allows gas to move either vertically or laterally, depending on whether the barometric pressure is higher or lower than the soil gas pressure. When barometric pressure is falling, landfill gas will tend to migrate out of the landfill into surrounding areas. As barometric pressure rises, gas may be retained in the landfill temporarily as new pressure balances are established.

12.6 BIOREACTORS

One of the major areas of research concerning landfill is the use of bioreactors. A *bioreactor* is formed under specific landfilling conditions. *Bioreactor landfilling* is a process in which water and air are circulated into a specially designed landfill, in order to cause accelerated biological decomposition of the waste material. The intention for this type of landfill operation is to maximize the generation of biogas, which is captured using a network of perforated pipes and burnt to generate electricity. Another desired outcome is the rapid stabilization of organic waste material (in order to minimize the length of time required to manage the landfill site, or to make use of the decomposed material as compost).

By adding and recalculating liquids in bioreactors, decomposition is accelerated under anaerobic conditions, increasing the production of landfill gas by 2 to 10 times, approximately half of which is methane with at least 23 times the warming potential as carbon dioxide. The result is to shift substantial volumes of methane production, which otherwise would not occur for decades hence, to the present.

In the landfill, which acts as a bioreactor, anaerobic digestion occurs insofar as the naturally occurring processes of anaerobic degradation are harnessed and contained. Anaerobic digestion has a long history dating back to the tenth century B.C.

The four stages key of anaerobic digestion are hydrolysis, acidogenesis, acetogenesis, and methanogenesis. These stages result from the biological treatment of organic waste by two key bacterial groups—acetogens and methanogens.

The *first stage* is the chemical reaction of hydrolysis, where complex organic molecules are broken down into simple sugars, amino acids, and fatty acids with the addition of hydroxyl groups. The *second stage* is the biological process of acidogenesis where a further breakdown by acidogens into simpler molecules, volatile fatty acids occurs, producing ammonia, carbon dioxide, and hydrogen sulfide as by-products. The *third stage* is the biochemical process of acetogenesis where the simple molecules from acidogenesis are further digested by acetogens to produce carbon dioxide, hydrogen, and mainly acetic acid. The *fourth stage* is the biochemical process of methanogenesis where methane, carbon dioxide, and water are produced by methanogens.

A simplified overall chemical reaction for the degradation of sugars produced by the hydrolysis of cellulose is:

$$C_6H_{12}O_6 \rightarrow 3CO_2 + 3CH_4$$

Thus, the desirable product of the landfill bioreactor is methane. Other products may (depending upon the composition of the waste material in the landfill) include (a) a solid fibrous material, which is spread without further treatment, or after composting (maturation), to provide organic matter for improvement of soil quality and fertility (improves soil structure and reduces summer irrigation demand) and (b) a liquid fraction which contains nutrients and can be spread as a fertilizer and sprayed on crops. If the solid and liquid fractions are not separated, the slurry can be spread on the soil.

Effective gas generation in bioreactors under saturated conditions of rapid differential settlement without a low permeable cover is impossible. The result of this practice, which significantly increases early gas generation without effective gas capture, is a significant near-term increase of potent greenhouse gas emissions into the atmosphere. This may have very serious implications for meteorological experts who warn that the earth is about to pass a climatic "tipping point."

The use of clay caps and advanced liner systems and bentonite is useful for the prevention of leachate and landfill gas release but a consequence of this is that moisture, that is required for the biodegradative processes, is often excluded. The result of this is that much of the waste can remain intact entombed inside the landfill, potentially for longer than the lifetime of the barriers. One method to reduce this effect, by enhancing and accelerating waste stabilization, is to operate the landfill as a bioreactor. The bioreactor landfill attempts to control, monitor, and optimize the waste stabilization process rather than contain the wastes as prescribed by most regulations (Reinhart et al., 2002). This is carried out in an aerobic environment as opposed to the normal situation in landfill sites, where conditions within the wastes are commonly anaerobic. Reinhart et al. (2002) has described the potential positive and negative effects that this type of landfill can pose (Table 12.4)

TABLE 12.4 Advantages and Disadvantages of the Landfill Bioreactor

Potential advantages

Rapid waste stabilisation—aerobic waste decomposition has been cited as a more rapid means of waste stabilisation
Improved gaseous emissions—methane is *not* a by-product of aerobic decomposition. Other odour-causing chemicals are also reduced.
Degradation of recalcitrant chemicals—may offer greater treatment for organic wastes and ammonia (which do not degrade or transform under anaerobic conditions)
Removal of moisture—the aerobic conditions act to strip moisture from the landfill, reducing the chance of leachate production

Potential disadvantages

Risk of fire and explosive gas mixtures—addition of air to landfills poses an increased potential for landfill fires.
Cost—increased costs may be incurred as a result of the requirement to pump air into the landfill, as well as the monitoring processes
Unknown gaseous emissions—although gaseous emissions such as methane may be reduced other hazardous chemicals may still be released. Also, nitrous oxide, a more potent greenhouse gas than methane, may be emitted.

Source: Reinhart, D. R., P. T. McCreanor, and T. Townsend: "The Bioreactor Landfill: Its Status and Future," *Waste Management and Research*, 20, 2002, pp. 172–86.

The future of the bioreactor may well be found in its use with an anaerobic system that is, a hybrid arrangement. For example, air could be added to the anaerobic processes after the degradation of waste has occurred, thus removing excess moisture from the landfill and fully composting the waste. The cycling of both conditions (aerobic and anaerobic) also offers the possibility of treating a greater range of chemicals such as the nitrification and denitrification of ammonia.

12.7 GASIFICATION OF LANDFILL WASTE

There is also the potential for the gasification (Chap. 8) of any organic waste material that is mined from the landfill as a solid or semisolid.

A waste gasification facility coverts the waste into a pressurized gas mixture which is then passed through a cleanup system to remove particulate matter and tar vapor, and finally used to turn turbines in a power plant. The remaining gas is a mixture of carbon monoxide, hydrogen, water vapor, and carbon dioxide. Finally, the particulates and tar are thermally treated (pyrolyzed) to produce carbon monoxide.

Any hydrogen produced that is not used as a constituents of synthesis gas (Chap. 7) can be removed and stored for use in hydrogen fuel cells. Waste gasification does produce some carbon dioxide that needs attention. Although there is the thought that although the carbon dioxide is produced from a renewable organic source, it has a short atmospheric residence time and therefore does not contribute to global warming. On this basis, carbon dioxide from biomass is acceptable but carbon dioxide from a fossil fuel is unacceptable.

The goal of using fuels from alternative sources is not only to supplement the dwindling supplies of fossil fuels but also to reduce the emissions of greenhouse gases. Carbon dioxide is a greenhouse gas—no matter what the source. To consider the carbon dioxide allowable because it is from a biomass-based process is, in any sense, playing with numbers and using a statistical means to justify the end.

Such thinking is not only irrational but is more than suspect—it is twisted logic!

12.8 POWER GENERATION

The gas generated in landfills can be tapped after the completion of the landfill, and when burnt as fuel has a calorific value about half that of natural gas. This is due to the fact that there is only about 60 percent combustible methane in the gas created and this has a calorific value of 37 GJ/t compared to 50 GJ/t for natural gas.

Generating electricity from landfill gas is clearly a better sustainable solution than merely trapping and not using it in any productive manner. It is a resource that should be employed. There is the added bonus for the site operators of extra generated financial revenue from the electricity sales. However, all of the environmental burdens still apply. The only offset is where electricity generation by nonrenewable sources can be reduced.

Large municipal or industrial landfills produce gas that can be tapped to generate electricity. Microorganisms that live in organic materials such as food wastes, paper, or yard clippings cause these materials to decompose. This produces landfill gas, typically comprises roughly of 60 percent methane and 35 percent carbon dioxide.

Landfill gas is collected from landfills by drilling "wells" into the landfills, and collecting the gases through pipes. Once the landfill gas is processed, it can be combined with natural gas to fuel conventional combustion turbines or used to fuel small combustion or combined cycle turbines. Landfill gas may also be used in fuel cell technologies, which use chemical reactions to create electricity, and are much more efficient than combustion turbines. However, use of the gas produced by landfills may reduce the harmful environment impacts that would otherwise result from landfill operations. Landfill gas electricity generation offers major air quality benefits where landfills already exist or where the decision to build the landfill has already been made.

Landfill gas power plants reduce methane emissions, a global climate change agent with 23 times the negative impact of carbon dioxide.

A landfill gas power plant burns methane that would otherwise be released into the atmosphere or burned off in a flaring process. Methane is a highly potent agent of global climate change, having about 23 times the negative impact on a pound-by-pound basis as carbon dioxide. Landfill gas combustion produces some carbon dioxide, but the impact of these emissions on global climate change is offset many times over by the methane emission reductions.

Landfill gas generators produce nitrogen oxides emissions that vary widely from one site to another, depending on the type of generator and the extent to which steps have been taken to minimize such emissions. Combustion of landfill gas can also result in the release of organic compounds and trace amounts of toxic materials, including mercury and dioxins, although such releases are at levels lower than if the landfill gas is flared.

There are few water impacts associated with landfill gas power plants. Unlike other power plants that rely upon water for cooling, landfill gas power plants are usually very small, and therefore pollution discharges into local lakes or streams are typically quite small.

12.9 REFERENCES

Barlaz, M. A. and R. K. Ham: "Leachate and Gas Generation," in *Geotechnical Practice for Waste Disposal,* D. E. Daniel (ed.), Chapman and Hall, London, U.K., 1993, pp. 113–36.

Barton, C. D., L. Paddock, C. S. Romanek, and J. C. Seaman: "Geochemistry of an Abandoned Landfill Containing coal Combustion Waste: Implications for Remediation," in *Chemistry of Trace Elements in Fly Ash,* K. S. Sajwan, A. K. Alva, and R. F. Keefer (eds.), Kluwer Academic/Plenum Publishers, London, England, 2003, pp. 105–41.

Crawford J. F. and P. G. Smith: *Landfill Technology,* Butterworths Scientific Press, London, U.K., 1985.

Hester, R. E. and R. M. Harrison (eds.): *Environmental and Health Impact of Solid Waste Management Activities,* The Royal Society of Chemistry, London, England, 2002.

Reinhart, D. R., P. T. McCreanor, and T. Townsend: "The Bioreactor Landfill: Its Status and Future," *Waste Management and Research,* 20, 2002, pp. 172–86.

Speight, J. G. (ed.):. *Lange's Handbook of Chemistry,* 16th ed., McGraw-Hill, New York, 2005, pp. 2.431.

Tchobanoglous G, H. Theisen, and S. Vigil: *Integrated Solid Waste Management, Engineering Principles and Management Issues,* McGraw-Hill, Inc. New York, 1993, pp. 381–417.

US EPA: *Compilation of Air Pollutant Emissions Factors, AP-42,* 5th ed., vol. 1, Stationary Point and Area Sources, U.S. Environmental Protection Agency, Washington, D.C., Jan., 1995, sec. 2.4— "Municipal Solid Waste Landfills," http://www.epa.gov/ttn/chief/ap42/ch02/ and http://www.atsdr.cdc.gov/HAC/landfill/PDFs/Landfill_2001_ch2mod.pdf.

Wisbrock, W. H.: "Landfill Gas to Methanol," in *Alcoholic Fuels,* S. Minteer (ed.), CRC-Taylor & Francis, Boca Raton, Fla., 2006, chap. 3.

APPENDIX A
DEFINITION AND PROPERTIES OF FUELS AND FEEDSTOCKS FROM DIFFERENT SOURCES

TABLE A.1 Properties of Selected Crude Oils

Crude oil	API gravity[*]	Specific gravity	Sulfur (% w/w)	Nitrogen (% w/w)
Alaska North Slope	26.2	0.8973	1.1	0.2
Arabian Light	33.8	0.8560	1.8	0.07
Arabian Medium	30.4	0.8740	2.6	0.09
Arabian Heavy	28.0	0.8871	2.8	0.15
Athabasca (Canada)	8	1.0143	4.8	0.4
Beta (California)	16.2	0.9580	3.6	0.81
Brent (North Sea)	38.3	0.8333	0.37	0.10
Bonny Light (Nigeria)	35.4	0.8478	0.14	0.10
Boscan (Venezuela)	10.2	0.9986	5.3	0.65
Ekofisk (Norway)	37.7	0.8363	0.25	0.10
Henan (China)	16.4	0.9567	0.32	0.74
Hondo Blend (California)	20.8	0.9291	4.3	0.62
Kern (California)	13.6	0.9752	1.1	0.7
Kuwait Export	31.4	0.8686	2.5	0.21
Liaohi (China)	17.9	0.9471	0.26	0.41
Maya (Mexico)	22.2	0.9206	3.4	0.32
Shengli (China)	13.8	0.9738	0.82	0.72
Tapis Blend (Malaysia)	45.9	0.7976	0.03	Nil
West Hackberry Sweet[†]	37.3	0.8383	0.32	0.10
West Texas Intermediate	39.6	0.8270	0.34	0.08
Xinjiang (China)	20.5	0.9309	0.15	0.35

[*] *API gravity* is related to specific gravity by the formula °API = 141.5 ÷ (specific gravity @ 60°F) − 131.5.
[†] Produced from a storage cavern in the U.S. Strategic Petroleum Reserve.
For more data, please see www.enbridge.com/pipelines/about/pdf/crudecharacteristics2006.pdf.

TABLE A.2 Properties of Tia Juana Crude Oil and the 650°F, 950°F, and 1050°F Residua

		Residua		
	Whole crude	650°F	950°F	1050°F
Yield, vol. %	100.0	48.9	23.8	17.9
Sulfur, wt. %	1.08	1.78	2.35	2.59
Nitrogen, wt. %		0.33	0.52	0.60
API gravity	31.6	17.3	9.9	7.1
Carbon residue, wt. %				
Conradson		9.3	17.2	21.6
Metals				
Vanadium, ppm		185		450
Nickel, ppm		25		64
Viscosity				
Kinematic				
At 100°F	10.2	890		
At 210°F		35	1010	7959
Furol				
At 122°F		172		
At 210°F			484	3760
Pour point, °F	−5	45	95	120

TABLE A.3 Properties of Selected Residua

Feedstock	Gravity API	Sulfur (wt. %)	Nitrogen (wt. %)	Nickel (ppm)	Vanadium (ppm)	Asphaltenes (heptane) (wt. %)	Carbon residue (conradson) (wt. %)
Arabian Light, >650°F	17.7	3.0	0.2	10.0	26.0	1.8	7.5
Arabian Light, >1050°F	8.5	4.4	0.5	24.0	66.0	4.3	14.2
Arabian Heavy, >650°F	11.9	4.4	0.3	27.0	103.0	8.0	14.0
Arabian Heavy, >1050°F	7.3	5.1	0.3	40.0	174.0	10.0	19.0
Alaska, North Slope, >650°F	15.2	1.6	0.4	18.0	30.0	2.0	8.5
Alaska, North Slope, >1050°F	8.2	2.2	0.6	47.0	82.0	4.0	18.0
Lloydminster (Canada), >650°F	10.3	4.1	0.3	65.0	141.0	14.0	12.1
Lloydminster (Canada), >1050°F	8.5	4.4	0.6	115.0	252.0	18.0	21.4
Kuwait, >650°F	13.9	4.4	0.3	14.0	50.0	2.4	12.2
Kuwait, >1050°F	5.5	5.5	0.4	32.0	102.0	7.1	23.1
Tia Juana, >650°F	17.3	1.8	0.3	25.0	185.0		9.3
Tia Juana, >1050°F	7.1	2.6	0.6	64.0	450.0		21.6
Taching, >650°F	27.3	0.2	0.2	5.0	1.0	4.4	3.8
Taching, >1050°F	21.5	0.3	0.4	9.0	2.0	7.6	7.9
Maya, >650°F	10.5	4.4	0.5	70.0	370.0	16.0	15.0

TABLE A.4 Properties of Maya Crude Oil and Athabasca Bitumen

	Maya (benchmark)	Orinoco Cerro Neqro	Athabasca bitumen	Athabasca DilBit*	Athabasca SynBit†
API gravity	22	8	8	21	20
% Sulfur	3.5	3.8	4.5	3.7	2.8
TAN	<0.5	3.3	3.5	2.4	1.8
Yields, LV%					
Naphtha, LPG	20	2	0	25	10
Distillate	22	17	14	15	26
Vacuum gas oil	20	26	34	24	37
Vacuum bottoms	38	55	52	36	27

*Blend of 68% Athabasca bitumen and 32% condensate.
†Blend of 52% Athabasca bitumen and 48% synthetic crude (future quality).
Source: Crandall, G. R.: "Non-Conventional Oil Market Outlook," Presentation to the International Energy Agency, Conference on Non-Conventional Oil, Prospects for Increased Production, Nov. 26, 2002.

TABLE A.5 General Properties of Unrefined (Left) and Refined (Right) Natural Gas

Relative molar mass	20–16
Carbon content, wt. %	73–75
Hydrogen content, wt. %	27–25
Oxygen content, wt. %	0.4–0
Hydrogen-to-hydrogen atomic ratio	3.5–4.0
Density relative to air @15°C	1.5–0.6
Boiling temperature, °C/1 atm	–162
Autoignition temperature, °C	540–560
Octane number	120–130
Methane number	69–99
Vapor flammability limits, vol. %	5–15
Flammability limits	0.7–2.1
Lower heating/calorific value, Btu	900
Methane concentration, vol. %	100–80
Ethane concentration, vol. %	5–0
Nitrogen concentration, vol %	15–0
Carbon dioxide concentration, vol. %	5–0
Sulfur concentration, ppm, mass	5–0

Source: http://www.visionengineer.com/env/alt_ng_prop.php.

TABLE A.6 Properties of Natural Gas Constituents and Selected Organic Liquids

	Molecular weight	Specific gravity	Vapor density air = 1	Boiling point,°C	Ignition point,°C	Flash point,°C
Methane	16	0.553	0.56	–160	537	–221
Ethane	30	0.572	1.04	–89	515	–135
Propane	44	0.504	1.50	–42	468	–104
Butane	58	0.601	2.11	–1	405	–60
Pentane	72	0.626	2.48	36	260	–40
Hexane	86	0.659	3.00	69	225	–23
Benzene	78	0.879	2.80	80	560	–11
Heptane	100	0.668	3.50	98	215	–4
Octane	114	0.707	3.90	126	220	13
Toluene	92	0.867	3.20	161	533	4
Ethyl benzene	106	0.867	3.70	136	432	15
Xylene	106	0.861	3.70	138	464	17

TABLE A.7 Classification of Coal

Class/group	Fixed carbon limits (dry, mineral-matter-free basis), %		Volatile matter limits (dry, mineral-matter-free basis), %		Gross calorific value limits (moist, mineral-matter-free basis)				Agglomerating character
					Btu/lb		MJ/kg		
	Equal to or greater than	Less than	Greater than	Equal to or less than	Equal to or greater than	Less than	Equal to or greater than	Less than	
Anthracitic									
Mela-anthracite	98	...	...	2	...	...	...	...	
Anthracite	92	98	2	8	...	...	...	...	Nonagglomerating
Semianthracite	86	92	8	14	...	...	...	...	
Bituminous									
Low volatile bituminous coal	78	86	14	22	...	...	...	...	
Medium volatile bituminous coal	69	78	22	31	...	...	...	...	
High volatile A bituminous coal	...	69	31	...	14,000	...	32.6	...	Commonly agglomerating
High volatile B bituminous coal	...	...	...	...	13,000	14,000	30.2	32.6	
High volatile C bituminous coal	...	...	...	...	11,500	13,000	26.7	30.2	
					10,500	11,500	24.4	26.7	Agglomerating

(Continued)

TABLE A.7 Classification of Coal (*Continued*)

Class/group	Fixed carbon limits (dry, mineral-matter-free basis), %		Volatile matter limits (dry, mineral-matter-free basis), %		Gross calorific value limits (moist, mineral-matter-free basis)				Agglomerating character
					Btu/lb		MJ/kg		
	Equal to or greater than	Less than	Greater than	Equal to or less than	Equal to or greater than	Less than	Equal to or greater than	Less than	
Subbituminous									
Subbituminous A coal	...	...	...	...	10,500	11,500	24.4	26.7	
Subbituminous B coal	...	...	...	...	9,500	10,500	22.1	24.4	Nonagglomerating
Subbituminous C coal	...	...	...	...	8,300	9,500	19.3	22.1	
Lignitic									
Lignite A	...	...	...	...	6,300	8,300	14.7	19.3	
Lignite B	...	...	...	...	...	6,300	...	14.7	

Source: ASTM D388

TABLE A.8 General Properties of Coal

Coal is a readily combustible rock containing more than 50% by weight of carbonaceous material, formed from compaction and indurations of variously altered plant remains similar to those in peat. Lignite increases in maturity by becoming darker and harder and is then classified as subbituminous coal. After a continuous process of burial and alteration, chemical and physical changes occur until the coal is classified as bituminous—dark and hard coal.

Bituminous coal ignites easily and burns long with a relatively long flame. If improperly fired bituminous coal is characterized with excess smoke and soot.

Anthracite is very hard and shiny.

Anthracite coal creates a steady and clean flame and is preferred for domestic heating. Furthermore, it burns longer with more heat than the other types.

Typical sulfur content in coal
Anthracite coal : 0.6–0.77 wt %
Bituminous coal : 0.7–4.0 wt %
Lignite coal : 0.4 wt %

Typical moisture content in coal
Anthracite coal: 2.8–16.3 wt %
Bituminous coal: 2.2–15.9 wt %
Lignite coal: 39 wt %

Typical fixed carbon content in coal
Anthracite coal: 80.5–85.7 wt %
Bituminous coal: 44.9–78.2 wt %
Lignite coal: 31.4 wt %

Typical bulk density of coal
Anthracite coal: 50–58 (lb/ft^3), 800–929 (kg/m^3)
Bituminous coal: 42–57 (lb/ft^3), 673–913 (kg/m^3)
Lignite coal: 40–54 (lb/ft^3), 641–865 (kg/m^3)

Typical ash content in coal
Anthracite coal: 9.7–20.2 wt %
Bituminous coal: 3.3–11.7 wt %
Lignite coal: 4.2 wt %

Source: http://www.engineeringtoolbox.com/classification-coal-d_164.html

TABLE A.9 Properties of Coal, Natural Gas, and Various Woody and Herbaceous Raw Materials

	Bituminous coal	Natural gas	Wood[a]	Bark	Willow	Forest residues[b]	Wood chips	Wood pellets	Cereal straw	Dedicated energy crops
Ash, d%	8.5–10.9	0	0.4–05	3.5–8	1.1–4.0	1–3	0.8–1.4	0.4–1.5	3–10	6.2–7.5
Moisture, w%	5–10	0	5–60	45–65	50–60	50–60	20–50	7–12	14–25	15–20
NCV, MJ/kg	26–28.3	48	18.5–20	18.0–23	18.4–19.2	18.5–20	19.2–19.4	16.2–19	16.5–17.4	17.1–17.5
Density, kg/m^3	1100–1500	N.A.[c]	390–640	320	120[d]	N.A.[e]	250–350 320–450[g]	500–780	100–170[f]	200[a]
Volatile matter, w%	25–40	100	>70	69.6–77.2	>70	>70	76–86	>70	70–81	>70
Ash melting point, T°C	1100–1400	–	1400–1700	1300–1700	N.A.	N.A.[h]	1000–1400	>1120	700–1000	700–1200
C, d%	76–87	75	48–52	48–52	47–51	48–52	47–52	48–52	45–48	45.5–46.1
H, d%	3.5–5	24	6.2–6.4	4.6–6.8	5.8–6.7	6.0–6.2	6.1–6.3	6.0–6.4	5.0–6.0	5.7–5.8
N, d%	0.8–1.5	0.9	0.1–0.5	0.3–0.8	0.2–0.8	0.3–0.5	<0.3	0.27–0.9	0.4–0.6	0.50–1.0
O, d%	2.8–11.3	0.9	38–42	24.3–42.4	40–46	40–44	38–45	≈40	36–48	41–44
S, d%	0.5–3.1	0	<0.05	<0.05	0.02–0.10	<0.05	<0.05	0.04–0.08	0.05–0.2	0.08–0.13
Cl, d%	<0.1	–	0.01–0.03	0.01–0.03	0.02–0.05	0.01–0.04	0.02	0.02–0.04	0.14–0.97	0.09
K, d%	0.003	–	0.02–0.05	0.1–0.4	0.2–0.5	0.1–0.4	≈0.02	N.A.	0.69–1.3	0.3–0.5
Ca, d%	4–12	–	0.1–1.5	0.02–0.08	0.2–0.7	0.2–0.9	≈0.04	N.A.	0.1–0.6	9

[a] Without bark.
[b] Coniferous trees with needles.
[c] Depends on the aggregate state (compression and temperature).
[d] Willow chips.
[e] Large variations are possible.
[f] Bales.
[g] The first range is for soft wood, the second range is for hard wood.
[h] Large variations are possible.

Source: B. Kavalov and S. D. Peteves, "Status and Perspectives of Biomass-to-Liquid Fuels in the European Union," European Commission, Directorate General Joint Research Centre (DG JRC) Institute for Energy, Petten, The Netherlands, 2005.

TABLE A.10 Properties of Cellulose, Lignin, and Lignocel

Ultimate (wt%, daf)

	Cellulose	Lignin	Lignocel
C	47.23	72.53	53.67
H	5.80	5.43	5.36
N	0.00	0.00	0.00
S	0.00	0.00	0.00
O	46.97	22.04	40.97
Cl	0.00	0.00	0.00

Proximate (wt%)

	Cellulose	Lignin	Lignocel
Moisture	4.30	2.61	9.45
Volatiles (daf)	84.65	76.66	76.45
Fixed Carbon (daf)	11.05	20.73	13.56
Ash (dry)	0.00	0.00	0.54

Source: Guo J.: "Pyrolysis of Wood Powder and Gasification of Wood-Derived Char," Technische Universiteit Eindhoven, Eindhoven, The Netherlands, 2004.

TABLE A.11 Fatty Acid Content of Some Vegetable Oils (%)

Vegetable oil	Palmitic 16:0	Stearic 18:0	Palmitoleic 16:1	Oleic 18:1	Linoleic 18:2	Ricinic 12-OH-oleic	Other acids
Tallow	29.0	24.5	–	44.5	–	–	–
Coconut oil	5.0	3.0	–	6.0	–	–	65.0
Olive oil	14.6	–	–	75.4	10.0	–	–
Groundnut oil	8.5	6.0	–	51.6	26.0	–	–
Cotton oil	28.6	0.9	0.1	13.0	57.2	–	0.2
Corn oil	6.0	2.0	–	44.0	48.0	–	–
Soybean oil	11.0	2.0	–	20.0	64.0	–	3.0
Hazelnut kernel	4.9	2.6	0.2	81.4	10.5	–	0.3
Poppy seed	12.6	4.0	0.1	22.3	60.2	–	0.8
Rapeseed	3.5	0.9	0.1	54.1	22.3	–	9.1
Safflower seed	7.3	1.9	0.1	13.5	77.0	–	0.2
Sunflower seed	6.4	2.9	0.1	17.7	72.8	–	0.1
Castor oil	–	3.0	3.0	3.0	1.2	89.5	0.3

Source: Pinto, A. C., L. N. N. Guarieiro, M.J.C. Rezende, N.M. Ribeiro, E.A. Torres, W.A. Lopes, et al.: "Biodiesel: An Overview," *J. Braz. Chem. Soc.*, 16, 2005, pp. 1313–30.

APPENDIX B
COMPARISON OF THE PROPERTIES OF GASEOUS FUELS FROM DIFFERENT SOURCES

TABLE B.1 General Properties and Description of Gaseous Fuels

	Molecular weight	Specific gravity	Boiling point (°F)	Ignition temperature (°F)	Flash point (°F)	Flammability limits in air (% v/v)
Methane	16.0	0.553	−258.7	900–1170	Gas	5.0–15.0
Ethane	30.1	0.572	−127.5	959	Gas	3.0–12.5
Ethylene	28.0		−154.7	914	Gas	2.8–28.6
Propane	44.1		−43.8	842	Gas	2.1–10.1
Propylene	42.1		−53.9	856	Gas	2.00–11.1
n-Butane	58.1	0.601	31.1	761	−76	1.86–8.41
Isobutane	58.1		10.9	864	−117	1.80–8.44
n-Butene	56.1	0.595	21.2	829	Gas	1.98–9.65
Isobutene	56.1		19.6	869	Gas	1.8–9.0

Types:
(A) Fuels found in nature:
Natural gas: (1) From crude oil reservoirs and (2) from natural gas reservoirs; most volatile fraction from the distillation of oil, petroleum gas (mainly butane, propane)
Methane from coal mines (Firedamp)
Methane from the digestion of biomass by bacteria (landfill gas, digester gas)
(B) Fuel gases made from solid fuel
Gases derived from coal
    (1) Pyrolysis, carbonization: the product of the heating of a solid or liquid to devolatilize the substance to form a char, coke, or charcoal
    (2) Coke oven gas
    (3) Gas from charcoal production
    (4) Gasification—The product of complete gasification of a solid or liquid feedstock—i.e., including the char. Only residue is ash.
Producer gas—low CV gas from partial combustion of coal in air
Blue or water gas—medium CV gas from gasification of coal with steam.
Carburetted water gas—medium-to-high CV gas. Town gas.
Gases derived from waste and biomass
From other industrial processes (blast furnace gas)
(C) Gases made from petroleum
Liquefied petroleum gas (LPG)
Refinery gases: formed by the cracking higher boiling fractions
Gases from residue or heavy oil gasification
(D) Gases from some fermentation process

TABLE B.2 Composition of Different Gaseous Fuels

	Natural gas	LPG	Coal producer gas	Wood producer gas
CO	–	–	20–30%	18–25%
H_2	–	–	8–20%	13–15%
CH_4	80–95%	–	0.5–3%	1–5%
C_2H_6	<6	–	Trace	Trace
> C_2H_6*	<4	100%	Trace	Trace
CO_2	<5	–	3–9%	5–10%
N_2	<5	–	50–56%	45–54%
H_2O	–	–	–	5–15%

*Contains hydrocarbons heavier than C_2H_6.

COMPARISON OF THE PROPERTIES OF GASEOUS FUELS FROM DIFFERENT SOURCES

TABLE B.3 Flash Points and Auto-Ignition Temperatures of Various Fuels

Substance	Flash point (°C)	Autoignition (°C)
Methane	−188	537
Ethane	−135	472
Propane	−104	470
n-Butane	−60	365
n-Octane	10	206
Isooctane	−12	418
n-Cetane	135	205
Methanol	11	385
Ethanol	12	365
Acetylene	Gas	305
Carbon monoxide	Gas	609
Hydrogen	Gas	400

Source: Bartok and Sarofim, *Fossil Fuel Combustion: A Source Book*, 1991, Wiley, by permission of John Wiley and Sons, Inc. (http://users.tkk.fi/~rzevenho/BR_Ch2.pdf.)

TABLE B.4 Composition of Synthesis Gas

	Syngas Composition,* Mole Percent	
Component	Heavy oil feed	Coke feed
CO	45.6	47.7
H_2	43.3	30.3
CO_2	8.2	17.9
H_2O	0.3	0.1
CH_4	0.4	0.01
Ar	1.0	0.8
N_2	0.5	1.3
H_2S	0.7	1.8
COS	0.0	0.02

*Composition also depends on gasifier licensor.
Source: Valerie Jue Francuz, F: "Gasification Workshop," Sep., 11–13, 2001. Indianapolis, Indiana.

TABLE B.5 Calorific Value of Gaseous Fuels

Gaseous fuels (dry)	Gross calorific value
Coal gas coke oven (benzene-free)	540 Btu ft^3, 20 MJ m^{-3}
Coal gas continuous vertical retort	480 Btu ft^3, 18 MJ m^{-3}
Coal gas low temperature	910 Btu ft^3, 34 MJ m^{-3}
Butane	3170 Btu ft^3, 118 MJ m^{-3}
Propane	2520 Btu ft^3, 94 MJ m^{-3}
Natural gas (North Sea)	1045 Btu ft^3, 39 MJ m^{-3}
Producer gas (from coal)	160 Btu ft^3, 6 MJ m^{-3}
Producer gas (from coke)	135 Btu ft^3, 5 MJ m^{-3}
Carburetted water gas	240 Btu ft^3, 9 MJ m^{-3}
Blue water gas blue	26 Btu ft^3, 1 MJ m^{-3}

Source: Herington, E. F. G.: Tables of Physical and Chemical Constants, Kaye Laby. (http://www.kayelaby.npl.co.uk/chemistry/3_11/3_11_4.html)

TABLE B.6 Composition (%) of Gaseous Fuels

Fuel	Carbon dioxide (CO_2)	Carbon monoxide (CO)	Methane (CH_4)	Butane (C_4H_{10})	Ethane (C_2H_6)	Propane (C_3H_8)	Hydrogen (H_2)	Hydrogen sulfide (H_2S)	Oxygen (O_2)	Nitrogen (N_2)
Carbon monoxide		100								
Coal gas	3.8	28.4	0.2				17.0			50.6
Coke oven gas	2.0	5.5	32				51.9		0.3	4.8
Digester gas	30		64				0.7	0.8		2.0
Hydrogen							100			
Landfill gas	47	0.1	47				0.1	0.01	0.8	3.7
Natural gas	0–0.8	0–0.45	82–93		0–15.8		0–1.8	0–0.18	0–0.35	0.5–8.4
Propane gas				0.5–0.8	2.0–2.2	73–97				

Source: http://www.engineeringtoolbox.com/chemical-composition-gaseous-fuels-d_1142.html.

APPENDIX C

COMPARISON OF THE PROPERTIES OF LIQUID FUELS FROM DIFFERENT SOURCES

TABLE C.1 General Properties of Liquid Products from Petroleum

	Molecular weight	Specific gravity	Boiling point (°F)	Ignition temperature (°F)	Flash point (°F)	Flammability limits in air (% v/v)
Benzene	78.1	0.879	176.2	1040	12	1.35–6.65
Diesel fuel	170–198	0.875			100–130	
Fuel oil no. 1		0.875	304–574	410	100–162	0.7–5.0
Fuel oil no. 2		0.920		494	126–204	
Fuel oil no. 4	198.0	0.959		505	142–240	
Fuel oil no. 5		0.960			156–336	
Fuel oil no. 6		0.960			150	
Gasoline	113.0	0.720	100–400	536	−45	1.4–7.6
n-Hexane	86.2	0.659	155.7	437	−7	1.25–7.0
n-Heptane	100.2	0.668	419.0	419	25	1.00–6.00
Kerosene	154.0	0.800	304–574	410	100–162	0.7–5.0
Neohexane	86.2	0.649	121.5	797	−54	1.19–7.58
Neopentane	72.1		49.1	841	Gas	1.38–7.11
n-Octane	114.2	0.707	258.3	428	56	0.95–3.2
Isooctane	114.2	0.702	243.9	837	10	0.79–5.94
n-Pentane	72.1	0.626	97.0	500	−40	1.40–7.80
Isopentane	72.1	0.621	82.2	788	−60	1.31–9.16
n-Pentene	70.1	0.641	86.0	569	—	1.65–7.70
Toluene	92.1	0.867	321.1	992	40	1.27–6.75
Xylene	106.2	0.861	281.1	867	63	1.00–6.00

TABLE C.2 Heat Content of Various Liquid Fuels

	kg/L	bbl/MT	MJ/bbl	MJ/kg	MJ/L
Petroleum products					
Premium unleaded gasoline	0.754	8.34	5255.1	43.81	33.05
Regular unleaded gasoline	0.740	8.50	5172.8	43.95	32.53
Automotive gas oil	0.843	7.46	5732.2	42.78	36.05
Marine diesel oil	0.850	7.40	5770.0	42.68	36.29
Light fuel oil	0.919	6.85	6041.5	41.36	38.00
Heavy fuel oil	0.948	6.63	6124.4	40.63	38.52
Bunker fuel oil	0.955	6.58	6137.4	40.40	38.60
Power station fuel oil	0.890	7.07	5717.0	40.40	35.96
Export fuel oil	0.940	6.69	6174.2	41.31	38.83
Lighting kerosene	0.788	7.98	5462.7	43.60	34.36
Jet fuel	0.803	7.84	5525.3	43.30	34.75
Aviation gasoline	0.716	8.78	5066.1	44.50	31.86
Blended heating oil	0.824	7.63	5686.1	43.40	35.76
Bitumen	1.028	6.12	6399.5	39.15	40.25
Liquid petroleum gas					
LPG60/40	0.534	11.78	3,876.3	45.65	24.38
General product LPG	0.536	11.73	3,890.9	45.66	24.47
Propane	0.508	12.38	3,713.5	45.98	23.36
Butane	0.572	11.00	4,118.2	45.28	25.90

Source: http://www.med.govt.nz/upload/28709/200507-1.pdf.

TABLE C.3 Properties of Distillate and Residual Liquid Fuels

Grade no.	1	2	4	5	6
	Kerosene	Distillate	Very light residual	Light residual	Residual
Color	Clear	Amber	Black	Black	Black
Specific gravity, at 16°C	0.825	0.865	0.928	0.953	0.986
Kinematic viscosity at 38°C (m²/s)	1.6×10^{-6}	2.6×10^{-6}	15×10^{-6}	50×10^{-6}	360×10^{-6}
Pour point (°C)	<−17	<−18	−23	−1	19
Flash point (°C)	38	38	55	55	66
Autoignition temperature (°C)	230	260	263	–	408
Carbon (%)	86.5	86.4	86.1	85.5	85.7
Carbon residue (%)	Trace	Trace	2.5	5.0	12.0
Hydrogen (%)	13.2	12.7	11.9	11.7	10.5
Oxygen (%)	0.01	0.04	0.27	0.3	0.38–0.64
Ash (%)	–	<0.01	0.02	0.03	0.04
HHV (MJ/kg)	46.2	45.4	43.8	43.2	42.4

Source: http://users.tkk.fi/~rzevenho/BR_ch2.pdf.

TABLE C.4 Comparison of Synthetic Crude Oil from Canada (Athabasca Tar Sand), Venezuela (Orinoco Bitumen), and Australia (Stuart Oil Shale) with West Texas Intermediate Crude Oil

	WTI (Benchmark)	Canadian synthetic*	Sincor synthetic[†]	Stuart oil shale[‡]
API Gravity	40	35	32	48
% Sulfur	0.3	0.09	0.07	0.01
Yields, LV%				
Naphtha, LPG	38	21	18	48
Distillate	31	40	40	37
Vacuum gas oil	21	39	42	15
Vacuum bottoms	10	0	0	0

Notes * Based on planned quality in 2005 from existing oil sands plants.
[†] Zuata Orinoco project.
[‡] Proposed for future commercial development by southern Pacific Petroleum.
Source: Crandall, G. R: "Non-Conventional Oil Market Outlook," Presentation to the International Energy Agency, Conference on Non-Conventional Oil, Prospects for Increased Production, Nov. 26, 2002.

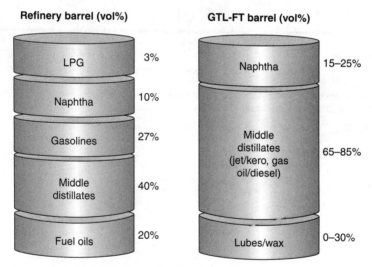

FIGURE C.1 Typical distribution of products from refining and gas-to-liquids processes. [*Source:* B. Kavalov and S. D. Peteves: "Status and Perspectives of Biomass-to-Liquid Fuels in The European Union," European Commission. Directorate General Joint Research Centre (DG JRC) Institute for Energy, Petten, The Netherlands, 2005.]

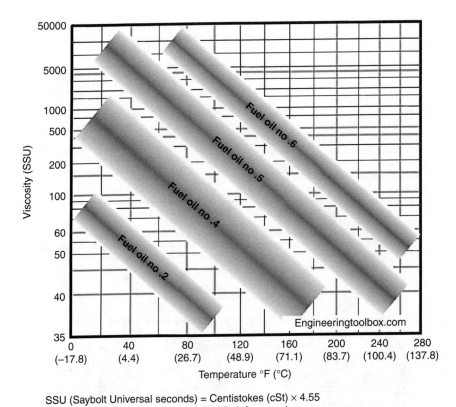

SSU (Saybolt Universal seconds) = Centistokes (cSt) × 4.55
4–5000 SSU is the maximum practical limit for pumping

FIGURE C.2 Viscosity of fuel oil. (*Source: www.engineeringtoolbox.com/viscosity-fuel-oils-d_1143.html.*)

TABLE C.5 Composition and Properties of Shale Oil Produced from United States Sources

	Gas combustion retorting process	Tosco retorting process	Union oil retorting process	Shell ICP process
Gravity, API	19.8	21.2	18.6	38
Pour point, °F	83.5	80	80	
Nitrogen (dohrmann), wt. %	2.14 ±0.15	1.9	2 (KJELDAHL)	1
Sulfur (X-ray F), wt. %	0.6999 ±0.025	0.9	0.9 (P BOMB)	0.5
Oxygen (neutron act.), wt. %	1.60.8	0.9	0.5	
Carbon, wt. %	83.92	85.1	84	85
Hydrogen, wt. %	11.36	11.6	12.0	13
Conradson carbon, wt.%	4.71	4.6	4.6	0.2
Bromine no.	33.2	49.5	Not available	
SBA wax, wt. %	8.1	Not available	6.9 (MEK)	
Viscosity, SSU.:				
100°F	270	108	210	
212°F	476	39	47	
Sediment, wt. %	0.042	Not available	0.043	
Ni, ppm.	6.4	6	4	1
V, ppm.	6.0	3	1.5	1
Fe, ppm.	108.0	100	55	9
Flash (O.C.), °F	240		192 (COC)	
Molecular weight	328		306 (Calculated)	
Distillation				TBP/GC
450° at vol. %	11.1	23	5	45
650° at vol. %	36.1	44	30	84
5 vol% at °F	378	200	390	226
10	438	275	465	271
20	529	410	565	329
30	607	500	640	385
40	678	620	710	428
50	743	700	775	471
60	805	775	830	516
70	865	850	980	570
80	935	920		624
90	1030			696
95	1099			756

Source: Johnson, H. R., P. M. Crawford, and J. W. Bunger: *Strategic Significance of America's Oil Shale Resource*, vol. 2: "Oil Shale Resources, Technology, and Economics," Office of Deputy Assistant Secretary for Petroleum Reserves, Office of Naval Petroleum and Oil Shale Reserves, U.S. Department of Energy, Washington, D.C., Mar. 2004.

TABLE C.6 Heat Content of Fuel Oil

Grade	Heating value, Btu/gal	Comments
Fuel oil no. 1	132,900–137,000	Small space heaters
Fuel oil no. 2	137,000–141,800	Residential heating
Fuel oil no. 4	143,100–148,100	Industrial burners
Fuel oil no. 5 (light)	146,800–150,000	Preheating in general required
Fuel oil no. 5 (heavy)	149,400–152,000	Heating required
Fuel oil no. 6	151,300–155,900	Bunker c

Source: http://www.engineeringtoolbox.com/fuel-oil-combustion-values-d_509.html.

TABLE C.7 Properties of Some Alternate Liquid Fuels

Fuel	Cetane number	Research octane number	Motor octane number	Density (lb/gal)	LHV (Btu/gal)	LHV (Btu/lb)	DEE (gal)	DEE (lb)
100% Ethanol	8	109	90	6.6	75,600	11,500	0.59	0.66
E85, 85% Ethanol		105	89	6.5	83,600	12,855	0.64	0.72
10% Ethanol/gasoline		96.5	86	6.1	111,000	18,000	0.86	1.39
10% Ethanol/diesel	45			7	123,000	17,500	0.95	1.01
100% Methanol	5	109	89	6.7	56,200	8,400	0.44	0.48
100% Soy methyl-ester	49			7.3	120,200	16,500	0.93	0.95
B100, 100% Biodiesel	54				117,000	15,800		
B20, 20% Biodiesel, 80% petrodiesel	46					18,100		
#2 Diesel	44			6.7–7.4	126,000–130,000	18,000–19,000	1	1
#1 Diesel	44			7.6	125,800	16,600	0.98	0.95
Gasoline		90–100	80–90	6	115,400	19,200	0.89	1.1
CNG—compressed natural gas	<0	>127	122			20,400		1.17
LNG—liquified natural gas	<0	>127	122	3.5	78,000	22,300	0.6	1.28
LPG—propane		109	96	4.2	83,600	19,900	0.65	1.14
DME—dimethyl ether	55–60			5.6	74,800	13,600	0.58	0.78

Source: http://www.engineeringtoolbox.com/alternative-fuels-d_1221.html.

TABLE C.8 Properties of Conventional and Alternative Fuels

Property	Gasoline	No. 2 diesel	Methanol	Ethanol	Propane	CNG	Hydrogen
Chemical formula	C_4 to C_{12}	C_3 to C_{25}	CH_3OH	C_2H_5OH	C_3H_8	CH_4	H_2
Physical state	Liquid	Liquid	Liquid	Liquid	Compressed gas	Compressed gas	Compressed gas or liquid
Molecular weight	100–105	~200	32.04	46.07	44.1	16.04	2.02
Composition (wt %)							
Carbon	85–88	84–87	37.5	52.2	82	75	0
Hydrogen	12–15	33–16	12.6	13.1	18	25	100
Oxygen	0	0	49.9	34.7	NA	NA	0
Main fuel source(s)	Crude oil	Crude oil	Natural gas, coal, or woody biomass	Corn, grains, or agricultural waste	Underground reserves	Underground reserves	Natural gas, methanol, and other energy sources
Specific gravity (60°F/60°F)	0.72–0.78	0.81–0.89	0.796	0.796	0.508	0.424	0.07
Density (lb/gal @ 60°F)	6.0–6.5	6.7–7.4	6.63	6.61	4.22	1.07	NA
Boiling temperature (°F)	80–437	370–650	149	172	–44	–259	–423
Freezing point (°F)	–40	–40–30	–143.5	–173.2	–305.8	–296	–435
Autoignition temperature (°F)	495	~600	867	793	850–950	1,004	1,050–1,080
Reid vapor pressure (psi)	8–15	0.2	4.6	2.3	208	2,400	NA

Source: Alternative Fuels Data Center, "Properties of Fuel," www.afdc.doe.gov/pdfs/fueltable.pdf, and "Fuel Comparison," www.afdc.doe.gov/fuel_comp.html, Aug. 2005.

TABLE C.9 Specifications for Diesel and Biodiesel (ASTM)

Property	Diesel	Biodiesel
Standard	ASTM D975	ASTM D6751
Composition	HC* (C10–C21)	FAME† (C12–C22)
Kin. viscosity (mm^2/s) at 40°C	1.9–4.1	1.9–6.0
Specific gravity (g/mL)	0.85	0.88
Flash point (°C)	60–80	100–170
Cloud point (°C)	−15 to 5	−3 to 12
Pour point (°C)	−35 to −15	−15 to 16
Water (vol%)	0.05	0.05
Carbon (wt%)	87	77
Hydrogen (wt%)	13	12
Oxygen (wt%)	0	11
Sulfur (wt%)	0.05	0.05
Cetane number	40–55	48–60
HFRR‡ (μm)	685	314
BOCLE§ scuff (g)	3,600	>7,000

*HC: hydrocarbons.
†FAME: fatty acid methyl esters.
‡HFRR: high frequency reciprocating rig.
§BOCLE: ball-on cylinder lubricity evaluator.

APPENDIX D
COMPARISON OF THE PROPERTIES OF SOLID FUELS FROM DIFFERENT SOURCES

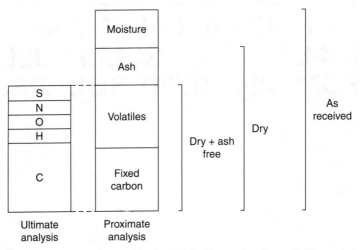

FIGURE D.1 Proximate analysis of solid fuels. (*Source: http://users.tkk.fi/~rzevenho/BR_ch2.pdf.*)

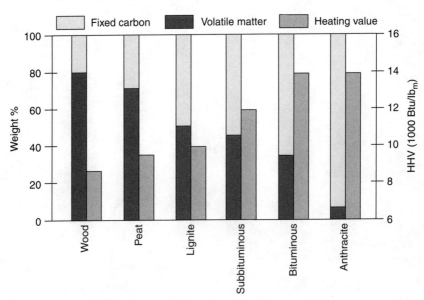

FIGURE D.2 Proximate analysis of wood and coal types. (*Source: http://users.tkk.fi/~rzevenho/BR_ch2.pdf.*)

COMPARISON OF THE PROPERTIES OF SOLID FUELS FROM DIFFERENT SOURCES

TABLE D.1 Proximate Analysis and Ultimate Analysis of Solid Fuels (Dry, Ash-Free)

	Fuel type				
	Wood	Peat	Lignite	Bituminous coal	Refuse-derived fuel
Proximate analysis (wt %)					
Volatile matter	81	65	55	40	85
Fixed carbon	19	35	45	60	15
Ultimate analysis (wt %)					
Hydrogen	6	6	5	5	7
Carbon	50	55	68	78	52
Sulfur	0.1	0.4	1	2	0.3
Nitrogen	0.1	1	1	2	0.6
Oxygen	44	38	25	13	40
Higher heating value (Btu/lb)	8700	9500	10,000	14,000	9,700

Source: http://users.tkk.fi/~rzevenho/BR_ch2.pdf.

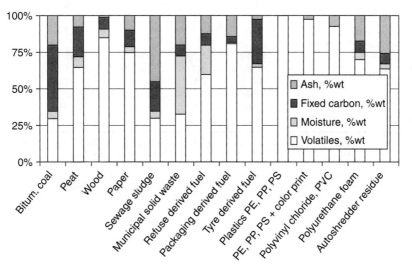

FIGURE D.3 Proximate analysis of fossil and renewable fuels. (*Source:* http://users.tkk.fi/~rzevenho/BR_ch2.pdf.)

TABLE D.2 Heat Content of Biomass Materials

Switchgrass/(Btu/lb)	7,341
Bagasse/(Btu/lb)	6,065
Rice hulls/(Btu/lb)	6,575
Poultry litter/(Btu/lb)	6,187
Solid wood waste/(Btu/lb)	6,000–8,000

Source: Wright, L., R. Boundy, R. Perlack, S. Davis, and B. Saulsbury: *Biomass Energy Data Book*, Office of Planning, Budget and Analysis, Energy Efficiency and Renewable Energy, United States Department of Energy, Contract No. DE-AC05-00OR22725, Oak Ridge National Laboratory, Oak Ridge, Tenn., 2006, p. 136.

TABLE D.3 Moisture, Ash, Heat Content, and Chemical Composition of Selected Biomass Fuels

Fuel type	Clean wood	Verge grass	Organic domestic waste	Demolition wood	Sludge
Moisture content/(%) of wet fuel	50	60	54	20	20
Ash content/(%) of dry fuel	1.3	8.4	18.9	0.9	37.5
LHV (as received)/(MJ/kg)	7.7	5.4	6.4	13.9	8.8
HHV (as received)/(MJ/kg)	9.6	7.4	8.3	15.4	9.9
Composition/(%) w/w (maf)					
C	49.10	48.70	51.90	48.40	52.50
H	6.00	6.40	6.70	5.20	7.20
O	44.30	42.50	38.70	45.20	30.30
N	0.48	1.90	2.20	0.15	6.99
S	0.01	0.14	0.50	0.03	2.74
Cl	0.10	0.39	0.3	0.08	0.19

Source: van Doorn, J.: "Characterization of Energy Crops and Biomass and Waste Streams," Netherlands Energy Research Foundation, Report ECN-C-95-047, Petten, The Netherlands, July, 1995.
Faaij, A., van Doorn, J., Curvers, A., Waldheim, L., Olsson, E., van Wijk, et al.: "Characterization and Availability of Biomass Waste and Residues in the Netherlands for Gasification," Department of Science, Technology, and Society, Utrecht University, Netherlands Energy Research Foundation, Termiska Processer AB Studsvik Sweden, Province of Noord-Holland, *Biomass and Bioenergy*, 12(4), 1997, pp. 225–40.

GLOSSARY

Alcohol: The family name of a group of organic chemical compounds composed of carbon, hydrogen, and oxygen. The molecules in the series vary in chain length and are composed of a hydrocarbon plus a hydroxyl group. Alcohol includes methanol and ethanol.

Alkylation: A process for manufacturing high-octane blending components used in unleaded petrol or gasoline.

Anaerobic digestion: Decomposition of biological wastes by microorganisms, usually under wet conditions, in the absence of air (oxygen), to produce a gas comprising mostly methane and carbon dioxide.

Annual removals: The net volume of growing stock trees removed from the inventory during a specified year by harvesting, cultural operations, such as, timber stand improvement, or land clearing.

API gravity: A measure of the lightness or heaviness of petroleum that is related to density and specific gravity.

$$°API = (141.5/\text{specific gravity @ } 60°F) - 131.5$$

Aromatics: A range of hydrocarbons which have a distinctive sweet smell and include benzene and toluene. These occur naturally in petroleum and are also extracted as a petrochemical feedstock, as well as for use as solvents.

Asphaltene (asphaltenes): The brown to black powdery material produced by treatment of petroleum, heavy oil, bitumen, or residuum with a low-boiling liquid hydrocarbon.

Barrel (bbl): the unit of measure used by the petroleum industry; equivalent to approximately 42 U.S. gallons or approximately 34 (33.6) Imperial gallons or 159 L; 7.2 bbl are equivalent to 1 metric ton of oil.

Barrel of oil equivalent (boe): The amount of energy contained in a barrel of crude oil, that is, approximately 6.1 GJ (5.8 million Btu), equivalent to 1700 kW.

Billion: 1×10^9

Biochemical conversion: The use of fermentation or anaerobic digestion to produce fuels and chemicals from organic sources.

Biodiesel: A fuel derived from biological sources that can be used in diesel engines instead of petroleum-derived diesel; through the process of transesterification, the triglycerides in the biologically derived oils are separated from the glycerin, creating a clean-burning, renewable fuel.

Bioenergy: Useful, renewable energy produced from organic matter—the conversion of the complex carbohydrates in organic matter to energy; organic matter may either be used directly as a fuel, processed into liquids and gasses, or be a residual of processing and conversion.

Bioethanol: Ethanol produced from biomass feedstocks; includes ethanol produced from the fermentation of crops, such as corn, as well as cellulosic ethanol produced from woody plants or grasses.

Biofuels: A generic name for liquid or gaseous fuels that are not derived from petroleum based fossils fuels or contain a proportion of non-fossil fuel; fuels produced from plants, crops such as sugar beet, rape seed oil, or reprocessed vegetable oils or fuels made from gasified biomass; fuels made from renewable biological sources and include ethanol, methanol, and biodiesel; sources include, but are not limited to: corn, soybeans, flaxseed, rapeseed, sugarcane, palm oil, raw sewage, food scraps, animal parts, and rice.

Biogas: A combustible gas derived from decomposing biological waste under anaerobic conditions. Biogas normally consists of 50 to 60 percent methane. See also landfill gas.

Biomass: Any organic matter that is available on a renewable or recurring basis, including agricultural crops and trees, wood and wood residues, plants (including aquatic plants), grasses, animal manure, municipal residues, and other residue materials. Biomass is generally produced in a sustainable manner from water and carbon dioxide by photosynthesis. There are three main categories of biomass—primary, secondary, and tertiary.

Biopower: The use of biomass feedstock to produce electric power or heat through direct combustion of the feedstock, through gasification and then combustion of the resultant gas, or through other thermal conversion processes. Power is generated with engines, turbines, fuel cells, or other equipment.

Biorefinery: A facility that processes and converts biomass into value-added products. These products can range from biomaterials to fuels such as ethanol or important feedstocks for the production of chemicals and other materials.

Biomass to liquid (BTL): The process of converting biomass to liquid fuels.

Bitumen: Also, on occasion, referred to as native asphalt, and extra heavy oil; a naturally occurring material that has little or no mobility under reservoir conditions and which cannot be recovered through a well by conventional oil well production methods including currently used enhanced recovery techniques; current methods involve mining for bitumen recovery.

Black liquor: Solution of lignin-residue and the pulping chemicals used to extract lignin during the manufacture of paper.

Bone dry: Having zero percent moisture content. Wood heated in an oven at a constant temperature of 100°C (212°F) or above until its weight stabilizes is considered bone dry or oven dry.

Bottoming cycle: A cogeneration system in which steam is used first for process heat and then for electric power production.

British thermal unit (Btu): A nonmetric unit of heat, still widely used by engineers; 1 Btu is the heat energy needed to raise the temperature of 1 lb of water from 60°F to 61°F at 1 atm pressure. 1 Btu = 1055 J (1.055 kJ).

Bunker: A storage tank.

Butanol: Though generally produced from fossil fuels, this four-carbon alcohol can also be produced through bacterial fermentation of alcohol.

Carbon dioxide (CO_2): A product of combustion that acts as a greenhouse gas in the earth's atmosphere, trapping heat and contributing to climate change.

Carbon monoxide (CO): A lethal gas produced by incomplete combustion of carbon-containing fuels in internal combustion engines. It is colorless, odorless, and tasteless.

Carbon sink: A geographical area whose vegetation and/or soil soaks up significant carbon dioxide from the atmosphere. Such areas, typically in tropical regions, are increasingly being sacrificed for energy crop production.

Catalyst: A substance that accelerates a chemical reaction without itself being affected. In refining, catalysts are used in the cracking process to produce blending components for fuels.

Cetane number: A measure of the ignition quality of diesel fuel; the higher the number the more easily the fuel is ignited under compression.

GLOSSARY

Closed-loop biomass: Crops grown, in a sustainable manner, for the purpose of optimizing their value for bioenergy and bioproduct uses. This includes annual crops such as maize and wheat, and perennial crops such as trees, shrubs, and grasses such as switch grass.

Cloud point: The temperature at which paraffin wax or other solid substances begin to crystallize or separate from the solution, imparting a cloudy appearance to the oil when the oil is chilled under prescribed conditions.

Coarse materials: Wood residues suitable for chipping, such as slabs, edgings, and trimmings.

Coking: A thermal method used in refineries for the conversion of bitumen and residua to volatile products and coke (see "Delayed coking" and "Fluid coking").

Conventional crude oil (conventional petroleum): Crude oil that is pumped from the ground and recovered using the energy inherent in the reservoir; also recoverable by application of secondary recovery techniques.

Cord: A stack of wood comprising 128 ft^3 (3.62 m^3); standard dimensions are 4 × 4 × 8 ft, including air space and bark. One cord contains approximately 1.2 U.S. tons (oven-dry) = 2400 lb = 1089 kg.

Cracking: A secondary refining process that uses heat and/or a catalyst to break down high molecular weight chemical components into lower molecular weight products which can be used as blending components for fuels.

Cropland: Total cropland includes five components: cropland harvested, crop failure, cultivated summer fallow, cropland used only for pasture, and idle cropland.

Cropland pasture: Land used for long-term crop rotation. However, some cropland pasture is marginal for crop uses and may remain in pasture indefinitely. This category also includes land that was used for pasture before crops reached maturity and some land used for pasture that could have been cropped without additional improvement.

Cull tree: A live tree, 5.0 in in diameter at breast height (d.b.h.) or larger that is nonmerchantable for saw logs now or prospectively because of rot, roughness, or species.

Cultivated summer fallow: Cropland cultivated for one or more seasons to control weeds and accumulate moisture before small grains are planted.

Delayed coking: A coking process in which the thermal reactions are allowed to proceed to completion to produce gaseous, liquid, and solid (coke) products.

Density: The mass (or weight) of a unit volume of any substance at a specified temperature (see also "Specific gravity").

Desulfurization: The removal of sulfur or sulfur compounds from a feedstock.

Diesel engine: Named after the German engineer Rudolph Diesel, this internal-combustion, compression-ignition engine works by heating fuels and causing them to ignite; can use either petroleum or bioderived fuel.

Diesel fuel: A distillate of fuel oil that has been historically derived from petroleum for use in internal combustion engines; also derived from plant and animal sources.

Diesel, Rudolph: German inventor famed for fashioning the diesel engine, which made its debut at the 1900 World's Fair; initially engine to run on vegetable-derived fuels.

Digester: An airtight vessel or enclosure in which bacteria decomposes biomass in water to produce biogas.

Direct-injection engine: A diesel engine in which fuel is injected directly into the cylinder.

Distillate: Any petroleum product produced by boiling crude oil and collecting the vapors produced as a condensate in a separate vessel, for example, gasoline (light distillate), gas oil (middle distillate), or fuel oil (heavy distillate).

Distillation: The primary distillation process which uses high temperature to separate crude oil into vapor and fluids which can then be fed into a distillation or fractionating tower.

Downdraft gasifier: A gasifier in which the product gases pass through a combustion zone at the bottom of the gasifier.

Dutch oven furnace: One of the earliest types of furnaces, having a large, rectangular box lined with firebrick (refractory) on the sides and top; commonly used for burning wood.

E85: An alcohol fuel mixture containing 85 percent ethanol and 15 percent gasoline by volume, and the current alternative fuel of choice of the U.S. government.

Effluent: The liquid or gas discharged from a process or chemical reactor, usually containing residues from that process.

Emissions: Substances discharged into the air during combustion, for example, all that stuff that comes out of your car.

Energy crops: Crops grown specifically for their fuel value; include food crops such as corn and sugarcane, and nonfood crops such as poplar trees and switch grass.

Energy balance: The difference between the energy produced by a fuel and the energy required to obtain it through agricultural processes, drilling, refining, and transportation.

Energy crops: Agricultural crops grown specifically for their energy value.

Energy-efficiency ratio: A number representing the energy stored in a fuel as compared to the energy required to produce, process, transport, and distribute that fuel.

Enhanced recovery: Methods that usually involve the application of thermal energy (e.g., steam flooding) to oil recovery from the reservoir.

Ethanol (ethyl alcohol, alcohol, or grain-spirit): A clear, colorless, flammable oxygenated hydrocarbon; used as a vehicle fuel by itself (E100 is 100 percent ethanol by volume), blended with gasoline (E85 is 85 percent ethanol by volume), or as a gasoline octane enhancer and oxygenate (10 percent by volume).

Feedstock: The biomass used in the creation of a particular biofuel (e.g., corn or sugarcane for ethanol, soybeans or rapeseed for biodiesel).

Fermentation: Conversion of carbon-containing compounds by microorganisms for production of fuels and chemicals such as alcohols, acids, or energy-rich gases.

Fiber products: Products derived from fibers of herbaceous and woody plant materials; examples include pulp, composition board products, and wood chips for export.

Fine materials: Wood residues not suitable for chipping, such as planer shavings and sawdust.

Flexible-fuel vehicle (flex-fuel vehicle): A vehicle that can run alternately on two or more sources of fuel; includes cars capable of running on gasoline and gasoline/ethanol mixtures, as well as cars that can run on both gasoline and natural gas.

Fluid coking: A continuous fluidized solids process that cracks feed thermally over heated coke particles in a reactor vessel to gas, liquid products, and coke.

Fluidized bed boiler: A large, refractory-lined vessel with an air distribution member or plate in the bottom, a hot gas outlet in or near the top, and some provisions for introducing fuel; the fluidized bed is formed by blowing air up through a layer of inert particles (such as sand or limestone) at a rate that causes the particles to go into suspension and continuous motion.

Fly ash: Small ash particles carried in suspension in combustion products.

Forest land: Land at least 10 percent stocked by forest trees of any size, including land that formerly had such tree cover and that will be naturally or artificially regenerated; includes transition zones, such as areas between heavily forested and nonforested lands that are at least 10 percent stocked with forest trees and forest areas adjacent to urban and built-up lands; also included are pinyon-juniper and chaparral areas; minimum area for classification of forest land is 1 acre.

Forest residues: Material not harvested or removed from logging sites in commercial hardwood and softwood stands as well as material resulting from forest management operations such as precommercial thinnings and removal of dead and dying trees.

Forest health: A condition of ecosystem sustainability and attainment of management objectives for a given forest area; usually considered to include green trees, snags, resilient stands growing at a moderate rate, and endemic levels of insects and disease.

Fossil fuel: Solid, liquid, or gaseous fuels formed in the ground after millions of years by chemical and physical changes in plant and animal residues under high temperature and pressure. Oil, natural gas, and coal are fossil fuels.

Fuel cell: A device that converts the energy of a fuel directly to electricity and heat, without combustion.

Fuel cycle: The series of steps required to produce electricity. The fuel cycle includes mining or otherwise acquiring the raw fuel source, processing and cleaning the fuel, transport, electricity generation, waste management, and plant decommissioning.

Fuel oil: A heavy residue, black in color, used to generate power or heat by burning in furnaces.

Fuel treatment evaluator (FTE): A strategic assessment tool capable of aiding the identification, evaluation, and prioritization of fuel treatment opportunities.

Fuel wood: Wood used for conversion to some form of energy, primarily for residential use.

Furnace: An enclosed chamber or container used to burn biomass in a controlled manner to produce heat for space or process heating.

Gasification: A chemical or heat process used to convert carbonaceous material (such as coal, petroleum, and biomass) into gaseous components such as carbon monoxide and hydrogen.

Gasifier: A device for converting solid fuel into gaseous fuel; in biomass systems, the process is referred to as pyrolytic distillation.

Gasohol: A mixture of 10 percent anhydrous ethanol and 90 percent gasoline by volume; 7.5 percent anhydrous ethanol and 92.5 percent gasoline by volume; or 5.5 percent anhydrous ethanol and 94.5 percent gasoline by volume.

Gas to liquids (GTL): The process of refining natural gas and other hydrocarbons into longer-chain hydrocarbons, which can be used to convert gaseous waste products into fuels.

Gel point: The point at which a liquid fuel cools to the consistency of petroleum jelly.

Genetically modified organism (GMO): An organism whose genetic material has been modified through recombinant DNA technology, altering the phenotype of the organism to meet desired specifications.

Grassland pasture and range: All open land used primarily for pasture and grazing, including shrub and brush land types of pasture; grazing land with sagebrush and scattered mesquite; and all tame and native grasses, legumes, and other forage used for pasture or grazing; because of the diversity in vegetative composition, grassland pasture and range are not always clearly distinguishable from other types of pasture and range; at one extreme, permanent grassland may merge with cropland pasture, or grassland may often be found in transitional areas with forested grazing land.

Grease car: A diesel-powered automobile rigged postproduction to run on used vegetable oil.

Greenhouse effect: The effect of certain gases in the earth's atmosphere in trapping heat from the sun.

Greenhouse gases: Gases that trap the heat of the sun in the earth's atmosphere, producing the greenhouse effect. The two major greenhouse gases are water vapor and carbon dioxide. Other greenhouse gases include methane, ozone, chlorofluorocarbons, and nitrous oxide.

Grid: An electric utility company's system for distributing power.

Growing stock: A classification of timber inventory that includes live trees of commercial species meeting specified standards of quality or vigor; cull trees are excluded.

Habitat: The area where a plant or animal lives and grows under natural conditions. Habitat includes living and nonliving attributes and provides all requirements for food and shelter.

Hardwoods: Usually broad-leaved and deciduous trees.

Heating value: The maximum amount of energy that is available from burning a substance.

Heavy (crude) oil: Oil that is more viscous that conventional crude oil, has a lower mobility in the reservoir but can be recovered through a well from the reservoir by the application of a secondary or enhanced recovery methods.

Hectare: Common metric unit of area, equal to 2.47 acres. 100 hectares = 1 km.

Herbaceous: Nonwoody type of vegetation, usually lacking permanent strong stems, such as grasses, cereals, and canola (rape).

Heteroatom compounds: chemical compounds that contain nitrogen and/or oxygen and/or sulfur and /or metals bound within their molecular structure(s).

Hydrocarbonaceous material: A material such as bitumen that is composed of carbon and hydrogen with other elements (hetero elements) such as nitrogen, oxygen, sulfur, and metals chemically combined within the structures of the constituents; even though carbon and hydrogen may be the predominant elements, there may be very few true hydrocarbons (qv).

Hydrocarbon compounds: Chemical compounds containing only carbon and hydrogen.

Hydrodesulfurization: The removal of sulfur by hydrotreating (qv).

Hydroprocesses: Refinery processes designed to add hydrogen to various products of refining.

Hydrotreating: The removal of heteroatomic (nitrogen, oxygen, and sulfur) species by treatment of a feedstock or product at relatively low temperatures in the presence of hydrogen.

Idle cropland: Land in which no crops were planted; acreage diverted from crops to soil-conserving uses (if not eligible for and used as cropland pasture) under federal farm programs is included in this component.

Incinerator: Any device used to burn solid or liquid residues or wastes as a method of disposal.

Inclined grate: A type of furnace in which fuel enters at the top part of a grate in a continuous ribbon, passes over the upper drying section where moisture is removed, and descends into the lower burning section. Ash is removed at the lower part of the grate.

Indirect-injection engine: An older model of diesel engine in which fuel is injected into a prechamber, partly combusted, and then sent to the fuel-injection chamber.

Indirect liquefaction: Conversion of biomass to a liquid fuel through a synthesis gas intermediate step.

Industrial wood: All commercial round wood products except fuel wood.

Joule (J): Metric unit of energy, equivalent to the work done by a force of 1 N applied over distance of 1 m (= 1 kg m^2/s^2). 1 J = 0.239 cal (1 cal = 4.187 J).

Kerosene: A light middle distillate that in various forms is used as aviation turbine fuel or for burning in heating boilers or as a solvent, such as white spirit.

Kilowatt (kW): A measure of electrical power equal to 1000 W. 1 kW = 3412 Btu/hr = 1.341 horsepower.

Kilowatt hour (kWh): A measure of energy equivalent to the expenditure of 1 kW for 1 hour. For example, 1 kWh will light a 100-W light bulb for 10 hours. 1 kWh = 3412 Btu.

Landfill gas: A type of biogas that is generated by decomposition of organic material at landfill disposal sites. Landfill gas is approximately 50 percent methane (see also "Biogas").

Lignin: Structural constituent of wood and (to a lesser extent) other plant tissues, which encrusts the walls and cements the cells together.

Live cull: A classification that includes live cull trees; when associated with volume, it is the net volume in live cull trees that are 5.0 in in diameter and larger.

Logging residues: The unused portions of growing-stock and non-growing-stock trees cut or killed logging and left in the woods.

M85: An alcohol fuel mixture containing 85 percent methanol and 15 percent gasoline by volume. Methanol is typically made from natural gas, but can also be derived from the fermentation of biomass.

Megawatt (MW): A measure of electrical power equal to 1 million watts (1000 kW).

Methanol: A fuel typically derived from natural gas, but which can be produced from the fermentation of sugars in biomass.

Million: 1×10^6

Mill residue: Wood and bark residues produced in processing logs into lumber, plywood, and paper.

Modified/unmodified diesel engine: Traditional diesel engines must be modified to heat the oil before it reaches the fuel injectors in order to handle straight vegetable oil. Modified, any diesel engine can run on veggie oil; without modification, the oil must first be converted to biodiesel.

Moisture content (MC): The weight of the water contained in wood, usually expressed as a percentage of weight, either oven-dry or as received.

Moisture content, dry basis: Moisture content expressed as a percentage of the weight of ovenwood, that is, [(weight of wet sample − weight of dry sample) / weight of dry sample] × 100.

Moisture content, wet basis: Moisture content expressed as a percentage of the weight of wood as-received, that is, [(weight of wet sample − weight of dry sample) / weight of wet sample] × 100.

MTBE: Methyl-tertiary-butyl ether is highly refined high octane light distillate used in the blending of petrol.

Nitrogen fixation: The transformation of atmospheric nitrogen into nitrogen compounds that can be used by growing plants.

Nitrogen oxides (NO_x): Products of combustion that contribute to the formation of smog and ozone.

Nonforest land: Land that has never supported forests and lands formerly forested where use of timber management is precluded by development for other uses; if intermingled in forest areas, unimproved roads and nonforest strips must be more than 120 ft wide, clearings, and so on, must be more than 1 acre in area to qualify as nonforest land.

Nonattainment area: Any area that does not meet the national primary or secondary ambient air quality standard established (by the Environmental Protection Agency) for designated pollutants, such as carbon monoxide and ozone.

Nonindustrial private: An ownership class of private lands where the owner does not operate wood processing plants.

Oil from tar sand: Synthetic crude oil (qv).

Oil mining: Application of a mining method to the recovery of bitumen.

OOIP (oil originally in place or original oil in place): The quantity of petroleum existing in a reservoir before oil recovery operations begin.

Open-loop biomass: Biomass that can be used to produce energy and bio-products even though it was not grown specifically for this purpose; include agricultural livestock waste, residues from forest-harvesting and crop-harvesting operations.

Oxygenate: A substance which, when added to gasoline, increases the amount of oxygen in that gasoline blend; includes fuel ethanol, methanol, and methyl-tertiary-butyl ether (MTBE).

Particulate emissions: Particles of a solid or liquid suspended in a gas, or the fine particles of carbonaceous soot and other organic molecules discharged into the air during combustion.

Pay zone thickness: The depth of a tar sand deposit from which bitumen (or a product) can be recovered.

Particulate: A small, discrete mass of solid or liquid matter that remains individually dispersed in gas or liquid emissions.

Photosynthesis: Process by which chlorophyll-containing cells in green plants convert incident light to chemical energy, capturing carbon dioxide in the form of carbohydrates.

Pour point: The lowest temperature at which oil will pour or flow when it is chilled without disturbance under definite conditions.

Primary oil recovery: Oil recovery utilizing only naturally occurring forces; recovery of crude oil from the reservoir using the inherent reservoir energy.

Primary wood-using mill: A mill that converts round wood products into other wood products; common examples are sawmills that convert saw logs into lumber and pulp mills that convert pulpwood round wood into wood pulp.

Process heat: Heat used in an industrial process rather than for space-heating or other housekeeping purposes.

Producer gas: Fuel gas high in carbon monoxide (CO) and hydrogen (H_2), produced by burning a solid fuel with insufficient air or by passing a mixture of air and steam through a burning bed of solid fuel.

Pulpwood: Round wood, whole-tree chips, or wood residues that are used for the production of wood pulp.

Pyrolysis: The thermal decomposition of biomass at high temperatures (greater than 400°F, or 204°C) in the absence of air; the end product of pyrolysis is a mixture of solids (char), liquids (oxygenated oils), and gases (methane, carbon monoxide, and carbon dioxide) with proportions determined by operating temperature, pressure, oxygen content, and other conditions.

Quad: One quadrillion Btu (10^{15} Btu) = 1.055 exajoules (EJ), or approximately 172 million barrels of oil equivalent.

Recovery boiler: A pulp mill boiler in which lignin and spent cooking liquor (black liquor) is burned to generate steam.

Refractory lining: A lining, usually of ceramic, capable of resisting and maintaining high temperatures.

Refuse-derived fuel (RDF): Fuel prepared from municipal solid waste; noncombustible materials such as rocks, glass, and metals are removed, and the remaining combustible portion of the solid waste is chopped or shredded.

Residues: Bark and woody materials that are generated in primary wood-using mills when round wood products are converted to other products.

Residuum (plural residua, also known as resid or resids): The nonvolatile portion of petroleum that remains as residue after refinery distillation; hence, atmospheric residuum, vacuum residuum.

GLOSSARY

Rotation: Period of years between establishment of a stand of timber and the time when it is considered ready for final harvest and regeneration.

Round wood products: Logs and other round timber generated from harvesting trees for industrial or consumer use.

Sandstone: A sedimentary rock formed by compaction and cementation of sand grains; can be classified according to the mineral composition of the sand and cement.

Saturated steam: Steam at boiling temperature for a given pressure.

Secondary oil recovery: Application of energy (e.g., water flooding) to recovery of crude oil from a reservoir after the yield of crude oil from primary recovery diminishes.

Secondary wood processing mills: A mill that uses primary wood products in the manufacture of finished wood products, such as cabinets, moldings, and furniture.

Specific gravity: The mass (or weight) of a unit volume of any substance at a specified temperature compared to the mass of an equal volume of pure water at a standard temperature.

Stand (of trees): A tree community that possesses sufficient uniformity in composition, constitution, age, spatial arrangement, or condition to be distinguishable from adjacent communities.

Steam turbine: A device for converting energy of high-pressure steam (produced in a boiler) into mechanical power which can then be used to generate electricity.

Straight vegetable oil (SVO): Any vegetable oil that has not been optimized through the process of transesterification.

Superheated steam: Steam which is hotter than boiling temperature for a given pressure.

Sustainable: An ecosystem condition in which biodiversity, renewability, and resource productivity are maintained over time.

Synthetic crude oil (syncrude): A hydrocarbon product produced by the conversion of coal, oil shale, or tar sand bitumen that resembles conventional crude oil; can be refined in a petroleum refinery.

Synthetic ethanol: Ethanol produced from ethylene, a petroleum by-product.

Tar sand (bituminous sand): A formation in which the bituminous material (bitumen) is found as a filling in veins and fissures in fractured rocks or impregnating relatively shallow sand, sandstone, and limestone strata; a sandstone reservoir that is impregnated with a heavy, extremely viscous, black hydrocarbonaceous, petroleum-like material that cannot be retrieved through a well by conventional or enhanced oil recovery techniques; (FE 76-4): The several rock types that contain an extremely viscous hydrocarbon which is not recoverable in its natural state by conventional oil well production methods including currently used enhanced recovery techniques.

Thermochemical conversion: Use of heat to chemically change substances from one state to another, for example, to make useful energy products.

Timberland: Forest land that is producing or is capable of producing crops of industrial wood, and that is not withdrawn from timber utilization by statute or administrative regulation.

Tipping fee: A fee for disposal of waste.

Ton (short ton): 2000 lb.

Tonne (Imperial ton, long ton, shipping ton): 2240 lb; equivalent to 1000 kg or in crude oil terms about 7.5 bbl of oil.

Topping cycle: A cogeneration system in which electric power is produced first; the reject heat from power production is then used to produce useful process heat.

Topping and back pressure turbines: Turbines which operate at exhaust pressure considerably higher than atmospheric (noncondensing turbines); often multistage with relatively high efficiency.

Transesterification: The chemical process in which an alcohol reacts with the triglycerides in vegetable oil or animal fats, separating the glycerin and producing biodiesel.

Traveling grate: A type of furnace in which assembled links of grates are joined together in a perpetual belt arrangement. Fuel is fed in at one end and ash is discharged at the other.

Trillion: 1×10^{12}

Turbine: A machine for converting the heat energy in steam or high temperature gas into mechanical energy. In a turbine, a high-velocity flow of steam or gas passes through successive rows of radial blades fastened to a central shaft.

Turn down ratio: The lowest load at which a boiler will operate efficiently as compared to the boiler's maximum design load.

Vacuum distillation: A secondary distillation process which uses a partial vacuum to lower the boiling point of residues from primary distillation and extract further blending components.

Viscosity: A measure of the ability of a liquid to flow or a measure of its resistance to flow; the force required to move a plane surface of area 1 square meter over another parallel plane surface 1 meter away at a rate of 1 meter per second when both surfaces are immersed in the fluid; the higher the viscosity, the slower the liquid flows.

Volatile organic compounds (VOCs): Name given to light organic hydrocarbons which escape as vapor from fuel tanks or other sources, and during the filling of tanks. VOCs contribute to smog.

Waste streams: Unused solid or liquid by-products of a process.

Waste vegetable oil (WVO): Grease from the nearest fryer which is filtered and used in modified diesel engines, or converted to biodiesel through the process of transesterification and used in any old diesel car.

Water-cooled vibrating grate: A boiler grate made up of a tuyere grate surface mounted on a grid of water tubes interconnected with the boiler circulation system for positive cooling; the structure is supported by flexing plates allowing the grid and grate to move in a vibrating action; ash is automatically discharged.

Watershed: The drainage basin contributing water, organic matter, dissolved nutrients, and sediments to a stream or lake.

Watt: The common base unit of power in the metric system; 1 W = 1 J/s, or the power developed in a circuit by a current of 1 A flowing through a potential difference of 1 V (1 W = 3.412 Btu/h).

Wheeling: The process of transferring electrical energy between buyer and seller by way of an intermediate utility or utilities.

Whole-tree harvesting: A harvesting method in which the whole tree (above the stump) is removed.

Yarding: The initial movement of logs from the point of felling to a central loading area or landing.

INDEX

absorption, gas analysis by, 91
acetogenesis, 355
acicular coke, 100
acid gases, 36. *See also* processing
acid rain, 164–165, 241
acid-catalyzed transesterification, 285–286
acidogenic digestate, 276
acids. *See* fatty acids
aerobic decomposition, 353–355, 362–363
afterdamp, 140
age of refuse, 349
agglomerating coal, 152
agitated gasifiers, 155
agitated reactors, 159
agricultural residues, 273–274, 280
air pollution from wastes, 326
air samplers, ambient, 358
Albanian tar sand deposits, 107
alcoholic fermentation, 231
alcohols. *See also* ethanol
   bioalcohol, 18–19
   from biomass, 244–247
   effect on biodiesel transesterification, 288
   oxygenated gasoline, 96
   producing from crops, 281–282
   properties, 370
   from waste, 338–340
alcoholysis, biodiesel, 284
alkali-catalyzed transesterification, 285
Alkazid process, 52
alkylation processes, 83–84, 93
alternative fuels, 1–3, 14–15, 241, 387–388.
   See also nonconventional fuel sources
aluminum chloride, 84
ambient air samplers, 358
amine process, 45–47, 49
ammonia, 346
amorphous carbon, 211

anaerobic digestion
   bioreactors, 362–363
   to create biogas, 243
   defined, 231
   energy crops, 275–278
   landfill mining, 359
   of waste, 18, 332, 353–355
analyses
   coal, 137, 144
   solid fuel, 392–393
animal charcoal, 320
animal waste, 277–278
anthracite, 8, 136, 373
anthropogenic pathways for landfill gas migration, 361
anticlinal natural gas reservoirs, 29–30
anticlinal tar sand traps, 104
antiknock quality of gasoline, 95–96
antioxidants in shale oil, 189
arsenic in shale oil, 192–194
artichokes, Jerusalem, 269
ash
   in coal, 373
   content of biomass, 235–236, 394
   as waste, 324
   from waste incineration, 331
as-mined coal, 141
asphalt, 7
associated gas, 31–32
Athabasca Oil Sands Project, Shell Canada Ltd., 123–124
Athabasca tar sand deposit, Canada, 105, 111, 115–116, 369
atmosphere, effects of carbon dioxide on, 164–165
atmospheric distillation, 68–69
auger mining of coal, 139

Australia
  oil shale, 174–176
  synthetic crude oil, 383
auto-ignition temperatures, 379
automotive gasoline, 92–95
autothermal reforming, 198, 206
aviation fuels, 95, 97

bacteria
  anaerobic digestion, 275–276
  in biophotolysis, 239
  degradation of organic matter by, 34–35
  fermentation of biomass, 232
  formation of landfill gas, 353–354
bagasse briquettes, 342
banded coals, 136
bark, 303
barometric pressure in landfill
    gas migration, 361
Bemolanga tar sand deposit, 107
beneficiation
  biomass feedstock, 227
  coal, 141
benzine. *See* gasoline
Bergius process, 9, 157
binders, briquette, 320, 341
bioalcohol, 18–19
biochemical conversion, 231
bioconversion
  biofuels from wood, 306
  biorefineries, 252–254
  future, 259–260
biodiesel. *See also* transesterification, biodiesel
  benefits of, 2
  as direct biofuel, 17
  feedstocks for, 283–284
  petroplants, 290–291
  production of, 247–248
  properties, 228, 288–289
  specifications, 389
  technical standards, 289
  thermal conversion, 256
  use of, 250–251, 289–290
  from waste, 336–338
biodiversity, 241, 327
bioenergy crops, 226
bioethanol, 19. *See also* ethanol
biofuels. *See also* solid fuels
  defined, 17, 221
  feedstock
    properties, 227–228
    types, 224–227

biofuels (*Cont.*):
  future, 259
  gaseous, 242–244
  liquid, 244–248
  overview, 223–224, 241–242
  solid, 248–249
  from synthesis gas, 249–250
  from wood
    gaseous, 307–310
    liquid, 310–313
    overview, 306–307
biogas
  anaerobic digestion, 231, 276
  from landfills, 351–352
  overview, 242–243
  versus syngas, 197
biogenic coalbed methane, 33
biomass, 221–264. *See also* biorefineries;
      energy crops; wood
  chemistry of, 228–231
  fuels
    feedstock properties, 227–228
    feedstock types, 224–227
    gaseous, 242–244
    liquid, 244–248
    overview, 223–224, 241–242
    solid, 248–249
    from synthesis gas, 249–250
  future, 259–260
  gasification, 23–24, 213
  overview, 16–18, 221–223
  processes
    biophotolysis, 239–240
    combustion, 236
    environmental issues, 240–241
    fermentation, 231–232
    gasification, 232–236
    overview, 231
    pyrolysis, 236–239
  production of synthetic fuels, 20
  properties, 394
  uses, 250–251
biomass ethanol, 19
biomass-to-liquids (BTL) production, 249–250
bio-oil, 248
biophotolysis, 239–240
bioprocesses, 25
bioreactors, 362–363
biorefineries
  bioconversion, 252–254
  biofuels from wood, 306
  ethanol production, 245

INDEX 405

biorefineries (*Cont.*):
  greenhouse gas production, 257
  other aspects, 257–259
  overview, 251–252
  thermal conversion, 254–257
biosyngas, 249–250
bitumen, 6–8, 108, 369. *See also* tar sand bitumen
bituminous coal, 8, 136, 373
bituminous sand. *See* tar sand bitumen
black liquor, 304–305
boilers, biomass, 271
boiling fractions, petroleum, 63, 65
boiling points, 40–41, 110–111
bone black, 320
bottom simulating reflector (BSR), 35
Brazilian oil shale, 176
Brent crude oil, 126
briquettes, 318–320, 341–342
British coal use, 141–142
BSR (bottom simulating reflector), 35
BTL (biomass-to-liquids) production, 249–250
bulk density
  biomass, 235
  coal, 373
  tar sand, 109–110
Bunker C oil, 99
burning, 325–326. *See also* combustion
burning springs, 27
butanol, 17, 19, 251, 282

caked coal, 152
Californian tar sand deposits, 106–107
calorific values
  biomass, 236
  gaseous fuels, 379
Canada
  oil shale occurrence and development, 176
  synthetic crude oil, 383
  tar sand bitumen
    future of use, 126–130
    mining, 115–116
    occurrence and reserves, 105
    primary conversion, 122–124
    properties, 111
    synthetic crude oil, 125–126
carbohydrates, 224–226, 229
carbon
  in coal, 373
  produced by steam reforming, 205
  rejection, 88, 120
  steam-methane reforming, 211–212

carbon dioxide. *See also* processing
  from biomass, 240
  effects of in atmosphere, 164–165
  from energy crops, 265–266
  in gasification, 199, 364
  in landfill gas, 346
  sequestration, 10
carbon monoxide, 149–150, 346
carbonate-washing process, 52–55
carbonization, 22, 163–164, 307, 319–320
carbon-neutral effects, 17, 23–24
casing head gas, 36
catalysts
  alcohols from waste, 339–340
  alkylation, 84
  biodiesel transesterification, 288
  coke, 100
  Fischer-Tropsch process, 215–217
  isomerization, 83
  methanation, 153
  steam-methane reforming, 209–212
catalytic conversion of syngas, 256
catalytic cracking processes, 74–77, 190
catalytic hydrocracking processes, 89–90
catalytic hydrotreating, 124
catalytic liquefaction of coal, 160
catalytic partial oxidation, 198, 206
catalytic polymerization, 84–85
catalytic reactors, 161
catalytic reforming, 80–82
catalytic transesterification, 284–286
CBM (coalbed methane), 33–34
CCTs (clean coal technologies), 165
cellulose, 229, 279–281, 301–302, 375
Central Pacific Minerals (CPM), 175
cereal straw, 303
cetane numbers, diesel fuel, 99
char, 152–153, 159, 237–238
charcoal, 164, 304–305, 314, 317–320
chemical ethanol, 19
chemicals
  biochemicals, 252–254
  oil recovery methods, 67
  treatment of shale, 193–194
  from wood, 300
chemistry
  biomass, 228–231, 394
  coal, 144, 150–153, 158–159
  gaseous fuels from waste, 333
  gasification to produce syngas, 199–200, 204–206
  landfill waste, 354–357
  tar sand bitumen, 110–115

China
  coal use in, 141
  oil shale, 174, 177
chips, wood, 303, 318
chitin, 230
clarain coal, 137
classifications
  coal, 136, 371–372
  landfill, 346–347
clathrates, 35
Claus sulfur recovery process, 55–57
clays, in catalytic cracking, 77
clean coal technologies (CCTs), 165
cleaning
  coal, 141
  natural gas. *See* processing
  syngas, 213
cleats, coal, 33
coal, 131–168
  versus biomass, 303
  classifications, 371–372
  clean coal technologies (CCTs), 165
  environmental aspects of utilization, 164–165
  Fischer-Tropsch processing, 25
  formation, 134–135
  gaseous fuels
    future, 157
    gaseous products, 148–150
    gasification processes, 153–155
    gasifiers, 146–148
    overview, 27–28, 145–146
    physicochemical aspects, 150–153
  gasification, 20, 23, 198–203
  liquid fuels
    future, 162–163
    liquefaction processes, 159–161
    overview, 157–158
    physicochemical aspects, 158–159
    products, 161–162
  mining, 137–141
  occurrence and reserves, 131–134
  overview, 8–10, 131
  preparation, 141
  properties, 373–374
  proximate analyses, 392
  relevant to petroleum, 7
  solid fuels, 163–164
  types, 135–137
  uses, 141–145
coalbed methane (CBM), 33–34
cobalt catalysts, Fischer-Tropsch process, 215–217

cocurrent fixed bed gasifiers, 147
co-firing biomass, 271–272
cogeneration, 271
coke, 100, 163–164
coking, 22, 71–74, 122–126
combined cycle
  gasifiers, 202–203
  IGCC plants, 145–146
  power generation from petroleum fractions, 206–207
combustion
  biomass, 225–226, 236
  direct, of energy crops, 270–272
  gas analysis by, 91
  in gasifiers, 200
  natural gas, 40, 43–44
  toe-to-heel air injection process, 119
  waste, 328–332
  wood, 304, 307
complete mix digesters, 277
composition
  biogas, 351–352
  gaseous fuels, 378–380
  natural gas, 35–36
  petroleum-based fuels, 4
  of raw materials, and thermal decomposition, 22
  shale oil, 385
  tar sand bitumen, 112–114
  of waste, and landfill gas, 348
  wood, 301–304
compound types in shale oil, 187
concentrated acid hydrolysis, 279
concentrations of landfill gas, uniform, 360
conditioning, hot water process, 116
construction wastes, 324
contaminants, natural gas, 45
continuous coal mining, 139
continuous-flow, stirred-tank reactors, 161
contour coal mining, 139
conventional coal mining, 139
conventional fuels, 4–5, 388
conventional gas, 31–33
conversion. *See also* bioconversion; thermal conversion; waste
  biochemical, 231
  tar sand bitumen
    in situ, 119
    overview, 122
    primary, 122–124
  yield, biodiesel transesterification, 287–288
cooking oil, waste, 336–338

coppicing, wood, 266–267
cordgrass, 268
corn, 225–226
countercurrent fixed bed gasifiers, 147
countercurrent hydrolysis, 279
covered lagoon digesters, 277
covers, landfill, 361
CPM (Central Pacific Minerals), 175
cracking
  catalytic, 74–77
  defined, 22
  gasoline, 93
  hydrocracking, 78–80
  relation to octane number, 96
  thermal, 71–74
  vacuum distillation to avoid, 69–70
creosote oil, 237
crop residues, 273–274
crops. *See* energy crops
crude oil, synthetic, 125–126. *See also* petroleum
cuts, petroleum. *See* fractions, petroleum
cyclic coal mining, 139
cycling process, natural gas, 30

damps in coal mines, 140
DAO (deasphalted oil), 85
DEA (diethanolamine), 46
dead crude oil, 31–32
deasphalted oil (DAO), 85
deasphalting, 85–86, 121, 125
debris, organic, 28
decomposition, 353–355. *See also* thermal decomposition
deep coal mining, 139
degradation. *See also* thermal decomposition
  aerobic, 355
  bacterial, of organic matter, 34–35
delayed coking, 72–73, 89, 122–123
demolition wastes, 324
dense medium separation of coal, 141
densification
  biomass feedstock, 235
  wood, 307
density
  natural gas, 37–40
  tar sand, 109–110
depletion, natural gas, 31
depolymerization, 336
desalting, 67–68
destructive distillation, oil shale, 11–12
destructive hydrogenation, 77

desulphurization, phosphate, 52
devolatilization. *See* pyrolysis
Devonian-Mississippian black shale, 11
dewatering, 67–68
dewaxing, 86–87
diesel fuel. *See also* biodiesel
  overview, 98–99
  produced by Fischer-Tropsch process, 217
  from shale oil, 189
  specifications, 389
  from wood, 312
Diesel, Rudolf, 284
diethanolamine (DEA), 46
diffusion, landfill gas, 360
digestates, 275–276
digester gas, 276–277. *See also* biogas
digesters, 277–278
digestion, anaerobic. *See* anaerobic digestion
diluents, 36
dilute acid hydrolysis, 279
dimethyl ethers of polyethylene glycol, 48
direct biofuels, 17, 250–251
direct coal liquefaction, 159–160, 163
direct combustion of energy crops, 270–272
direct hydrogenation of coal, 9
direct liquefaction
  of coal, 9
  hydrothermal, 275
direction of wind in landfill gas migration, 361
dissolved natural gas, 31–32
distillate fuels, 98, 190, 383
distillation
  destructive, of oil shale, 11–12
  gas analysis by, 91
  petroleum, 68–70, 111
  tar sand bitumen, 111, 115, 121–122
domestic fuel oil, 99
domestic waste, 323, 325–326. *See also* waste
downdraft gasifiers, 147, 272
dredged soil, 324
drift mines, coal, 138
dry ash gasifiers, 154–155
dry box process, 49–51
dry gas, 32, 36
drying
  biomass feedstock, 235
  firewood, 315
dump sites, 347
durain coal, 137
dusts in coal mines, 140

E85 fuel blend, 19
earth kilns, 317
easter oil shale, United States, 180–181
ebullating bed processes, 89–90, 161
economics
  oil shale retorting, 184
  tar sand bitumen, 127–128
electricity generation. *See* power generation
electronic waste, 323–324, 326
electrostatic desalting units, 68
elemental composition
  biomass, 235
  tar sand bitumen, 112–113
Elk Hill field project, 13–14
emissions. *See also* greenhouse gases
  fossil fuel, 43–44
  landfill gas power generation, 364
  waste incineration, 330–332
energy
  natural gas, 40
  oil recovery, 66
  from wood, 304–306
energy crops, 265–295
  biodiesel
    catalytic transesterification, 284–286
    effect of reaction parameters on conversion yield, 287–288
    feedstocks for, 283–284
    noncatalytic supercritical methanol transesterification, 286–287
    overview, 282–283
    properties, 288–289
    technical standards, 289
    transesterification, 284
    uses, 289–290
  creating biogas from, 243
  ethanol, 278–281
  hydrocarbons, 290–291
  other alcohols, 281–282
  versus other biomass sources, 303
  overview, 265–266
  processes
    anaerobic digestion, 275–278
    direct combustion, 270–272
    gasification, 272–274
    pyrolysis, 274–275
  types of, 225–226, 266–270
enhanced oil recovery (EOR) methods, 3, 66–67
entrained bed gasification of coal, 153, 155
entrained flow gasifiers, 148
entrained-flow reactors, 159, 161
entrained-solids process, 201

entrapment, tar sand, 104
environment
  biomass processes, 236, 240–241
  coal
    liquefaction, 163
    mining, 139–141
    utilization of, 164–165
  effect of oil shale exploitation, 13–14
  lignocellulosic materials, 257
  natural gas, 43–44, 59
enzymatic hydrolysis, 252–254, 280
enzyme-catalyzed transesterification, 285–286
EOR (enhanced oil recovery) methods. *See* enhanced oil recovery methods
Estonian oil shale, 173–174, 177–178
ethanol
  biochemical conversion, 231
  creating from biomass, 244–246
  from crops, 278–281
  overview, 17–19
  properties, 228
  use of, 251
  from waste, 338–339
  from wood, 311–313
ethers, 96, 312
ethyl alcohol. *See* ethanol
Euphorbiaceae plants, 291
Eureka process, 125
Europe
  coal reserves in, 133
  tar sand deposits, 107–108
ex situ retorting processes, oil shale, 12
explosive properties of natural gas, 40–43
extractable constituents of wood, 302
extraction processes, tar sand bitumen, 120
extruded charcoal, 320

facilities, landfill gas, 350–351
fast pyrolysis, 274–275
fats, 224–225, 283
fatty acids, 287, 338, 375
FCC (fluid catalytic cracking), 76, 89, 121
feedstocks
  for biodiesel, 283–284
  biofuel
    properties, 227–228
    types, 224–227
  biomass, 222–223, 234–235, 394
  catalytic cracking, 75
  for ethanol production, 278, 280–281
  gasification to produce syngas, 212–213, 249–250

INDEX

feedstocks (*Cont.*):
  hydrocracking processes, 78
  properties from different sources, 367–375
fermentation
  alcoholic, 231
  biomass, 231–232
  formation of landfill gas, 355
  for propanol and butanol, 282
ferric oxide, 49–50
Ferrox process, 50
fibers, plant, 224
field separation, petroleum, 67–68
fine charcoal, 320
fire points of natural gas, 42
fireplaces, 315–316
firewood, 302, 304, 315–316
Fischer assay, 12–13
Fischer-Tropsch process
  alcohols from waste, 339
  liquefaction of coal, 9–10, 158, 160–161
  liquid fuels from waste, 335
  production of synthetic fuels, 24–25
  thermal conversion, 256
  use of synthesis gas in, 213–218
fixed bed gasifiers
  for biomass, 232–233
  cocurrent, 147
  countercurrent, 147, 272
  for biomass, 232–233
fixed bed processes
  catalytic cracking, 76
  coal gasification, 153–155
fixed bed reactors, Fischer-Tropsch, 215–216
fixed carbon content of coal, 137
fixed incinerators, 329–330
fixed kilns, 317
fixed-bed catalytic reactors, 161
flammability
  gaseous fuels, 92
  natural gas, 40–43
  oil shale, 13
flash points, 41–42, 379
flexicoking, 74, 89, 125
fluid catalytic cracking (FCC), 76, 89, 121
fluid coking, 72–74, 89, 123, 126
fluidized bed technologies
  combustors, 271
  Fischer-Tropsch reactors, 215–216
  gasification, 147, 153, 201, 233–234
  hydroretort process, 191–192
  incinerators, 330

fluidized bed technologies (*Cont.*):
  liquefaction reactors, 161
  for oil shale retorting, 184
fluidized reactors, 159–160
food wastes, 323–324
forest residues, 280, 303–304. *See also* wood
fossil fuels, 7, 43–44, 393. *See also* coal; natural gas; petroleum
fractionation, 5, 7
fractions. *See also* gasification
  from biomass pyrolysis, 237
  coal, 141
  petroleum, 63–65, 68–70, 99
  shale oil, 189–191
  tar sand bitumen, 104, 113–114
Franklin stoves, 314
free fatty acids, 287, 338
free radicals in shale oil, 188–189
freeze walls, Shell in situ process, 186
froth flotation of coal, 141
fuel oils, 97–100, 385–386
fuel sources, 1–26
  conventional, 4–5
  gaseous fuels from different, 377–380
  liquid fuels from different, 381–389
  nonconventional
    bioalcohol, 18–19
    biomass, 16–18
    coal, 8–10
    gas hydrates, 16
    natural gas, 14–15
    oil shale, 10–14
    overview, 5–6
    tar sand bitumen, 6–8
  overview, 1–4
  properties of fuels and feedstocks from, 367–375
  solid fuels from different, 391–394
  synthetic fuels
    overview, 19–21
    production of, 22–25
fuel wood, 314–318
fugitive emissions, 44
furnaces
  direct combustion, 270
  pipe still, 68–69
Fushun, China oil shale, 177
fusion coal, 137

gas. *See also* biogas; gaseous fuels; gasoline; greenhouse gases; landfill gas; natural gas; synthesis gas; water gas

gas. (*Cont.*):
  digester, 276–277
  from coal
    high heat-content, 150
    low heat-content, 149
    medium heat-content, 149–150
    primary gasification, 152
    reactions, 151
  in coal mines, 140
  digester, 276–277
  gas well, 32
  injection for oil recovery, 66
  liquefied petroleum, 91–92
  mixed, 91
  producer, 100, 200, 308–310
  reversion, 80
  stranded, 217
Gas Combustion retort process, 12
gas hydrates, 16, 34–35
gas oils, cracking, 71
gaseous fuels
  biofuels, 242–244
  coal
    future, 157
    gasification processes, 153–155
    gasifiers, 146–148
    physicochemical aspects, 150–153
    products, 148–150
    underground gasification, 156
  petroleum, 90–92
  properties of different sources, 377–380
  from waste, 332–334
  from wood, 307–310
gasification
  biomass, 232–236
  coal
    gaseous products, 148–150
    gasifiers, 146–148
    overview, 142–143, 145–146
    primary, 152
    processes, 153–155
    to produce syngas, 198–203
    secondary, 152
    underground, 156
  energy crop, 272–274
  of feedstocks to produce syngas, 212–213, 249–250
  landfill waste, 363–364
  for methanol production, 281–282
  overview, 23–24

gasification (*Cont.*):
  of petroleum fractions to produce syngas
    chemistry, 204–206
    overview, 203–204
    processes, 206–209
  thermal conversion, 256–257
  waste, 332–334, 338–339
  wood, 307–310
gasifiers
  biomass, 232–234
  coal
    agitated, 155
    dry ash, 154–155
    overview, 146–148
    to produce syngas, 202–203
  energy crop, 272–273
  processes within, 199–200
  wood, 309–310
gasoline
  isomerization, 82–83
  mixed with ethanol, 19
  natural, 31
  overview, 92–96
  reforming, 80–82
  from shale oil, 188
gasoline-type jet fuel, 97
gas-to-liquid (GTL) processes, 14–15, 21, 384
generation of power. *See* power generation
German oil shale, 178
Giammarco-Vetrocoke process, 53
glycerol, 290
glycol, 32
gober gas, 352–353
grain alcohol. *See* ethanol
grain crops, 265
grasses, 267–270
grate combustors, 271
gravity
  specific, 37
  steam-assisted drainage, 117–119
  tar sand bitumen, 114
greases, 224–225
Green River oil shale, 11, 179–181
greenhouse gases
  biofuels, 260
  biorefineries, 257
  effects in atmosphere, 165
  energy crop gasification, 273
  methane, 16, 18
  natural gas emissions, 43–44
groundwater levels and landfill gas migration, 361

GTL (gas-to-liquid) processes, 14–15, 21, 384
guard beds, 54
gum, 211

harvesting
  hemp, 270
  short rotation coppice, 267
  timber, 315
hazardous waste landfill sites, 346, 359
hazards
  landfill gas, 349–350
  from wastes, 327
HDN (hydrodenitrogenation), 194
heat energy in oil shale processes, 192–193
heat, latent, 41
heat values
  biomass, 394
  biomass feedstock, 228
  fuel oils, 385
  gases, 92
  liquid fuels, 382
  natural gas, 40
heating, 313–314, 318–319. *See also* retorting
heavy fuel oil, 98
heavy metals, 356
heavy oil, 4, 6, 88–90
heavy residue gasification, 206–207
hemicelluloses, 229, 238, 301
hemp, 270
herbaceous biomass, 303, 374
heteroatomic species, 104
high heat-content (high-Btu) gas, 150
high-temperature Fischer-Tropsch (HTFT) process, 215–216
H-Oil process, 89–90
Hot Potassium Carbonate process, 52–53
hot water process, 116–117
HTFT (high-temperature Fischer-Tropsch) process, 215–216
Hubbert peak oil theory, 1
hybrid gasification process, 207
hydrates, gas, 16
hydrocarbonization, coal pyrolysis, 159
hydrocarbons
  compound types in shale oil, 187
  fuels from
    liquefied petroleum gas, 91–92
    overview, 290–291
    from waste, 335–336
  gasification of, 207
  gasoline, 92–93, 95–96

hydrocarbons (*Cont.*):
  kerosene, 97
  resources containing, 103–104
  resources producing, 7
hydroconversion processes, residuum, 121
hydrocracking, 78–80, 190–192
hydrodenitrogenation (HDN), 194
hydrogasification of coal, 153
hydrogen
  addition processes, 88–89, 120
  atmosphere, thermal decomposition in, 23
  biophotolysis, 239
  coal liquefaction, 159–160
  in gaseous products from coal, 149–150
  gasification of petroleum fractions to produce
    chemistry, 204–206
    overview, 203–204
    processes, 206–209
    steam reforming processes, 209–212
  in landfill gas, 346
  liquid fuels from waste, 336
  produced by catalytic reforming, 81
  in shale oil, 194
  from tar sand bitumen, 125
hydrogen fluoride, 84
hydrogen sulfide, 49–51. *See also* processing
hydrogenation, 9, 19, 57, 77–80
hydrolysis, 57, 252, 279–280, 355
hydroprocesses
  petroleum, 77–80
  shale oil, 188, 190–191
hydroretorting shale oil, 191–192
hydroskimming refineries, 65
hydrothermal liquefaction, direct, 275
hydrotransport of oil sands ore, 116
hydrotreating, 78–79, 121, 124, 186–187
hypro process, 207

ICP process, Shell, 193
IFPEXOL-2 process, 52
IFPEXOL-1 process, 51–52
IGCC (Integrated Gasification Combined Cycle), 145–146, 202
immobility of tar sand bitumen, 114
impurities in natural gas, 45
in situ processes
  combustion in oil recovery, 67
  gasification of coal, 156
  oil shale retorting, 12–13, 184–186, 193
  tar sand bitumen recovery, 117–120
incineration of waste, fuels from, 328–332. *See also* combustion

indirect gasifiers, 234
indirect liquefaction of coal, 9–10, 160–162
industrial fuel oil, 99
industrial use of fuel wood, 317
industrial waste, 323, 325–326. *See also* waste
inert waste landfill sites, 347
injection of gas for oil recovery, 66
inorganic reactions in landfill gas formation, 356
Integrated Gasification Combined Cycle (IGCC), 145–146, 202
intermolecular hydrogenation, 19
intramolecular hydrogenation, 19
iron catalysts, Fischer-Tropsch process, 215–217
iron sponge process, 49–51
isobutane, 83–84
isobutyl alcohol, 96
isomerization processes, 82–83, 93
isoparaffins, 83–84
Israeli oil shale, 178

Jerusalem artichoke, 269
jet fuel, 97, 189
Jijuntun Formation of oil shale, China, 177
Jordanian oil shale, 178–179

Karrick process, 9, 25
Kentucky tar sand deposits, United States, 107
kerogen
  defined, 169
  in oil shale, 10–11, 171–172
  relevant to petroleum, 7
  reserves, 173
  thermal decomposition, 181–182
kerosene, 97–98
kerosene-type jet fuel, 97
ketone, 86–87
kilns, charcoal, 317–318
knocking, gasoline, 95–96
kohlentype, 137
Koppers-Totzek Process, 155, 201

landfill gas
  basic information regarding, 347–351
  biogas, 243, 351–352
  bioreactors, 362–363
  formation of
    inorganic reactions, 356
    landfill leachate, 356–357
    landfill mining, 358–359
    monitoring, 357–358
    organic reactions, 355–356
    overview, 353–355

landfill gas (*Cont.*):
  gasification of landfill waste, 363–364
  gober gas, 352–353
  landfill classification, 346–347
  migration, 359–361
  overview, 345–346
  power generation, 364–365
  renewable natural gas, 352
landfill gas (LFG), 18. *See also* natural gas
landfill leachate, 356–357
landfill mining and reclamation (LFMR), 358–359
landfills, 327–328, 345–347
landslides and gas hydrates, 16
latent heat, 41
LC-Fining process, 89–90
leachate, landfill, 356–357
leaks, natural gas, 44
lean amine solution, 47
lean gas, 32, 36–37
LEL (lower explosive limit) of natural gas, 42
LFG (landfill gas), 18. *See also* natural gas
LFMR (landfill mining and reclamation), 358–359
light-range fuels, 189–191
lignin, 229–231, 238, 301–302, 375
lignite, 8, 135–136, 373
lignocellulose
  bioconversion, 252–254, 260
  defined, 229
  environment, 257
  properties, 301–302, 375
  thermoconversion, 254–257
liners, landfill, 347
lipids, 283
liquefaction
  coal, 8–10, 144, 159–161
  direct hydrothermal, 275
liquefied natural gas, 32–33
liquefied petroleum gas (LPG), 91–92
liquid fraction of biomass pyrolysis, 237
liquid fuels
  alternate, 387
  biofuels, 244–248
  coal
    future, 162–163
    liquefaction processes, 159–161
    physicochemical aspects, 158–159
    products, 161–162
  Fischer-Tropsch processing, 24–25
  oil shale
    in situ technologies, 184–186
    mining, 182–184
    overview, 181

INDEX 413

liquid fuels, oil shale (*Cont.*):
   retorting, 182–184
   thermal decomposition, 181–182
   overview, 1–2
   petroleum
      fuel oil, 98–100
      gasoline, 92–96
      kerosene and related fuels, 97–98
      properties, 382
      produced from natural gas, 14–15
   properties of different sources, 381–389
   from waste
      biodiesel, 336–338
      ethanol, 338–339
      hydrocarbon fuels, 335–336
      other alcohols, 339–340
      overview, 334
   from wood, 307, 310–313
liquid redox sulfur recovery processes, 56
liquids, organic, 370
lithotype, coal, 137
logs, 318
longevity, coal, 134
longwall coal mining, 139
low heat-content (low-Btu) gas, 149
low temperature carbonization (LTC), 9
lower explosive limit (LEL) of natural gas, 42
low-temperature Fischer-Tropsch (LTFT) process, 215–216
LPG (liquefied petroleum gas), 91–92
LR (Lurgi-Ruhrgas) process, 12
lumber. *See* wood
lump charcoal, 320
Lurgi process, 154–155
Lurgi-Ruhrgas (LR) process, 12

maceral coal groups, 136
Madagascar tar sand deposits, 107
man-made pathways in landfill gas migration, 361
manure, 277–278, 352–353
marine sediments, 16
marsh gas. *See* natural gas
Maya Crude Oil, 369
McMurray-Wabasca tar sand deposits, Canada, 105
MDEA (methyldiethanolamine), 46
medium heat-content (medium-Btu) gas, 149–150
membrane processing systems, 54–55
mercaptans, 36

mercury, 36
metal oxide processes, 49–51
metals, heavy, 356
methanation, coal, 153
methane. *See also* natural gas
   anaerobic digesters to trap, 277–278
   in biogas, 351–352
   boiling point and density, 40
   in coal mines, 140
   coalbed, 33–34
   gas hydrates, 16
   in gaseous products from coal, 149–150
   as greenhouse gas, 18
   in landfill gas, 345–346, 348–349
   landfill gas power generation, 364
methane hydrate, 34–35
methanogenesis, 355–356
methanogenic digestate, 276
methanol
   defined, 19
   indirect synthesis, 9–10
   processes based on, 51–52
   producing from crops, 281–282
   Rectisol process, 48
   supercritical, 286–287
   thermal conversion, 256
   from wood, 311–313
methyl esters. *See* biodiesel
methyl tertiary-butyl ether (MTBE), 96, 311–312
methyldiethanolamine (MDEA), 46
middle-distillate range fuels, 189–192
migration, landfill gas, 348, 359–361
mineralogy, tar sand bitumen, 109
mining
   coal, 137–141
   landfill, 358–359
   oil shale, 182–184
   tar sand bitumen, 115–116
miscanthus, 267–268
miscible oil recovery methods, 67
Missouri tar sand deposits, United States, 107
mixed gas, 91
modified in situ processes
   oil shale retorting, 186
   tar sand bitumen extraction, 120
moisture
   in biomass feedstock, 227–228, 234, 394
   in coal, 373
   effect on biodiesel transesterification, 287
   landfill gas migration, 361
molar ratio in biodiesel transesterification, 288
molecular sieve process, 54

molten salt processes, 154–155
monitoring landfill gas, 357–358
Moroccan oil shale, 179
motor fuels, 241
motor octane numbers, 95
moving bed catalytic cracking process, 76
moving bed hydroprocessing reactors, 191
moving grate incinerators, 329–330
MSW (municipal solid waste), 243, 323. *See also* landfill gas; waste
MTBE (methyl tertiary-butyl ether), 96, 311–312
multi-tubular fixed bed reactors, 215–216
municipal solid waste (MSW), 243, 323. *See also* landfill gas; waste

natural gas
  versus biomass, 303
  composition, 35–36
  conventional gas, 31–33
  and environment, 59
  Fischer-Tropsch processing, 10
  formation and occurrence, 28–31
  gaseous fuels, 90–91
  history of, 27–28
  for hydrogen production, 203
  overview, 14–15, 27
  processing
    carbonate-washing and water-washing, 52–55
    hydrogenation and hydrolysis, 57
    metal oxide, 49–51
    methanol based, 51–52
    olamine, 45–47
    overview, 44–45
    physical solvent, 47–49
    sulfur recovery, 55–56
  producing methanol from, 281
  properties
    density, 37–40
    environmental, 43–44
    general, 370
    heat of combustion, 40
    versus other fuels, 374
    overview, 37
    volatility, flammability, and explosive, 40–43
  pyrolysis, 207–208
  unconventional gas, 33–35
  uses, 57–59
natural gas liquids (NGLs), 30–31
natural gasoline, 31
natural pathways in landfill gas migration, 361

Naval Petroleum and Oil Shale Reserves, 13–14
needle coke, 100
neutralization of free fatty acids, 338
New Albany shale, United States, 181
New Brunswick oil shale, Canada, 176
New Mexico tar sand deposits, United States, 107
NGLs (natural gas liquids), 30–31
nitrogen, 148, 164–165, 346
nitrogen oxide, 207
NMOCs (non-methane organic compounds), 346
No. 1 fuel oil, 98–99
No. 2 fuel oil, 99
No. 6 fuel oil, 99
nonassociated gas, 32
nonbanded coals, 136
noncatalytic reactors, 161
noncatalytic supercritical methanol transesterification, 286–287
nonconventional fuel sources
  bioalcohol, 18–19
  biomass, 16–18
  coal, 8–10
  gas hydrates, 16
  natural gas, 14–15
  need for, 1–4
  oil shale, 10–14
  overview, 5–6
  tar sand bitumen, 6–8
nondestructive hydrogenation, 77
non-methane organic compounds (NMOCs), 346
nonpoint sources of emissions, 44
Nova Scotia oil shale, Canada, 176
Nuclear Utility Services (NUS) Corporation, 9

Occidental vertical modified in situ process, 186
ocean biophotolysis, 239
ocean floor sediments, 16
oceanic methane hydrate reservoir, 35
octane ratings
  gasoline, 95–96
  liquid fuels from wood, 311–312
  reforming, 80–82
odorants, natural gas, 36, 43, 91
Officina tar sand deposit, Venezuela, 107
oil. *See also* biodiesel; heavy oil; petroleum; vegetable oils
  cracking gas, 71
  creosote, 237
  fuel, 97–100
oil sand. *See* tar sand bitumen

oil shale
  future, 192–195
  history, 173–174
  liquid fuels
    in situ technologies, 184–186
    mining, 182–184
    overview, 181
    retorting, 182–184
    thermal decomposition, 181–182
  occurrence and development
    Australia, 175–176
    Brazil, 176
    Canada, 176
    China, 177
    Estonia, 177–178
    Germany, 178
    Israel, 178
    Jordan, 178–179
    Morocco, 179
    overview, 174–175
    Russian Federation, 179
    Thailand, 179
    United States, 179–181
  origin, 172–173
  overview, 10–14, 169–172
  refining, 186–192
  relevant to petroleum, 7
olamine processes, 45–47
olefins, 83–85
once-through Rectisol process, 51
open core gasifiers, 148, 272–273
open-pit mining
  coal, 138
  oil shale, 182
ore transportation, oil sands, 116
organic debris, 28
organic liquids, 370
organic matter, bacterial degradation of, 34–35
organic reactions in formation of landfill gas, 355–356
organic sediments, 7, 104
organic waste. *See* landfill gas
oxidation processes. *See also* partial oxidation processes
  Claus sulfur recovery, 55–56
  formation of landfill gas, 356
  gasification of petroleum fractions, 208–209
  tar sand bitumen, 110, 113
oxides, 164–165
oxygen
  in landfill gas, 346, 348–349
  in shale oil products, 188–189
  in tar sand bitumen, 113

oxygenated fuels, 96, 241
oxygen-blown Claus process, 56

paraffins, 82–84. *See also* kerosene
partial coking, 125
partial oxidation processes
  gasification of petroleum fractions, 208–209
  hydrogen production with, 205–206
  syngas, 198
particle size and thermal decomposition, 23
particulates
  removal systems, syngas, 213
  from wood fires, 316
Patos oil field, Albania, 107
PDA (propane deasphalter asphalt), 85
peat, 135
pellets, 303, 318–319
perennial energy crops, 267–270
permeability
  landfill gas, 360–361
  tar sand, 110
Petrobrás (Petróleo Brasileiro), 12, 176
petrochemicals, 90
petrol. *See* gasoline
petroleum, 63–100. *See also* refining
  Brent crude oil versus syncrude, 126
  conventional fuel, 4–5
  gaseous fuels, 90–92
  gasification of fractions to produce syngas
    chemistry, 204–206
    overview, 203–204
    processes, 206–209
  history, 63–65
  liquid fuels
    fuel oil, 98–100
    gasoline, 92–96
    kerosene and related fuels, 97–98
  properties, 382
  and natural gas formation, 28–29, 31–32
  overview, 1–4
  products, 90–100
  properties, 368–369
  recovery, 66–67
  shale oil compared to, 188
  solid fuels, 100
  synthetic, 8
  tar sand bitumen relevant to, 7
petroplants, 290–291
Petrosix process, 176
phosphates, 52, 85
photosynthesis, 229, 239

physical solvent processes, 47–49
physicochemical aspects
  gaseous fuels from coal, 150–153
  liquid fuels from coal, 158–159
  tar sand bitumen, 110–115
pile burners, 271
pipe still furnaces, 68–69
pipelines, natural gas, 58–59
pitch, 7, 104
plant materials, 134–135. *See also* biomass; energy crops
platinum catalysts, catalytic reforming, 82
plug-flow digesters, 277–278
point sources of emissions, 44
pollution from wastes, 326
polyethylene glycol, dimethyl ethers of, 48
polyforming, 80
polymer flooding, 67
polymeric sugars, 229
polymerization processes, 84–85, 93
polysaccharides, 229
porosity of tar sand, 109–110
portable steel kilns, 318
potassium phosphate, 52
pour points of tar sand bitumen, 114
power generation
  from coal, 142–144
  combined cycle, 206–207
  from energy crops, 271, 273
  landfill gas, 364–365
  from waste, 325–326, 329–331
precautions, firewood, 315–316
pressure
  barometric and soil gas, 361
  landfill gas, 349, 360
  natural gas reservoirs, 30
  in steam-assisted gravity drainage, 117–118
  thermal decomposition, 22
  vapor, 41
pretreatment
  bioconversion, 252
  coal, 152
  thermal conversion, 254
  waste, 328
primary biomass feedstocks, 222
primary conversion of tar sand bitumen, 122–124
primary gasification of coal, 152
primary oil recovery, 66
process selectivity, 45
processes. *See individual processes by name*
processing
  in biorefineries, 258–259
  coal, 9–10, 152–153
  Fischer-Tropsch, 24–25
  natural gas
    carbonate-washing and water-washing, 52–55
    hydrogenation and hydrolysis, 57
    metal oxide, 49–51
    methanol based, 51–52
    olamine, 45–47
    overview, 14–15, 44–45
    physical solvent, 47–49
    removing impurities, 30–31
    sulfur recovery, 55–56
  oil shale kerogen, 11–12
  to produce synthetic fuels, 20–21
producer gas, 100, 200, 308–310
production
  associated gas, 31–32
  biodiesel, 247–248
  bitumen from tar sand, 8
  coalbed methane, 33–34
  ethanol, 279–281
  hydrogen from tar sand bitumen, 125
  oil shale kerogen, 10–11
  petroleum-based fuels, 1–2
  synthetic fuels, 19–20
  of synthetic fuels, 22–25
propane, 85–86
propane deasphalter asphalt (PDA), 85
propanol, 282
properties
  alternate liquid fuels, 387
  biodiesel, 288–289, 389
  biogas, 242
  biomass, 223, 234–235
  conventional and alternative fuels, 388
  diesel, 389
  fuels and feedstocks from different sources, 367–375
  gaseous fuels from different sources, 377–380
  kerosene, 98
  liquid fuels from different sources, 381–389
  natural gas
    density, 37–40
    environmental, 43–44
    heat of combustion, 40
    overview, 37
    volatility, flammability, and explosive, 40–43
  shale oil, 385

properties (*Cont.*):
  solid fuels from different sources, 391–394
  tar sand bitumen
    chemical and physical, 110–115
    and structure, 108–110
  wood, 301–304
proximate analyses
  coal, 137, 144
  solid fuel, 392–393
P-series fuels, 246–247
purification of oil for transesterification, 338
pyroligneous acid, 237
pyrolysis
  biomass, 234, 236–239
  coal, 159–160
  defined, 22
  energy crops, 274–275
  gasification to produce syngas, 199, 207–208
  oil shale, 11–12
  thermal conversion, 254–256
  thermal depolymerization, 336
  wood, 316–317
pyrolytic carbon, 211

Queensland oil shale, Australia, 174

rain, acid, 164–165, 241
ranks, coal, 136
reactions
  biodiesel transesterification, 287–288
  coal, 151–152
  Fischer-Tropsch process, 215
  landfill waste, 354–356
  temperature of, and thermal decomposition, 22
reactors
  coal liquefaction, 161
  coal pyrolysis, 159
  Fischer-Tropsch, 215–216
  gaseous fuels from waste, 333–334
  moving bed hydroprocessing, 191
recoverable reserves of coal, 131–133
recovery
  petroleum, 66–67
  tar sand bitumen, 116–120
Rectisol process, 47–48, 51
recycle solvents in coal solvent extraction processes, 160
redox sulfur recovery processes, 56
reed canary grass, 268
reed plants, 269
refineries, 63–65. *See also* biorefineries

refinery gas, 203–204
refining. *See also* processing; upgrading
  biomass
    bioconversion, 252–254
    greenhouse gas production, 257
    other aspects, 257–259
    overview, 251–252
    thermal conversion, 254–257
  demand for hydrogen, 203–206
  oil shale, 186–192
  petroleum
    alkylation processes, 83–84
    catalytic processes, 74–77
    deasphalting, 85–86
    dewatering and desalting, 67–68
    dewaxing, 86–87
    distillation, 68–70
    hydroprocesses, 77–80
    isomerization processes, 82–83
    polymerization processes, 84–85
    reforming processes, 80–82
    versus shale oil, 189–190
    thermal processes, 70–74
  products, 161–162, 384
  underground, 119
reformate, 80
reforming processes, 80–82, 93, 198, 203–206
refuse. *See* landfill gas; waste
regeneration, ferric oxide, 49–50
rejection, carbon, 88–89
renewable energy, 16–17
renewable fuels, 393
renewable materials. *See* biomass
renewable natural gas, 352
research octane numbers, 95
research, wood, 299–300
reserves
  coal, 131–134
  natural gas, 31
  oil shale, 173
  tar sand bitumen, 104–108
reservoirs
  methane hydrate, 35
  natural gas, 29–30
  petroleum, 66
  shale gas, 35
Resid FCC (residuum fluid catalytic cracking), 89, 121
residua
  properties, 110–111, 368–369
  refining, 88–90
  visbreaking, 71

residual fuel oil, 98–99
residual liquid fuels, 383
residual oil, 66–67
residual waste, 323
residue gas, 36
residues
    agricultural, 273–274
    energy crop, 225–226
    for ethanol production, 280
    forest, 303–304
    waste, 324
residuum fluid catalytic cracking (Resid FCC), 89, 121
residuum hydroconversion processes, 121
residuum hydrotreating processes, 121
retort oil. *See* shale oil
retorting
    oil shale
        defined, 12–13
        in situ, 184–186
        mining and, 182–184
        overview, 172
        refining, 190–192
    Petrosix process, 176
    to produce syngas, 200–201
reversion, gas, 80
rich gas, 31–32, 37
road asphalt, 7
room and pillar mining, 139, 183
rotary-kiln incinerators, 330
rubbish, 323–324
Rumanian tar sand deposits, 108
run-of-mine coal, 141
Russian Federation oil shale, 179

Sabatier process, 23
saccharides, 229
safety, coal mine, 139–141
SAGD (steam-assisted gravity drainage), 117–119
sands, tar. *See* tar sand bitumen
sandstone reservoirs, 6
sanitary landfill sites, 346–347
Sasol company, 9, 216–217
sawdust briquettes, 342
SCM (supercritical methanol), 286–287
SCOT (Shell Claus off-gas treating) process, 57
Scottish oil shale, 173
screening criteria for enhanced oil recovery methods, 3
seasoned firewood, 315
secondary biomass feedstocks, 222
secondary gasification of coal, 152

secondary oil recovery, 66
secondary upgrading of tar sand bitumen, 124
sediments
    marine, 16
    organic, 7, 104
selectivity, process, 45
Selenizza tar sand deposit, Albania, 107
Selexol process, 47–48
separation
    coal, 141
    hot water process, 116
    petroleum field, 67–68
    processes, tar sand bitumen, 120–122
sequestration, carbon dioxide, 10
shaft mines, 139
shale gas, 35
shale oil, 169, 172, 385. *See also* oil shale
shale streaks in tar sand deposits, 118
Shell Canada Ltd., 123–124
Shell Claus off-gas treating (SCOT) process, 57
Shell Gasification Process, 208
Shell ICP process, 193
Shell in situ process, 185–186
shift conversion of coal, 153
short rotation coppice (SRC), 267
shortwall coal mining, 139
simple hydrogenation, 77
simple incinerators, 329
simple sugars, 229
sizing of biomass feedstock, 235
slope coal mines, 138
slurry processes, 51
slurry reactors, Fischer-Tropsch, 215–216
smoke, wood, 316–317
SNG (synthetic natural gas), 150
soil, dredged, 324
soil gas pressure in landfill gas migration, 361
solid fuels
    biofuels, 248–249
    coal, 163–164
    petroleum, 100
    properties of from different sources, 391–394
    synthetic, 20
    from waste, 340–342
    from wood
        charcoal, 319–320
        fuel wood, 314–318
        logs and wood chips, 318
        overview, 313–314
        pellets, 318–319
solvent deasphalting, 85–86, 121

solvent dewaxing processes, 86–87
solvent extraction processes, 160
solvents, physical, 47–49
sorghum, 269–270
sorption-enhanced methane-steam reforming, 210–211
sour gas, 36. *See also* processing
sour synthetic crude oils, 126
sources, fuel. *See* fuel sources
South Africa
    Fischer-Tropsch process, 216–217
    indirect liquefaction of coal, 9
Southern Pacific Petroleum (SPP), 175
special wastes, 324
specialty plant products, 224
specific gravity, 37, 114
specifications, diesel and biodiesel, 389
speed of wind in landfill gas migration, 361
spent shale, 13, 183–184, 193
sponges, iron, 49–51
SPP (Southern Pacific Petroleum), 175
springs, burning, 27
SRC (short rotation coppice), 267
standard test methods, natural gas, 38
starch, 229
steam
    direct combustion of energy crops, 271
    in gasification processes, 201–202
    oil recovery methods, 67
    reforming processes, 198, 203–205, 209–212
steam-assisted gravity drainage (SAGD), 117–119
steam-naphtha reforming, 208–209
steel kilns, 318
stirred-tank reactors, 161
storage, firewood, 315
stove oil, 99
stoves, wood, 314–316
stranded gas, 217
stratigraphic tar sand traps, 104
straw, 226, 273–274, 303
Stretford process, 50
structural tar sand traps, 104
Stuart oil shale deposit, Australia, 175–176
sub-bituminous coal, 8, 136
substitute natural gas. *See* synthetic natural gas
subsurface monitoring for landfill gas, 358
subsurface retorting, 12–13
sugar beets, 225–226, 291
sugar crops, 265
sugarcane, 225–226, 291
sugars, 229, 278–280, 306
sulfate, 356

sulfides, 346
Sulfinol process, 49
sulfur
    in coal, 373
    in gasoline from shale oil, 188
    hydrogenation and hydrolysis processes, 57
    metal oxide processes, 49–50
    recovery processes, 55–56
sulfur dioxide emissions, 165
sulfuric acid, 84, 279
Suncor mining and processing plant, 115–116, 122–123, 125–130
supercritical methanol (SCM), 286–287
surface coal mining, 138–139
surface monitoring for landfill gas, 358
surface retorting, 183
surfactant flooding, 67
swamp gas. *See* natural gas
sweep efficiency, oil recovery, 66–67
sweet gas, 36. *See also* processing
sweet sorghum, 225–226, 269–270
sweet synthetic crude oils, 126
switchgrass, 226, 268–269
syncrude (synthetic crude oil). *See* synthetic crude oil
Syncrude Canada mining and processing plant, 115–116, 123, 125–128
synfuel (synthetic fuel). *See* synthetic fuel
synthesis, Fischer-Tropsch. *See* Fischer-Tropsch process
synthesis gas (syngas), 197–219
    biofuels from, 249–250, 256–257
    from coal
        chemistry, 199–200
        gasifiers, 202–203
        overview, 145, 198–199
        processes, 201–202
    composition, 379
    defined, 20–21
    Fischer-Tropsch processing, 9–10, 24, 160–162, 213–218
    gasification, 23, 212–213
    overview, 92, 197–198
    from petroleum fractions
        chemistry, 204–206
        overview, 203–204
        processes, 206–209
    steam-methane reforming, 209–212
    use of, 240
    from waste, 335, 339–340
synthetic crude oil (syncrude), 8, 119, 125–126, 383

synthetic fuel (synfuel). *See also*
   nonconventional fuel sources
  defined, 1–3
  overview, 19–21
  produced from natural gas, 14–15
  production of, 22–25
synthetic natural gas (SNG), 150

tail gas-treating process, 56–57
tailings, hot water process, 116–117
tar, 7, 85, 104, 237
tar sand bitumen
  bitumen recovery, 116–120
  defined, 4
  future of use, 126–130
  mining technology, 115–116
  occurrence and reserves, 104–108
  overview, 6–8
  properties
    chemical and physical, 110–115
    structure and, 108–110
  synthetic crude oil, 125–126
  upgrading
    hydrogen production, 125
    other processes, 124–125
    overview, 120–122
    primary conversion, 122–124
    secondary, 124
TCLP (toxicity characteristic leaching procedure), 357
TDP (thermal depolymerization), 336
technical standards for biodiesel, 289
temperatures
  auto-ignition, 379
  effect on biodiesel use, 289–290
  and formation of natural gas, 28–29
  landfill gas
    migration, 361
    production of, 349
  reaction
    biodiesel transesterification, 288
    thermal decomposition, 22
    tar sand bitumen immobility, 114
tertiary biomass feedstocks, 222
test methods, natural gas, 38
Texaco Gasification Process, 209
Texas tar sand deposits, United States, 107
THAI (toe-to-heel air injection) process, 119
Thai oil shale, 179
The Oil Shale Corporation (TOSCO) process, 12, 184

thermal conversion, 228, 231, 254–257
thermal cracking, 70–74, 120, 191
thermal deasphalting, 125
thermal decomposition, 22–23, 163–164, 181–182, 249
thermal depolymerization (TDP), 336
thermal oil recovery methods, 67
thermal polymerization, 84
thermal reforming, 80
thermogenic coalbed methane, 33
thiols, 36
Tia Juana Crude Oil, 368
tight gas reservoirs, 35
timber. *See* wood
time, effect of on biodiesel transesterification, 288
tire pyrolysis, 336
toe-to-heel air injection (THAI) process, 119
topping refineries, 65
TOSCO (The Oil Shale Corporation) process, 12, 184
towers, distillation, 68–70
toxicity characteristic leaching procedure (TCLP), 357
traditional fuels, 241
transesterification, biodiesel
  alcoholysis, 284
  catalytic, 284–286
  effect of reaction parameters on conversion yield, 287–288
  noncatalytic supercritical methanol, 286–287
  use of, 290
  from waste oil, 337–338
traps
  natural gas, 29–30
  tar sand, 104
treatment plant wastes, 324
trees. *See* wood
Tremblador tar sand deposit, Venezuela, 107
trickle-bed reactors, 161
Trinidad tar sand deposits, 107–108
turbines, combined cycle, 202–203
two-stage Rectisol process, 51

UEL (upper explosive limit) of natural gas, 42
Uinta Basin tar sand deposits, United States, 105–106
ultimate analyses
  coal, 137, 144
  solid fuel, 393
ultimate composition, tar sand bitumen, 112–113
unconventional gas, 33–35
underground gasification of coal, 156

INDEX

underground mining
  coal, 139
  oil shale, 182–183
underground refining, 119
uniform concentrations, landfill gas, 360
Union Oil retorting process, 12
United States
  oil shale, 169–171, 174–175, 179–181
  tar sand bitumen, 103, 105–107, 128–129
Unocal company, 184, 193
updraft gasifiers, 147, 309–310
upgrading
  heavy oil, 88–90
  shale oil, 186–192
  tar sand bitumen
    hydrogen production, 125
    other processes, 124–125
    overview, 120–122
    primary conversion, 122–124
    secondary upgrading, 124
upper explosive limit (UEL) of natural gas, 42
urban waste briquettes, 342
Utah tar sand deposits, United States, 105–106

vacuum distillation, 69–70
vapor density, 39–40
vapor pressure, 41–42
vapor-assisted extraction (VAPEX), 119
vegetable oils, 17, 224, 247, 375
vegetal matter, 134–135
Venezuela
  synthetic crude oil, 383
  tar sand bitumen occurrence and reserves, 107
ventilation, coal mine, 140
viscosity, 114, 117, 386
viscosity breaking (visbreaking), 71–72
vitamin coal, 137
vitrinite, 136
volatile matter content of biomass, 235
volatile yield, thermal decomposition, 22
volatility of natural gas, 40–43
Voyageur Project, Suncor, 129

waste, 323–344. *See also* landfill gas
  anaerobic digestion of, 243, 277–278
  as biomass feedstock, 18, 227
  domestic, 325–326
  effects of, 326–328
  energy crop gasification, 273
  fuels from incineration, 328–332
  future, 342–343
  gaseous fuels, 332–334

waste (*Cont.*):
  gasification of to produce syngas, 212–213
  industrial, 325–326
  liquid fuels
    biodiesel, 336–338
    ethanol, 338–339
    hydrocarbon fuels, 335–336
    other alcohols, 339–340
    overview, 334
  overview, 323–325
  solid fuels, 340–342
  woody, 304
waste gas shift reaction, 153
water
  in biodiesel, 290
  in landfill gas, 349
  in oil shale operations, 194
  pollution from wastes, 326
water gas, 100, 198, 200
water-gas shift reaction, 205
water-washing process, 52–55
waxes, removing from petroleum, 86–87
Wellman Galusha process, 155
wells, landfill gas, 350
West Texas Intermediate Crude Oil, 383
wet gas, 30, 36
wet oxidation processes, 56
whisker carbon, 211
willow, 273, 303
wind in landfill gas migration, 361
Winkler process, 201
wood, 297–321. *See also* solid fuels
  bioconversion, 252–254
  as biomass feedstock, 226–227
  composition, 301–304
  energy crops, 265–266
  energy from, 304–306
  fuels from
    gaseous, 307–310
    liquid, 310–313
    overview, 306–307
  history, 297–300
  properties, 301–304, 374
  proximate analyses, 392
  pyrolysis of, 237–239
  short rotation coppice, 267
wood alcohol. *See* methanol
workup, biodiesel, 338

yeasts, 282
Young, James, Dr., 173

ABOUT THE AUTHOR

James G. Speight, Ph.D., D.Sc., is a fuels consultant and visiting professor at both the University of Utah and the University of Trinidad and Tobago. He is recognized internationally as an expert in the characterization, properties, and processing of conventional and synthetic fuels. Dr. Speight is the author, editor, or compiler of more than 30 books related to fossil fuel processing and environmental issues, including *Lange's Handbook of Chemistry*, Sixteenth Edition; *Chemical Process and Design Handbook*; and *Perry's Standard Tables and Formulas for Chemical Engineers*, all available from McGraw-Hill.